SOLAR HEATING and COOLING

SOLAR HEATING and COOLING

ACTIVE AND PASSIVE DESIGN

Second Edition

JAN F. KREIDER
Consulting Engineer

FRANK KREITH
Solar Energy Research Institute

HEMISPHERE PUBLISHING CORPORATION Washington New York London

McGRAW-HILL BOOK COMPANY New York St. Louis San Francisco
Auckland Bogotá Hamburg Johannesburg London Madrid Mexico Montreal
New Delhi Panama Paris São Paulo Singapore Sydney Tokyo Toronto

SOLAR
HEATING
and COOLING:
Active and Passive Design
Second Edition

1 2 3 4 5 6 7 8 9 0 B C B C 8 9 8 7 6 5 4 3 2 1

This book was set in Press Roman by Hemisphere Publishing Corporation. The editors were Jeremy Robinson, Valerie Ziobro, and Mary Dorfman; the designer was Sharon Martin DePass; the production supervisor was Miriam Gonzalez; and the typesetter was Sandra F. Watts. BookCrafters, Inc. was printer and binder.

Library of Congress Cataloging in Publication Data

Kreider, Jan F., date
Solar heating and cooling.

Bibliography: p.
Includes index.
1. Solar heating. 2. Solar air conditioning.
I. Kreith, Frank. II. Title.
TH7413.K73 1982 697′.78 81-6842
ISBN 0-07-035486-3 AACR2

Contents

Preface

During the six years since the publication of the first edition of *Solar Heating and Cooling*, many changes have taken place in solar technology. In writing this new edition, we have attempted to incorporate the important new developments that have occurred in the interim and to update the material the architect, builder, or mechanical engineer needs to properly design a solar system for heating and cooling a building or providing domestic hot water.

Foremost among the developments that have taken place is a fuller recognition of the costs and limitations of active solar energy systems. Experience with numerous heating systems for buildings in different parts of the country has shown that in many cases the passive approach can be more suitable than the installation of an active heating or cooling system. For this reason, passive design criteria have been added to this new edition.

However, even when the installation of an active space-heating system is not cost-effective, the development of improved hardware for solar domestic hot-water systems and incentives such as tax rebates, cost sharing, and low-interest loans have made the installation of domestic hot-water systems, both in new buildings and as retrofits, economically attractive in most parts of the country. More emphasis therefore has been placed on designing, sizing, and estimating the cost of solar domestic hot-water systems, as well as solar heating of swimming pools.

In addition to approaches to reducing the costs and fuel requirements for thermal conditioning of buildings, the technologies associated with solar ponds, photovoltaics, and wood stove heating are nearing maturity. Solar ponds are probably not suitable for private residences, but experience in Israel and the

United States has shown that salt-stabilized ponds may be suitable for heating, and possibly also cooling, applications in multi-family housing, hotels, or cooperative or district building heating. Therefore a separate section dealing with the design and sizing of solar ponds has been included.

Although solar photovoltaic power is not yet cost-effective, the price of photovoltaic cells has decreased by a factor of 10 during the past few years. With current emphasis on research and development in photovoltaics, it is expected that the price will decrease further. In the near future it is likely that photovoltaic installations will be attractive for isolated locations where electric power is unavailable. Sizing and costing of photovoltaic power systems have therefore been included in this edition.

Wood stoves are an old tradition in the United States, but current use of wood stoves for residential heating has raised numerous questions regarding their efficiency and required size. We found to our surprise that technical knowledge about wood stoves is not extensive, but we have tried to give a basic understanding of their operation and sizing in this edition.

As in the previous edition, we have not dwelt on theory, but have taken a pragmatic approach suitable for the practitioner who wishes to use solar energy.

We learned from the first edition that some mistakes are unavoidable despite careful attempts to eliminate them in proof. We therefore would appreciate it if any reader who discovers an error or misprint would let us know.

Finally, it should be apparent that solar technology is rapidly changing; therefore, we suggest that the reader who wishes to keep up-to-date in the field continue to follow the solar literature as it appears in the United States and Europe, for example, in *Solar Engineering, Solar Age,* and the journals *Solar Energy* and *Solar Energy Engineering*. In this way it should be possible to continue updating the information base in this book and apply the latest knowledge in the field to the new homes that will be constructed in the years to come.

Jan F. Kreider
Frank Kreith

Preface to the First Edition

One of the great maladies of our time is the way sophistication seems to be valued above common sense.

NORMAN COUSINS

By the end of this decade solar-powered systems for heating and cooling of buildings and providing hot service water will be commercially available and will be competitive with fossil fuels and electrical systems in many parts of the United States. By 1985 solar heating and cooling systems could supply about one-third of the 30 percent total United States energy consumption now used for residential and commercial space heating and cooling. By the turn of the century solar energy could provide 20 percent of the energy needs of the United States.

Until a few years ago, proposals for using the sun's energy on a significant scale were likely to be received with skepticism. During the last several years, however, the increasing cost of conventional fuels, the political uncertainties related to the supply of petroleum, and the problems associated with electric-energy generation from nuclear sources have modified this skepticism dramatically.

Although solar energy is the most abundant form of energy available, it is also one of the most dilute and intermittent forms and therefore requires different methods of collection and utilization than forms of energy widely used heretofore. There have been predictions, however, that with current technology the solar energy impinging on 4 percent of the land area of the United States would be sufficient to provide the projected energy needs of the country in the year 2000; by comparison, about 15 percent of the land area of the United States is used today for agricultural purposes. The successful utilization of solar energy on a large scale depends upon the design of systems and a thorough consideration of their economics. In contrast to gas or electric heating systems, which can be easily standardized, no single solar system is satisfactory for all tasks. The design and size of a solar system must, therefore, be matched to its task: It should not only meet technical requirements but should also involve—and resolve satisfactorily—the impact on construction economics, system amortization, and building esthetics.

One significant but generally unrecognized reason the use of solar energy has progressed so slowly is that conventional engineering education has not produced courses that deal with this interdisciplinary field involving principles of physics, chemistry, engineering, architecture, meteorology, and astronomy. Because of this education gap, the authors began to offer solar-energy seminars for architects, engineers, contractors and planners. These seminars presented, in a comprehensive and practical way, the available knowledge of use and economics of solar energy. The seminar material, expanded and supported with examples and related information, forms the basis of this book.

The emphasis in this book is on heating and cooling methods that can be employed economically in the near term. By extension, the principles are also applicable to engineering analysis of solar thermal-power applications, crop drying, and solar distillation. Photochemical, photosynthetic, and photovoltaic processes are not considered in depth; the indirect solar technologies of wind power and ocean-thermal-gradient utilization are likewise not included. This book attempts to describe practical systems, either those of more classical design with long, successful histories of use or those resulting from the latest generation of solar research that seem to have considerable promise because of their use of new concepts, materials, and manufacturing techniques. The reader must recognize that the field is rapidly changing; to remain current one must follow developments on an almost daily basis.

It is recognized that the audience for this book is not homogeneous. Therefore, throughout the book, which may serve as a textbook or a practical handbook, a dual method of presentation is used. Results of analysis are presented in analytical (equation) form and in graphical or tabular form. The seasoned engineer will find the analytical approach more appropriate in his designs by computer, while the architect and builder may find the graphical presentations more useful.

The material is divided into five chapters and eight appendixes. The five chapters contain—in this order—introductory information on solar and conventional energy-use concepts and requirements, fundamental principles of heat transfer and the nature of solar radiation, practical and efficient methods of collecting solar energy, and detailed quantitative descriptions of the practical systems for heating or cooling by means of solar energy together with an analysis of their economics. The appendixes contain extensive tables of reference data, including heat-transfer properties of materials, geographical and seasonal variation of sunshine and climate in the United States, and economic amortization schedules. The appendixes also contain information on the legal implications of solar-energy use for buildings and several typical solar-heated building designs. In keeping with the goal of this book to provide information in a readily usable form, the units of measurement are those endorsed by the American National Standards Institute (ANSI)—those used in practice in the United States and Canada—not those units used in the Système International d'Unités (SI) system.

This book draws on many sources for information, much of which dates from a small group of solar pioneers who were years before their time. Wherever possible, the authors have tried to indicate the specific sources from which the information was derived, but certain information could not be ascribed specifically to a single source. The authors have tried to give credit where credit is due; if the authors have failed to cite an original source, it is an oversight and not an intention. This book is not intended to be a compendium of literature on solar-energy applications; however, a section of general references has been included that can serve as a starting point for a thorough examination of the literature.

The notes upon which this book is based were developed as a part of the program of the Center for Management and Technical Programs at the University of Colorado, Boulder. The authors would like to express their appreciation to the Center for its help and cooperation and to the many authors and publishers who gave permission to use drawings or other materials. The authors wish to acknowledge the patience and helpfulness of Mr. Rolla Rieder during the typing of the manuscript. The authors also wish to acknowledge the contribution of Ms. Honey Sauberman, whose careful editing significantly improved the clarity, continuity, and organization of this book.

Jan F. Kreider
Frank Kreith

SOLAR HEATING and COOLING

1 Introduction: Why Solar Energy?

Eyes, though not ours, shall see
Sky-high a signal flame,
The sun returned to power above
A world but not the same

CECIL DAY LEWIS

Strictly speaking, all forms of energy are derived from the sun. However, our most common forms of energy—fossil fuels—received their solar input eons ago and have changed their characteristics so that they are now in a highly concentrated form. Since it is apparent that these stored, concentrated energy forms are now being used at such a rapid rate that they will be depleted in the not-too-distant future, we must begin to supply a large portion of our energy needs not from stored, but from incoming solar energy as soon as possible.

ENERGY—CONTEMPORARY SOURCES AND USAGE PATTERNS

Energy is defined in classical thermodynamics as the capacity to do work. From a practical point of view, it is the basic ingredient for all industrialized societies. In the United States energy is currently derived from four primary sources: petroleum, natural gas and natural gas liquids, coal, and wood. The supplies of these common energy sources, except for wood, are finite. Their lifetime is estimated to range from 35 years for natural gas to 200 years for coal. As current energy sources become exhausted an energy gap will develop, exacerbated by the synergistic effects of population growth and increased dependence on energy. After nonrenewable energy sources are consumed in what some authors call this "fossil fuel age," humanity must turn to longer-term, permanent energy sources. The two most significant of these are nuclear and solar energy. Nuclear energy requires highly technical and costly means for its safe and reliable utilization and may have undesirable side effects. Solar energy, on the other hand, shows promise of becoming a dependable energy source without new requirements of a highly technical and specialized nature for its widespread utilization. In addition, there appear to be no significant polluting effects from its use [1].

Figure 1.1 illustrates the time span of the fossil fuel age in which we live. In this figure, the rate of energy use is plotted as a function of time, using a prediction based on a careful analysis of all available fossil fuel resources by Hubbert [1]. The resulting curve has been called the "energy pimple" [2]. It is apparent that our fossil fuel consumption is increasing at a precarious rate and that the time when the world must begin to reduce the rate of consumption of nonrenewable energy is not far away.

The United States uses more than 35 percent of the world's fossil fuels, but has only about 6 percent of the world's population and 20 percent of the world's fossil fuel supply. Much of the energy produced in the United States is lost through waste. Total U.S. energy use, classified according to business sectors, is illustrated in Fig. 1.2. Approximately 34 percent of the energy is used directly for heating and cooling residential and commercial buildings. In 1978 President Jimmy Carter proclaimed as a national goal that by the year 2000 at least 20 percent of total U.S. energy needs would be supplied by the sun [3]. Since, as shown in Table 1.1, a large fraction of our consumption is in heating water for domestic use and in heating and cooling buildings, these end uses of energy must constitute a large fraction of the solar market for the end of this century.

It should be noted, however, that solar energy use and energy conservation are two sides of the same coin. Therefore, for proper solar designs, buildings must first conserve energy to reduce energy demands. The American Institute of Architects (AIA) estimates that by 1990, with proper conservation in buildings, it will be possible to save more energy than can be produced from U.S. domestic oil, nuclear power, or natural gas (domestic and imported) [4]. Conservation is the first step in reducing fossil fuel energy usage in buildings, whether the

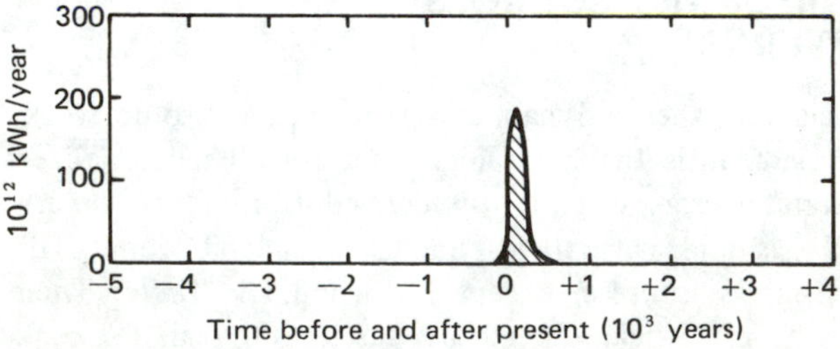

Figure 1.1 Historical perspective on world fossil fuel consumption: the energy pimple. (*Adapted from [1, 2].*)

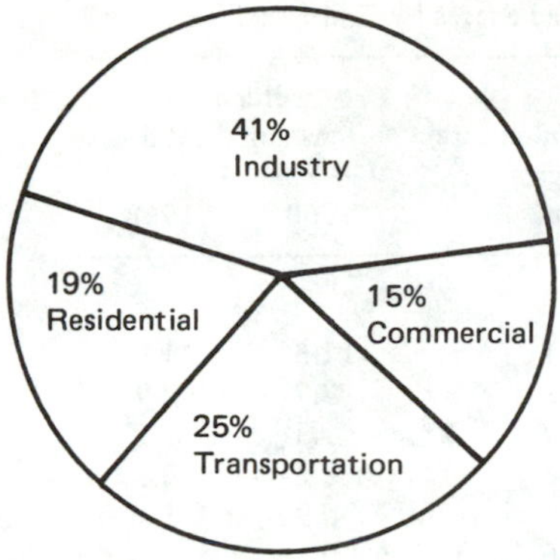

Figure 1.2 Approximate division of U.S. total energy use in four sectors. (*From [4].*)

building uses conventional or renewable alternative energy sources. As the next step solar technologies should be applied, as appropriate, to further reduce use of fossil fuels.

Table 1.1 indicates that of the total national energy consumption, 19 percent is used in residential buildings, 15 percent in commercial buildings, 41 percent in industrial processes, and 25 percent in transportation. The major uses of energy in buildings are in space heating, air conditioning, and service-hot-water supply. Table 1.1 shows the division of energy use in terms of consumption in trillion British thermal units (Btu) and percentage of the national total, as well as the annual growth rate. Space heating of residences accounts for 11 percent of the total national energy consumption, while space heating for commercial occupancy represents an additional 7 percent. Air conditioning represents 2.5 percent of the total consumption but has an annual growth rate of about 16 percent in residences.

The energy losses that determine the effectiveness of heating or air conditioning in buildings are essentially the same. Unnecessary energy losses are caused by inadequate insulation, excessive ventilation, high rates of air infiltration from outside, and excessive fenestration. Building heat losses have recently been recognized by the federal government as a major cause of fuel resource waste. In its *Minimum Property Standards* (1965), the Federal Housing Administration (FHA) of the U.S. Department of Housing and Urban Development (HUD) permitted heat loss of 2000 Btu/(1000 $ft^3 \cdot °F$) per day in single-family residences. The standard required by HUD Operation Breakthrough in 1970, however, reduced this figure to 1500, and the newly implemented *Minimum Property Standards* requires it to be less than 1000. The reduction required is to be achieved principally by improved thermal insulation and reduction of air infiltration. Approximately 40 percent of space-heating fuel can be saved through thicker or more effective insulation and improved draft control in commercial buildings. It appears at present technologically and economically feasible to reduce heat losses in buildings to approximately 700 Btu/(1000 $ft^3 \cdot °F$) per day

Table 1.1 Total Fuel Energy Consumption in the United States by End Use

End use*	Consumption (trillion Btu) 1960	1968	Annual rate of growth (%)	Percent of national total 1960	1968
		Residential			
Space heating	4,848	6,675	4.1	11.3	11.0
Water heating	1,159	1,736	5.2	2.7	2.9
Cooking	556	637	1.7	1.3	1.1
Clothes drying	93	208	10.6	0.2	0.3
Refrigeration	369	692	8.2	0.9	1.1
Air conditioning	134	427	15.6	0.3	0.7
Other	809	1,241	5.5	1.9	2.1
Total	7,968	11,616	4.8	18.6	19.2
		Commercial			
Space heating	3,111	4,182	3.8	7.2	6.9
Water heating	544	653	2.3	1.3	1.1
Cooking	93	139	4.5	0.2	0.2
Refrigeration	534	670	2.9	1.2	1.1
Air conditioning	576	1,113	8.6	1.3	1.8
Feedstock	734	984	3.7	1.7	1.6
Other	145	1,025	28.0	0.3	1.7
Total	5,742	8,766	5.4	13.2	14.4
		Industrial			
Process steam	7,646	10,132	3.6	17.8	16.7
Electric drive	3,170	4,794	5.3	7.4	7.9
Electrolytic processes	486	705	4.8	1.1	1.2
Direct heat	5,550	6,929	2.8	12.9	11.5
Feedstock	1,370	2,202	6.1	3.2	3.6
Other	118	198	6.7	0.3	0.3
Total	18,340	24,960	3.9	42.7	41.2
		Transportation			
Fuel	10,873	15,038	4.1	25.2	24.9
Raw materials	141	146	0.4	0.3	0.3
Total	11,014	15,184	4.1	25.5	25.2
National total	43,064	60,526	4.3	100.0	100.0

*Electric-utility consumption is allocated to each end use.
Source: Adapted from [8] with permission.

through the use of proper design, increased insulation, and reduction of unnecessary ventilation and infiltration. Figure 1.3 illustrates the reduction in building heat loss that can be realized by adding storm windows and increasing ceiling and wall insulation. Appendix 12 contains an extensive list of energy conservation methods for buildings and presents the relative efficiency of each in several climatic zones of the United States.

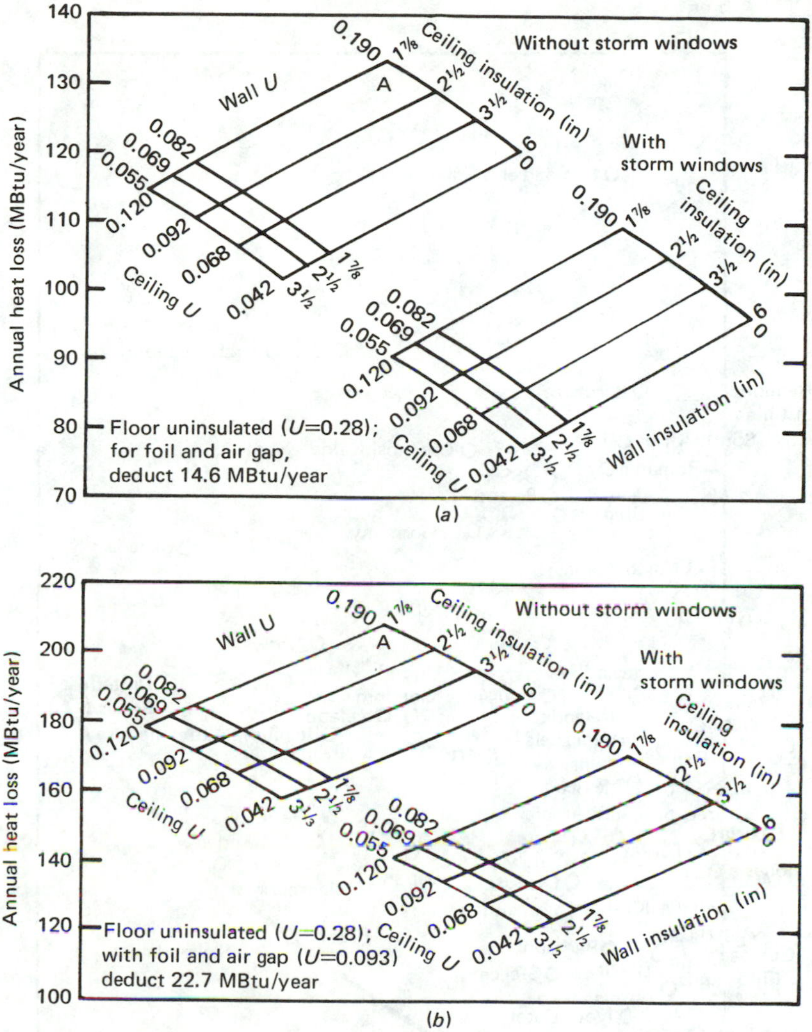

Figure 1.3 Annual heat loss from model homes with various weights of insulation in (*a*) New York and (*b*) Minneapolis. Abbreviations: 1 MBtu = 10^6 Btu; $U = \text{Btu}/(\text{h} \cdot \text{ft}^2 \cdot {}^\circ\text{F})$. A quantitative estimate of the saving in energy for home heating that is technologically feasible may be obtained as follows. The upper point *A* in (*a*) and (*b*) may be assumed to represent the approximate state of insulation and storm window sealing in approximately 90 percent of housing built before minimum property standards were issued. Heat losses from these houses can be reduced approximately 45 percent by application of heavy ceiling insulation and side wall insulation and by installation of storm windows. Thus installing insulation and storm windows in present housing units could reduce current fuel consumption for space heating of residences by approximately 40 percent. (*Adapted with permission from Moyers [9].*)

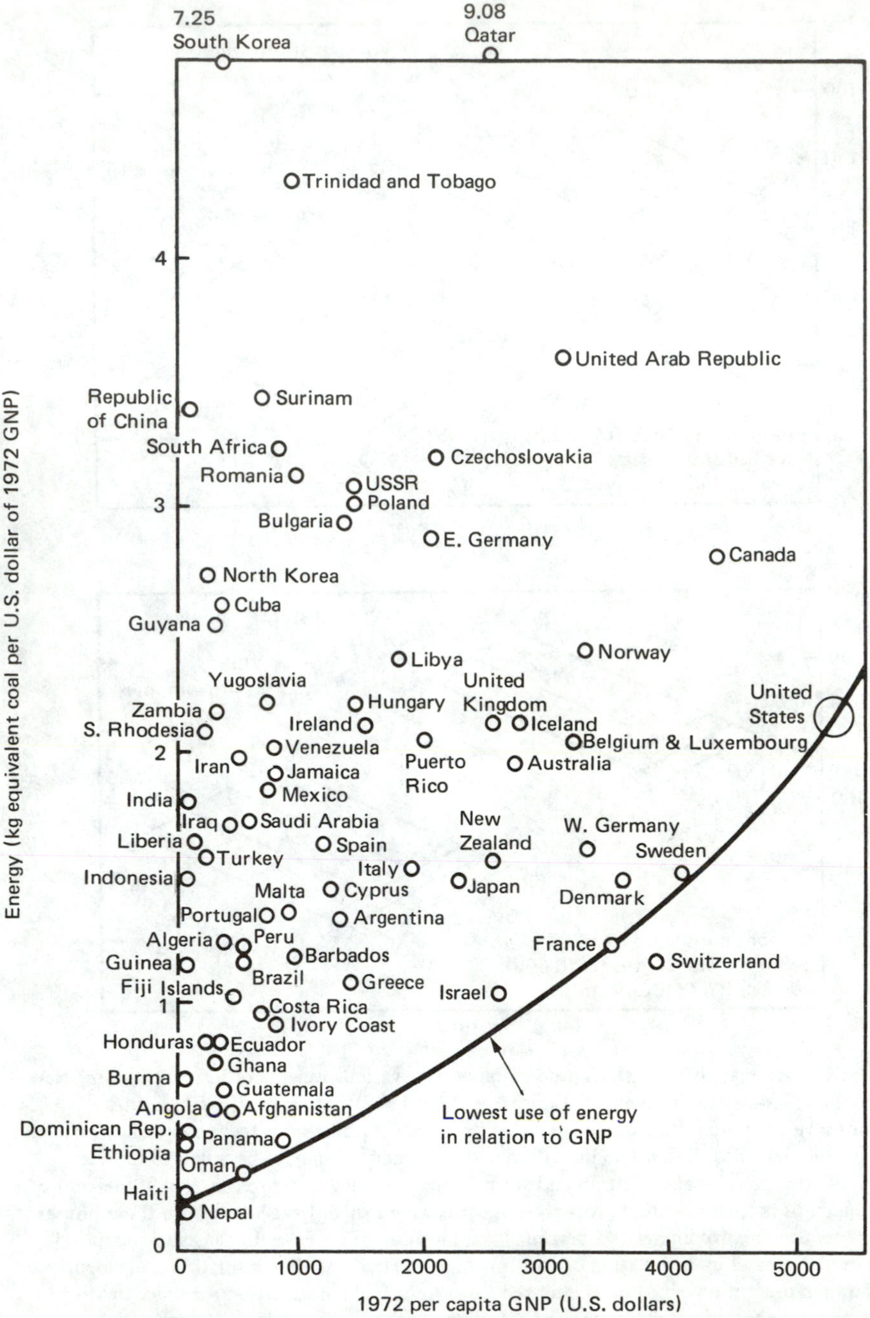

Figure 1.4 Worldwide energy consumption per unit of GNP compared with GNP per capita. (*From F. Felix, "Using Electricity More Efficiently,"* Electrical World, *December 1, 1974, p. 59. Copyright 1974, McGraw-Hill, Inc. All rights reserved.*)

Felix [5] compiled energy use data for more than 200 countries and colonies; per capita energy use for a number of these countries is shown in Fig. 1.4. Although it is true that the United States, with less than 6 percent of the world's population, consumes about one-third of the world's energy, the energy use per dollar of gross national product (GNP) lies on the lower bound for all countries. It seems that below this lower bound, a given level of GNP cannot be sustained. This curve shows that the United States is among the most efficient energy consumers in the world. The present U.S. standard of living cannot be maintained solely by increased energy conservation; new sources of supply are required. Solar energy is one of the most important such new energy sources.

THE NATURE OF SOLAR ENERGY

Solar energy is the world's most abundant permanent source of energy. The amount of solar energy intercepted by the planet earth is 170 trillion kW, an amount 5000 times greater than the sum of all other inputs (terrestrial nuclear, geothermal, and gravitational energies and lunar gravitational energy). Of this amount, about 30 percent is reflected to space, 47 percent is converted to low-temperature heat and reradiated to space, and 23 percent powers the evaporation-precipitation cycle of the biosphere; less than $\frac{1}{2}$ percent is represented in the kinetic energy of the wind and waves and photosynthetic storage in plants. The amount of the sun's energy intercepted by the earth is only a tiny fraction–one-thousandth of one-millionth–of the total energy released by the conversion of 4 million tons of hydrogen per second to helium in the sun.

Although it is abundant, solar energy impinging on the earth's atmosphere is relatively dilute–approximately 430 Btu/(h·ft^2). Traversing the earth's atmosphere, it is further diluted by attenuation, local weather phenomena, and air pollution. Moreover, solar energy is received only intermittently at any point on the earth. The solar energy that arrives at the surface of the earth is in two forms: direct radiation and diffuse radiation. *Direct radiation* is collimated and capable of casting a shadow; *diffuse radiation* is dispersed, or reflected, by the atmosphere and not collimated. In considering how solar energy can best be used, the ratio of direct to diffuse radiation becomes important, as will be shown in Chap. 2. The ratio of direct to diffuse radiation varies with time and location. For example, in the Four Corners states, the ratio of direct to diffuse radiation is on the order of 5, but for a large city it may be only on the order of 2. The amount of direct radiation diminishes as air pollution increases. The nature of the sun's energy reaching the surface of the earth and its temporal, seasonal, and geographic variations are described in detail in Chap. 2.

On the average, the radiation striking the earth's surface is on the order of 100 to 200 Btu/(h·ft^2). Because of the dilute nature of solar radiation, large collection areas are required, and the initial cost of installing heating and cooling equipment using solar energy is larger than that for other contemporary fuels. However, with properly designed and constructed equipment, solar heating and cooling of homes and buildings is, at current power costs, less expensive than electric climate control under most conditions and at most locations in the United States.

HISTORICAL PERSPECTIVE

The first person known to have used the sun's energy on a large scale is Archimedes, who reputedly set fire to an attacking Roman fleet at Syracuse in 212 B.C. "by means of a burning glass composed of small square mirrors moving every way upon hinges . . . so as to reduce [the Roman fleet] to ashes at the distance of a bowshot." Serious studies of the sun and its potential began in the seventeenth century, when Galileo and Lavoisier utilized the sun in their research. By 1700 diamonds had been melted, and by the early 1800s heat engines were operating with energy supplied by the sun. In the early twentieth century solar energy was used to power water distillation plants in Chile and irrigation pumps in Egypt. By 1930 Robert Goddard had applied for five patents on various solar devices to be used in his project to send a rocket to the moon. Most of these projects, however, were considered curiosities, because they were ahead of their time.

By the 1920s and 1930s practical use was being made of the sun's energy in California–for solar service-hot-water heaters. Devices similar to solar water heaters were also used to heat buildings in the United States. The first building to be practically heated with converted solar service-hot-water heaters was constructed at the Massachusetts Institute of Technology in 1938. Some 20 other experimental building heating projects–an average of about 1 per year–were completed between 1938 and 1960. Performance data recorded for a number of these projects are used in solar-heated building design today.

The space age gave solar energy its first postwar boost. The success of solar cells in powering service modules of the National Aeronautics and Space Administration (NASA) in terrestrial orbit and lunar excursions led some engineers to propose other uses for solar energy in the space program. Solar thermal power plants were designed and prototypes built; other exotic devices were proposed–even some that would use the small pressure exerted by sunlight to propel interplanetary vehicles. Although none of these projects was directed at providing large-scale terrestrial solar conversion economically, some of this NASA technology has found use in practical, applied projects for economical solar heating and cooling in the 1970s.

The impending shortages of fossil fuels and the government response of initiating federal funding for solar research and development have resulted in the first significant efforts in the United States directed toward practical use of the sun for building heating and cooling. The first federal funding of solar research in the 1970s amounted to \$1.7 million in fiscal year 1972. By 1975 the figure was \$50 million; by 1980 it was more than \$1000 million. Although these are small amounts in the federal energy research and development budget, they have provided the impetus required to move solar energy from the hobbyist/inventor stage to the first level of practical, widespread use.

CONVERSION OF SOLAR ENERGY TO HEAT

Solar energy is transmitted from the sun through space to the earth by electromagnetic radiation. It must be converted to heat before it can be used in practical heating or cooling systems. Since solar energy is relatively dilute when it reaches the earth, a system used to convert it to heat on a practical scale must be relatively large. Solar energy collectors, the devices used to convert the sun's radiation to heat, usually have a surface that efficiently absorbs radiation and converts this incident flux to heat, which raises the temperature of the absorbing material. Part of this energy is then removed from the absorbing surface by means of a heat transfer fluid, which may be either liquid or gaseous.

Since 1900 at least 50 solar collector designs have been demonstrated as functional. These designs are separated into two generic classes: concentrating and nonconcentrating. *Nonconcentrating*, or flat-plate, *collectors* intercept solar radiation on a metal or glass absorber plate from which heat is transferred and used in the thermal application. Since the temperature of the absorber plate is greater than that of the environment, unrecoverable heat losses occur from the entire absorbing surface of the collector to the environment. Consequently, 100 percent collector efficiency cannot be realized in practice.

Concentrating collectors attempt to reduce heat loss by using an absorber area smaller than the area that intercepts the sun's rays—the aperture area. This performance improvement is accomplished by reflecting the sun's rays from the large aperture area to the small absorbing area by use of shaped mirrors or other reflecting surfaces. Since only the direct or collimated portion of solar radiation is amenable to effective concentration, most concentrators must move to track the sun and cannot collect as much diffuse radiation as flat-plate collectors. In Chap. 3, where several collector designs are analyzed in detail, it will be shown that these two negative aspects of concentrators can be offset by their greater efficiency, particularly in sunny regions of the world.

Since solar energy is available only during daylight hours and during periods when the sun is not significantly obscured by clouds, a means of providing continuous heat from this intermittent source is required. In nearly all applications a form of thermal energy storage is used for this purpose. Three practical storage media have shown acceptable performance. Pebble or rock beds and water have been the two most widely used means of sensible heat storage (storage in a material by virtue of a temperature rise). The first is used with air-cooled solar conversion devices; the second, with liquid-cooled devices. The third storage method uses the latent heat of materials undergoing a phase change. Such materials can store more energy per unit volume than sensible heat storage materials, but to date they have shown unacceptable reliability in repeated freeze-thaw cycles. Since the heating and cooling loads on a building do not occur in phase with useful solar collection, all practical solar building thermal systems require storage. The size and cost of storage are considered in detail in Chap. 4.

This book deals primarily with the design and economic analysis of heating and cooling systems for buildings. Solar energy can, of course, be applied for other end uses, and there are other solar technologies [2] besides those treated here. The following section is a brief overview of the most important solar technologies and their applications.

OVERVIEW OF SOLAR TECHNOLOGIES

Solar energy technologies can be divided into two broad classifications: natural collection systems and technological conversion systems (see Fig. 1.5).

In *natural collection technologies* the total biosphere–earth, wind, and water–provides free collectors. Since no collectors need to be built, the energy costs for natural systems are determined by the converter, e.g., the wind turbine. In *technological conversion systems* the amount of collectable energy is determined by the amount of solar energy per unit area of collector at a given geographic location.

Solar Heating and Cooling

Solar energy now makes its biggest impact in domestic water heating and space heating of buildings. Two distinct but complementary approaches are used to convert sunlight to heat–passive and active systems.

A *passive system* may be defined as one in which the energy and mass flow is by natural means (involving conduction, convection, radiation, and evaporation). An *active system* is one in which all the mass flow is by forced means (involving fans or pumps). A *hybrid system* is one in which at least one of the significant flows is by natural means and at least one by forced means. These topics will be treated in detail in later chapters.

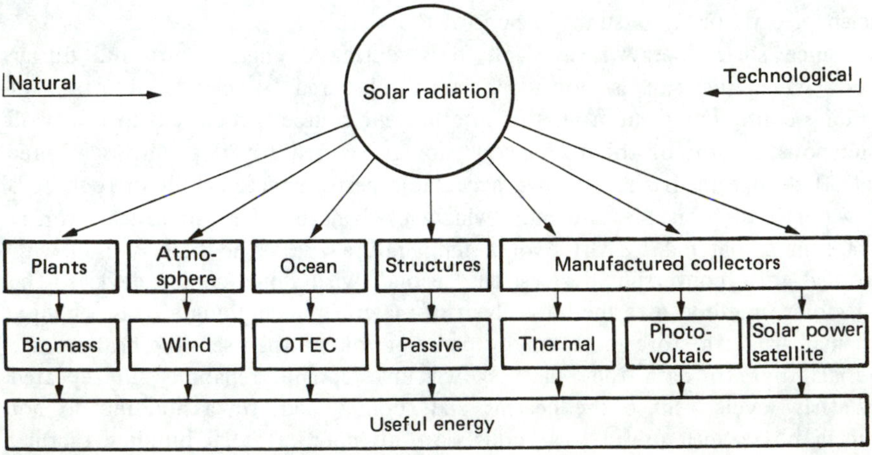

Figure 1.5 Solar energy technologies.

Biomass

Biomass is material derived from growing organisms–wood, corncobs, or seaweed. It is a direct and natural form of solar energy and can be produced where the sun shines on land and oceans. Direct combustion of wood has been the principal form of energy for most of human existence. In the United States wood provided 75 percent of the energy 100 years ago, and Sweden currently obtains 8 percent of its total energy from wood, while Finland relies on wood for a full 17 percent of its total energy.

Gasifiers can generate from organic residues a low-energy gas that can substitute for gas and oil in existing equipment. Digesters that convert cattle manure from large feedlots into pipeline-quality gas are already operating on a commercial scale, and the by-products from these digesters can be used for fertilizer and even cattle feed.

Although gas is a useful form of energy, liquid fuels are even more important. Pyrolysis (high-temperature heating in the absence of oxygen) of biomass can produce liquid fuel. Biomass is also an attractive source of alcohols that can be used as fuels. Alcohols that can be produced from biomass are methanol, wood alcohol derived from wood or municipal wastes, and ethanol, grain alcohol reduced by fermentation from agricultural products [6].

Alcohols are in some ways superior to gasoline as fuel; they have higher octane numbers and they can burn cleaner than gasoline. One promising use of alcohol is as a mixture with gasoline called gasohol.

Wind Energy

A fraction of the solar radiation incident on the earth is converted by the atmosphere into the kinetic energy of winds. A wind turbine can convert the kinetic energy of moving air into mechanical energy. Wind energy provided power for transportation at sea from the beginning of civilization and for agricultural purposes in Western Europe as early as the eleventh century. Wind power was also used extensively in the United States until recent times. More than 6 million small windmills (each less than 1 kW) were used in this country before the 1950s to pump water and to generate electricity on farms and in small communities.

Modern designs for using wind power cover a wide range of sizes and technologies. Deployment plans range from units for single homes to plants with the capacity of small fossil fuel plants (see Fig. 1.6).

Ocean Thermal Energy Conversion

Ocean thermal energy conversion (OTEC) is an indirect form of solar energy utilization. OTEC systems tap the thermal gradients between various ocean depths. In a closed cycle, surface water heated by the sun vaporizes a fluid such as ammonia. The pressurized vapor drives a turbine to produce electricity, which is then transmitted to land via cable. The vapor is condensed by colder water from the deep regions and again converted to a liquid. It is then ready to begin the cycle again.

Figure 1.6 Large wind turbine: MOD-1, 2000 kW rated capacity. (*Courtesy of NASA*.)

The amount of thermal energy in the ocean is large, but converting it requires the development of efficient, durable, and cheap heat exchange equipment that can withstand rough and corrosive conditions. The first demonstration plant began operation in 1980 [2], and the first commercial plant is expected to be built by 1990 (see Fig. 1.7).

Another method of utilizing the oceans for energy generation is through harnessing ocean waves and tidal currents. Systems to accomplish this are being developed in Japan and the United Kingdom, where prototype installations have been built. Cost estimates and projections in areas of the world favorable for this type of solar energy use look promising.

Industrial Process Heat

The possibility of utilizing a solar system year-round, as in solar domestic water heating, makes the solar industrial process heating market attractive. Approximately 41 percent of the energy used in the United States each year goes to industry. At least 60 percent of this, or about 25 percent of our energy requirement, goes to processes such as heating cleanup water, cooking, curing, steam heating, and air drying.

Numerous demonstration projects have been built for end uses such as drying lumber in Mississippi, drying raisins in California, providing steam for bleaching surgical gauze, and producing heat for concentrating orange juice. The technology for these applications is close to commercialization.

Conversion of Solar Energy to Electricity

The problem of converting solar energy to electricity has several possible solutions. One is *solar thermal conversion*, in which various types of solar collectors may be used to generate steam, which drives a turbine generator to produce electricity.

The most promising type of solar collecting system is the power tower, or

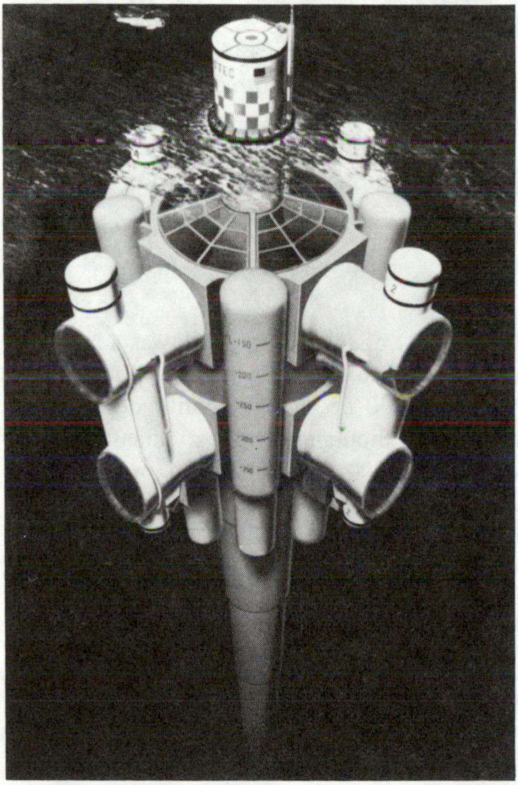

Figure 1.7 OTEC system. This is an artist's concept of a structure proposed by Lockheed's Ocean Systems Division for an OTEC system. The structure is a platform with crew quarters and maintenance facilities. Attached around the outside are turbine generators and pumps. It is 250 ft in diameter and 1600 ft long and weighs about 300,000 tons. The structure is designed to send 160,000,000 W of power ashore to distribution networks. This is enough power to supply a city of 100,000 people. (*Courtesy of Lockheed Aircraft Corp.*)

central receiver. This is a large tower surrounded by a field of tracking mirrors (heliostats), which concentrate the sun's rays onto a boiler located in the top of the tower (see Fig. 1.8).

Solar thermal conversion has the potential for centralized production of electricity, but for small-scale distributed installation—e.g., for individual houses—direct conversion with photovoltaic cells is much more promising.

Photovoltaic cells, or solar cells, convert solar radiation directly into electricity. They are a direct spin-off of the satellite and space program. This simple, solid-state device holds the promise of long operating life with little need for servicing. It is expected that by 1985 solar cells could contribute greatly to a wide range of applications that consume electricity.

Solar cells are connected and placed in a sealed glass or plastic unit called a module. A number of these modules, each of which might deliver 12 W, are then positioned in a rigid frame, called an array, to provide a specific power rating (Fig. 1.9). But since solar cells deliver power only during daylight hours, a photovoltaic system requires storage (usually a battery system) if power is to be supplied at night or in cloudy weather. Also, the power requires conditioning by an inverter to change the dc current to ac current and to regulate and keep the voltage output constant.

One of the most immediate applications for this technology may be in providing power in remote areas where there is no existing power grid.

ENERGY SUPPLIED BY SOLAR TECHNOLOGIES

In 1978 a domestic policy review (DPR) of solar energy [6] was conducted by an interagency solar energy policy committee with significant public participation. More than 100 officials representing 30 or more executive departments and government agencies participated. Several thousand people attended meetings all over the United States and more than 2000 individuals and organizations submitted oral or written comments.

Some of the principal conclusions are presented in Table 1.2, which shows the expected distribution of U.S. energy sources for different assumed costs of oil. As the price of oil escalates, not only the consumption of oil but also the consumption of energy overall is expected to decrease. Unfortunately, the actual cost of oil in 1980 exceeded the highest DPR estimate of $32 for 2000. It is therefore not clear how one can extrapolate the results and predictions of the DPR. However, Table 1.3 shows the DPR predictions for the amounts of energy that will be displaced by solar technologies in 2000, as well as the technical limits for these amounts.

The amount of solar energy used will depend as much on government energy policies and incentives as on developments in solar technologies. A wide range of government programs could be initiated to increase solar energy use.

Figure 1.8 Five-megawatt solar thermal central receiver test facility located at Sandia National Laboratory, Albuquerque, New Mexico. [*Courtesy of Solar Energy Research Institute (SERI) and Sandia National Laboratory.*]

Figure 1.9 Solar cell array. This flat-plate solar cell array supplying 1.6 kWe peak power is installed atop the Museum of Science and Technology, Chicago, Illinois. (*Courtesy of SERI.*)

The DPR grouped these programs into three basic policy options representing three levels of support:

1 Continue existing federal programs.
2 Expand current efforts in select program areas.
3 Dramatically increase federal support with solar energy use as a national goal.

Option 2 is estimated to cost $2.5 billion more than option 1 between 1980 and 1985, and option 3 to cost approximately $40 billion more than option 2 for the same period.

The results of the DPR show that, irrespective of the kind of future envisioned, solar energy will contribute an increasing share of our energy needs, and between now and the end of the century a large part of the solar contribution will be in the area of heating domestic water and buildings by active and passive means. Similar conclusions were reached by studies conducted by Stanford Research Institute, using various levels of assumed growth in energy consumption [7, 8].

Solar energy is, in the long term, the only inexhaustible energy resource. It is the only energy source capable of supplying humanity's increasing energy demands—which are increasingly difficult to meet by sources used previously. Although comparatively dilute, the energy reaching the earth from the sun far exceeds the energy requirements of the world's population. The first large-scale thermal application of solar energy in this century will be for heating and cooling residential and commercial buildings. It is the purpose of this book to provide the architect, engineer, and builder with the tools required to design and construct active and passive heating and cooling systems.

Table 1.2 Primary Energy Supply in the Base Case, Assuming Three Different Prices for Imported Oil in the Year 2000

	Energy supplied (quadrillion Btu's)			
	1977	2000		
Source	$14.50/bbl*	$18/bbl	$25/bbl	$32/bbl
Oil	36.9	44.0	32.1	22.8
Gas	19.6	20.2	18.0	14.5
Coal	14.2	43.0	38.5	31.5
Nuclear	2.7	17.0	15.0	13.0
Solar	4.2	7.3	9.9	12.7
Other†	–	0.5	0.5	0.5
Total	77.6	132.0	114.0	95.0

*Landed price of imported oil.
†Includes geothermal and other nonsolar renewable energy sources.
Source: Reuyl et al. [7].

Table 1.3 Energy Displaced by Solar Technologies in 1977 and 2000*

Energy end use or technology	1977	2000: Base case $25/bbl	2000: Base case $32/bbl	2000: Maximum practical	2000: Technical limit
Residential/commercial heating, hot water, cooling	Small	0.9	1.3	2.0	3.8
Passive design	Small	0.2	0.3	1.0	1.7
Industrial and agricultural†	–	1.0	1.4	2.6	3.5
Hydropower	2.4‡	3.9	4.0	4.3	4.5
(High head)	(2.4)	(3.5)	(3.5)	(3.5)	(3.5)
(Low head)	(Small)	(0.4)	(0.5)	(0.8)	(1.0)
Biomass	1.8	3.1	4.4	5.4	7.0
Solar thermal electric	–	0.1	0.2	0.4	1.5
Wind	–	0.6	0.9	1.7	3.0
Photovoltaics§	–	0.1	0.2	1.0	2.5
OTEC	–	–	–	0.1	1.0
Solar power satellite	–	–	–	–	–
Total	4.2	9.9	12.7	18.5	28.5

*Numbers in this table represent amount of conventional energy that can be displaced by solar energy rather than amount of energy actually delivered by solar systems. The DPR estimated ranges for solar penetration in the year 2000; figures given are midpoints of these ranges.
†Includes process heat, on-site electricity, and heating and hot water.
‡Energy displaced by existing dams in years of normal rainfall is 3.0 quads (1 quad = 10^{15} Btu).
§Penetration of photovoltaics is dependent on substantial cost reductions.
Source: [6].

REFERENCES

1. M. K. Hubbert, *World Energy Resources, 10th Commonwealth Mining and Metallurgical Congr.*, Ottawa, Canada, 1974; see also "Measurement of Energy Resources," *J. Dyn. Syst. Meas. Control*, vol. 101, 1974, pp. 16–30.
2. R. L. Bailey, *Solar Electrics,* Ann Arbor Science, Ann Arbor, Mich., 1980.
3. Executive Office of the President, *The National Energy Plan,* Washington, D.C., 1978.
4. G. Franta (ed.), *Solar Architecture*, Arizona State Conference of American Institute of Architects, Forum 75, Washington, D.C., 1975.
5. F. Felix, "Using Electricity More Efficiently," *Electr. World*, Dec. 1, 1974, p. 59.
6. U.S. Department of Energy, *Domestic Policy Review of Solar Energy,* Response memorandum to the President of the United States, TID-22834, Government Printing Office, Washington, D.C., Feb. 1979.
7. J. S. Reuyl, W. W. Harman, R. C. Carlson, M. D. Levine, and J. G. Witwer, *Solar Energy in America's Future*, Rep. DSE-115/2, Stanford Research Institute, Menlo Park, Calif., 1977.
8. Stanford Research Institute, *Patterns of Energy Consumption in the U.S.,* Government Printing Office, Washington, D.C., 1972.
9. J. C. Moyers, *The Value of Thermal Insulation in Residential Construction: Economics and Conservation of Energy,* Oak Ridge National Laboratory, Oak Ridge, Tenn., 1971.

2 Solar Radiation Calculations

In this chapter, the motion of the sun is described from the point of view of a terrestrial observer. Although the sun is the star about which the earth rotates, we will use the convention that the sun moves about the earth since the resulting geometric relations are simpler and easier to understand. After describing the motion of the sun and its relation to shading of buildings and collectors, we will discuss the amount of solar radiation striking a surface at various common orientations. Since most of the methods of estimation used later in the book require only a single calculation per month, monthly-averaged total solar radiation calculations will be presented. The method described can be used for any location in the world.

INTRODUCTION

Three types of solar radiation are usable by either active or passive terrestrial systems. The most significant type of radiation for solar thermal processes is called beam radiation. Beam radiation is the solar radiation that travels from the sun to a point on the earth with negligible change in direction. It is the type of sunlight that casts a sharp shadow, and on a sunny day it can be as much as 80 percent of the total sunlight striking a surface.

The second type of solar radiation is diffuse or scattered sunlight. This is sunlight that comes from all directions in the sky dome other than the direction of the sun. It is produced by scattering of sunlight by atmospheric components such as particulates, water vapor, and aerosols. On a cloudy day the sunlight is 100 percent diffuse.

The third type of radiation that is sometimes present at the glazing of a solar collector or a window is reflected radiation. This is either diffuse or direct radiation reflected from the foreground onto the solar aperture. The amount of reflected radiation varies significantly with the nature of the foreground, being relatively higher for a light-colored environment near the collector and relatively lower for a dark-colored environment. For solar collectors mounted on the roof of a building, for example, the amount of reflected radiation may be relatively small. A qualitative feel for the importance of reflected radiation can be obtained by standing at the location of a solar collector and estimating the percentage of the collector's field of view that consists of the foreground. In a later section, foreground reflectances are tabulated.

Calculation of diffuse, direct, and reflected radiation is one of the first tasks of the solar designer in establishing the feasibility of the solar resource for heating or cooling. The U.S. National Weather Service and other meteorological services around the world do not measure these three radiation components. Therefore a method of decomposing measured solar radiation into its three components is required. The standard method for doing this is presented in a later section of this chapter.

The motion of the sun is important in determining the angle at which beam radiation strikes a surface. Therefore the first topic dealt with quantitatively has to do with the seasonal and daily variation of the location of the sun relative to a viewer on the earth. Most solar textbooks use a mixture of Greek and Roman letters to denote the angles that are used in calculating the precise location of the sun. We will use an abbreviated form of the name of the angle itself to enable the reader to better relate the equations to the angles involved. The next section deals with solar geometry in detail, including several examples.

One of the principal concerns of a solar designer, in addition to calculating the amount of solar radiation on a surface, is to ensure that the collector surface is illuminated during the periods of the year when solar heat is required. To make this assessment, the possibility of system aperture shading must be evaluated. Shading may be included by design in order to reduce the amount of solar radiation on a surface when not needed, or it may be a characteristic of the site on which a solar building is to be located. For a solar building in an urban environment the possibility of shading is significant and is one of the first things a designer must consider. Surrounding buildings, trees, and fences cannot be relocated to improve access to solar radiation; therefore the designer must review these characteristics early in order to position the building properly for solar gain. Access to sunlight is not guaranteed by law in most areas of the world, although in some cases local ordinances permit a solar homeowner to purchase solar easements to ensure access to sunlight. Solar access law is surveyed in [1] and Chap. 11 of this book.

THE SUN'S MOTION

When viewed from a location on the earth, the sun appears to move in a circular orbit about the earth. Although this view is physically incorrect, it is convenient for purposes of terrestrial solar design, and the results are as accurate as those obtained by assuming correctly that the earth moves about the sun.

The sun may be assumed to be located on the celestial sphere. This is an imaginary sphere centered at the earth's center and located at a sufficiently large distance from the earth that the locations of the sun and other stars can be considered to be given by single points. A body moving on a sphere has two degrees of freedom in its motion. We will specify the two degrees of freedom by angles, and two angles are sufficient information to locate the sun on the celestial sphere at any time.

Solar Altitude Angle

Figure 2.1 shows the solar altitude angle ALT and azimuth angle AZM. The altitude angle is measured upward from the local horizontal plane to a line between the observer and the sun. The azimuth angle is measured in the horizontal plane between the due-south direction and the projection of the sun-earth line onto the horizontal plane. Azimuth angles have a sign convention, as do other solar angles, but for our purposes the sign associated with the solar angles need not be considered.

Figure 2.2 shows the solar altitude and azimuth angles at three times of year for a location at latitude 40°N. The maximum solar angle occurs at noon in all seasons of the year. In winter the noon sun is only $26\frac{1}{2}°$ above the horizon, whereas in summer it is $73\frac{1}{2}°$ above the horizon. Because of this large variation in altitude angle from season to season, a solar collector must be tilted up from the horizontal plane to face the sun during the season when the peak requirement for solar heat exists. If the requirement is uniform for the entire year, with no seasonal peak, then a tilt angle that maximizes year-round collection will be used. This tilt angle is typically equal to the local latitude. In subsequent chapters, the proper tilt angle for solar collectors for various tasks is given.

The numbers inside the circles in Fig. 2.2 are the azimuth angles of the sun at various times of day. The times shown in the figure are in solar time.* It can

*Solar time is based on the due-south position of the sun, which is defined as solar noon. Solar time differs from clock time as described on p. 26.

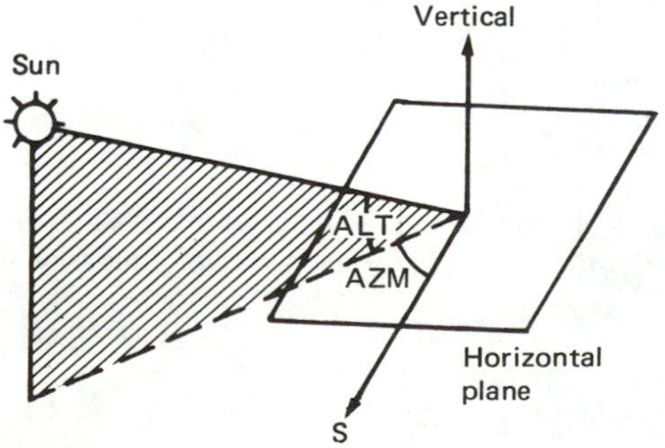

Figure 2.1 Diagram showing solar altitude angle ALT and solar azimuth angle AZM. (*Adapted from [2].*)

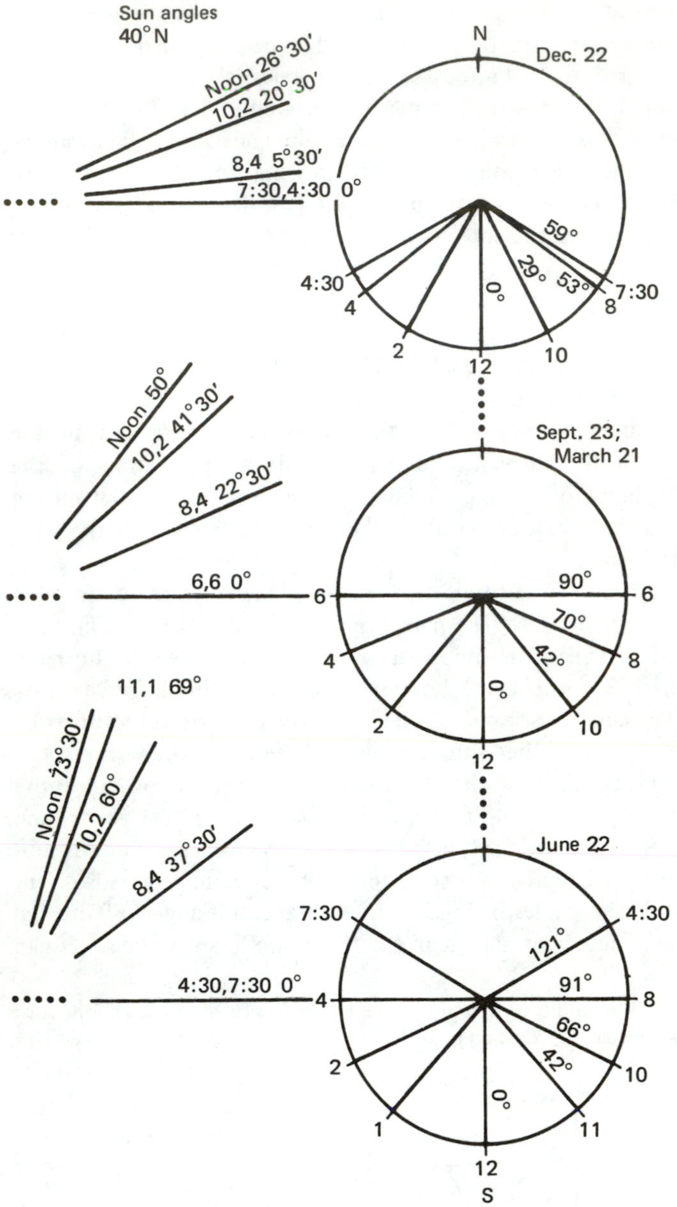

Figure 2.2 Solar elevation and azimuth angles for the solstices and equinoxes for latitude 40° N. (Left) Altitude angles for selected hours of the day; (right) azimuth angles for the same hours of the day.

be seen that the sun rises in winter to the south of due east and sets to the south of due west. The sun rises in the east and sets in the west exactly on only 2 days a year, the equinoxes, when there are 12 hours of day and 12 hours of night. Between the equinoxes, in spring and summer, the sun rises and sets to the north of the east-west line. Figure 2.2 applies for only one latitude; for other latitudes the specific angles will be different but the general relation of summer to winter sun angles is the same.

Solar altitude and azimuth angles can be calculated from simple spherical trigonometry equations, which are not developed in this book but are given directly below. For the development of these equations, the reader may refer to [2]. The altitude angle measured above the horizon is given by Eq. (2.1):

$$\text{ALT} = \sin^{-1}\,[\cos(\text{DEC})\cos(\text{LAT})\cos(\text{HOUR}) + \sin(\text{DEC})\sin(\text{LAT})] \qquad (2.1)$$

The altitude angle depends on three fundamental angles, the solar declination DEC, the latitude LAT, and the solar hour angle HOUR. Each of these angles will be described in turn.

The solar declination angle is a measure of the variation of the position of the sun on a seasonal basis. The declination angle is the same as the latitude at which the sun is directly overhead on a given day. These values of latitude lie between the Tropics of Cancer and Capricorn and have a variation of $\pm 23\frac{1}{2}°$ during a year. Alternatively, the declination can be thought of as the angle between the overhead point for an observer on the equator and the position of the sun at noon on a given day. This is equivalent to the earlier definition. The solar declination angle varies slowly and can be considered constant for a specific day. Figure 2.3 is a plot of average solar declination for a year. The maximum

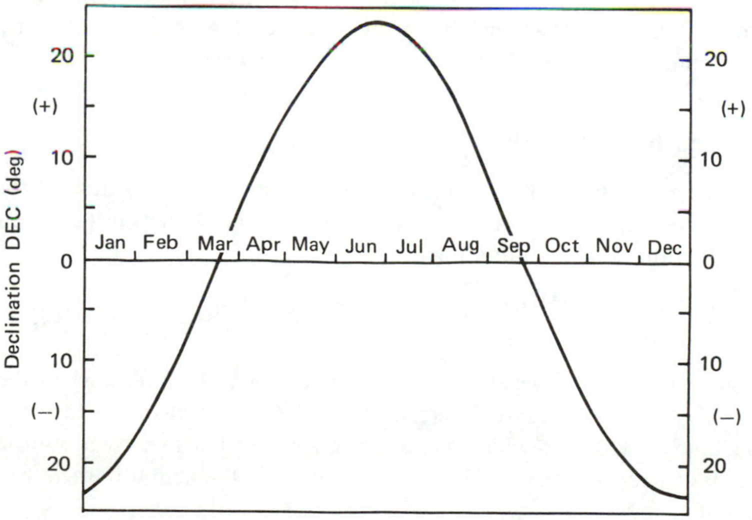

Figure 2.3 Plot of average solar declination in degrees for a year. Solar declination varies from $-23\frac{1}{2}$ to $+23\frac{1}{2}°$.

value of the declination, $+23\frac{1}{2}°$, occurs on June 21 and the minimum value occurs on December 21. The declination can be calculated with an accuracy of a few percent from Eq. (2.2):

$$\text{DEC} = -23.45° \cos [0.986 (\text{DAY} + 10.5)] \tag{2.2}$$

where DAY is the day number counted from January 1.

The hour angle is defined as the number of hours between solar noon and the time of interest multiplied by the constant 15°/h. The value of this constant is determined by the rate at which the sun appears to move around the earth, namely 360° in 24 h or 15° per hour. In calculating the hour angle it is important to use solar time and not clock time. Furthermore, standard time rather than daylight time must be used.

The third angle used in calculating the solar altitude angle is the latitude LAT. Latitude for a particular site of interest can be read from an atlas or from App. 6 of this book. In the conterminous United States, the latitude varies between 25°N and approximately $47\frac{1}{2}$°N. By convention, latitudes in the Southern Hemisphere are given a minus sign.

Equation (2.1) can also be used to calculate the time of day at which the sun rises or sets. Sunrise and sunset are defined as the points at which the altitude angle of the sun is zero, i.e., the center of sun has passed below the local horizontal plane. Equation (2.1) can be solved for the hour angle at which this occurs, as shown in Eq. (2.3), where HOURSET is the sunset hour angle,

$$\text{HOURSET} = \cos^{-1} [-\tan (\text{DEC}) \tan (\text{LAT})] \tag{2.3}$$

To determine the time at which sunset occurs, HOURSET, expressed in degrees, is divided by 15. The result is the number of hours after local solar noon at which the sun sets. To find the hour of sunrise, this same number of hours is subtracted from noon, as shown below in an example. Figure 2.4 shows the positions of the sunrise and sunset angles for winter and summer for a location at 40°N.

Solar Azimuth Angle

The solar azimuth angle, measuring the relation of the sun to due south, depends on the same three angles as the solar altitude angle. Equation (2.4) can be used to calculate the solar azimuth angle:

$$\text{AZM} = \sin^{-1} \left[\frac{\cos (\text{DEC}) \sin (\text{HOUR})}{\cos (\text{ALT})} \right] \tag{2.4}$$

The azimuth angle will be greater than 90° for some hours of the day when the length of day is greater than 12 h. Therefore Eq. (2.4) must also be evaluated relative to the time of year for which a calculation is being made. Near sunrise and sunset on days between March 21 and September 21 the azimuth angle will be greater than 90°.

Appendix 5 gives solar altitude and azimuth angles for several latitudes in the Northern Hemisphere. Hourly values of the angles, calculated from Eqs. (2.1)

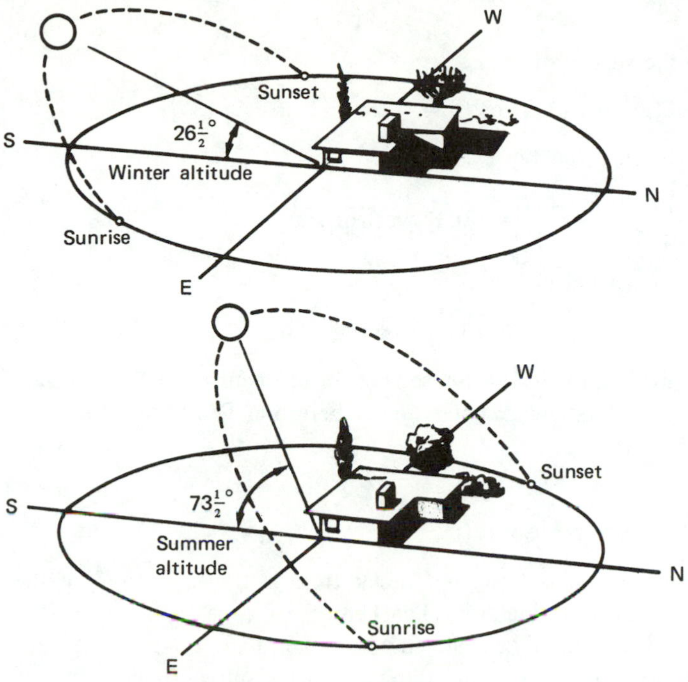

Figure 2.4 Sun angle diagram showing relative locations of summer and winter sunrise and sunset for location at 40° N. In winter, shadows are longer and possible shading problems more severe.

and (2.4), are given for one day of each month of the year. For latitudes other than those given in the table, interpolation can be used for rough calculations; the basic equations should be used for final design studies.

EXAMPLE

For a location at 37°N, find the altitude angle and the azimuth angle at 3 p.m. solar time on February 20. Also find the times at which the sun rises and sets on this date.

SOLUTION

For February 20 the day number is

$$DAY = 31 + 20 = 51$$

so the declination is

$$DEC = -23.45 \cos [0.986(51 + 10.5)] = -11.5°$$

The hour angle is

$\text{HOUR} = 3 \times 15 = 45°$

Then from Eq. (2.1) the solar altitude is

$\text{ALT} = \sin^{-1} [\cos(-11.5) \cos(37) \cos(45) + \sin(-11.5) \sin(37)] = 25.7°$

The azimuth angle is found from Eq. (2.4):

$$\text{AZM} = \sin^{-1} \left[\frac{\cos(-11.5) \sin(45)}{\cos(25.7°)} \right] = 50.3° \text{ west of south}$$

To find the sunset hour see Eq. (2.3):

$\text{HOURSET} = \cos^{-1} [-\tan(-11.5) \tan(37)] = 81.2°$

Convert to hours by dividing by 15°/h. Sunset is seen to occur at 81.2/15 = 5.4 h after solar noon or at 5:24 standard solar time. Therefore, February 20 has 10 h, 48 min of daylight.

Other Topics in Solar Geometry

The relation between solar time and clock time is useful in finding the position of the sun at a specific location. The United States is divided into four time zones, each of which has a central, standard meridian of longitude. For example, in the eastern time zone the standard meridian is 75°W. For the central, mountain, and Pacific zones, the standard meridian increases by 15° per zone as one moves westward. If a location under study is precisely on one of the standard meridians, then solar time and clock time are almost identical except for a small correction given by the equation of time, accounting for certain irregularities in the earth's orbit. The equation of time is described in [2].

For a location that is not on a standard meridian, a correction of 4 min of clock time is required for each degree of difference in longitude between the site and the standard meridian. For example, the standard meridian for the mountain time zone is 105°W. If a location under study is at 106°W, the sun will arrive at this location 4 min after it arrives at the standard meridian. Therefore a correction of 4 min must be made to the hour angle that would be calculated from clock time, as opposed to the true solar hour angle. Hence if clock time is 3 p.m. in the mountain time zone, the solar time at 106°W is 2:56. Other corrections to hour angle calculations need not concern us in this book.

Another angle that is useful in solar design is the incidence angle INC. This is the angle between beam radiation from the sun and a line constructed perpendicular to an irradiated surface. The incidence angle is zero if the surface is perpendicular to the direct rays of the sun; it is 90° if the surface is parallel to the rays from the sun. Equation (2.5) can be used to calculate the incidence angle for south-facing planar surfaces.

$$\text{INC} = \cos^{-1} [\cos(\text{DEC}) \cos(\text{LAT} - \text{TILT}) \cos(\text{HOUR}) + \sin(\text{DEC}) \sin(\text{LAT} - \text{TILT})] \qquad (2.5)$$

where TILT is the surface tilt angle measured from the horizontal. For surfaces that do not face south and for surfaces that move (e.g., concentrating collector apertures) other incidence angle equations apply. These are summarized in [2].

At times of year when the day is longer than 12 h, the sun may "set" on the collector plane before it passes below the horizon. In these cases the period of collection is controlled by sunrise and sunset at the collector plane rather than at the horizontal plane. The hour angle at which collector sunrise and sunset occur can be found by the same approach used for Eq. (2.3). However, instead of solving Eq. (2.1), one uses Eq. (2.5).

Sunpath Diagrams

Solar altitude and azimuth angles may be shown in a useful graphic way by means of a sunpath diagram. This is a plot in polar coordinates of values of the altitude and azimuth angles for a given latitude versus declination and time of day. Figure 2.5 shows a sunpath diagram for 45°N. Values of solar altitude angle are indicated by concentric circles on the diagram and azimuth angles are measured from due south in 10° increments. The declination and hour angle are shown, respectively, by approximately horizontal and approximately vertical lines.

By referring to Fig. 2.5, several uses of the sunpath diagram can be described. First, the values of the altitude and azimuth angles can be read directly. As an example, consider September 10, when the declination is +5°. At 10:00 a.m. the sun is at the position shown by the heavy dot in Fig. 2.5. This corresponds to an azimuth angle of approximately 42° east and an altitude angle of approximately 43° above the horizontal plane.

The sunpath diagram is used for determining the times of sunrise and sunset. By referring once again to the declination line +5° corresponding to September 10, sunrise is seen to occur at approximately 5:40 a.m. and sunset at approximately 6:20 p.m. The length of day can also be determined from the sunpath diagram; it is twice the sunset time, or 12 h, 40 min.

A fourth use of the sunpath diagram is described in the next section, on shading calculations. This is probably the most common application of the diagram for the solar designer. Sunpath diagrams for six values of latitude are contained in App. 3.

Other versions of sunpath diagrams are available. For example, a rectangular coordinate system is used in [4]. However, the circular coordinate system is better suited to the type of shading calculations described in the next section.

SHADING CALCULATIONS

During the early stages of solar design, it is necessary to assess the effect of possible shading of a solar collector of either the active or passive type by surrounding features on site. In addition, for some solar systems, especially those of the passive type, shading is introduced as a necessary feature of the design.

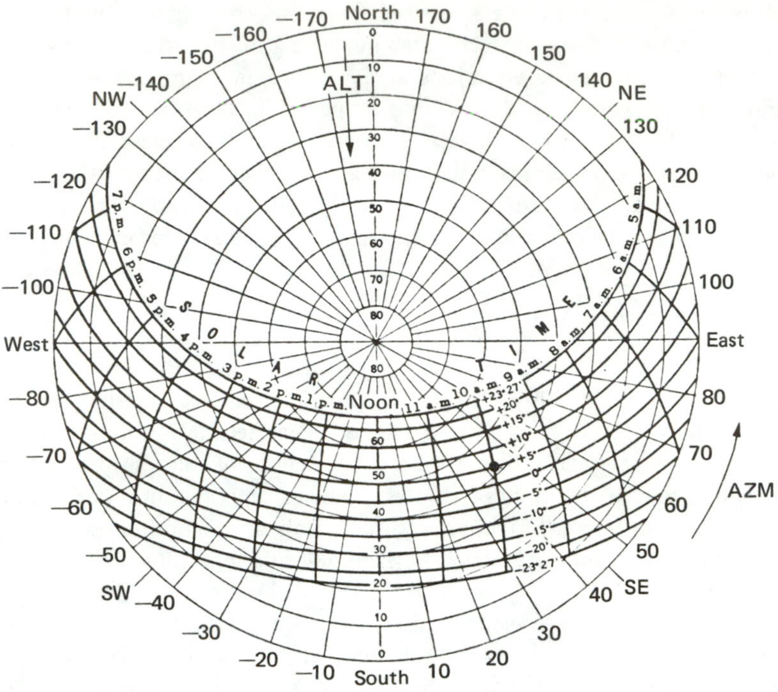

Declination	Approximate dates
+23°27′	June 22
+20°	May 21, July 24
+15°	May 1, Aug. 12
+10°	Apr. 16, Aug. 28
+5°	Apr. 3, Sept. 10
0°	Mar. 21, Sept. 23
−5°	Mar. 8, Oct. 6
−10°	Feb. 23, Oct. 20
−15°	Feb. 9, Nov. 3
−20°	Jan. 21, Nov. 22
−23°27′	Dec. 22

Figure 2.5 Sunpath diagram for 45° N, showing azimuth and altitude angles and their relation to declination and time of day. (*Adapted from [2]*.)

The sunpath diagram can be used for both purposes in determining the time of year and hours of day when a particular feature of the environment may cast a shadow on a solar collector.

To make quantitative shading calculations, an angle called the profile angle is required. The profile angle P is shown in Fig. 2.6. It is the projection of the solar altitude angle ALT onto another plane ($ABCD$) located at an azimuth angle AZM. The profile angle can be calculated from the simple equation:

$$P = \tan^{-1}\left[\frac{\tan(\text{ALT})}{\cos(\text{AZM})}\right] \tag{2.6}$$

where AZM is the azimuth angle between the plane in which the profile angle is measured and the direction of the sun. This azimuth angle is the same as the solar azimuth angle described earlier only for projection of the altitude angle onto the north-south plane. If a plane *ABCD* is used other than the north-south plane, the azimuth angle in Eq. (2.6) will differ from the solar azimuth angle by the azimuth angle difference between plane *ABCD* and the north-south plane.

Use of the profile angle in solving one common shading problem is illustrated in Fig. 2.7. A north-south cross section of a sawtooth array of collectors on a flat roof is shown. It is desired to space the collectors in such a way that they do not shade one another on the shortest day of the year, when shadows are longest. A typical rule is that the shadow line, shown as a dashed line in Fig. 2.7, should intersect the bottom of the collector to the north at 10:00 a.m. and 2:00 p.m. on December 21. This will allow 4 h of uninterrupted solar collection on the shortest day of the year, although there will be some shading before 10:00 a.m. and after 2:00 p.m. Since the sun spends only a few days in the position near winter solstice, it is not necessary to size the distance *D* between the collectors to ensure no shading whatsoever on December 21. Economic constraints will dictate that some shading be tolerated in order to fit more collector area on a given amount of roof area. Widely spaced collectors

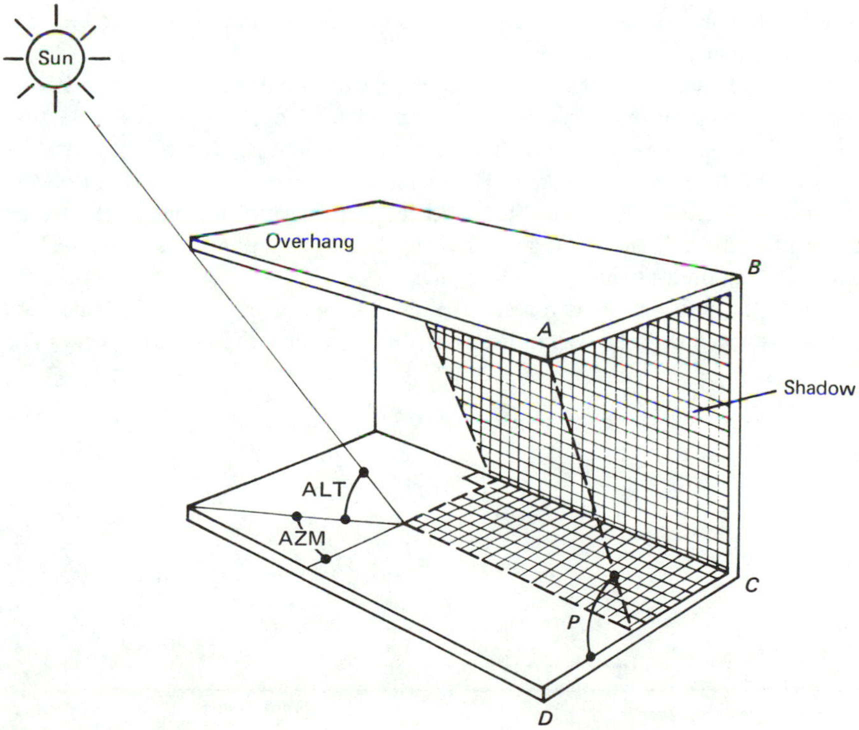

Figure 2.6 Diagram showing profile angle *P* as projection of solar altitude angle ALT onto plane *ABCD*. (*Adapted from [2]*.)

require additional piping runs and may consume more roof area than is available for a project. By using the profile angle equation, it is easy to show that the distance D between two successive collector rows should be approximately three times the face length L_c of the collector to meet the shading criterion given above. This rule applies for latitude 40° N and for collectors positioned for winter heating, i.e., at a tilt angle of 15° plus the latitude.

The distance between collectors can be calculated from

$$D = L_c \frac{\sin(\text{TILT}) + P)}{\sin(P)} \tag{2.7}$$

In Eq. (2.7), the profile angle P to be used corresponds to that for the hour and the day at which the designer has determined that the shadow line should just reach the foot of the next northern row of collectors. On December 21 at an hour angle of 30°, this profile angle is approximately 18° for 40°N.

Shadow Angle Protractor

The shadow angle protractor shown in Fig. 2.8 is similar to the sunpath diagram. However, instead of solar altitude angle, the solar profile angle P is plotted for various values of the azimuth angle. The sunpath diagram can be superimposed over the shadow angle protractor to show when certain profile angle values occur. This is a particularly useful feature when attempting to evaluate shading phenomena on solar collectors.

Use of the shadow angle protractor is relatively simple. The lines that cast a shadow on a solar collector aperture are identified from plan and section drawings of the solar building site. The sections required are those that show the shadow-casting lines *in an end view*. For example, in Fig. 2.7 the section drawing has been constructed such that the upper edge of the collector, which is the shadow-casting line, is shown in an end view, that is, as a point. As many section drawings are required as shadow-casting lines.

From the sections constructed, the governing values of profile angle are determined as well as solar azimuth angles during which shading can occur. The

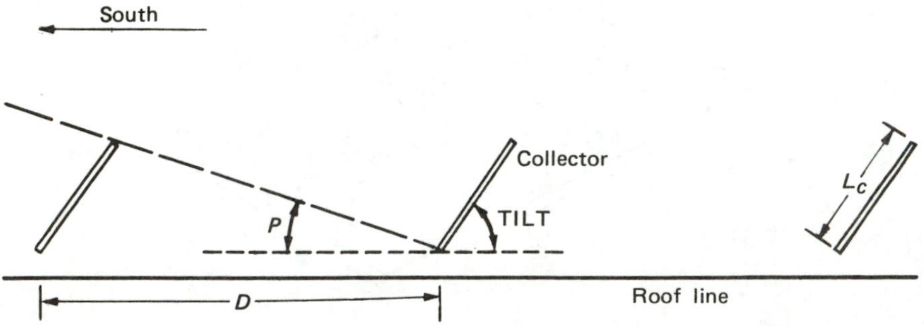

Figure 2.7 Sawtooth collector arrangement showing limiting value of profile angle P for no shading of one collector by another.

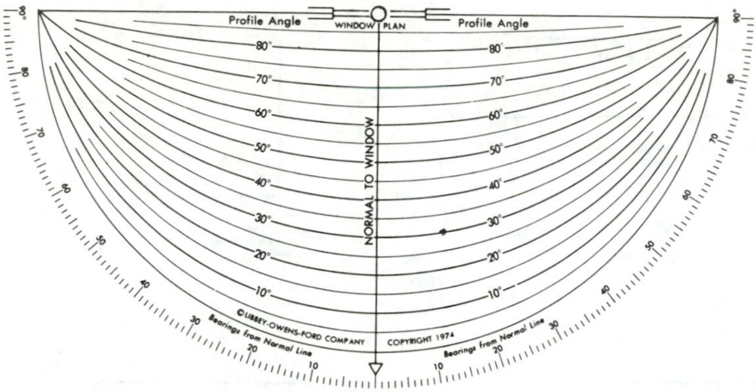

Figure 2.8 Shadow angle protractor diagram. Protractor is a graphic construction of profile angles for various values of solar azimuth angle. (*Courtesy of Libbey-Owens-Ford Glass Co.*)

profile and azimuth angles are plotted on the shadow angle protractor, which is then superimposed on a sunpath diagram to determine the time when shading will occur. The example below illustrates the use of the sunpath diagram and shadow angle protractor. The same method is used for all shading calculations, one section being required for each building line that can cast a shadow on the collector.

EXAMPLE

By using the sunpath diagram and shadow angle protractor, find the months of the year and hours of the day when the overhang shown in Fig. 2.9 will shade point *J* at the base of the vertical wall. The vertical wall might be a window and the overhang to be designed shades the window in summer. Use the angles and dimensions shown in the figure to establish proper profile angles and azimuth angles to prepare the shadow map.

SOLUTION

Three edges of the overhang, *AB*, *BC*, and *CD*, cast shadows on point *J* at different hours of the day and times of the year. Therefore three cross sections of the overhang are required. These are shown in Fig. 2.9, *b* to *d*.

Beginning with shadow-casting line segment *AB* shown in cross section in Fig. 2.9*b*, the shadow angle protractor is oriented so that the line of symmetry (called "normal to window" in Fig. 2.8) faces west. In this position, the shadow angle protractor can be used to prepare the shadow map of line *AB*. Shading will occur for azimuth angles between due west and 30° south of west. These are shown by two solid lines on the shadow angle protractor in Fig. 2.10. Likewise, shading will occur for profile angles greater than 30°. Consequently, for the triangular wedge shown in Fig. 2.10, shading will occur. The triangle is

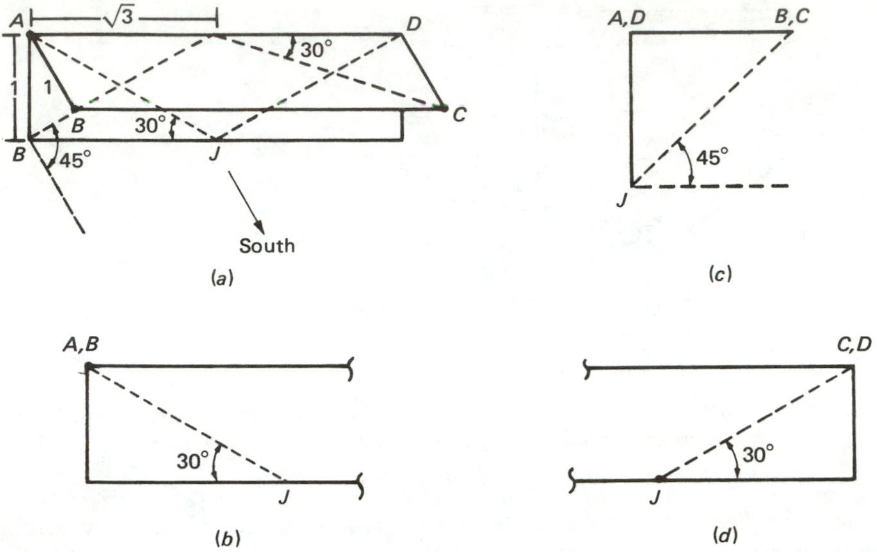

Figure 2.9 Example problem involving overhang. (*a*) isometric sketch; (*b*) section showing line segment *AB* in end view; (*c*) section showing line segment *BC* in end view; (*d*) section showing line segment *CD* in end view.

formed by the limiting values of profile and azimuth angles that must coexist for shading of point *J* to occur. The shadow map shown in Fig. 2.10 is superimposed on the sunpath diagram for 35°N in Fig. 2.11. Points *A* and *B* are labeled in this composite diagram.

The same exercise is now repeated for line segment *BC*. The cross section involved is shown in Fig. 2.9*c*, with governing values of the azimuth angle ranging between 60° east of south and 60° west of south. The limiting value of the profile angle for shading of point *J* to occur is 45°. The shadow map for line

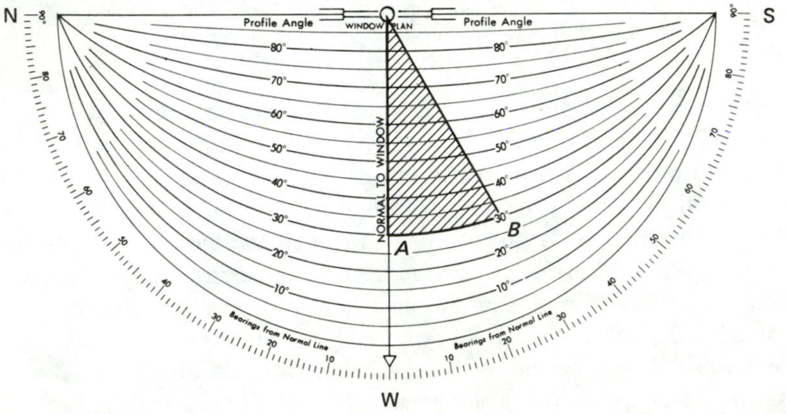

Figure 2.10 Shadow map for line segment *AB*. (*Courtesy of Libbey-Owens-Ford Glass Co.*)

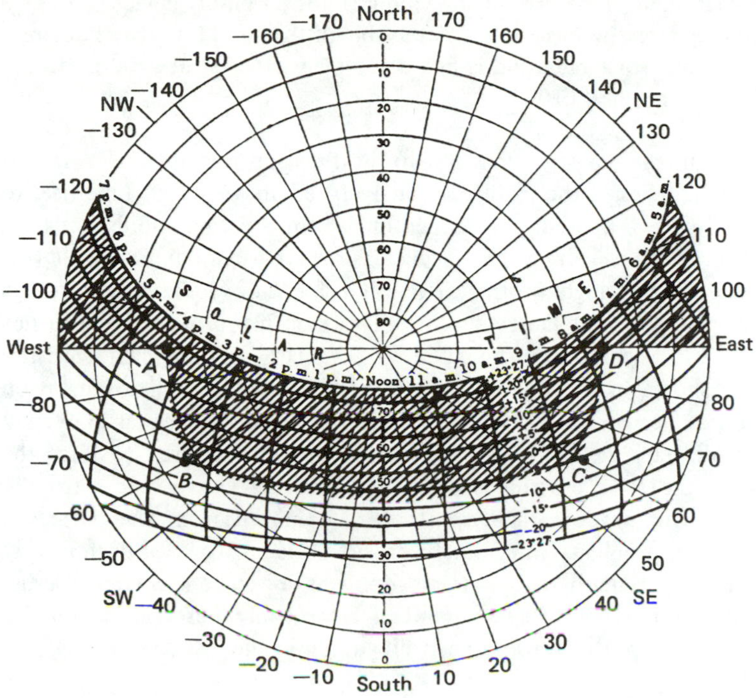

Figure 2.11 Shadow map for overhang *ABCD* superimposed on sunpath diagram for 35°N. Months and hours when shading occurs can be read directly from this figure.

segment *BC* is constructed in the same way as for segment *AB* except that the shadow angle protractor is now faced due south. This direction is determined by the plane in which the cross section is drawn. In each case, the line of symmetry of the shadow angle protractor is placed parallel to the cross-sectional plane for the given line segment being evaluated. The shadow map for segment *BC* is then superimposed on the 35°N sunpath diagram as shown in Fig. 2.11. Points *B* and *C* are labeled in the figure.

Finally, line segment *CD* is plotted on a shadow angle protractor. It is the mirror image of line segment *AB* and is shown plotted on the composite diagram in Fig. 2.11.

The cross-hatched area in Fig. 2.11 bounded by points *A, B, C,* and *D* is the shadow map for the overhang shown in Fig. 2.9*a*. In addition to the shadows cast by the overhang, the plane containing point *J* will be in a shadow whenever the sun is north of the east-west line. Two triangular additions must be made to the composite drawing of Fig. 2.11 to account for this phenomenon. These are shown by the triangular, shaded regions above the east-west line.

The composite diagram can be used to determine when shading of point *J* will take place. It will occur between 3:00 and 4:00 p.m. for the period of the year when the declination is greater than −5°, i.e., between approximately March 8 and October 6. As indicated in the diagram, shading does not occur at precisely the same time each day. Therefore this overhang is not an on-off device but a device that requires a few weeks to completely block the sun at a given hour of the day. For example, solar radiation is not completely blocked at 4:00 p.m. until the declination reaches +10°.

To carry out the construction shown in the previous example, several relatively simple sketches of the feature to be analyzed are required. Likewise, to analyze shading of a proposed new collector by an existing building would require a site plan and at least two section drawings. An approach that is equivalent to the shadow angle protractor one but does not require any drawings has been developed by Solar Pathways, Inc. This firm has produced the device shown in Fig. 2.12, which can be used to evaluate shading characteristics immediately. As shown in the figure, the reflection of the solar environment on the glass dome can be used to sketch the shadow map on a clear plastic overlay of the sunpath diagram included with the device. A pencil is used to trace the characteristics of the solar environment reflected on the glass dome onto the horizontal plane. This shadow map can then be removed from the device to evaluate any shading problems immediately on site. It is most useful during early stages of design, when potential site shading problems are to be evaluated. Design of overhangs and vertical fins for sun control is probably best carried out by using the shadow angle protractor approach illustrated in the last example.

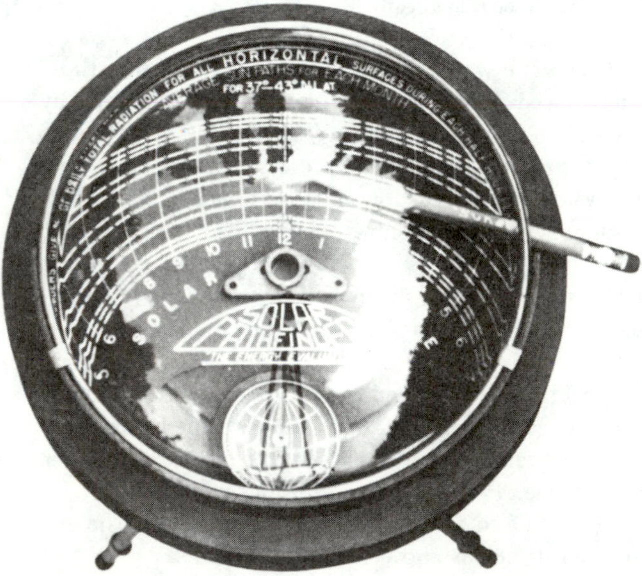

Figure 2.12 Solar pathfinder shading device manufactured by Solar Pathways, Inc. Device can be used to evaluate shading problems on site without making any drawings. (*Courtesy of Solar Pathways, Glenwood Springs, Colorado.*)

QUANTITATIVE ANALYSIS OF SOLAR RADIATION

In the preceding sections we have discussed the position of the sun and the determination of whether a surface receives radiation. However, the precise amount of solar radiation striking a surface has yet to be determined. In principle this is a very complicated calculation, since meteorological factors, solar geometry, and receiving surface characteristics must all be known before the amount of solar radiation striking a given surface can be calculated. However, for system design it is necessary to calculate the amount of solar radiation on various surfaces on a monthly basis only. Therefore certain meteorological variables can be replaced by their average values over a month and the calculations are considerably simplified.

In this section, a method of calculating solar radiation on a monthly basis on any south-facing surface is described. It is presented as a five-step recipe, which can be used with the tables presented below and in App. 6. If the reader prefers to read tabulated solar radiation values for various surface orientations, App. 6 presents radiation values on horizontal surfaces, surfaces with tilt equal to the latitude, the latitude $+15°$ and vertical surfaces for more than 200 cities. Also presented are monthly heating degree-days, which can be used to predict solar heating system performance, and monthly average ambient temperatures, used for a similar purpose. If the location for which a system is to be designed is not represented in App. 6, local solar radiation can be used in the five-step process delineated below. If solar radiation data of a historical nature do not exist, those for a nearby site with similar climate characteristics can be used.

Five-Step Recipe for Calculating Tilted Surface Insolation

Most weather services in the United States and elsewhere measure the sum of beam and diffuse radiation on a horizontal surface. Such records exist for hundreds of locations around the world and have been averaged over each month of the year. Therefore total horizontal solar radiation data on a monthly basis are generally available. Although they may not be available for a specific project in a specific site, they will in all likelihood be available for a site nearby.

The key problem in calculating solar radiation on a tilted surface is determining the relative amounts of beam and diffuse radiation contained in the horizontal total measured by the meteorological services. Collectors are tilted so that they will capture more beam radiation. Since diffuse radiation comes from all directions, there is no preferred orientation for capturing diffuse radiation. However, beam radiation comes from a well-defined location, namely the sun, whose position varies significantly between summer and winter. The collector used to capture solar radiation is positioned in such a way that solar heat production is maximized for the season of peak energy requirement.

In the case of solar heating, collectors are typically tilted at an angle 15° greater than the local latitude in order to favor the winter sun. Passive solar systems typically use a vertical aperture, since they are incorporated in walls of buildings in many cases. Solar water-heating systems, having a relatively uniform energy demand year-round, use collectors tilted at approximately the local latitude. In this section, we show how horizontal radiation measurements can be converted to these various angles of tilt.

To use the five-step recipe, the following information is required:

1 Local latitude.
2 Collector tilt angle.
3 Monthly extraterrestrial radiation EXTERR.
4 Monthly terrestrial radiation TOTH.
5 Beam, diffuse, and reflected radiation tilt factors.

The five-step process is described below, followed by an example.

Step 1. Insolation Data

Monthly average horizontal total radiation data for the site should be collected. These data are preferably long-term averages measured over a period greater than 10 years. Extraterrestrial radiation is required for each month for the same latitude. Monthly extraterrestrial radiation is the average radiation striking a horizontal surface located in exactly the same location, but in the absence of the atmosphere. Amounts of extraterrestrial radiation can be calculated very accurately, since no intervening meteorological or atmospheric conditions need to be taken into account; their calculation is described in [2]. Table 2.1 gives values of monthly averaged extraterrestrial radiation for latitudes between 20 and 65°N for each month of the year.

Step 2. Calculate Clearness Index CLEAR

The clearness is the ratio of terrestrial solar radiation TOTH to extraterrestrial radiation EXTERR. Physically, it is the average atmospheric transmittance to solar radiation for the month in question.

Step 3. Calculate Horizontal and Diffuse Components of Radiation

In this step, the total horizontal direct terrestrial solar radiation TOTH is decomposed into its direct component DIRH and its diffuse component DIFFH. At this step short-term calculations are complex. However, over the long term it has been found that a relatively simple expression, given by Eq. (2.8), relates the diffuse radiation to the clearness index and the total horizontal radiation [3].

$$\text{DIFFH} = (1.0 - 1.097\ \text{CLEAR})\ \text{TOTH} \tag{2.8}$$

After calculating the diffuse component from Eq. (2.8), one can use Eq. (2.9) to calculate the direct or beam component DIRH by simple difference; this calculation is carried out for each month of the year.

$$\text{DIRH} = \text{TOTH} - \text{DIFFH} \tag{2.9}$$

Table 2.1 Average Extraterrestrial Radiation on a Horizontal Surface in SI Units and in English Units Based on a Solar Constant of 1.353 kW/m^2 or 429 Btu/(h·ft^2)

Latitude (deg)	Jan.	Feb.	Mar.	Apr.	May	June	July	Aug.	Sept.	Oct.	Nov.	Dec.
					Watt-hours per square meter per day							
20	7,415	8,397	9,552	10,422	10,801	10,868	10,794	10,499	9,791	8,686	7,598	7,076
25	6,656	7,769	9,153	10,312	10,936	11,119	10,988	10,484	9,494	8,129	6,871	6,284
30	5,861	7,087	8,686	10,127	11,001	11,303	11,114	10,395	9,125	7,513	6,103	5,463
35	5,039	6,359	8,153	9,869	10,995	11,422	11,172	10,233	8,687	6,845	5,304	4,621
40	4,200	5,591	7,559	9,540	10,922	11,478	11,165	10,002	8,184	6,129	4,483	3,771
45	3,355	4,791	6,909	9,145	10,786	11,477	11,099	9,705	7,620	5,373	3,648	2,925
50	2,519	3,967	6,207	8,686	10,594	11,430	10,981	9,347	6,998	4,583	2,815	2,100
55	1,711	3,132	5,460	8,171	10,358	11,352	10.825	8,935	6,325	3,770	1,999	1,320
60	963	2,299	4,673	7,608	10,097	11,276	10,657	8,480	5,605	2,942	1,227	623
65	334	1,491	3,855	7,008	9,852	11,279	10,531	8,001	4,846	2,116	544	97
					Btu's per square foot per day							
20	2,346	2,656	3,021	3,297	3,417	3,438	3,414	3,321	3,097	2,748	2,404	2,238
25	2,105	2,458	2,896	3,262	3,460	3,517	3,476	3,316	3,003	2,571	2,173	1,988
30	1,854	2,242	2,748	3,204	3,480	3,576	3,516	3,288	2,887	2,377	1,931	1,728
35	1,594	2,012	2,579	3,122	3,478	3,613	3,534	3,237	2,748	2,165	1,678	1,462
40	1,329	1,769	2,391	3,018	3,455	3,631	3,532	3,164	2,589	1,939	1,418	1,193
45	1,061	1,515	2,185	2,893	3,412	3,631	3,511	3,070	2,410	1,700	1,154	925
50	797	1,255	1,963	2,748	3,351	3,616	3,474	2,957	2,214	1,450	890	664
55	541	991	1,727	2,585	3,277	3,591	3,424	2,826	2,001	1,192	632	417
60	305	727	1,478	2,407	3,194	3,567	3,371	2,683	1,773	931	388	197
65	106	472	1,219	2,217	3,116	3,568	3,331	2,531	1,533	670	172	31

Source: Adapted from [2].

Step 4. Calculate Tilt Factors

The tilt factor is the ratio of a particular type of radiation on a surface tilted at any angle to that on a horizontal surface. After completion of step 3, all horizontal components of radiation are known. After multiplying by the tilt factors evaluated in this step, the radiation amounts are known for each type on tilted surfaces. Beam, diffuse, and reflected radiation tilt factors can be calculated from equations given in [2]; values are summarized in Tables 2.2 and 2.3 for south-facing surfaces.

Table 2.2 Monthly Beam Radiation Tilt Factors BTILT for Various Latitudes and Collector Tilts

Latitude (deg)	Jan.	Feb.	Mar.	Apr.	May	June	July	Aug.	Sept.	Oct.	Nov.	Dec.
					(*a*) TILT = LAT − 15°							
25	1.22	1.15	1.08	1.01	0.97	0.95	0.96	1.00	1.05	1.13	1.20	1.24
30	1.38	1.26	1.14	1.03	0.96	0.93	0.95	1.00	1.10	1.22	1.35	1.42
35	1.59	1.40	1.21	1.06	0.96	0.92	0.94	1.02	1.15	1.33	1.54	1.66
40	1.88	1.58	1.31	1.10	0.97	0.92	0.94	1.04	1.22	1.48	1.80	2.00
45	2.31	1.83	1.43	1.15	0.98	0.92	0.95	1.07	1.31	1.69	2.17	2.51
50	2.99	2.18	1.60	1.21	1.00	0.92	0.96	1.11	1.43	1.97	2.75	3.34
55	4.19	2.71	1.82	1.28	1.02	0.93	0.97	1.17	1.58	2.37	3.71	4.92
					(*b*) TILT = LAT							
25	1.47	1.31	1.14	0.98	0.87	0.83	0.85	0.93	1.07	1.25	1.43	1.53
30	1.66	1.43	1.20	1.00	0.87	0.81	0.84	0.94	1.12	1.35	1.60	1.74
35	1.91	1.58	1.28	1.02	0.87	0.81	0.83	0.96	1.17	1.46	1.82	2.02
40	2.25	1.79	1.38	1.06	0.88	0.80	0.84	0.98	1.24	1.65	2.12	2.42
45	2.76	2.06	1.51	1.11	0.89	0.80	0.84	1.01	1.34	1.87	2.56	3.03
50	3.54	2.46	1.68	1.17	0.90	0.81	0.85	1.05	1.45	2.17	3.22	4.01
55	4.93	3.05	1.91	1.24	0.92	0.81	0.86	1.10	1.61	2.62	4.32	5.87
					(*c*) TILT = LAT + 15°							
25	1.63	1.38	1.12	0.88	0.73	0.66	0.69	0.81	1.02	1.29	1.56	1.71
30	1.83	1.50	1.18	0.90	0.72	0.65	0.68	0.82	1.06	1.39	1.74	1.94
35	2.10	1.66	1.25	0.92	0.72	0.64	0.68	0.83	1.12	1.52	1.98	2.25
40	2.47	1.87	1.35	0.95	0.73	0.64	0.68	0.85	1.18	1.69	2.30	2.69
45	3.01	2.16	1.48	1.00	0.74	0.64	0.68	0.88	1.27	1.92	2.77	3.34
50	3.86	2.57	1.65	1.05	0.75	0.64	0.69	0.91	1.38	2.23	3.47	4.41
55	5.34	3.18	1.88	1.12	0.77	0.65	0.70	0.95	1.53	2.68	4.64	6.42
					(*d*) TILT = 90°							
25	1.34	0.95	0.55	0.21	0.05	0.01	0.02	0.13	0.40	0.82	1.24	1.47
30	1.59	1.12	0.66	0.29	0.11	0.05	0.08	0.21	0.50	0.97	1.47	1.74
35	1.90	1.33	0.80	0.38	0.18	0.11	0.14	0.29	0.62	1.15	1.75	2.10
40	2.31	1.59	0.95	0.48	0.25	0.17	0.21	0.37	0.75	1.37	2.11	2.58
45	2.90	1.92	1.13	0.59	0.33	0.24	0.28	0.47	0.89	1.64	2.62	3.28
50	3.79	2.37	1.35	0.70	0.41	0.31	0.35	0.57	1.06	2.00	3.36	4.40
55	5.32	3.02	1.63	0.84	0.49	0.38	0.43	0.67	1.26	2.49	4.57	6.47

Table 2.3 Diffuse and Reflected Radiation Tilt Factors, DTILT and RTILT

TILT (deg)	DTILT	RTILT
0	1.00	0.00
10	0.99	0.01
20	0.97	0.03
30	0.93	0.07
40	0.88	0.12
50	0.82	0.18
60	0.75	0.25
70	0.67	0.33
80	0.59	0.41
90	0.50	0.50

The beam radiation tilt factor BTILT depends on latitude and surface tilt angle and is shown in Table 2.2 for several values of each. The diffuse and reflected surface tilt factors DTILT and RTILT depend primarily on surface tilt and are given in Table 2.3 for angles between 0 and 90°.

Step 5. Calculate Total Collector Plane Insolation

The total collector plane insolation is the sum of the beam, diffuse, and reflected sunlight components. Each is calculated by multiplying the tilt factor

Table 2.4 Solar Reflectance Values REFL for 15 Characteristic Surfaces

Surface	Average reflectivity
Snow (freshly fallen or with ice film)	0.75
Water surfaces (relatively large incidence angles)	0.07
Soils (clay, loam, etc.)	0.14
Earth roads	0.04
Coniferous forest (winter)	0.07
Forests in autumn, ripe field crops, plants	0.26
Weathered blacktop	0.10
Weathered concrete	0.22
Dead leaves	0.30
Dry grass	0.20
Green grass	0.26
Bituminous and gravel roof	0.13
Crushed rock surface	0.20
Building surfaces, dark (red brick, dark paints, etc.)	0.27
Building surfaces, light (light brick, light paints, etc.)	0.60

Source: Hunn and Calafell [5]. Reprinted with permission from *Solar Energy*, vol. 19, B. D. Hunn and D. O. Calafell, "Determination of Average Ground Reflectivity for Solar Collectors," Copyright 1977, Pergamon Press, Ltd. See also List [6].

by the respective horizontal insolation component. For reflected radiation it is assumed that horizontal beam and diffuse radiation are reflected similarly; therefore the horizontal solar flux reflected is the total horizontal radiation. In addition to the tilt factor and radiation component, the reflected component term will include the reflectance of the material immediately in the foreground of the collector from which sunlight is reflected onto the collector. Reflectances are summarized in Table 2.4. Equation (2.10) expresses the calculation of total collector plane insolation TOTCOL as the sum of the three terms described above.

$$\text{TOTCOL} = \underbrace{\text{DIRH} \times \text{BTILT}}_{\text{Direct}} + \underbrace{\text{DIFFH} \times \text{DTILT}}_{\text{Diffuse}} + \underbrace{\text{TOTH} \times \text{REFL} \times \text{RTILT}}_{\text{Reflected}} \tag{2.10}$$

where REFL is the foreground reflectance. The five-step method is illustrated in the following example.

EXAMPLE

How much sunlight strikes a solar collector tilted at 55° in Denver, Colorado (40°N) in February? Assume that the ground is covered with day-old snow with a reflectance of 0.70. Use the five-step recipe.

SOLUTION

Step 1. From App. 6, TOTH = 1127 Btu/(ft^2·day) and from Table 2.1 EXTERR = 1769 Btu/(ft^2·day).

Step 2. The clearness index is CLEAR = 1127/1769 = 0.64.

Step 3. The diffuse and beam components are

$$\text{DIFFH} = (1 - 1.097 \times 0.64) \times 1127 = 336 \text{ Btu/(ft}^2\cdot\text{day)}$$

The remaining flux is direct radiation:

$$\text{DIRH} = 1127 - 336 = 791 \text{ Btu/(ft}^2\cdot\text{day)}$$

Step 4. Tilt factors are read from Tables 2.2 and 2.3, interpolated as needed.

$$\text{BTILT} = 1.87$$

$$\text{DTILT} = 0.79$$

$$\text{RTILT} = 0.21$$

Step 5. Total flux TOTCOL is found from Eq. (2.10):

$$\text{TOTCOL} = (791 \times 1.87) + (336 \times 0.79) + (1127 \times 0.7 \times 0.21)$$

$$\text{TOTCOL} = 1910$$

Therefore the average solar flux in February on the 55° surface in Denver is 1910 Btu/(ft^2·day).

Clear-sky radiation on an hourly basis has been tabulated by the American Society of Heating, Refrigerating and Air-Conditioning Engineers (ASHRAE). These values are not useful for solar system performance design, since solar systems respond to actual monthly solar radiation, including all climatic and meteorological effects. Clear-sky radiation tabulations are useful for estimating peak cooling loads, or for quickly estimating the effect of surface tilt on the amount of solar radiation intercepted. However, for long-term performance predictions for actual systems one must use real data, not synthesized clear-sky data.

Many vertical passive collectors use overhangs to restrict the amount of beam radiation striking vertical passive apertures in summer. The overhang effect can be calculated on a monthly basis with an equation similar to Eq. (2.10); however, two additional terms describing the overhang are required. Calculation of solar flux on vertical surfaces with overhangs is described in [7], where passive systems are discussed in detail.

REFERENCES

1. G. B. Hayes, *Solar Access Law*, Ballinger, Cambridge, Mass., 1979.
2. F. Kreith and J. F. Kreider, *Principles of Solar Engineering*, Hemisphere, Washington, D.C., 1978.
3. J. K. Page, *Proc. U.N. Conf. New Sources Energy*, vol. 4, 1964, p. 378.
4. R. Bennett, *Sun Angles for Design*, Robert Bennett, Bala Cynwyd, Pa., 1978.
5. B. D. Hunn and D. O. Calafell, "Determination of Average Ground Reflectivity for Solar Collectors," *Solar Energy*, vol. 19, 1977, p. 87.
6. R. J. List, *Smithsonian Meterological Tables*, 6th ed., Smithsonian Institution, Washington, D.C., 1949, pp. 442–443.
7. J. F. Kreider, *The Solar Heating Design Process*, McGraw-Hill, New York, 1981.

3 Solar Collectors

Solar collectors are heat exchangers that use solar radiation to heat a working fluid, usually a liquid or air. They can be classified conveniently in three groups. The first group consists of flat-plate collectors, which use no optical concentration of sunlight. They are generally stationary and their outlet temperature capability is below 200°F (95°C). The second group are focusing collectors, which must "track" the sun and can generally utilize only the direct radiation. They are capable of producing high temperatures. Intermediate between these are nonimaging concentrators, which do not produce a well-defined focal spot and are capable of achieving temperatures up to about 350°F (175°C). These systems require no tracking if their concentration ratio (ratio of aperture to absorber area) is below 1.7 and only seasonal adjustments at concentration ratios up to 5. Each of these collector types has an appropriate set of applications, and choosing the correct collector for a given application is an important task of the solar engineer. For heating and cooling, flat-plate and stationary nonimaging collectors are best [1, 2].

FLAT-PLATE COLLECTORS

Flat-plate collectors are commonly used for space heating, swimming pool heating, and domestic hot-water applications. A typical flat-plate collector consists of the following components (see Fig. 3.1):

Absorber Plate

This is usually made of copper, steel, or plastic. The surface is covered with a flat black material of high absorptance. If the absorber plate is made of copper

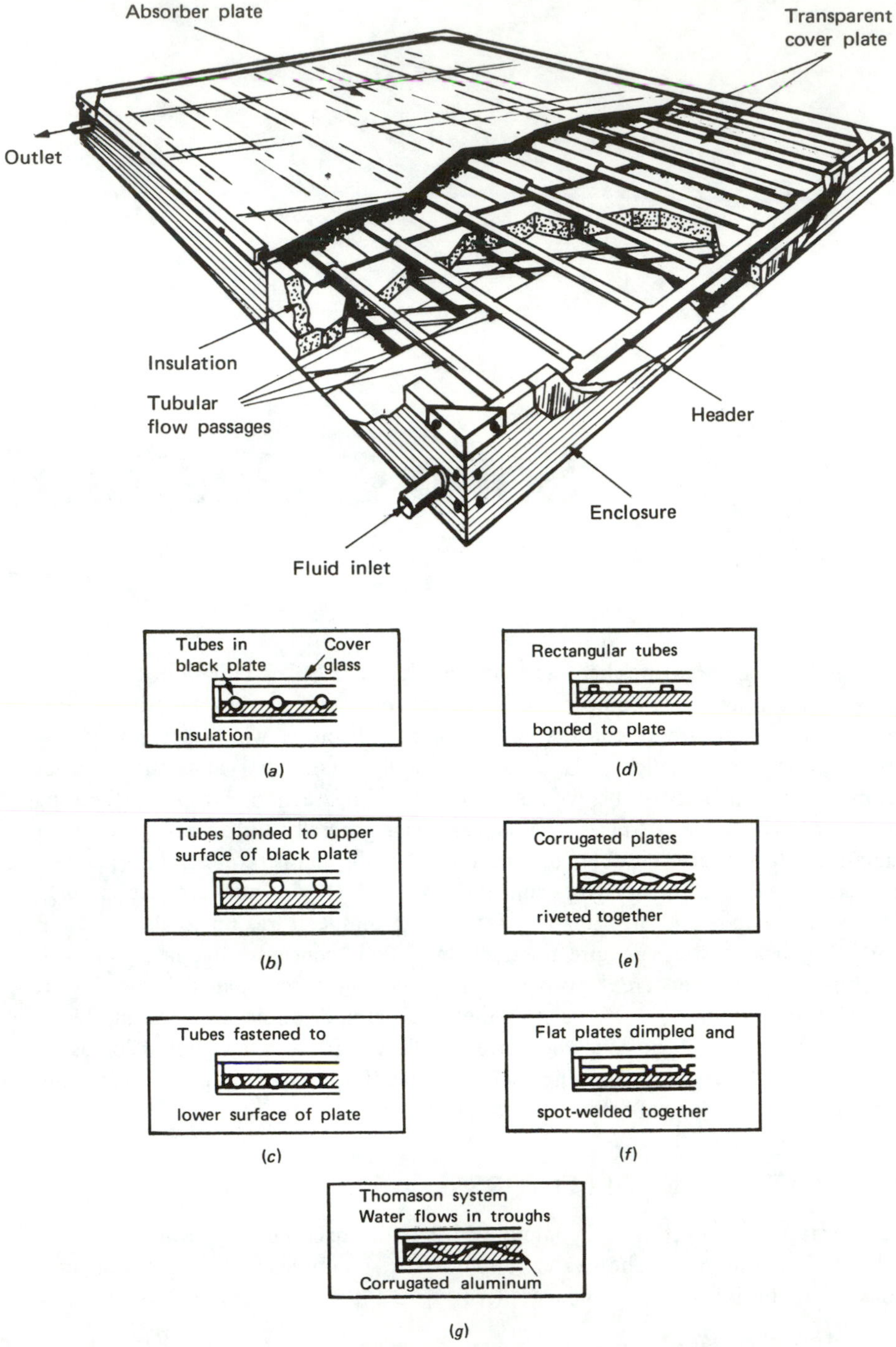

Figure 3.1 Cross sections of typical liquid flat-plate collectors.

or steel, it is possible to apply a selective coating that maximizes the absorptance of solar energy and minimizes the radiation emitted by the plate.

Flow Passages

The flow passages conduct the working fluid through the collector. If the working fluid is a liquid, the flow passage is usually a tube that is attached to or is an integral part of the absorber plate. If the working fluid is air, the flow passage should be below the absorber plate to minimize heat losses, which are excessive if the solar-heated air is in contact with the outer glazing. In air collectors it is often useful to increase the heat transfer surface area by means of fins attached to the absorber plate or metal screens that promote turbulence.

Cover Plate

To reduce convective and radiative heat losses from the absorber, one or two transparent covers are generally placed above the absorber plate. An exception is a collector used for swimming pool heating, where only a small temperature rise above ambient is required. Cover plates may be made from glass or transparent sheets of plastic.

Insulation

Insulating materials such as fiberglass are placed at the back and sides of the collector to reduce parasitic heat losses.

Enclosure

The collector is enclosed in a box that holds the components together, protects them from the weather, and facilitates installation of the collector on a roof or appropriate frame.

Typical cross sections of liquid and air flat-plate collectors are shown in Figs. 3.1 and 3.2. Aluminum is not suitable for liquid collectors because

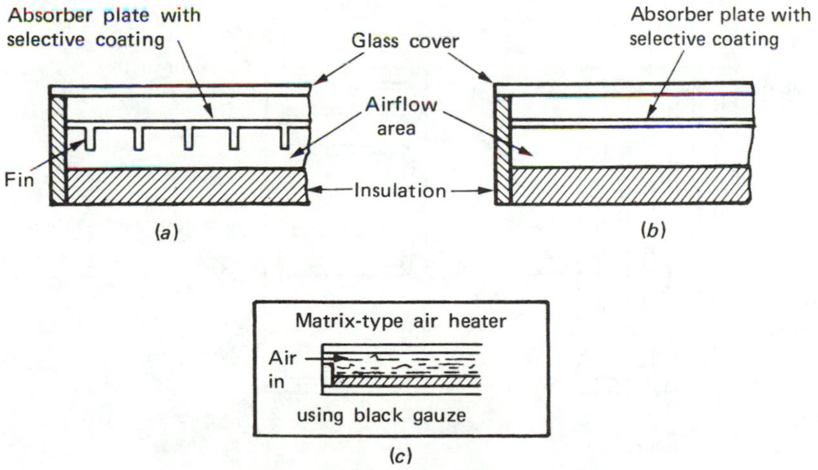

Figure 3.2 Cross sections of typical air flat-plate collectors. (*a*) With fins; (*b*) without fins; (*c*) matrix type. (*Adapted from [3]*.)

electrolytic corrosion would occur between copper tubes in the distribution system and an aluminum collector. It may be used in air systems.

Absorber Plate/Flow Passages

In a liquid collector the flow passage generally does not extend over the entire absorber plate surface, and heat must therefore be conducted along the absorber plate to reach the fluid. In such an installation the tubes are spaced several inches apart; the absorber surface between them acts like a fin, with radiant energy impinging on the surface, being conducted toward the tubes, and being transferred from the tube inner surface by convection to the liquid working fluid. Copper, which has high conductivity and is corrosion-resistant, is the best material for absorber plates, but because copper is expensive steel is also widely used.

Tube spacing is determined by a trade-off between the fin efficiency of the collector plate and the cost of the tubes. For a copper plate 0.02 in (0.05 cm) thick with $\frac{1}{2}$-in (1.25 cm) tubes spaced 6 in (15 cm) apart in good thermal contact with the copper, the fin efficiency is better than 97 percent. That is, compared to a plate with infinitely high conductivity, the effect of conducting heat along the copper plate toward the tubes reduces the heat collection by less than 3 percent.

Flow tubes in a liquid collector must be thermally bonded to the collector plate in such a way that there is good thermal conductance. Soldered or welded contacts reduce the contact resistance between tube and plate; clamped contacts have excessive thermal contact resistance and should be avoided. The flow tubes can be routed through the collector in parallel paths from an inlet to an outlet header (Fig. 3.3*a*), or a single tube can be routed in serpentine fashion (Fig.

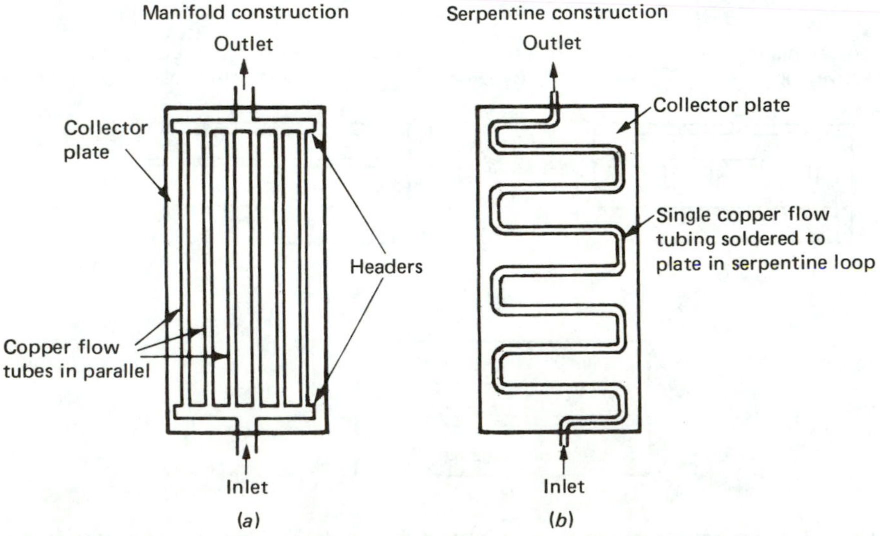

Figure 3.3 Flow-tube routing. (*a*) With headers; (*b*) in a serpentine loop.

3.3*b*). The latter arrangement eliminates the possibility of header leaks and ensures uniform flow. However, it also increases the pressure drop, and it is not suitable for a system using drain-down protection because the curved flow passages cannot be drained.

In air collectors the fluid generally contacts the rear surface of the entire absorber plate, as shown in Fig. 3.2. Because of the low thermal conductivity of air, the heat transfer between the absorber and the working fluid is considerably less than for a liquid system. The heat transfer is often enhanced by use of fins or corrugations or by creating turbulence; however, these measures increase the fan power required and therefore often result in excessive parasitic losses. Air systems have the advantage of eliminating freezing and boiling problems, and although leaks in air systems are more difficult to plug and detect, they are not as troublesome as leaks in liquid systems.

The surface of the absorber plate determines how much of the incident solar radiation is absorbed and how much is emitted at a given temperature. Flat black paint, which is widely used as a coating, has an absorptance of about 95 percent for incident shortwave solar radiation. It is durable and easy to apply. However, it radiates a very large fraction of blackbody radiation at long wavelengths and is therefore inferior to selective surface coatings from the point of view of thermal efficiency. Unfortunately, selective coatings cannot be simply painted on an absorber plate. They require electroplating and can therefore be applied only to metals. One of the most popular selective surfaces is black chrome, which has an absorptance of 90 percent in the solar wavelength range and an emittance of only about 10 percent in the long-wavelength range. Selective surfaces can improve collector performance appreciably, but they also increase the cost of the collector and often deteriorate with time. Table 3.1 is a summary of absorption coating characteristics, including the maximum temperature at which they can be used.

Cover Plates

A cover plate for a collector should have a high transmittance for solar radiation and should not deteriorate with time. The material most commonly used is glass. A $\frac{1}{8}$-in- (0.32-cm-) thick sheet of window glass (iron content, 0.12 percent) transmits 85 percent of solar energy at normal incidence. If the iron content is reduced to about 0.01 percent, to make so-called water-white glass, the transmittance increases to about 92 percent. All glass is practically opaque to long-wavelength radiation emitted by the absorber plate. If properly tempered, glass has high durability and negligible deterioration due to exposure to ultraviolet radiation, even over very long periods.

Some plastic materials can be used for collector glazing. They are cheaper and lighter than glass and, because they can be used in very thin sheets, they often have a higher transmittance. However, they are not as durable as glass and they often degrade with exposure to ultraviolet radiation or high temperatures. For example, Tedlar (duPont de Nemours), which has a very high transmittance, cannot be used for an inner glazing in a two-cover collector because it degrades

Table 3.1 Characteristics of Absorptive Coatings

Property or material	Absorptance* α	Emittance ϵ	$\frac{\alpha}{\epsilon}$	Breakdown temperature [°F (°C)]	Comments
Black chrome	0.87–0.93	0.1	~9		
Alkyd enamel	0.9	0.9	1		Durability limited at high temperatures
Black acrylic paint	0.92–0.97	0.84–0.90	~1		
Black inorganic paint	0.89–0.96	0.86–0.93	~1		
Black silicone paint	0.86–0.94	0.83–0.89	1		Silicone binder
PbS/silicone paint	0.94	0.4	2.5	662 (350)	Has a high emittance for thicknesses > 10 μm
Flat black paint	0.95–0.98	0.89–0.97	~1		
Ceramic enamel	0.9	0.5	1.8		Stable at high temperatures
Black zinc	0.9	0.1	9		
Copper oxide over aluminum	0.93	0.11	8.5	392 (200)	
Black copper over copper	0.85–0.90	0.08–0.12	7–11	842 (450)	Patinates with moisture
Black chrome over nickel	0.92–0.94	0.07–0.12	8–13	842 (450)	Stable at high temperatures
Black nickel over nickel	0.93	0.06	15	842 (450)	May be influenced by moisture at elevated temperatures
Ni-Zn-S over nickel	0.96	0.97	14	536 (280)	
Black iron over steel	0.90	0.10	9		

*Dependent on thickness and vehicle-to-binder ratio.
Source: U.S. Department of Housing and Urban Development [4].

at temperatures that may be reached by the second cover plate. Although two cover plates reduce the heat loss from the collector plates, each transparent cover reduces the radiation incident on the collector by at least 10 percent. Therefore, single-glazed collectors will outperform double-glazed collectors at low and medium temperatures. A single-glazed collector with a selective surface is generally the best choice for temperatures up to 200°F (~100°C) and shows the highest thermal performance. Table 3.2 lists the physical characteristics of the materials most commonly used for collector covers.

Enclosure/Insulation

The collector enclosure is usually made from steel, aluminum, or fiberglass. The frame should shade as little of the absorber plate as possible throughout the

Table 3.2 Characteristics of Cover Plate Materials[a]

Test	Polyvinyl fluoride[b]	Polyethylene tereph-thalate or polyester[c]	Polycar-bonate[d]	Fiberglass-reinforced plastics[e]	Methyl methacry-late[f]	Fluorinated ethylene-propylene[g]	Ordinary clear lime glass (float) (0.10–0.13% iron)	Sheet lime glass (0.05–0.06% iron)	Water-white glass (0.01% iron)
Solar transmission, %	92–94	85	82–89	77–90	89	97	85	87	85–91
Maximum operating temperature, °F	227	220	250–270	200°F produces 10% transmission loss	180–190	248	400	400	400
Tensile strength, lb/in²	13,000	24,000	9,500	15,000–17,000	10,500	2,700–3,100	1,600 annealed, 6,400 tempered	1,600 annealed, 6,400 tempered	1,600 annealed, 6,400 tempered
Thermal expansion coefficient, in/(in·°F) × 10^6	24	15	37.5	18–22	41.0	8.3–10.5	4.8	5.0	4.7–8.6
Elastic modulus, lb/in² × 10^6	0.26	0.55	0.345	1.1	0.45	0.5	10.5	10.5	10.5
Thickness, in	0.004	0.001	0.125	0.040	0.125	0.002	0.125	0.125	0.125
Weight for above thickness, lb/ft²	0.028	0.007	0.77	0.30	0.75	0.002	1.63	1.63	1.63
Greatest load, lb/ft² :ft²	–	–	–	–	–	–	–	–	30:30 annealed, 100:28 tempered
Length of life, years	In 5 years retains 95% of total transmission	4	–	7–20	–	–	–	–	–

[a]Values were obtained from the following sources: Grimmer and Moore [5], Kobayashi and Sargent [6], Scoville [7], Clarkson and Herbert [8], *Modern Plastics Encyclopedia* [9], and Toenjes [10].
[b]For example, Tedlar.
[c]For example, Mylar.
[d]For example, Lexan, Merlon.
[e]For example, Kalwall's Sunlite.
[f]For example, Lucite, Plexiglas, Acrylite.
[g]For example, Teflon.
Source: U.S. Department of Housing and Urban Development [4].

day, and the enclosure box should be well sealed to keep out water. Gasket materials in the enclosure should prevent condensation on the inner surfaces of the cover plates and should be able to withstand stagnation temperature without outgassing. (Stagnation is the no-flow condition during which the absorber rises in temperature to a point where heat losses from it equal absorbed insolation.)

Insulation is used in the enclosure to prevent heat losses from the absorber back and the sides. The insulation material should be protected from water and should be of closed-cell construction. It should also be nonflammable. Characteristics of some insulation materials are presented in Table 3.3.

Enclosures should be lightweight and have points of attachment to fasten the collector to a roof or a frame. In some cases, planar reflectors are used in front of flat-plate collectors to increase the amount of solar radiation that impinges on the absorber plate. In such installations special care should be taken to integrate the planar reflectors with the enclosure and provide a means of runoff for rain and snow.

COLLECTORS FOR HIGHER TEMPERATURES

As the temperature requirement increases, the efficiency of a flat-plate collector falls off. Other collectors are often chosen, particularly for supplying higher temperature water to operate an absorption chiller. Examples of some advanced collector designs are discussed below.

Evacuated-Tube Collectors

Convective heat losses can be reduced by evacuating the space around the absorber surface. Since such a vacuum would cause a typical flat-plate collector to collapse, this technique is used in conjunction with a tubular design. There are various types of evacuated-tube collectors on the market.

Table 3.3 Characteristics of Insulation Materials

Material	Density (lb/ft^3)	Thermal conductivity at 200°F (95°C) [$Btu/(h \cdot ft^2 \cdot °F \cdot in)$]	Temperature limits °F	Temperature limits °C
Fiberglass with organic binder	0.6	0.41	350	175
	1.0	0.35	350	175
	1.5	0.31	350	175
	3.0	0.30	350	175
Fiberglass with low binder	1.5	0.31	850	450
Ceramic fiber blanket	3.0	0.4 (at 400°F)	2300	1450
Mineral fiber blanket	10.0	0.31	1200	920
Calcium silicate	13.0	0.38	1200	920
Urea-formaldehyde foam	0.7	0.20 (at 75°F)	210	100
Urethane foam	2–4	0.20	250–400	120–205

Source: U.S. Department of Housing and Urban Development [4].

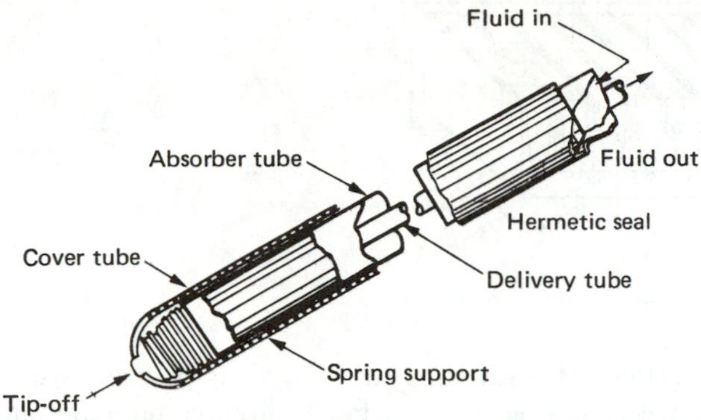

Figure 3.4 Sketch of a concentric glass evacuated-tube collector.

One design, shown in Fig. 3.4, uses three concentric glass tubes. Fluid flows into the annular space between the inner and second tubes and back out the inner tube. The annulus between the second tube and the third (outer) tube is evacuated, and the outside of the second tube has a selective coating [11].

The design in Fig. 3.5 uses only two concentric glass tubes. The space between them is evacuated, and the outer surface of the inner tube has an absorptive coating. A metal fin conforms to the inside surface of the inner tube, and to this a copper U-tube, which carries the fluid, is attached. The U-tube accommodates thermal expansion, and glass breakage will not result in a leak.

Various shapes of reflectors can be placed behind the evacuated tubes to provide a small amount of concentration. Evacuated tubes can collect both direct and diffuse solar radiation and do not require tracking. Because of the extremely high stagnation temperatures—over 600°F (316°C)—that evacuated-tube collectors can reach, thermal shocking with cold water has sometimes been a problem.

Compound Parabolic Concentrator

A compound parabolic concentrator (CPC) collector, as shown in Fig. 3.6, consists of many troughs, each comprised of two half-parabolic reflectors with an

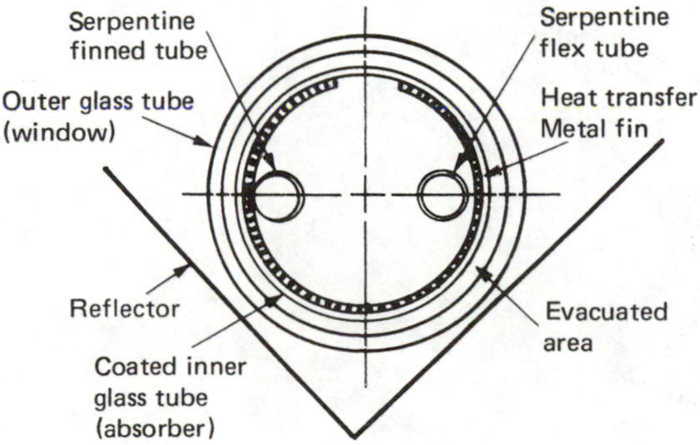

Figure 3.5 Cross section of an evacuated-tube collector with copper tube and V-shaped reflector.

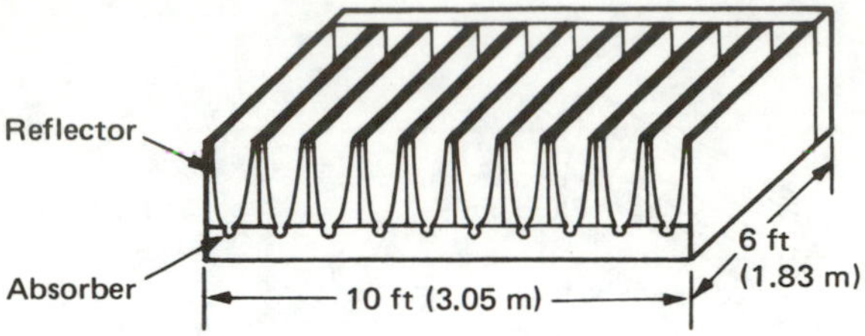

Figure 3.6 A CPC collector module.

absorber surface at the bottom of each trough. A CPC is a nonimaging concentrator with a wide acceptance angle and low concentration ratio; it essentially "funnels" the radiation rather than focuses it. If oriented east-west, troughs with a low concentration ratio can be stationary. At most they require only seasonal adjustment. CPCs collect a considerable amount of diffuse as well as direct radiation. Unlike the other concentrators, they need not have a high specular reflective surface and can thus better tolerate dust and degradation. A disadvantage is the relatively large amount of reflector area required, but truncated designs can be used without seriously degrading performance [12].

Tracking Parabolic Troughs

Another way to reduce heat loss is to concentrate the solar energy onto a smaller absorber surface. The most common concentrator is a parabolic reflector trough with an absorber pipe located along the focal line, as shown in Fig. 3.7. The collectors can be oriented east-west or north-south; in the latter case they are usually tilted from the horizontal. In most cases, concentration ratios are high enough that continuous tracking is required, even with east-west troughs. Since most diffuse radiation falls outside the concentrator's acceptance angle, only direct radiation can be collected. As a result, a concentrating collector receives less of the available sunlight than a flat-plate collector.

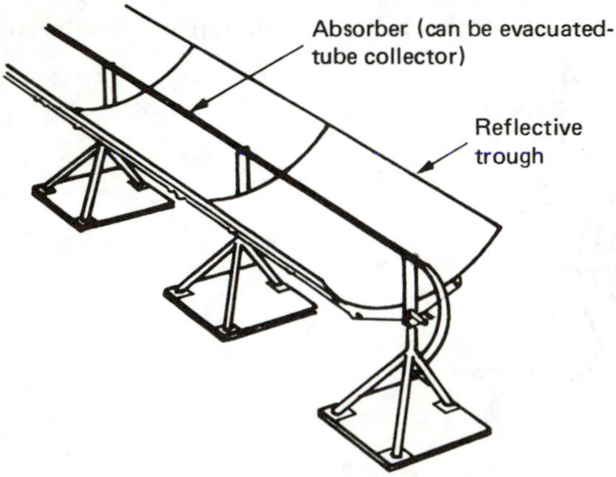

Figure 3.7 Tracking parabolic trough collector.

COLLECTOR PERFORMANCE

The thermal performance of a collector can be calculated from a first-law energy balance. For a simple flat-plate collector an instantaneous energy balance, according to the first law of thermodynamics, is

Useful energy collected = energy absorbed by the collector plate − heat loss to surroundings

In the form of an algebraic equation, the energy balance can be written as

$$q_u = I_c A_c (\tau\alpha)_s - U_c A_c (T_c - T_a) \tag{3.1}$$

where q_u = rate of useful energy gain, Btu/h (W)
A_c = area of collector that absorbs solar radiation, ft^2 (m^2)
I_c = global insolation on plane of collector, Btu/(h·ft^2) (W/m^2)
τ_s = net solar transmittance of glazing
α_s = solar absorptance of collector plate
U_c = overall heat loss coefficient, Btu/(h·ft^2·°F) [W/(m^2·°C)]
T_c = average collector plate surface temperature, °F (°C)
T_a = ambient air temperature, °F (°C)

The key to the thermal performance of a solar collector is the value of the overall energy loss coefficient U_c. This coefficient determines how much of the incoming solar energy is lost to the surroundings for a given average plate surface temperature T_c. Thus to improve the performance of a solar collector, it is necessary either to reduce the overall energy loss coefficient or to reduce the area from which energy is lost. The first measure requires either placing a second glazing over the absorber or surrounding the absorber with an evacuated space. The second approach, reducing the area from which heat loss occurs for a given aperture, requires concentration. Both approaches are used in improving collector performance.

Thermal Efficiency

Although it is possible to calculate the overall energy loss coefficient analytically, it is customary to determine its value by performing experiments. From these experiments, it is also possible to determine the thermal efficiency of the collector η_c, defined as the ratio of the useful energy collected or delivered by the collector to the total insolation on the collector aperture. In equation form, the collector efficiency can be written

$$\eta_c \equiv \frac{q_u}{A_c I_c} \eta_o - \frac{U_c(T_c - T_a)}{I_c} \tag{3.2}$$

where η_o is the optical efficiency of the collector, equivalent to $(\tau\alpha)_s$ in Eq. (3.1). For a simple flat-plate configuration, the optical efficiency equals the product of the net cover transmittance times the solar absorber absorptance. Typical values of τ_s range from 0.75 to 0.90 and values of α_s range from 0.90 to 0.95.

For concentrating collectors, the optical efficiency is given by

$$\eta_o = \rho_s \tau_s \alpha_s \delta \tag{3.3}$$

where ρ_s is the solar reflectance of the concentrator reflector surface and δ is the intercept factor. Reflectances depend on the material used for the concentrator surface and range from 0.80 to 0.92. The intercept factor depends on the accuracy of construction and the geometric size and relation of the receiver and the reflector; it varies from about 0.9 to 1.0, depending on the concentration ratio and the receiver area.

Heat Removal Factor

Because it is difficult to measure the collector plate temperature, T_c, this is usually eliminated from the efficiency relation by defining a collector heat removal efficiency factor F_R. This factor is defined as the useful energy collected divided by the energy that would be collected if the entire collector plate were at the inlet fluid temperature $T_{f,i}$. In algebraic form this relation is [13, 14]

$$F_R = \frac{I_c(\tau\alpha)_s - U_c(T_c - T_a)}{I_c(\tau\alpha)_s - U_c(T_{f,i} - T_a)} \tag{3.4}$$

and the useful energy delivered by the collector becomes

$$q_u = A_c F_R \left[I_c(\tau\alpha)_s - U_c(T_{f,i} - T_a)\right] \tag{3.5}$$

Figure 3.8 shows qualitatively the efficiency for a flat plate collector, a nontracking CPC with a concentration ratio of 1.5 and an evacuated absorber, and a tracking parabolic trough plotted against the difference between the inlet temperature of the collector and the environmental temperature divided by the appropriate insolation. Examination and interpretation of the efficiency curves of Fig. 3.8*a*, which are plotted as instantaneous η_c versus $(T_{f,i} - T_a)/I_c$, show that their slope is $-U_c F_R$ and their intercept is $F_R(\tau\alpha)_s$. Thus the collector heat removal efficiency factor and the overall energy loss coefficient can easily be calculated from an experimentally determined collector efficiency curve if the insolation, the transmittance of the cover plate, and the solar absorptance of the collector plate are known.

The efficiency η_a of a tracking parabolic trough collector with a 20 to 1 concentration ratio (20X) is shown in Fig. 3.8*b* as a function of the temperature difference $(T_{f,i} - T_a)$ divided by the beam insolation per unit aperture area I_b. The efficiency is less than for a flat plate collector at small values of $(T_{f,i} - T_a)/I_f$. But as a result of concentration, the absorber area is smaller than for a flat plate collector and the heat loss is therefore reduced at larger temperature differences. Thus, the slope $U_c F_R$ is smaller than for a flat plate collector and the efficiency does not decrease appreciably as the operating temperature increases.

Figure 3.9 shows peak efficiency curves for three flat-plate liquid collec-

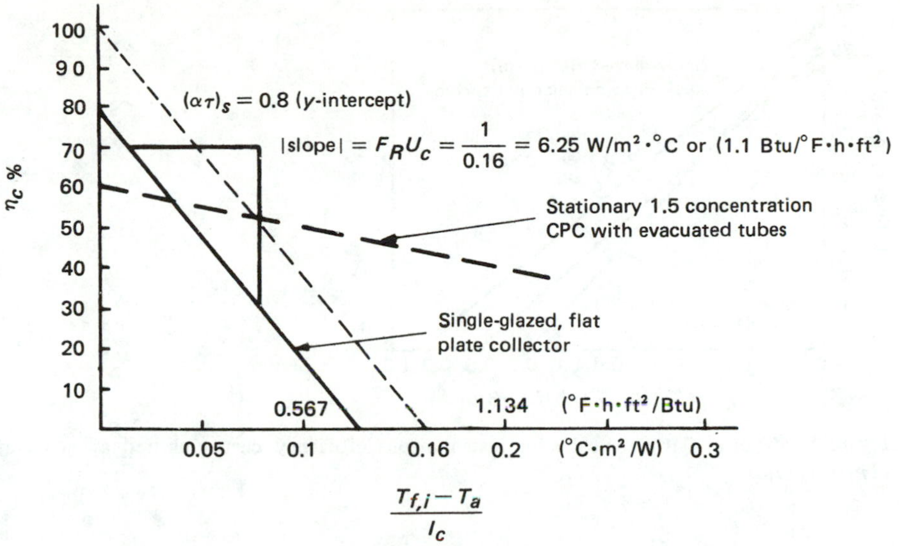

(*a*)

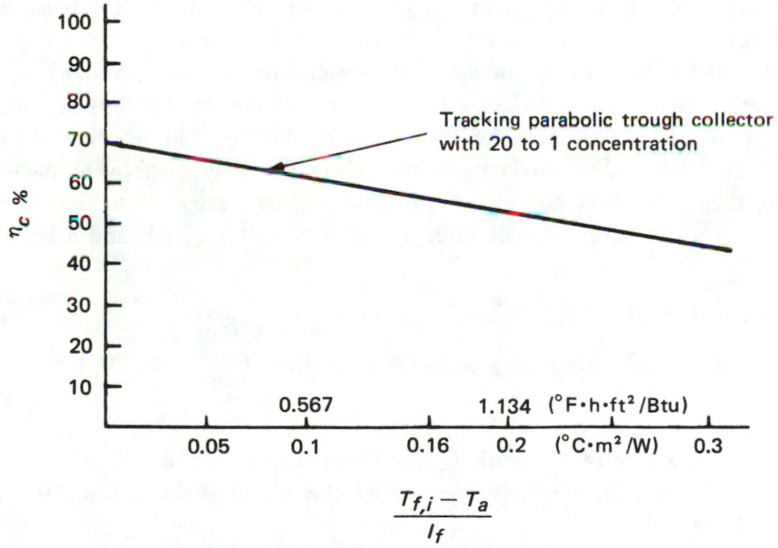

(*b*)

Figure 3.8 (*a*) Typical instantaneous efficiencies η_c for a flat plate and a CPC (1.5X) collector vs. $(T_{f,i} - T_a)/I_c$. (*b*) Typical instantaneous efficiency for a tracking parabolic trough with 20 to 1 concentration.

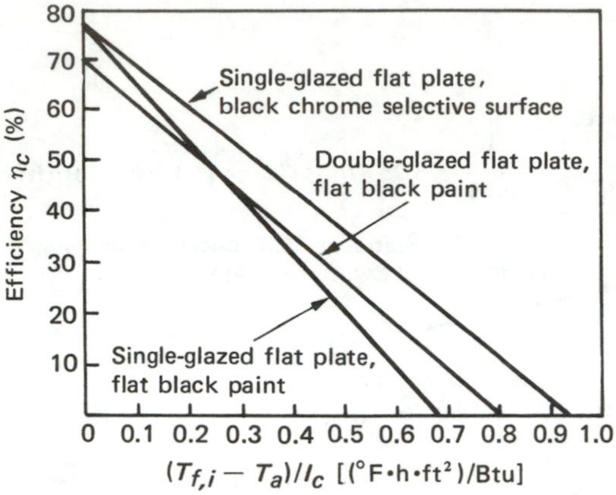

Figure 3.9 Typical flat-plate collector instantaneous efficiency curves plotted as η_c versus $(T_{f,i} - T_a)/I_c$.

tors: two with flat black absorbers (one with a single cover and the other with two covers) and the other with a black chrome selective surface and a single cover. Note that for large values of $(T_{f,i} - T_a)/I_c$, corresponding to the collector supplying water at a much higher than ambient temperature, the two-cover collector has a higher efficiency than the single cover collector. This is because the second cover reduces heat loss, resulting in a smaller U_c and a reduced slope. At small values of $(T_{f,i} - T_a)/I_c$, however, the single-cover collector outperforms the two-cover model because it has a higher transmittance τ and thus a larger value of $F_R(\tau\alpha)_s$. It is easy to take this one step further and see that for a swimming pool collector, which heats water only a few degrees above ambient temperature, it is often best not to use any cover. Using a selective surface with a single cover has the advantages of both a high intercept $F_R(\tau\alpha)_s$ and a lower heat loss.

Collector Testing

The useful thermal output of a collector q_u is given by

$$q_u = \dot{m} c_p (T_{\text{out}} - T_{\text{in}}) \tag{3.6}$$

where $\dot{m}$ = mass flow rate of the working fluid, lb/h (kg/s)

c_p = specific heat at constant pressure of the working fluid, Btu/(lb·°F) [kJ/(kg·C)]

T_{in} = average fluid inlet temperature to collector, °F (°C)

T_{out} = average fluid outlet temperature from collector, °F (°C)

In a test, the mass flow rate of the working fluid is usually measured with a nozzle or an orifice meter when air is used as the working fluid and with an orifice meter or a turbine flowmeter when water is used. The temperature of the fluid entering the collector and the temperature of the fluid leaving it can be measured with a thermocouple or a resistance thermometer.

Insolation incident on the collector aperture is measured with an instrument called a *pyranometer* (Fig. 3.10). A pyranometer measures the hemispheric or total insolation coming from the surroundings above the plane of the instrument. When dealing with a concentrating collector, the beam or direct insolation rather than the hemispheric insolation is sought. This is measured with a *pyrheliometer*, which shades all but the solar disk from the view of the measuring device. A simpler approach is often taken, however, and instead of measuring the beam radiation directly, the sun is shaded from view by use of a shadow band on a pyranometer. Insolation from the surroundings, except that directly from the sun, is thus determined. The beam radiation is then evaluated by subtracting from the total hemispheric insolation I_c the diffuse insolation of the surroundings I_d, or

$$I_b = I_c - I_d \tag{3.7}$$

In evaluating the instantaneous efficiency of a collector, the total insolation is used for collectors with a concentration ratio less than 2, such as flat plates and nonmoving CPCs. For collectors with concentration greater than about 2, such as parabolic troughs, the beam insolation is used to evaluate the efficiency according to the relation

$$\eta_b = \frac{q_{\text{out}}}{I_b} \tag{3.8}$$

Figures 3.11 and 3.12 show the complete test configurations, instruments, and auxiliary equipment for liquid and air collectors, respectively. Details on testing and evaluating the thermal efficiencies of collectors with concentration less than 2 are given in ASHRAE Standard 93-77 [15].*

*Available from the American Society of Heating, Refrigerating and Air-Conditioning Engineers, Inc., 345 East 47th Street, New York 10017.

Figure 3.10 Solar pyranometer manufactured by Eppley Laboratory. (*Courtesy of Eppley Laboratory.*)

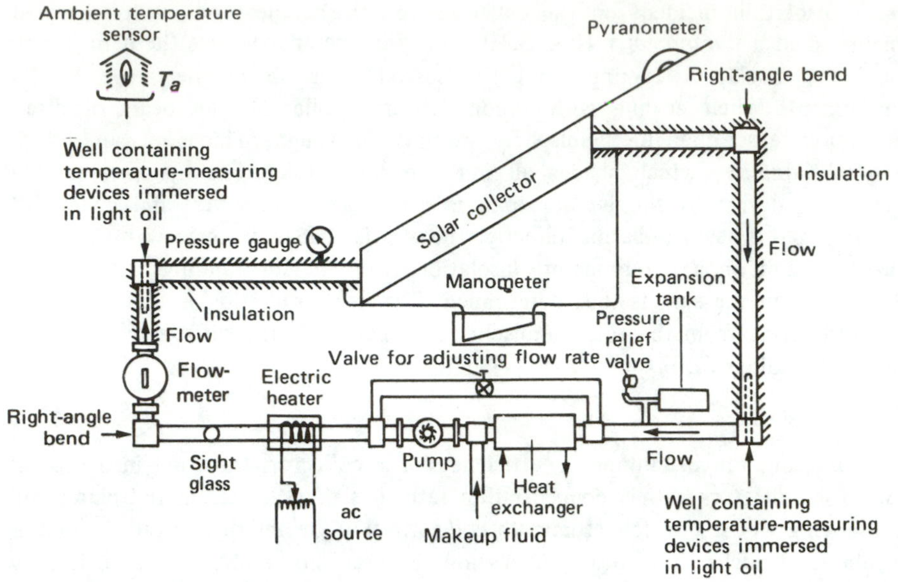

Figure 3.11 Test configuration for liquid collectors. (*From [16].*)

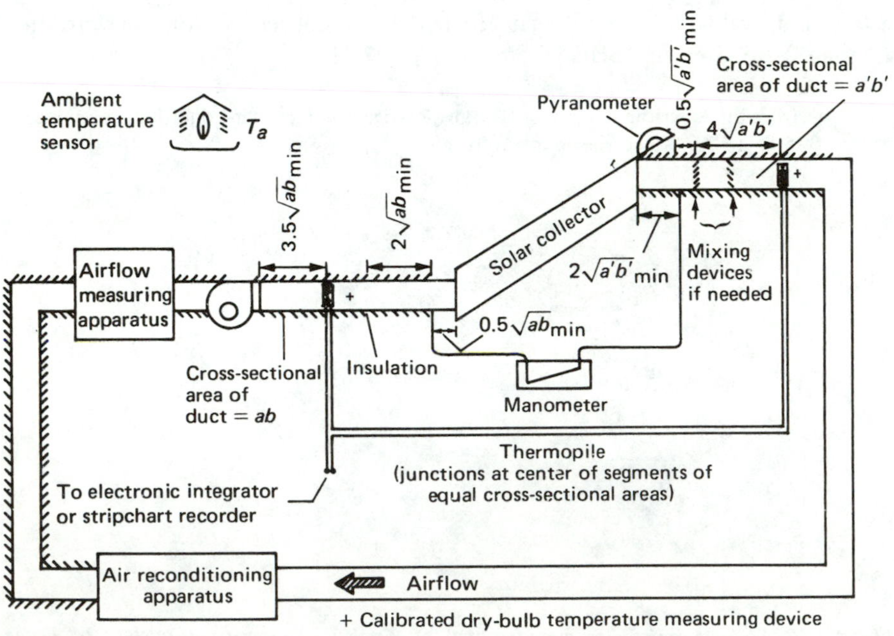

Figure 3.12 Test configuration for air collectors. (*From [16].*)

Practical Hints for Collector Selection and Use

On the basis of experience, the following precautions are recommended:

1 Obtain and thoroughly review collector efficiency and stagnation test results.

2 Consider costs of maintenance, such as replacing plastic glazings every several years. Be sure the collector array installation provides access for maintenance.

3 Do not install the collector until flow can be provided, or cover the collector to protect it against stagnation damage unless it is designed for high-temperature stagnation.

4 Protect collectors, especially the evacuated-tube type, from thermal shock.

5 Make sure the collector is capable of withstanding design pressures without leaking. Check the city water pressure.

6 Provide adequate freeze protection.

7 Ensure that the collector is properly handled during delivery and installation.

8 Be sure header pipe sizes are adequate to provide uniform flow in the collector array. In large arrays, flow balancing valves should be considered.

REFERENCES

1. F. Kreith and J. F. Kreider, *Principles of Solar Engineering*, Hemisphere, Washington, D.C., 1978.
2. J. F. Kreider, *The Solar Heating Design Process,* McGraw-Hill, New York, 1981.
3. U.S. Energy Research and Development Administration, *Solar Design Handbook,* Government Printing Office, Washington, D.C., 1977.
4. U.S. Department of Housing and Urban Development, *Intermediate Minimum Property Standards Supplement, Solar Heating and Domestic Hot Water Systems,* Government Printing Office, Washington, D.C., 1977.
5. D. P. Grimmer and S. W. Moore, "Practical Aspects of Solar Heating: A Review of Materials Use in Solar Heating Applications," presented at the Society for Advancement in Materials Process Engineering Meeting, Albuquerque, N.M., Oct. 14–16, 1975.
6. T. Kobayashi and L. Sargent, "A Survey of Breakage-resistant Materials for Flat-Plate Solar Collector Covers," presented at the U.S. Section of the International Solar Energy Society Meeting, Fort Collins, Colo., Aug. 20–23, 1974.
7. A. E. Scoville, "An Alternate Cover Material for Solar Collectors," presented at the International Solar Energy Society Congress and Exposition, Los Angeles, Calif., July 1975.
8. C. W. Clarkson and J. S. Herbert, "Transparent Glazing Media for Solar Energy Collectors," presented at the U.S. Section of the International Solar Energy Society Meeting, Fort Collins, Colo., Aug. 20–23, 1974.
9. *Modern Plastics Encyclopedia*, McGraw-Hill, New York, 1975–1976.
10. R. B. Toenjes, "Integrated Solar Energy Collector Final Summary Report," LA-6143-MS, Los Alamos Scientific Laboratory, Los Alamos, N.M., Nov. 1975.
11. G. R. Mather, Jr., "ASHRAE 93-77 Instantaneous and All-Day Tests of the Sunpak™ Evacuated Tube Collector," *J. Solar Energy Eng*., vol. 102, 1980, pp. 294–304.
12. A. Rabl, N. B. Goodman, and R. Winston, "Practical Design Considerations for CPC Solar Collectors," *Solar Energy,* vol. 22, 1979, pp. 343–382.

13. R. W. Bliss, "The Derivation of Several 'Plate Efficiency Factors' Useful in the Design of Flat-Plate Solar Heat Collectors," *Solar Energy,* vol. 3, 1959, pp. 55–64.
14. H. C. Hottel and B. B. Woertz, "The Performance of Flat-Plate Solar Heat Collectors," *Trans. ASME*, vol. 64, 1942, pp. 91–104.
15. "Methods of Testing to Determine the Thermal Performance of Solar Collectors," ASHRAE Standard 93-77, Am. Soc. for Heating, Refrigeration, and Air-Conditioning Engineering, New York, 1977.
16. J. E. Hill, E. R. Streed, G. E. Kelly, J. L. Geist, and T. Kaisuda, "Development of Proposed Standards for Testing Solar Collectors and Thermal Storage Devices," NBS Tech. Note 899, National Bureau of Standards, Gaithersburg, Md., 1976.

4 Thermal Energy Storage

Solar systems can also provide heat when the sun is not shining. On sunny days, a properly sized system should be able to collect more energy than is needed to meet the daytime heating load, and the excess energy can then be stored for later use. The most common heat storage systems use rock beds or water tanks; these types of storage systems are treated in this chapter [1, 2].

Figure 4.1 represents a typical air-based system that uses a rock bed for storage, and Fig. 4.2 a typical liquid-based system that uses water for storage. Many variations of these space-heating systems are possible, but in any storage system it is always necessary to have (1) a heat storage material, (2) a container, and (3) provisions for adding and removing heat.

Heat capacity is the ability of a material to store sensible heat. In the English system of units it is measured in terms of the number of British thermal units required to raise the temperature of 1 lb of the material by 1°F. Water has a heat capacity of 1 Btu/(lb·°F). Most other materials have lower heat capacities; rock, for example, has a heat capacity of 0.21 Btu/(lb·°F). The volumetric heat capacity is the quantity of heat that can be stored per unit volume for every degree of temperature change.

Table 4.1 gives the heat capacities and volumetric heat capacities of some common storage materials. The voids referred to in the table are the spaces between individual pieces of piled-up rock or other loose material. The ratio of the volume of these spaces to the total volume of the rock bed is called the void fraction. Experiments have shown that loose materials pack with a void fraction of 30 to 40 percent. Such materials behave as if their thermal capacity were reduced by the same percentage.

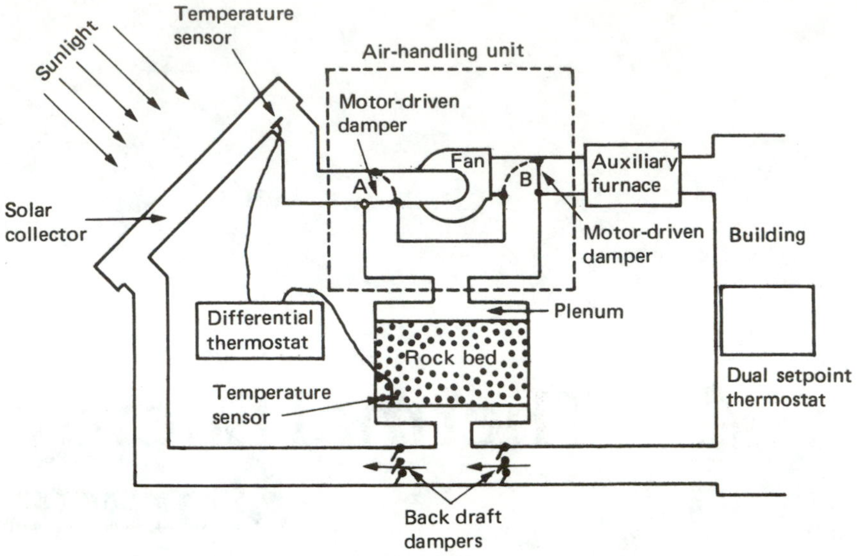

Figure 4.1 Typical air-heating system with rock bed storage.

Inspection of Table 4.1 shows that 1 ft^3 of water can store 62.4 Btu for every degree of temperature rise. A cubic foot of rock packed to a 30 percent void fraction can store only 24.3 Btu per degree of temperature rise. Therefore to store the same quantity of heat over the same temperature range, a rock bed would have to have a volume 2.6 times greater than that of a water tank.

GENERAL INFORMATION

The daily temperature range of a storage system is a function of its size. Both air- and liquid-based systems typically operate over a daily temperature

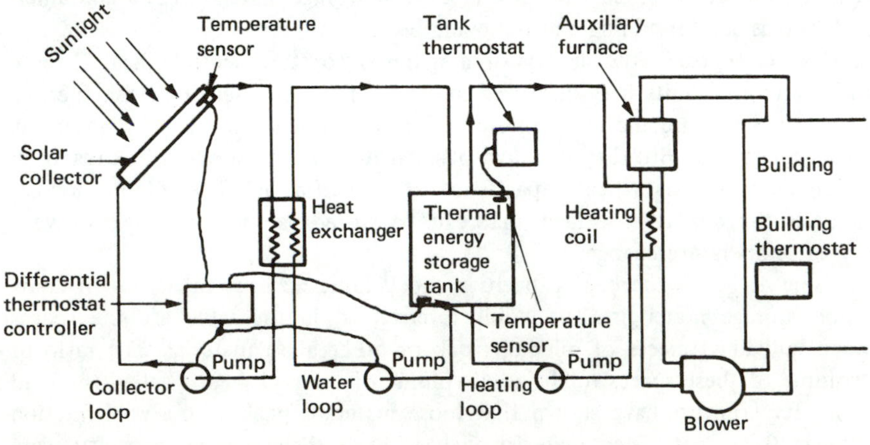

Figure 4.2 Typical liquid-heating system with tank storage and heat exchanger.

Table 4.1 Sensible Heat Storage Materials*

Material	Density (lb/ft³)	Heat capacity [Btu/(lb·°F)]	Volumetric heat capacity [Btu/(ft³·°F)]	
			No voids	30% voids
Water	62.4	1.00	62.4	–
Scrap iron	489	0.11	53.8	37.7
Scrap aluminum	168	0.22	36.96	25.9
Scrap concrete	140	0.27	27.8	26.5
Rock	167	0.21	34.7	24.3
Brick	140	0.21	29.4	20.6

*Data from Cole et al. [2].

range of less than 60°F on a sunny winter day. The exact range is highly variable from system to system, season to season, and day to day. Factors that influence the daily temperature range include the amount of sunshine available, the size of the storage device, the heat capacity of the storage material, the demand for heat, the type of system, the way the system is connected to the load, and the temperature limitations of materials in the system.

The daily temperature swing $T_{max} - T_{min}$ (°F) is related to the amount of usable heat Q (Btu) stored in the device by the equation [3]

$$Q = mc_p(T_{max} - T_{min}) \tag{4.1}$$

where m is the mass of the storage material (lb) and c_p is the heat capacity of the storage medium [Btu/(lb·°F)].

EXAMPLE

How much water is required to store 400,000 Btu from solar collectors on a sunny winter day if the daily temperature swing is limited to 40°F?

SOLUTION

Solving Eq. (4.1) for m gives

$$m = \frac{Q}{c_p(T_{max} - T_{min})} = \frac{400{,}000 \text{ Btu}}{1 \text{ [Btu/(lb·°F)]} \times 40\text{°F}} = 10{,}000 \text{ lb} \tag{4.2}$$

Since 1 gal of water weighs 8.34 lb, the amount of water required is 1200 gal.

As a rule, the storage capacity for residential space-heating or domestic water-heating systems should be about 10 to 15 Btu per degree Fahrenheit per square foot of collector. Dividing this amount by the heat capacity from Table 4.1 yields the mass of storage material required. For rock, the mass required is about 50 to 75 lb/ft² [4]; for water, it is 10 to 15 lb/ft². Alternatively, dividing the 10 to 15 Btu/(°F·ft^2) by the volumetric heat capacity yields the required volume of storage material. Since a rock bed usually has a void fraction of about 30 percent, the required storage volume is about 0.41 to 0.62 ft^3 of rock per square foot of collector (for convenience, these numbers are usually rounded upward to 0.50 to 0.75 ft^3/ft^2). About 0.16 to 0.24 ft^3 of water per square foot of collector is required. Converted to gallons, the volume of water needed is 1.25 to 2.0 gal/ft^2.

It is sometimes not possible to install an optimum-sized storage device because of space limitations or, in liquid-based systems, tank availability. Figure 4.3 shows what happens when storage capacity is reduced. If the storage capacity is less than about 10 Btu/(°F·ft^2) the fraction of the heating load supplied by solar energy is less than it should be and some of the heat collected is wasted. If the storage capacity is greater than about 15 Btu/(°F·ft^2) it will not significantly increase the percentage of the heating load supplied by solar energy, but heat losses and costs will be excessive.

For collector-to-storage heat losses to be minimized, the storage device should be as close as possible to the solar collectors. In practice, storage containers usually have to be located in basements, crawl spaces, garages, or outdoors; Tables 4.2 and 4.3 summarize the advantages and disadvantages of these locations.

TEMPERATURE STRATIFICATION

Temperature stratification occurs when some parts of the storage material are hotter than other parts. In theory, a solar system with perfectly stratified storage can perform 5 to 10 percent more efficiently than a thermally mixed system, if the system is designed to take advantage of stratification.

Stratification is relatively easy to achieve in a rock bed if the solar-heated

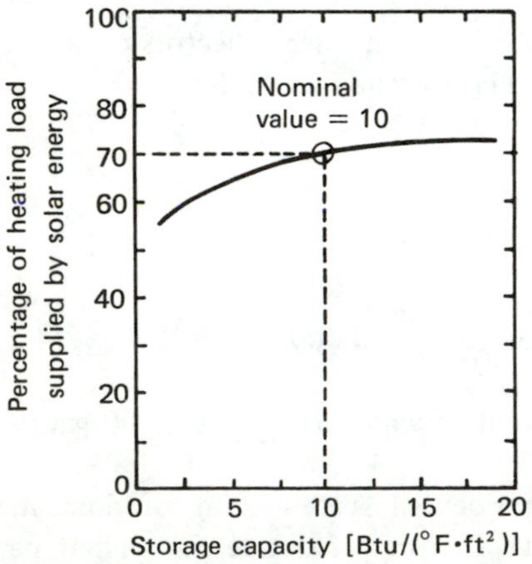

Figure 4.3 Effect of storage capacity on percentage of heating load supplied by solar energy for a constant heating load and collector area. (*From [9].*)

Table 4.2 Advantages and Disadvantages of Storage Locations

Utility room or basement	Unheated garage	Crawl space	Outdoors, above grade	Outdoors, below grade
		Advantages		
Insulation requirement is minimal Insulation is protected from weather Thermal losses contribute to building heat in winter Leaks are easily detected Access for repairs is relatively easy	Insulation is protected from weather Leaks are easily detected Access for repairs is easy Steel or fiberglass-reinforced plastic tanks can be installed in existing garage	Insulation is protected from weather Thermal losses may contribute to building heat in winter	Access is easy Thermal losses do not add to air-conditioning load Storage unit does not reduce living space	Thermal losses do not add to air-conditioning load Storage unit does not reduce living space
		Disadvantages		
Living space is reduced Thermal losses add to air-conditioning load in summer Leaks may damage building interior Steel or fiberglass-reinforced plastic tanks are difficult to install in existing building	Garage space is reduced Extra insulation is required Freeze protection is required in most of the United States Leaks may damage garage Thermal losses cannot be recovered	Thermal losses may add to air-conditioning load in summer Access is difficult for retrofit or repairs Shape of tank may require extra insulation	Extra insulation is required Weather protection is required Thermal losses cannot be recovered Freeze protection is required in most of the United States Vermin may burrow into insulation	Access for repairs is difficult Ground water may cause several types of problems Thermal losses cannot be recovered Vermin may burrow into insulation Careful design is required to ensure sufficient net positive suction head for pump

Source: Cole et al. [2].

Table 4.3 Storage Locations: Applicability and Special Requirements

Storage location	Applicability: New building	Applicability: Retrofit	Special requirements: Weather-proof insulation	Extra insulation	Freeze protection	Protection from ground water	Long-lifetime components
Utility room or basement	Yes	X*	No	No	No	No	Yes
Unheated garage	Yes	Yes	No	Yes	Yes	No	No
Crawl space	Yes	X	No	Yes	No	No	Yes
Outdoors, above grade	Yes	Yes	Yes	Yes	Yes	No	No
Outdoors, below grade	Yes	Yes	Yes	Yes	X	Yes	Yes

*Items marked X must be determined by the individual situation.
Source: Cole et al. [2].

air enters at the top of the bed and flows downward. Thus the rock bed is hot at the top but relatively cool at the bottom. Since rocks are stationary, this temperature differential is easily maintained, and the collector inlet air is at a lower average temperature than it would be if there were no stratification. Temperature stratification from top to bottom of a rock bed compensates for the lower efficiency of air-type collectors compared with liquid-type collectors.

Temperature stratification can also be achieved in water tanks, but pumping the water to the collector or the heat exchanger tends to mix the hotter and cooler water. Even natural convection from a coil-in-tank heat exchanger can upset stratification. Thus if a water tank is used for thermal energy storage, one should not count on much performance advantage from stratification.

HEAT LOSSES FROM STORAGE AND TRANSFER DUCTS

A storage device should be well insulated to minimize heat loss. The U.S. Department of Housing and Urban Development *Intermediate Minimum Property Standards* specifies that the storage device should be insulated so that losses during a 24-h period do not exceed 2 percent of the storage capacity. However, minimum standards do not necessarily produce the most cost-effective system. Because the energy-saving benefits of insulation are so great, storage devices should, if possible, have more insulation than the minimum specified.

The thermal resistance to the loss of heat from a storage device—called the R value—is an indication of the effectiveness of insulation. To determine the R value of insulation that will limit the heat loss to 2 percent in 12 h, find the entry in Table 4.4 or 4.5 that corresponds most closely to the shape and size of the storage device. This entry is a multiplier to be used in the equation

$$\text{R value} = \text{multiplier from table} \times (\text{storage temperature} - \text{ambient temperature}) \tag{4-3}$$

For storage temperature, use the average maximum temperature (°F) expected in the storage device on an average January day. For ambient temperature, use the average temperature (°F) of the storage device's surroundings on an average January day.*

The R value obtained will have the units of degrees Fahrenheit × square feet × hours per Btu, which are the units commonly used by insulation manufacturers in the United States.

*Assume an ambient temperature of about 65°F for storage devices in heated areas. For unheated locations, find out the average January outdoor temperature from the nearest U.S. Weather Bureau office. The ASHRAE *Handbook of Fundamentals* [5] recommends that underground temperatures may be assumed to be the same as water temperatures at depths of 30 to 60 ft. These temperatures may be obtained from Collins [6]. See also Carson [7].

Table 4.4 Insulation Multipliers for Horizontal and Vertical Cylindrical Water Tanks*

Size (gal)	Multiplier [(ft^2·h)/Btu]				
	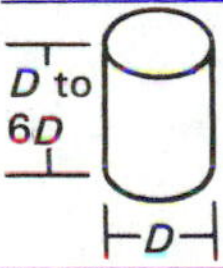	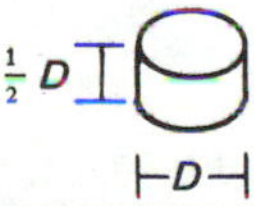	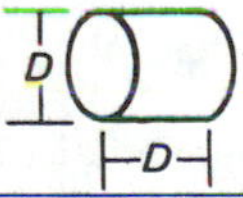	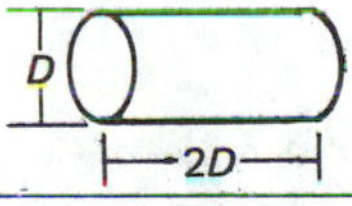	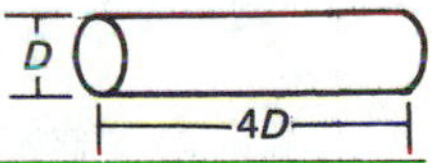
80	0.47†,‡	0.53†			
120	0.41	0.47			
250	0.32	0.36	0.28	0.29	0.33
500	0.26	0.29	0.22	0.23	0.26
750	0.22	0.25	0.19	0.20	0.23
1000	0.20	0.23	0.17	0.18	0.21
1500	0.18	0.20	0.15	0.16	0.18
2000	0.16	0.18	0.14	0.14	0.16
3000	0.14	0.16	0.12	0.13	0.14
4000	0.13	0.14	0.11	0.11	0.13
5000	0.12	0.13	0.10	0.11	0.12

*Table values are for a 2% loss in 12 h with a daily temperature range of 60°F. To determine the required R value of insulation, multiply the table value by the difference (°F) between the storage temperature and the ambient temperature.

†The R value of the bottom insulation is assumed to be half of that on the top and sides for tanks specified by the second and third columns.

‡ The second column is applicable to all tanks with a height 1 to 6 times the diameter. Insulation multipliers for most domestic hot-water tanks can be found in this column.

Source: Cole et al. [2].

Table 4.5 Insulation Multipliers for Rock Beds*

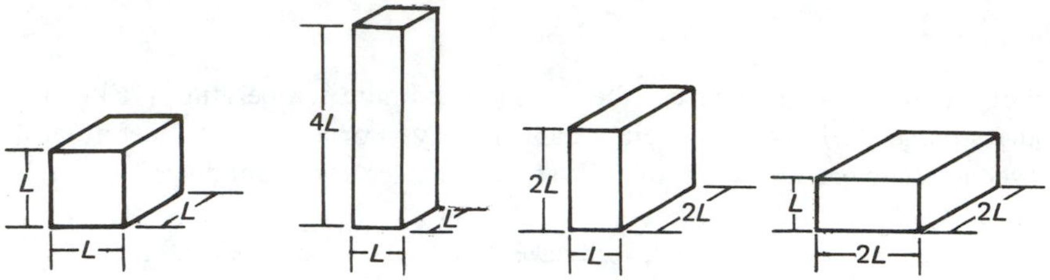

Volume of rock† (ft^3)	Multiplier [(ft^2·h)/Btu]: $L \times L \times L$	Multiplier: $L \times L \times 4L$	Multiplier: $L \times 2L \times 2L$	Multiplier: $2L \times 2L \times L$
100	0.82	0.83	0.80	0.97
150	0.72	0.72	0.70	0.85
200	0.65	0.66	0.64	0.76
300	0.57	0.57	0.56	0.67
400	0.52	0.52	0.51	0.61
500	0.48	0.48	0.47	0.56
600	0.45	0.46	0.44	0.53
800	0.41	0.41	0.40	0.48
1000	0.38	0.38	0.37	0.45

*Table values are for a 2% loss in 12 h with an assumed daily temperature range of 60°F. To obtain the required R value for the side and top insulation, multiply the table value by the difference (°F) between the storage temperature and the ambient temperature. The R value for the bottom insulation is assumed to be half of that on the top and sides.

†The insulation is assumed to cover both plenums, but the volumes and shapes given are for the rocks only.

Source: Cole et al. [2].

Table 4.6 R Values and Densities of Common Building and Insulating Materials

Material	Density (lb/ft^3)	R value [(°F·ft^2·h)/Btu per in thickness]
Acoustic tile	18.0	2.53
Asbestos-cement board	120.0	0.22
Brick:		
Common	120.0	0.20
Face	130.0	0.11
Cellulose fill	–	3.70
Cement (mortar or plaster with sand)	116.0	0.20
Concrete, heavyweight	140.0	
Dried aggregate		0.11
Undried aggregate		0.08
Concrete, lightweight	30.0	1.11
Concrete block, heavyweight		
4 in	101.0	0.18
6 in	85.0	0.15
8 in	69.0	0.13
12 in	76.0	0.11

Table 4.6 R Values and Densities of Common Building and Insulating Materials (*continued*)

Material	Density (lb/ft^3)	R value [(°F·ft²·h)/Btu per in thickness]
Concrete block, medium weight		
4 in	76.0	0.28
6 in	65.0	0.23
8 in	53.0	0.18
12 in	58.0	0.18
Concrete block, lightweight		
4 in	65.0	0.33
6 in	55.0	0.30
8 in	45.0	0.25
12 in	49.0	0.19
Fiberglass		3.85
Gypsum or plasterboard	50.0	0.90
Gypsum plaster		
Lightweight aggregate	45.0	0.63
Sand aggregate	105.0	0.18
Hardboard		
Medium-density siding	40.0	1.53
Medium-density other	50.0	1.37
High-density standard tempered	55.0	1.22
High-density service tempered	63.0	1.00
Insulation board		
Sheathing	18.0	2.63
Shingle backer	18.0	2.52
Nail base sheathing	25.0	2.28
Mineral board, preformed	–	3.47
Mineral wool/fiber		
Batt	–	3.33
Fill	–	3.09
Particleboard		
Low density		1.85
Medium density		0.11
High density		0.08
Underlayment		0.46
Polystyrene, expanded		4.17
Polyurethane, expanded	1.0	6.26
Urea formaldehyde	0.7	4.17
Roof insulation, preformed	16.0	2.78
Wood, soft (fir, pine, etc.)	32.0	1.25
Wood, hard (maple, oak, etc)	45.0	0.91

Source: Groven and Hirsch [10].

Table 4.6 gives R values per inch of thickness of typical insulators. Most storage units have several insulation layers, including the walls. The R value of multilayer insulation is the sum of the R values of the individual layers, as shown in Fig. 4.4.

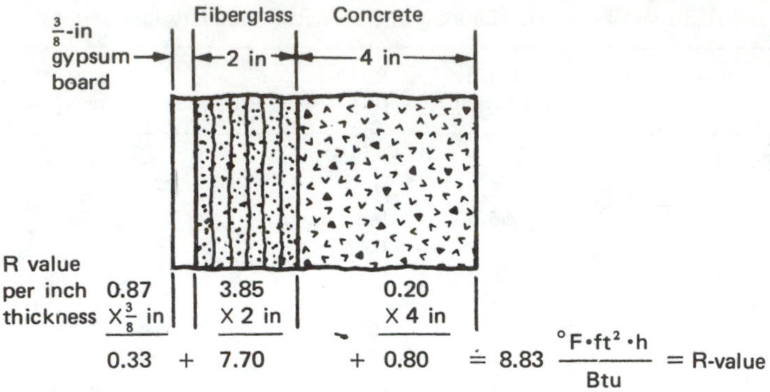

Figure 4.4 R value of multilayer insulation.

EXAMPLE

Calculate the HUD-specified maximum heat loss rate and the corresponding amount of fiberglass insulation needed on a 1500-gal cylindrical water tank that will be installed in a heated basement. The tank is horizontal, 10 ft long, and 5 ft in diameter. Assume that the maximum temperature of the tank on an average January day will be 160°F and that the ambient temperature in the basement in January will be 65°F.

SOLUTION

From Table 4.4 the appropriate multiplier for a water tank of this size and shape is 0.16. The insulation R value that will limit the tank's losses to 2 percent in 12 h will be

$$\text{R value} = 0.16(160 - 65) = 0.16(95) = 15\ (^\circ\text{F}\cdot\text{ft}^2\cdot\text{h})/\text{Btu}$$

The total insulation around the tank should therefore have an R value of 15 (°F·ft²·h)/Btu.

The thickness corresponding to this value of R is obtained next. From Table 4.6 the R value per inch of thickness for fiberglass is 3.85 (°F·ft²·h)/(Btu·in). Hence the thickness t is

$$t = \frac{\text{R}}{\text{R}/t} = \frac{15\ (^\circ\text{F}\cdot\text{ft}^2\cdot\text{h})/\text{Btu}}{3.85\ (^\circ\text{F}\cdot\text{ft}^2\cdot\text{h})/(\text{Btu}\cdot\text{in})} = 3.9\ \text{in}$$

A 4-in thickness is commercially available and will ensure compliance with HUD minimum standards.

The surface area A of the tank is

$$A = \pi DL + \frac{\pi D^2}{4} = 3.14\left(5 \times 10 + \frac{25}{4}\right) = 176.6\ \text{ft}^2$$

Then the heat loss per hour q in January will be

$$q = \frac{\Delta T}{R} A = \frac{160°F - 65°F}{15\ (°F \cdot ft^2 \cdot h)/Btu}\ 176.6\ ft^2 = 1118\ Btu/h$$

Heat is also lost as the fluid is moved into and out of storage. These losses include

1 Losses between the collector and the storage unit (charging losses)
2 Losses between the storage unit and the heating load (discharge losses)

To minimize these losses, (1) the ducting or piping from collector to storage must be well insulated and have weather protection and (2) the piping or ductwork from storage to load must be as short as possible and well insulated. It is recommended that insulation having an R value of 4, i.e., R-4, be used for pipes with diameters less than 1 in and R-6 insulation for pipes with diameters 1 to 4 in.

HEAT EXCHANGERS

Heat exchangers are devices that transfer heat from one fluid to another while preventing mixing of the two fluids. Hot fluid flows on one side of a metal barrier and heats a cold fluid flowing on the other side. For heat to be transferred, there must be a temperature difference between the fluids across the barrier. This temperature difference leads to a loss of overall system efficiency each time a heat exchanger is used. Heat exchangers typically used in solar energy systems are shown in Figs. 4.5 through 4.10.

In a liquid system, the collector must be protected against freeze-up in the winter. If antifreeze is used for protection a liquid-to-liquid heat exchanger must be installed, as shown in Fig. 4.2, to separate the heat transfer fluid from the water in storage, since antifreeze is too expensive to be used as a storage medium. Because there must be a temperature difference from the collector side to the storage side of the heat exchanger, the collector must operate at a higher,

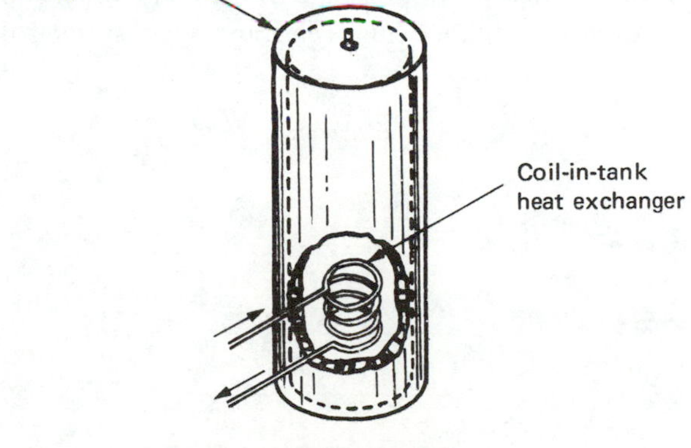

Figure 4.5 Coil-in-tank heat exchanger.

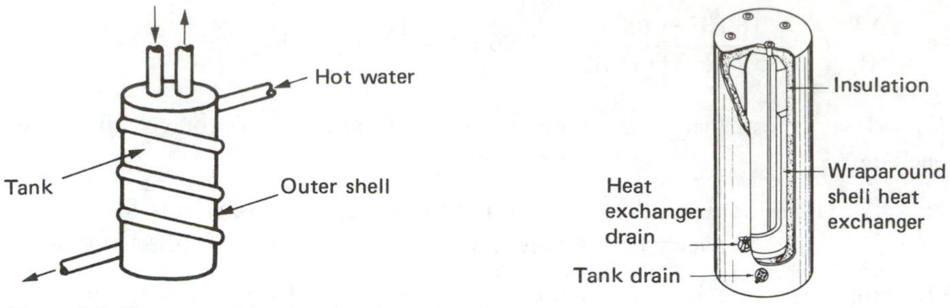

Figure 4.6 Wraparound heat exchanger and tank.

less efficient temperature than it would in a system without a heat exchanger between collector and storage. Thus the heat exchanger imposes a performance penalty on the system.

A collector-to-storage heat exchanger can be a simple coiled tube immersed in the storage tank (Fig. 4.5) or wrapped around the outside of the tank (Fig. 4.6). These types of heat exchangers rely on natural convection to move the water inside the tank past the heat exchange surface. If the tank is large, say several hundred gallons, natural convection is an inefficient means of transferring heat. Shell-and-tube heat exchangers (Fig. 4.7) are often used in this case, with a pump circulating the water between the tank and the heat exchanger.

In the simple heating system shown in Fig. 4.2, one heat exchanger (liquid-to-air) is needed to transfer heat to the building. This is often a finned-tube unit (Fig. 4.8) inserted in an air duct. Less frequently used are baseboard convectors, radiant heating coils, and individual fan-coil units.

Solar-heated domestic hot water requires a heat exchanger to separate the potable, or drinkable, hot water from either the nonpotable storage fluid or the collector fluid. Heat exchangers for use with potable water are subject to special safety requirements, as discussed in [2].

Another method of protecting collectors is to drain them whenever there is danger of freezing weather. A system employing this method, known as drain-down, is one of the most efficient solar collection systems available. Details of the drain-down system and its many variations are available in systems design manuals such as the International Telephone and Telegraph solar systems design manual [8].

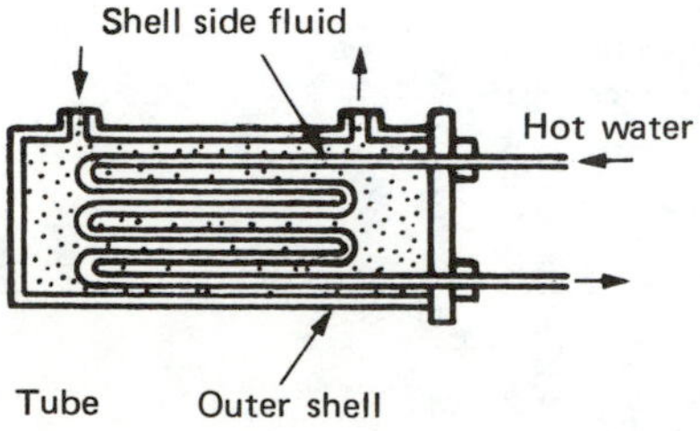

Figure 4.7 Shell-and-tube heat exchanger.

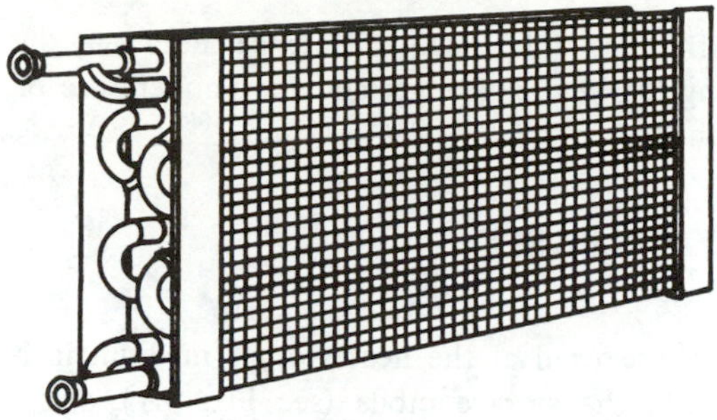

Figure 4.8 Typical liquid-to-air or air-to-liquid heat exchanger. (*Courtesy of Bohn Heat Transfer Division, Gulf-Western Manufacturing Co., Danville, Ill.*)

A drain-down system must be totally foolproof. Pipes must be pitched and collectors selected to ensure that all of the water will drain when it should. A single failure can ruin the collectors. Many designers prefer to use antifreeze in the collectors rather than risk this loss. The designer must decide whether to pay the penalties of lower performance and higher first cost for an antifreeze system in return for eliminating the risk of an expensive failure.

Air-based heating systems (Fig. 4.1), in contrast to liquid-based systems, do not need separate heat exchangers between the collectors and the heating load. The rock bed is both the storage device and the storage-to-load heat exchanger. On the other hand, if air-based collector systems are used to heat domestic water, they must have an air-to-liquid heat exchanger. Such a heat exchanger usually consists of finned water tubing in the air-handling duct (Fig. 4.8), similar to the finned-tube heat exchanger used in air ducts of liquid-based systems.

Using heat exchangers imposes a penalty on the solar space-heating system. Just as a collector-to-storage heat exchanger forces the collector to operate at a higher temperature, a storage-to-load heat exchanger forces the storage system to operate at a higher temperature than would be required if the heat exchanger could be eliminated.

To calculate how efficiently a system with heat exchangers performs, it is necessary to determine the penalty imposed on the system by the heat exchangers. This penalty is called *heat exchanger effectiveness*; it is defined as the actual rate of heat exchange divided by the rate of heat exchange of a perfect, infinitely large heat exchanger.

Since there is no perfect, infinitely large heat exchanger, the designer's task is to choose the size of heat exchanger that will minimize the overall cost of the system. This relatively complex task is described in [3]. Alternatively, most heat exchanger manufacturers can select a heat exchanger of the proper size if the following information about the system is supplied:

1 Physical characteristics of the two fluids in the heat exchanger
2 Amount of heat to be transferred (Btu/h)
3 Flow rates (gal/min) on both sides of the heat exchanger

4 Approach temperature difference (°F), i.e., difference between the temperature of the hot fluid entering the heat exchanger and the temperature of the heated fluid leaving the heat exchanger

THERMAL DESIGNS OF AIR-BASED STORAGE SYSTEMS

Rocks, in a suitable container, are used as the heat storage medium in a solar air space-heating system [4]. In the storage mode (see Fig. 4.1), solar-heated air is forced into the top of the container (the plenum) and then passes evenly down through the rock bed, heating the rocks. Cool air is drawn off at the bottom and returned to the collectors. When space heating from storage is required, the airflow is reversed, and warm air is delivered from the top damper B of the rock bed into the building (damper B open and damper A closed in Fig. 4.1). Thus the rock bed functions as both the storage medium and the heat exchanger between the solar collectors and the load. When heating without storage is required, the solar-heated air goes directly into the building.

Uniform distribution of air in the rock bed is important; with nonuniform distribution hot air bypasses some of the rocks and does not heat them. Uniform air distribution is achieved by designing the rock bed for a vertical airflow through the rocks with a pressure drop of at least 0.15 in of water and providing a sufficiently large plenum chamber. The cross-sectional area of each plenum should be at least 10 percent of the cross-sectional area of the rock bin; i.e., the height × width of the plenum should be at least 10 percent of the width × length of the rock bin. The rocks in a rock bed should be round and screened to ensure uniform size. Rocks conforming to the "Standard Specifications for Concrete Aggregates," ASTM C33 of the American Society for Testing and Materials, are satisfactory for use in rock beds.

Leaks can be a problem in air-based systems because they can degrade performance to the point where the system becomes uneconomical to operate. Air often leaks through seams in the corners of the rock bin, through joints between ductwork and the rock bin, and from dampers that do not seal properly.

The real performance of the rock bed depends on several interrelated factors: the rock bed face velocity, the size of the rocks, the heat transfer relaxation length, and the pressure drop across the rock bed. The *face velocity* describes the quantity of air moving through the rock bed. To compute it, divide the total airflow rate by the rock bed's cross-sectional area. Rock bed storage systems are usually designed for face velocities ranging from 10 to 60 ft/min. Increasing the face velocity increases the pressure drop across the bed.

As long as the rocks are clean, approximately round, and of uniform size, their size can be specified by their average diameter. Rock size primarily affects the relation between the heat transfer relaxation length and the pressure drop. For a given face velocity, smaller rocks give shorter heat transfer relaxation lengths but larger pressure drops. For larger rocks the reverse is true.

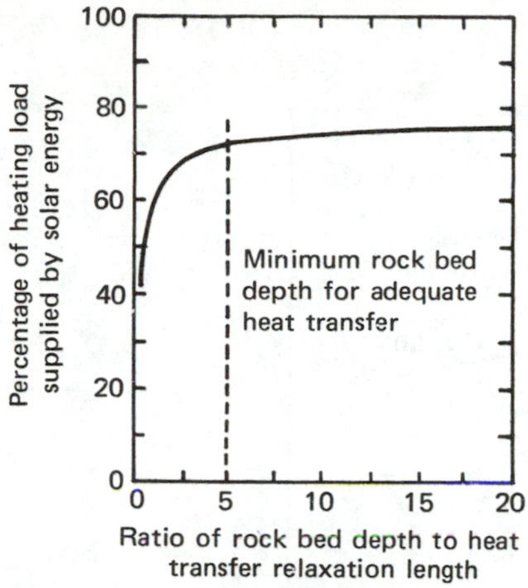

Figure 4.9 Effect of rock bed depth on system performance. (*From [2].*)

The *heat transfer relaxation length* is the ratio of the input heat to the heat that can be absorbed per foot of depth of the rock bed. For design purposes, it specifies the minimum rock bed depth needed for effective temperature stratification.

The effect of rock bed depth on system performance is illustrated in Fig. 4.9, in which the horizontal axis represents the rock bed depth expressed as a multiple of the heat transfer relaxation length, the vertical axis represents the percentage of the heating load supplied by solar energy, and the volume of rock and the size of the solar collector are taken as constant. Numerical study of a large number of cases showed that temperature stratification and adequate heat transfer will occur as long as the rock bed depth is more than five times the heat transfer relaxation length [2, 9]. Most rock bed designs easily exceed this depth, but this should be checked in every design to ensure stratification and adequate heat transfer.

For heat to be transferred from the air to the rocks, the flow path should be sufficiently long. But as the flow path increases, the pressure drop also increases. A pressure drop of at least 0.15 in of water is necessary to ensure uniform airflow in the rock bed, but to avoid excessive pumping power requirements, the pressure drop should be less than 0.30 in of water. The pressure gradient is the pressure drop divided by the length of the rock bed.

The relation among face velocity, rock size, relaxation length, and pressure gradient is shown in the rock bed performance map in Fig. 4.10. The vertical axis represents the pressure gradient across the rock bed, and the horizontal axis represents the heat transfer relaxation length corresponding to a given rock diameter and face velocity.

Rock bed storage must be designed according to the following systems constraints:

1 Rock volume of 0.50 to 0.75 ft^3 per square foot of collector
2 Airflow of 1 to 3 ft^3/min per square foot of collector

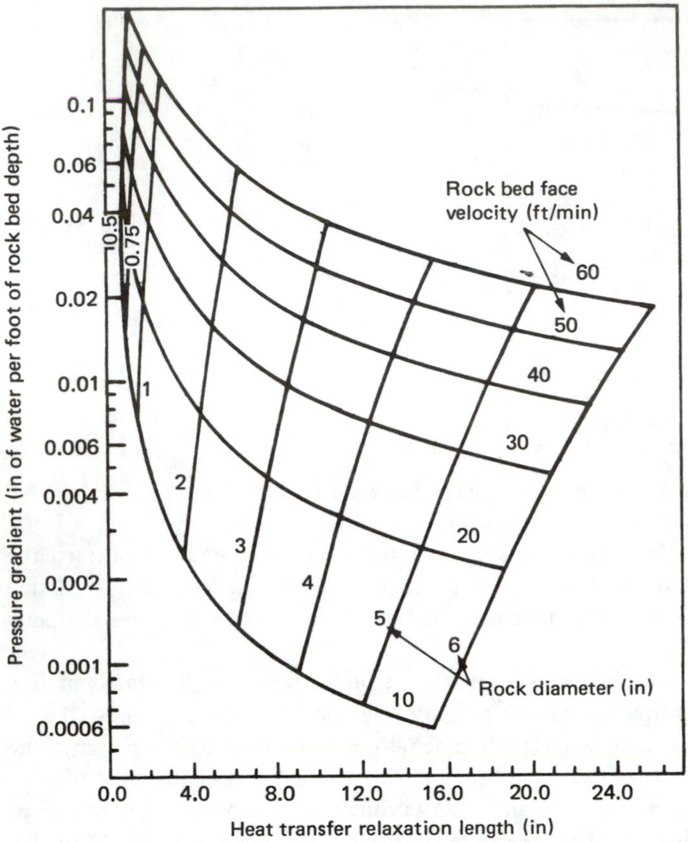

Figure 4.10 Rock bed performance map. Data from [11].

3 Rock bed pressure drop greater than 0.15 but less than 0.30 in of water
4 Rock bed length restriction (e.g., ceiling height in a basement)
5 Availability of rock sizes

The rock bed performance map can be used to adjust either the face velocity or the rock diameter to achieve a suitable pressure drop in the space available for the storage unit.

EXAMPLE

A design using 500 ft^2 of collector on a residence is subject to the following constraints:

1 Airflow rate of 2 $ft^3/(min \cdot ft^2)$, or 1000 ft^3/min.

2 0.50 to 0.75 ft^3 of rock per square foot of collector, or 250 to 375 ft^3 of rock.

3 Pressure drop across the rock bed of 0.15 to 0.30 in of water.

4 Rock bed length at least five times the heat transfer relaxation length.

5 Rock bed location in a basement that has a 7-ft floor-to-ceiling height. Allowing 8 in for insulation and 6 in for each plenum leaves 5 ft 4 in for the rocks.

Under these constraints, calculate maximum and minimum values for the rock bed cross-sectional area, face velocity, and pressure gradient and the maximum value of the heat transfer relaxation length. Also select suitable rock sizes and estimate the weight of the rocks.

SOLUTION

The maximum and minimum values, when plotted on the rock bed performance map, help to configure the system. The rock bed cross-sectional area is its volume divided by its depth:

$$\text{Minimum cross-sectional area} = \frac{250\ \text{ft}^3}{5.33\ \text{ft}} = 46.9\ \text{ft}^2$$

$$\text{Maximum cross-sectional area} = \frac{375\ \text{ft}^3}{5.33\ \text{ft}} = 70.3\ \text{ft}^2$$

The face velocity is the airflow rate divided by the rock bed cross-sectional area:

$$\text{Minimum face velocity} = \frac{1000\ \text{ft}^3/\text{min}}{70.3\ \text{ft}^2} = 14.2\ \text{ft/min}$$

$$\text{Maximum face velocity} = \frac{1000\ \text{ft}^3/\text{min}}{46.9\ \text{ft}^2} = 21.3\ \text{ft/min}$$

The pressure gradient is calculated as follows:

$$\text{Minimum pressure gradient} = \frac{0.15\ \text{in of water}}{5.33\ \text{ft}} = 0.028\ \text{in of water per foot}$$

$$\text{Maximum pressure gradient} = \frac{0.30\ \text{in of water}}{5.33\ \text{ft}} = 0.056\ \text{in of water per foot}$$

Finally, the maximum heat transfer relaxation length is the rock bed depth divided by 5, or

$$\frac{64\ \text{in}}{5} = 12.8\ \text{in}$$

Figure 4.11 shows the maximum and minimum face velocities and pressure gradients and the maximum heat transfer relaxation length plotted on the rock bed performance map. The lines drawn on the performance map define the acceptable design region for this example. The designer is free to base the design on any point within the acceptable design region.

Rock diameters of $\frac{1}{2}$ to $\frac{3}{4}$ in are satisfactory. For this design arbitrarily choose $\frac{3}{4}$-in rock and a 20 ft/min face velocity for the minimum pressure drop. The cross-sectional area of the rock bed is

$$\frac{1000\ \text{ft}^3/\text{min}}{20\ \text{ft/min}} = 50\ \text{ft}^2$$

A cross-sectional area of 50 ft^2 can be obtained by making the box 8 ft long and 6 ft 3 in wide on the inside.

The volume of rock will be 8 ft × 6 ft 3 in × 5 ft 4 in = 267 ft^3. This amounts to 0.53 ft^3 of rock per square foot of collector. Recall that the density of rock is about 167 lb/ft^3 and allow for a 30 percent void fraction. The weight of the rock then is

$$267\ \text{ft}^3 \times 167\ \text{lb/ft}^3 \times 0.70 = 31{,}212\ \text{lb} = 15.6\ \text{tons}$$

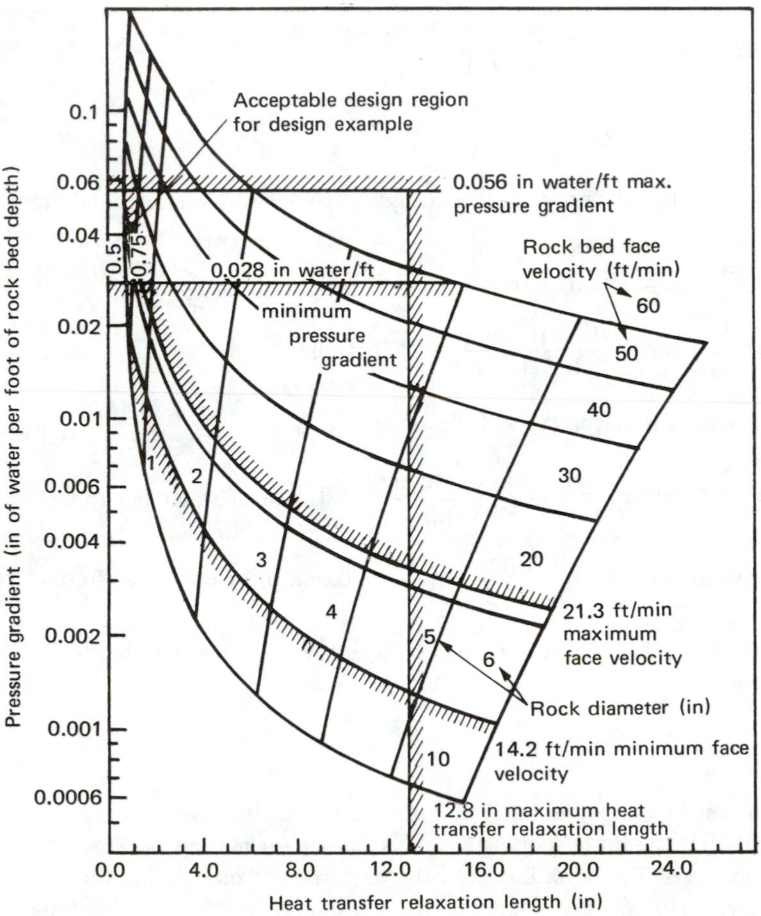

Figure 4.11 Rock bed performance map for design example. (*From [2]*.)

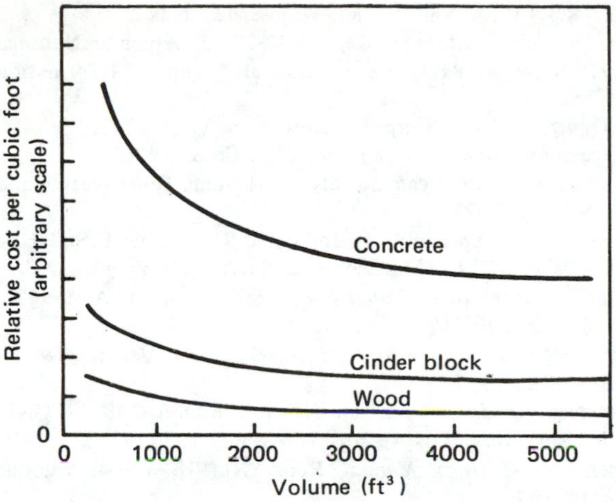

Figure 4.12 Relative cost of rock bed containers, excluding insulation. (*From [12].*)

To monitor performance two temperature sensors should be located in the rock bed, one 6 in below the top of the rocks and the other 6 in above the bottom. Low temperature readings by both sensors indicate that little heat remains in the rock bed, and the auxiliary heater must supply space heat. High temperature readings by both sensors indicate that the rock bed is fully charged. A high temperature reading at the top and a low temperature reading at the bottom indicate that the rock bed is partially charged. Extreme care must be taken to avoid damaging the temperature sensors while filling the rock bin. It is a good idea to install a spare temperature sensor, especially at the bottom of the rock bed, where a temperature sensor would be difficult to replace.

Rock Bed Cost

The cost of a rock bed varies widely from region to region. In 1978 the cost of rocks at the quarry typically ranged from $3 to $10 per ton. In locations far from suitable quarries, delivery is an extra expense.

The cost of the rock container depends on the type used. Wooden containers are the least expensive, followed by cinder block and concrete. Figure 4.12 shows the relative cost per cubic foot of these three types of containers. The cost per cubic foot is lower for larger containers than for smaller ones, because the volume goes up faster than the surface area, and it is the surface area that determines the amount of material and labor involved in construction of the container.

REFERENCES

1. C. Wyman, J. Castle, and F. Kreith, "A Review of Collector and Energy Storage Technology for Intermediate Temperature Applications," *Solar Energy,* vol. 24, 1980, pp. 517–540.

2. R. L. Cole, K. J. Nield, R. R. Rohde, and R. M. Wolosewicz (eds.), *Design and Installation Manual for Thermal Energy Storage,* Rep. ANL-79-15, Argonne National Laboratory, Argonne, Ill, 1979 (also available from National Technical Information Service, Springfield, Va.).
3. F. Kreith, *Principles of Heat Transfer*, 3d ed., Harper & Row, New York, 1973.
4. Solaron Corp., *Application Engineering Manual,* Commerce City, Colo., 1976.
5. ASHRAE, *Handbook of Fundamentals,* American Society of Heating, Refrigerating and Air-Conditioning Engineers, New York, 1972.
6. W. D. Collins, "Temperature of Water Available for Industrial Use in the U.S.," U.S. Geological Survey Water Supply Paper 520 F, U.S. Geological Survey, Washington, D.C.
7. J. E. Carson, *Soil Temperatures and Weather Conditions,* 16th ed., Rep. ANL-6470, Argonne National Laboratory, Argonne, Ill., 1970.
8. International Telephone and Telegraph Corp., *Solar Heating System Design Manual,* Bulletin TESE-576, Morton Grove, Ill., 1976.
9. D. Balcomb et al., *Solar Heating Handbook for Los Alamos,* Rep. UC-5967/CONF-75027-1, Los Alamos Scientific Laboratory, Los Alamos, N.M., May 1975.
10. R. M. Graven and P. R. Hirsch, *DOE-1 Users' Manual,* Rep. ANL/ENG-77-04, Argonne National Laboratory, Argonne, Ill, 1977.
11. R. V. Dunkle and W. M. Ellul, "Randomly-Packed Particulate Bed Regenerators and Evaporative Cooling," *Mech. Chem. Eng. Trans. Inst. Eng. Aust.,* vol. MC8, no. 2, 1972, pp. 117–121.
12. U.S. Energy Research and Development Administration, Division of Solar Energy, *Inter-Technology Corporation Technology Summary, Solar Heating and Cooling,* ERDA Rep. C00/2688-76-10, Washington, D.C., 1976.

5 Solar Energy Economics

This chapter describes various economic matters concerned with the design and sizing of solar heating and cooling systems. Solar energy is expensive energy, and the costs and benefits of solar systems must be carefully analyzed if the capital investment is to be used most effectively. Sections in this chapter deal with the ideas of engineering economics and how they are applied to solar systems, a simplified optimization method for sizing solar systems, and, finally, a more detailed optimization method that includes all economic parameters of interest to the system owner.

INTRODUCTION

The principal reason for using solar thermal systems for heating and cooling is to reduce the energy costs associated with operating buildings or other thermal processes. Therefore an economic analysis must be carried out to determine whether a particular solar system is economically advantageous for a particular project.

The benefit associated with owning a solar system is the value of the fossil or nuclear energy saved by substituting solar heat for these forms of heat. For each type of solar thermal system a somewhat different performance prediction method is used to calculate the energy delivery. The specifics of these methods are described in Chaps. 6 through 8.

The cost of the solar system associated with the benefit realized by fuel saving depends on the initial capital cost of the system as well as the recurring operating and maintenance costs distributed throughout the lifetime of the system. Solar systems are different from other energy systems from this perspective, since the principal investment in a solar system occurs initially, whereas the costs for fossil or nuclear energy are distributed throughout the lifetime of the heating system. To reconcile these two different types of cash flows and compare them on an impartial basis, the principles of engineering economics are used. This engineering discipline can compare cash flows that occur at any time during an economic period of analysis, reduce them to the same cash flow basis, and determine which of several system options is the most cost-effective. The next section describes engineering economics as it applies to solar heating and cooling systems.

Solar systems differ from conventional systems in another way that has an important effect on their relative economics. Solar system operation is subject to the law of diminishing returns. This is due to the nonlinear nature of the interactions of collectors and storage and delivery devices. The physics of these component interactions is such that a solar system does not produce proportionately more energy savings as the size of the system is increased. For example, if the collector area on a building is doubled, the amount of useful solar heat produced and delivered to the building will not be doubled. Figure 5.1 illustrates this concept in schematic form. Therefore, providing a 100 percent solar heating or cooling system for a building will be quite costly, since the last few percent of the annual heating load becomes very expensive in terms of initial capital investment. Usually, however, the solar system provides most but not all of a given thermal load, with the remaining demand provided by an auxiliary or backup system based on a conventional fuel. The precise amount of backup energy used is determined by an economic trade-off between solar energy and fossil energy costs. The cost of solar energy is defined as the amount invested in the solar system, prorated on an annual basis, and divided by the annual energy production. The cost is expressed in units of dollars per MMBtu (MMBtu = 1

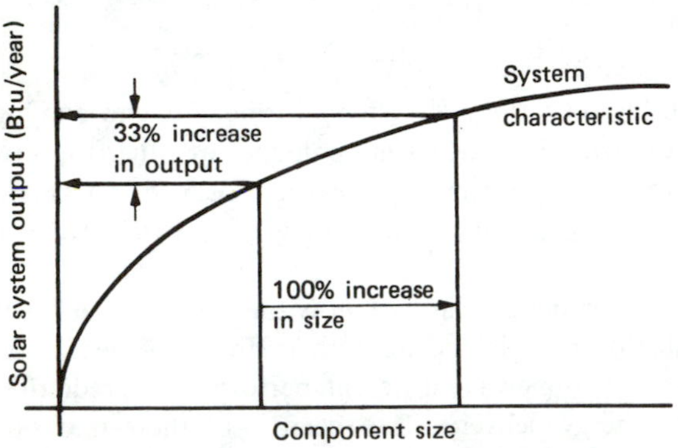

Figure 5.1 Characteristic curve exhibiting the law of diminishing returns, which applies to all components of solar thermal systems. For the example shown, a 100 percent increase in the component size results in only a 33 percent increase in the output of the system. A further increase in size would produce a still smaller increase in output.

million Btu) or per kilowatt-hour. If the cost of solar energy so measured is less than that for competing energy sources, there is an economic incentive for the use of solar energy. This is equivalent to requiring that the ratio of solar system costs to solar system benefits (dollar value of fuel saving) be less than unity. Of course, the cost and benefits must be measured in comparable cash flow terms, as described in the next section.

Another consequence of the law of diminishing returns is that the sizing of solar systems is based on economic criteria rather than on meeting a peak energy demand, as is the case for nonsolar energy resources. The economic optimum is established by examining a range of solar system sizes, calculating their costs and benefits, and determining the size that produces the maximum benefit for the system owner. Calculating costs and benefits for a range of solar system sizes is inherently more complicated than sizing a fossil fuel boiler or vapor compression air-conditioner to handle a peak heating or cooling load.

Therefore a new type of expertise required of a solar system designer is the ability to size systems on the basis of engineering economics rather than on a specific thermal requirement. Of course, the thermal requirement must be known before the economic sizing step; that is, the optimal sizing adds a new activity in each step of the design process. Specifically, as the design is refined, costs become more accurate, and recalculation of the optimal size at two or three points in the design process is carried out. The first calculation is done during the schematic design phase on the basis of estimated loads and rough cost estimates. During design development and construction document phases, cost estimates will improve, as will estimates of thermal demands. Therefore in progressively later stages of the design the optimal size of the solar system will be progressively more accurately determined.

Other factors may enter into the sizing of a solar system. In some cases, the system owner may arbitrarily decide that a particular heating or cooling system is to provide a specific annual solar heating or cooling fraction, e.g., 75 or 80 percent. In this case the economic optimization procedure described in this chapter need not be carried out, and the performance prediction methods described in Chaps. 6 through 8 can be used directly. Several calculations of estimated performance are made for several system sizes, and the size that produces the required annual solar contribution is identified without reference to economic criteria.

Fuel price inflation affects the value of solar system benefits in future years. As the value of displaced fossil or nuclear energy increases, the size of a solar system that can be economically purchased to save this energy will also increase. Therefore projected inflation in energy prices is an important variable in determining optimum system size. However, since systems must be sized before they are built and cannot be continuously increased in size as increased fuel prices may justify, it is necessary to estimate fuel inflation during the facility design process. Several guidelines for estimating fuel price inflation will be presented later, and it will be shown that this is one of the most sensitive inputs to an economic analysis.

Available budget may be a constraint on the size of a solar system. The optimum or least-cost system may require a greater initial investment than is available for a specific application. In that case the maximum budget amount should be invested to produce the maximum savings. Although this savings may not be the theoretical maximum possible if a larger system were to be purchased, it will be the maximum savings for a system owner given the budget constraint. Another constraint that may limit the size of a solar system is the area available for collectors. For example, if collectors are to be mounted on the flat roof of a building, the area of the building and shading considerations described in Chap. 2 will determine the maximum area of collector that can be installed. This may again limit the savings realized by the owner to some value less than the theoretical optimum. It will be seen in later sections that the solar cost curve has a fairly broad minimum; therefore, if the absolute theoretical optimum cannot be achieved, it is likely that most of the life-cycle savings associated with a system size can still be realized if restrictions on budget or area limit the array size.

Since solar heating systems are expensive, it is important that they be accurately sized. More care must be taken in the design of solar systems than has historically been taken in the design of nonsolar systems. Errors in nonsolar system design can be compensated for by burning more fossil fuel or consuming more electricity, since the economic penalties of reduced plant efficiency may not be significant and may be difficult for an owner to discover. However, incorrectly sized solar systems have significant initial cost penalties for the owner, and precise sizing based on economic optimization methods is essential. It is most important that solar systems not be designed by rule of thumb. This method was advocated by some early system designers, and it invariably produces systems that are either too large or too small for the economic environment in which they operate. A rule of thumb relating collector size to floor area, for example, is entirely inadequate.

The following section describes the features of engineering economics that are important for economic optimization of solar thermal systems. The next two sections describe a simplified and then a more detailed and comprehensive economic optimization method. The simplified method is probably adequate for most residential heating and cooling designs, where the accuracy of economic parameters does not warrant the more detailed approach. In commercial and larger residential applications, the detailed method can be used for more precise sizing, although it requires the specification of many more economic parameters.

ENGINEERING ECONOMICS APPLIED TO SOLAR SYSTEMS

The key idea in engineering economics is that money has a time value. That is, a given amount of money or a given cash flow will be worth less in the future than it is today. The relation between present and future cash flows is embodied in the discount rate. The discount rate is a measure of the time value of money and is related to the interest rate paid on home mortgages, for example. Our purpose in introducing several ideas from engineering economics is to enable the reader to compare solar and nonsolar energy sources on an impartial and rational basis. Comparison of different types of cash flows, initial versus recurring, requires that the discount rate be applied year by year.

The application of engineering economics to energy systems of various types has frequently been called life-cycle costing. In life-cycle costing the energy costs of the building are added to the costs of purchase and installation, maintenance, repair, and insurance for the solar system investment. Associated with each of these costs is the cost of money, i.e., the interest or discount rate.

Discounted Cash Flow Factors

The most common method of applying engineering economics to solar energy systems is to reduce all future cash flows to their present value or present worth. This subject is treated in many economics textbooks, e.g., [1]. Solar system economic analysis can also be carried out in terms of future dollars with the same results.

The following simple example of the idea of present worth illustrates the basic assumption of engineering economics: that it is the present purchasing power of a cash flow that is important. A \$500 payment made today has the same value as a \$550 payment made a year from today if the discount rate is 10 percent. In other words, if the 10-percent time value of money expressed on an annual basis is factored out of the \$550 future cash flow, the present value is seen to be \$500. The assumption is that the investor is indifferent to the \$500 cash flow today or to the \$550 cash flow a year from today.

Expressed in equation form, a cash flow occurring N years from today can be reduced to its present value as

$$P = \frac{F}{(1 + i)^N} \tag{5.1}$$

where P is the present worth or present value, F is the future cash flow, and i is the discount rate. The factor relating present to future cash flows is called the present-worth factor PWF(i, N):

$$\text{PWF}(i, N) = \frac{1}{(1 + i)^N} \tag{5.2}$$

The present-worth factor is the ratio of present to future cash flow and is used to reduce a single payment or cash flow in the future to present dollars. For example, it can be used to relate payments made in the future for natural gas or fuel oil to their equivalent present-value sum. This present-value sum can then be compared to the initial cost of a solar system to determine whether the solar system is economically viable. Of course, the rate of inflation of fuel prices would also need to be included in this calculation, as described shortly.

Table 5.1 shows present-worth factors for various discount rates up to 25 percent and for economic periods of analysis up to 20 years. For values of the

Table 5.1 Present-Worth Factors*

	i (%)									
N	0	2	4	6	8	10	12	15	20	25
1	1.0000	0.9804	0.9615	0.9434	0.9259	0.9091	0.8929	0.8696	0.8333	0.8000
2	1.0000	0.9612	0.9246	0.8900	0.8173	0.8264	0.7972	0.7561	0.6944	0.6400
3	1.0000	0.9423	0.8890	0.8396	0.7938	0.7513	0.7118	0.6575	0.5787	0.5120
4	1.0000	0.9238	0.8548	0.7921	0.7350	0.6830	0.6355	0.5718	0.4823	0.4096
5	1.0000	0.9057	0.8219	0.7473	0.6806	0.6209	0.5674	0.4972	0.4019	0.3277
6	1.0000	0.8880	0.7903	0.7050	0.6302	0.5645	0.5066	0.4323	0.3349	0.2621
7	1.0000	0.8706	0.7599	0.6651	0.5835	0.5132	0.4523	0.3759	0.2791	0.2097
8	1.0000	0.8535	0.7307	0.6274	0.5403	0.4665	0.4039	0.3269	0.2326	0.1678
9	1.0000	0.8368	0.7026	0.5919	0.5002	0.4241	0.3606	0.2843	0.1938	0.1342
10	1.0000	0.8203	0.6756	0.5584	0.4632	0.3855	0.3220	0.2472	0.1615	0.1074
11	1.0000	0.8043	0.6496	0.5268	0.4289	0.3505	0.2875	0.2149	0.1346	0.0859
12	1.0000	0.7885	0.6246	0.4970	0.3971	0.3186	0.2567	0.1869	0.1122	0.0687
13	1.0000	0.7730	0.6006	0.4688	0.3677	0.2897	0.2292	0.1625	0.0935	0.0550
14	1.0000	0.7579	0.5775	0.4423	0.3405	0.2633	0.2046	0.1413	0.0779	0.0440
15	1.0000	0.7430	0.5553	0.4173	0.3152	0.2394	0.1827	0.1229	0.0649	0.0352
16	1.0000	0.7284	0.5339	0.3936	0.2919	0.2176	0.1631	0.1069	0.0541	0.0281
17	1.0000	0.7142	0.5134	0.3714	0.2703	0.1978	0.1456	0.0929	0.0451	0.0225
18	1.0000	0.7002	0.4936	0.3503	0.2502	0.1799	0.1300	0.0808	0.0376	0.0180
19	1.0000	0.6864	0.4746	0.3305	0.2317	0.1635	0.1161	0.0703	0.0313	0.0144
20	1.0000	0.6730	0.4564	0.3118	0.2145	0.1486	0.1037	0.0611	0.0261	0.0115

*For interest rates *i* of 0 to 25 percent and periods of analysis *N* of 1 to 20 years.

discount rate or life-cycle period *N* not contained in the table, Eq. (5.2) can be used to evaluate the present-worth factor.

EXAMPLE

Assume that opportunity A generates benefits equal to \$100, \$150, and \$200 at the end of years 1, 2, and 3, respectively. Assume that opportunity B yields benefits of \$225 in year 2 and \$225 in year 3. Therefore over the 3-year life-cycle period both opportunities yield total benefits of \$450. However, the timing of the benefits received is different. By using the idea of the present-worth factor, compare the two benefit cash flows in terms of their present value.

SOLUTION

Step 1

Compute the present-worth factors from Eq. (2.2), using an interest rate of 10 percent:

$$\text{Year 1: PWF} = \frac{1}{1 + 0.10} = 0.9091$$

$$\text{Year 2: PWF} = \frac{1}{(1 + 0.10)^2} = 0.8264$$

$$\text{Year 3: PWF} = \frac{1}{(1 + 0.10)^3} = 0.7513$$

Step 2

Now compute the present worth of investment opportunities A and B by multiplying the prescribed cash flows by the appropriate present-worth factors. The annual benefit amounts expressed in present dollars are tabulated below.

		Annual benefit ($)		Present worth ($)	
Year	PWF	A	B	A	B
1	0.9091	100	0	90.91	0
2	0.8264	150	225	123.96	185.24
3	0.7513	200	225	150.26	169.04
Total		450	450	365.13	354.98

The effect of the time value of money is to reduce cash flows in future years relative to those in years closer to the present. For example, opportunity B, with cash flows of $225 in both years 2 and 3, has a relatively smaller present worth cash flow in year 3 than in year 2.

The comparison of investment opportunities A and B summarized in the table above indicates that opportunity A produces a greater present worth than opportunity B, although the total life-cycle cash flows on an undiscounted basis are identical.

Capital Recovery Factor–Series of Payments

The present-worth factor described above applies to a single payment made at some point in the future. It is common in the analysis of solar systems to encounter a uniform series of payments made over the life cycle of the system. This occurs, for example, in the repayment of a loan or mortgage for purchase of a solar system. A fixed or level payment is made each year to pay off the initial loan used to purchase the system. The relation between the annual payment and the initial loan amount is the capital recovery factor. Specifically, the capital recovery factor is the ratio of the annual (or monthly) payment to the total sum that must be repaid. In equation form, the capital recovery factor $\text{CRF}(i, N)$ is given by

$$\text{CRF}(i, N) = \frac{i}{1 - (1 + i)^{-N}} \tag{5.3}$$

In this book annual cash flows are assumed in the analysis of solar systems since performance predictions are generally made on an annual time scale. Monthly values of present-worth factors and capital recovery factors can be used, although the difference between them and annual quantities is small.

Table 5.2 shows capital recovery factors for discount rates up to 25 percent and life-cycle periods up to 20 years. If the discount rate is zero, as it is for some public projects using capital funds, the capital recovery factor is simply the reciprocal of the life-cycle period of analysis N. For example, the capital recovery factor for a 20-year period would be $\frac{1}{20}$ or 0.05.

EXAMPLE

A solar system is purchased as part of a 30-year home mortgage at 9 percent interest. If the solar system cost \$10,000, what is the annual payment on the loan and what is the total amount of interest paid over the 30-year period?

SOLUTION

The relation between the initial payment of \$10,000 and the annual payment is the product of \$10,000 × CRF(0.09, 30). Since this value is not in Table 5.2, Eq. (5.3) must be used:

$$\text{CRF}(0.09, 30) = \frac{0.09}{1 - (1 + 0.09)^{-30}} = 0.09734$$

The annual payment A is

$$A = 10{,}000 \times \text{CRF}(0.09, 30) = \$973.40$$

Over 30 years the total payment is

$$30 \times 973.40 = \$29{,}200$$

of which \$10,000 is the loan principal. The remaining payment is for interest totaling \$19,200.

Value of the Discount Rate

The discount rate used in discounted cash flow analyses is well known for most commercial and public projects but is not well defined for most homeowners. In the case of businesses, the discount rate is the same as the minimum acceptable return on investment. It usually differs from the interest rate, which would be used on a loan to purchase a solar system. The discount rate would apply to a firm's internal capital funds used for investment in an energy-saving system.

For public projects, the discount rate used would typically be the interest paid on bonds used to purchase a facility. For example, if a solar system were to be part of a public library purchased by bonds, the discount rate would be the

Table 5.2 Capital Recovery Factors*

	i (%)									
N	0	2	4	6	8	10	12	15	20	25
1	1.00000	1.02000	1.04000	1.06000	1.08000	1.10000	1.12000	1.15000	1.20000	1.25000
2	0.50000	0.51505	0.53020	0.54544	0.56077	0.57619	0.59170	0.61512	0.65455	0.69444
3	0.33333	0.34675	0.36035	0.37411	0.38803	0.40211	0.41635	0.43798	0.47473	0.51230
4	0.25000	0.26262	0.27549	0.28859	0.30192	0.31547	0.32923	0.35027	0.38629	0.42344
5	0.20000	0.21216	0.22463	0.23740	0.25046	0.26380	0.27741	0.29832	0.33438	0.37184
6	0.16667	0.17853	0.19076	0.20336	0.21632	0.22961	0.24323	0.26424	0.30071	0.33882
7	0.14286	0.15451	0.16661	0.17914	0.19207	0.20541	0.21912	0.24036	0.27742	0.31634
8	0.12500	0.13651	0.14853	0.16101	0.17401	0.18744	0.20130	0.22285	0.26061	0.30040
9	0.11111	0.12252	0.13449	0.14702	0.16008	0.17364	0.18768	0.20957	0.24808	0.28876
10	0.10000	0.11133	0.12329	0.13587	0.14903	0.16275	0.17698	0.19925	0.23852	0.28007
11	0.09091	0.10218	0.11415	0.12679	0.14008	0.15396	0.16842	0.19107	0.23110	0.27349
12	0.08333	0.09156	0.10655	0.11928	0.13270	0.14676	0.16144	0.18148	0.22526	0.26845
13	0.07692	0.08812	0.10014	0.11296	0.12652	0.14078	0.15568	0.17911	0.22062	0.26454
14	0.07143	0.08260	0.09467	0.10758	0.12130	0.13575	0.15087	0.17469	0.21689	0.26150
15	0.06667	0.07783	0.08994	0.10296	0.11683	0.13147	0.14682	0.17102	0.21388	0.25912
16	0.06250	0.07365	0.08582	0.09895	0.11298	0.12782	0.14339	0.16795	0.21144	0.25724
17	0.05882	0.06997	0.08220	0.09544	0.10963	0.12466	0.14046	0.16537	0.20944	0.25576
18	0.05556	0.06670	0.07899	0.09236	0.10670	0.12193	0.13794	0.16319	0.20781	0.25459
19	0.05263	0.06378	0.07614	0.08962	0.10413	0.11955	0.13576	0.16134	0.20646	0.25366
20	0.05000	0.06116	0.07358	0.08718	0.10185	0.11746	0.13388	0.15976	0.20536	0.25292

*For interest rates *i* of 0 to 25 percent and periods of analysis *N* of 1 to 20 years.

rate of interest paid on the bonds plus any administrative charges. In the case of grants made by the federal government to state and local governments, a discount rate of zero might apply, depending on the type of bookkeeping system used by the state or municipality. Since grant funds are not repaid, their discount rate is typically taken to be zero.

In the case of homes, the discount rate is bracketed by the homeowner's historical earning ability with other investments and the interest rate that would be paid on a loan for the solar system. A homeowner able to invest in a solar system would expect a return commensurate with his or her prior investment history. This rate of return would be the upper limit for the discount rate. The lower limit would be represented by the owner's willingness to pay interest on a solar system loan in order to reduce heating bills.

If the homeowner is unable to specify a discount rate to be used for the purpose of sizing a solar system, the interest rate on a solar loan is usually taken to be equivalent to the discount rate. Historically, interest rates in real terms are approximately 2 percent. That is, the real cost of money, after subtracting inflation, is 2 percent per year. This rule has applied in the United States for the past several decades [1]. The fluctuation that occurs in market interest rates is usually well correlated with the inflation rate of the economy as measured by the consumer price index or the gross national product (GNP) deflator compiled by the federal government. The actual time value of money is the so-called market interest rate reduced by the current value of the consumer price index or GNP deflator.

Interest Fraction in Annual Payments

As indicated by the preceding example, part of the capital recovery factor accounts for interest payments on a loan. Since interest is deductible for private and corporate taxpayers, it is important to know the fraction of each year's payment on a solar loan that consists of interest. Table 5.3 shows the interest fraction on an annual basis for various mortgage rates and mortgage terms up to 20 years. The equation used to calculate the values in this table is relatively complicated and is not reproduced here, but it can be found in [2].

EXAMPLE

For a \$10,000 solar system purchased at 10% interest find the fraction of interest paid in the first year's payment, the tenth year's payment, and the last (20th) year's payment.

SOLUTION

For an interest rate of 10 percent, Table 5.3 shows that 85.1 percent of the first year's payment, 65.0 percent of the tenth year's payment, and 9.1 percent of the last year's payment are interest. Taking these percentages of the annual payment of \$973.40 yields the interest payments for these years, which are \$828.36,

Table 5.3 Fraction of Interest and Yearly Mortgage Payment for Specified Year*

Years remaining on mortgage	Annual mortgage interest rate (%)										
	7	$7\frac{1}{2}$	8	$8\frac{1}{2}$	9	$9\frac{1}{2}$	10	$10\frac{1}{2}$	11	$11\frac{1}{2}$	12
20	0.742	0.765	0.785	0.804	0.822	0.837	0.851	0.864	0.876	0.887	0.896
19	0.723	0.747	0.768	0.788	0.806	0.822	0.836	0.850	0.862	0.874	0.884
18	0.704	0.728	0.750	0.770	0.788	0.805	0.820	0.834	0.847	0.859	0.870
17	0.683	0.708	0.730	0.750	0.769	0.786	0.802	0.817	0.830	0.843	0.854
16	0.661	0.686	0.708	0.729	0.748	0.766	0.782	0.798	0.812	0.825	0.837
15	0.638	0.662	0.685	0.706	0.725	0.744	0.761	0.776	0.791	0.805	0.817
14	0.612	0.637	0.660	0.681	0.701	0.719	0.737	0.753	0.768	0.782	0.795
13	0.585	0.609	0.632	0.654	0.674	0.693	0.710	0.727	0.742	0.757	0.771
12	0.556	0.580	0.603	0.624	0.644	0.663	0.681	0.698	0.714	0.729	0.743
11	0.525	0.549	0.571	0.592	0.612	0.631	0.650	0.667	0.683	0.698	0.713
10	0.492	0.515	0.537	0.558	0.578	0.596	0.614	0.632	0.648	0.663	0.678
9	0.456	0.478	0.500	0.520	0.540	0.558	0.576	0.593	0.609	0.625	0.639
8	0.418	0.439	0.460	0.479	0.498	0.516	0.533	0.550	0.566	0.581	0.596
7	0.377	0.397	0.417	0.435	0.453	0.470	0.487	0.503	0.518	0.533	0.548
6	0.334	0.352	0.370	0.387	0.404	0.420	0.436	0.451	0.465	0.480	0.493
5	0.287	0.303	0.319	0.335	0.350	0.365	0.379	0.393	0.407	0.420	0.433
4	0.237	0.251	0.265	0.278	0.292	0.304	0.317	0.329	0.341	0.353	0.364
3	0.184	0.195	0.206	0.217	0.228	0.238	0.249	0.259	0.269	0.279	0.288
2	0.127	0.135	0.143	0.151	0.158	0.166	0.174	0.181	0.188	0.196	0.203
1	0.065	0.070	0.074	0.078	0.083	0.087	0.091	0.095	0.099	0.103	0.107

*This table can be used for any mortgage period from 1 to 20 years for the interest rates shown.

$632.71, and $88.58. Thus as time goes on, more of the annual payment goes to repaying the principal and less to paying the interest.

Effect of Inflation

Inflation of fuel prices, which has exceeded that of the general economy in the past decade, causes the value of energy savings in future years to be progressively higher. That is, although a solar system may save roughly the same amount of Btu's per year, the dollar value of that energy progressively increases owing to fuel price escalation.

Discounted cash flows can be modified to include the effect of inflation of fuel prices. As shown in [3], a quantity called the levelized cost of fuel can be used to include both discount and inflation effects on future fuel prices. The levelized cost of fuel is a uniform amount that, if paid during each year of the life-cycle period, would result in the same cash flows as the real cost of fuel inflating at a specific rate j throughout the same period. The levelized cost of fuel $\overline{C}_f$ can be calculated from

$$\bar{C}_f = C_{f,0}\,\frac{\text{CRF}(i, N)}{\text{CRF}(i', N)} \tag{5.4}$$

where $C_{f,0}$ is the unit cost of fuel in year zero, and the capital recovery factor in the denominator is based on an inflation-adjusted discount rate i' given by

$$i' = \frac{i - j}{1 + j} \tag{5.5}$$

In Eq. (5.5), j is the annual fuel inflation rate and i is the discount rate as used in previous sections.

The value of the fuel price inflation rate j has been estimated by the U.S. Department of Energy and is shown in Table 5.4 for coal, fuel oil, gas, and electricity. The regional variation in the expected inflation rate for electricity is governed by the type of fuel used in different areas of the country. Note that i' as calculated from Eq. (5.5) can be positive, negative, or zero, depending on the relative magnitudes of i and j. For negative values of i', Eq. (5.3) should be used to determine the capital recovery factor in the denominator, since Table 5.2 does not include negative discount rates.

EXAMPLE

If the cost of oil is $C_{f,0}$ = \$36.50 per barrel today, what is the levelized cost over 15 years if the discount rate i = 10 percent and the price of oil increases at 5 percent per year?

SOLUTION

Find i' and the capital recovery factors from Eq. (5.3):

$$i' = \frac{0.1 - 0.05}{1.05} = 0.0476$$

$$\text{CRF}(0.0476, 15) = 0.0947$$

$$\text{CRF}(0.10, 15) = 0.1315$$

Then from Eq. (5.4)

$$\bar{C}_f = 36.50\,\frac{0.1315}{0.0947} = \$50.68 \text{ per barrel}$$

SIMPLIFIED ECONOMIC ANALYSIS OF SOLAR HEATING AND COOLING SYSTEMS

The accuracy of economic parameters and solar prediction methods for some situations does not warrant a detailed economic analysis including all possible cash flow terms that may occur over the life cycle of the system. The simplified economic method presented in this section is usually adequate for all

Table 5.4 U.S. Department of Energy Differential Inflation Rates for Fuels and Electricity

Coal	5%		
Fuel oil	8%		
Gas, natural or LPG*	10%		
Electricity, by region:			
New England	6.9%	East North Central	5.6%
Connecticut		Illinois	
Maine		Indiana	
Massachusetts		Michigan	
New Hampshire		Ohio	
Rhode Island		Wisconsin	
Vermont		West North Central	5.6%
Middle Atlantic	5.9%	Iowa	
New Jersey		Kansas	
New York		Minnesota	
Pennsylvania		Nebraska	
Delaware		North Dakota	
South Atlantic	5.8%	South Dakota	
District of Columbia		Missouri	
Florida		West South Central	7.5%
Georgia		Arkansas	
Maryland		Louisiana	
North Carolina		Oklahoma	
South Carolina		Texas	
Virginia		Mountain	5.7%
West Virginia		Arizona	
East South Central	5.6%	Colorado	
Alabama		Idaho	
Kentucky		Montana	
Mississippi		Nevada	
Tennessee		New Mexico	
Pacific	7.3%	Utah	
California		Wyoming	
Oregon			
Washington			

*LPG, liquefied petroleum gas.
Source: [4].

residential applications as well as for the schematic design of larger systems for commercial, industrial, and institutional projects.

The parameters included in this first-order economic approach include the initial cost of the solar system reduced by any applicable tax credits, the interest paid on a solar loan, if any, the cost and inflation rate in fuel prices, and the length of the period of economic analysis. Terms not included in the simplified approach are maintenance, operating costs, salvage value, depreciation, and tax savings on interest payments for solar systems. For many systems the tax savings roughly offset the costs listed above and the loss of accuracy due to using the simplified approach for preliminary design is relatively small.

The Cost Equation

The total cost of heating or cooling a particular building by use of both solar and nonsolar (backup) energy is

$$C_{tot} = C_s + C_f \tag{5.6}$$

where C_{tot} = total heating or cooling task cost expressed on an annual basis
C_s = annual cost of the solar system
C_f = annual cost of backup fuel

By using the ideas of capital recovery factors and levelized fuel cost developed above, Eq. (5.6) can be expressed in terms of other quantities as shown in Eq. (5.7):

$$C_{tot} = (C_{s,\text{initial}} - \text{ITC})\text{CRF}(i, N) + (1 - \bar{f}_s)L\,\bar{C}_f \tag{5.7}$$

where $C_{s,\text{initial}}$ = total initial investment in the solar system above that required for a nonsolar system
ITC = total state and federal tax credit
$\bar{f}_s$ = annual solar heating or cooling fraction
L = annual heating or cooling load
$\bar{C}_f$ = levelized cost of backup energy

In words, the two terms on the right in Eq. (5.7) represent the annual payment for owning the solar system and the annual payment for backup energy. The quantity $(1 - \bar{f}_s)L$ is the amount of energy that must be provided by the backup system. This energy amount multiplied by the levelized cost of fuel, which includes inflation effects, represents the life-cycle energy payment on an annualized basis.

Although Eq. (5.7) is expressed in units of dollars or other currency per year, an equivalent equation can be written on a total present-worth basis. That is, instead of dealing with annualized quantities, the total present worth of the solar system and backup energy purchases can be used to evaluate the total cost of heating or cooling a particular building. The present-worth equation can be derived from Eq. (5.7) simply by dividing out CRF(i, N) from both terms. By referring to Eq. (5.4), it is seen that the CRF term is present in the formula for levelized cost. Therefore, costs calculated on a total life-cycle basis and on an annual basis are related by the capital recovery factor only. In this book annualized costs are used, since most firms and homeowners prepare their economic statements and tax returns on an annual basis. In addition, most solar performance prediction methods calculate energy delivery and solar fraction $\bar{f}_s$ on an annual basis. However, economic optimizations using annual or total life-cycle costs will always result in the same solar system size to minimize the total cost.

Calculation of the annual solar fraction $\bar{f}_s$ will be treated in Chaps. 6 through 8 for both active and passive heating systems. A prediction method suitable for solar air-conditioning systems is not yet available in simplified form, and detailed computer models must be used to calculate the annual solar fraction for solar cooling systems as a function of system size.

Not only is the second term of Eq. (5.7) related to system size through the annual solar fraction, but the first term of the equation is also related to system size since total initial cost varies with collector, storage, and heat exchanger size. The variation of total annual cost with collector area is the subject of the next section, where methods of optimization are described.

Tax Credits

Investment tax credits for solar systems are available from the federal government and from some state governments. The National Energy Act passed in 1980 gives an investment tax credit for residential active solar systems of 40 percent of the first $10,000 spent for solar hardware. This represents additional expenditures for solar systems not present in a nonsolar building. Therefore, for example, a backup heating system would not be included in the solar tax credit since it would be part of a nonsolar building. For commercial applications, the investment tax credit is 15 percent of the initial solar system cost. According to the 1980 act, passive heating and cooling systems do not fully qualify for all tax reductions.

Many states offer tax credits for solar and energy conservation systems. These vary from a few percent up to relatively large tax credits, even exceeding the federal credit. The highest tax credit on a state level in 1980 was that offered by Colorado, which permitted an additional 30 percent of the extra cost of *active and passive* solar heating systems to be deducted from state tax payments. Since payments for state income taxes are typically relatively small in comparison with a 30 percent tax credit, the Colorado credit can be recovered over a 5-year period if the homeowner's tax due is less than the credit due in the first year. State tax credits are continuously changing, and private and corporate solar system purchasers should determine the current credits for the state where the system is to be installed. A complete tabulation of state tax credits is not presented here owing to their rapid evolution.

Variation for Commercial Systems

The treatment of backup energy costs for businesses is somewhat different from that indicated in Eq. (5.7). Since operating costs are deductible business expenses, the second term of the equation is multiplied by one minus the effective tax rate to determine the cost after taxes for backup energy. In equation form, the annualized cost of solar heating or cooling for income-producing businesses is

$$C_{\text{tot}} = (C_{s,\text{initial}} - \text{ITC})\,\text{CRF}(i, N) + (1 - \bar{f}_s)L\bar{C}_f(1 - T_{\text{inc}}) \tag{5.8}$$

Of course, the income tax credit ITC for businesses differs from that for homeowners, as described above. In Eq. (5.8) T_{inc} is the combined effective rate for state and federal tax payments, including any deductions from state taxes for federal tax payments and vice versa.

Typical Solar System Costs

For the simplified economic analyses treated in this section, the cost of solar system components and their installation can be expressed on a cost-per-

Table 5.5 Active Solar System Cost Estimates for Commercial Buildings, Based on 1978 Dollars

Subsystem	Fraction of total solar cost
Collectors and supports	35%
Storage and heat exchangers	20%
Piping, controls, electrical, and installation	45%
Subsystem component	**Cost per square foot of collector**
Collectors	$ 5–20
Nonselective	$ 5–10
Selective	$ 10–20
Collector support	$ 3–10
Heat exchangers	$ 0.40–0.80
Collector fluid	$ 0.15–0.20
Storage tank and insulation	$ 2–5
Piping, insulation, expansion tanks, valves	$ 3–6
Pumps	$ 0.40–1.00
Controls and electrical	$3000–5000 per installation
Type of system	**Installed cost per square foot of collector**
Building service hot water (BSHW) only	$20–35
Space and BSHW heating	$25–50
Space heating and cooling	$35–65

Source: U.S. Department of Energy [6].

square-foot-of-collector basis. Table 5.5 shows typical system costs expressed in 1978 dollars for all major components of solar heating systems in commercial buildings. The average cost of solar heating systems in the late 1970s was between $35 and $50 per square foot of collector, depending on location. In regions of the country where solar construction is active, competition between contracting firms will push the cost of solar systems toward the lower limit of this range. In other regions, where solar construction is still a novelty, higher costs should be expected.

In a period of sustained inflation, costs estimated during the early stages of solar system design should be increased by an inflation index to produce costs applicable at the start of construction, when equipment is actually purchased. These inflation rates can be determined by a solar contractor or from standard manuals used in the heating, ventilating, and air-conditioning (HVAC) engineering profession, such as the mechanical and electrical cost data manual produced annually by Robert S. Means Co., Kingston, Mass., or McGraw-Hill's Dodge Reports. Other construction cost indices are available for some large metropolitan areas.

Various levels of cost estimates beyond that embodied in the simple per-square-foot approach of Table 5.5 will be required. For schematic design and preliminary feasibility studies, the data in the table or quotations from local installers are satisfactory. As additional cost refinement is required in subsequent stages of design, however, quotations based on specifications should be used to determine the optimum cost of the system. It is not possible here to give detailed prices of components of solar systems. Quotations from contractors should be used instead, since prices vary from location to location and year to year as inflation and competition in the industry manifest their offsetting effects.

SOLAR SYSTEM OPTIMIZATION

We saw in the previous section that the cost of heating or cooling a building with solar energy depends on the size of the system used, since both terms in Eq. (5.7) vary with system size. It is the purpose of optimization to find the point at which the sum of the two cost terms–solar and backup–is a minimum. This is the design point for the solar system, and it is the most advantageous system size for the owner, since the gross savings in fuel, less solar system costs, are a maximum at the point where the total net heating or cooling cost is a minimum.

Figure 5.2 is a plot of the solar cost equation, Eq. (5.7), and shows the first and second terms as well as their total. The solar system cost, expressed in terms of dollars per square foot, increases approximately linearly with solar collector area. Collector area is the index of size most commonly used for solar heating and cooling systems, since the sizes of most other components are keyed by well-established rules to the size of the collector. Therefore identification of

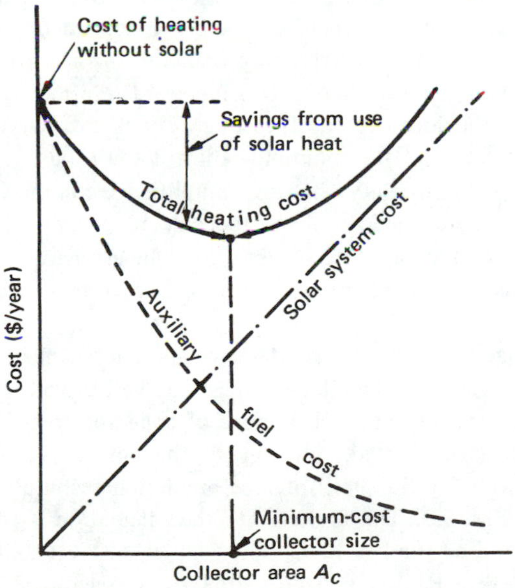

Figure 5.2 Annual solar system cost curve formed by adding auxiliary fuel cost to solar system annual cost. The total is the annual cost of performing a given solar heating task.

the optimum collector area immediately determines the optimum size of all other system components except for the load heat exchanger [2].

The auxiliary fuel cost curve shown in Fig. 5.2 is a monotonically decreasing curve that is not linear, but follows the law of diminishing returns. Therefore doubling collector area does not reduce the amount of auxiliary fuel required by a factor of 2. The relative reduction in auxiliary fuel required is always less than the percent increase in solar system size.

If the solar system cost and auxiliary fuel cost curves are added, their sum is the total heating cost curve in Fig. 5.2. The minimum point of this curve specifies the least-cost system size. It is the purpose of the optimization step required in solar system design to identify this point. Recall that the optimization and economic sizing methodology is not required for nonsolar systems, since they are typically sized to provide 100 percent of the heating load. The law of diminishing returns dictates that solar systems will not provide 100 percent of the energy requirement, hence the optimization calculation requirement.

Optimization Procedure

To construct the two curves shown in Fig. 5.2, the performance of a solar system applied to the specific project in question must be determined. To plot the auxiliary fuel cost curve, the annual solar fraction $\bar{f}_s$ must be calculated; the method of calculation depends on the application and will be described in later chapters. The levelized fuel cost is also a factor in the auxiliary fuel cost curve.

To plot the solar system cost curve, the cost of all solar components in dollars per square foot is required; the slope of this curve is given by the total of all costs expressed on this basis. In some cases the solar system investment includes a second type of cost, which is called a fixed cost and is independent of solar system size. The source of fixed costs would be, for example, the controls and piping needed for even the smallest solar system considered. The effect of fixed costs in Fig. 5.2 is to change the intercept of the solar system cost curve from the origin to a point above the origin. The minimum point of the total cost curve is unaffected. Therefore, for simplicity in the simplified economics methodology, the effect of fixed costs need not be considered insofar as optimization is concerned. Of course, the building owner's total investment for the system includes both fixed costs and costs that depend on collector area—the so-called variable costs.

Nearly all methods for calculating $\bar{f}_s$ require an iterative approach, since it is not possible to identify a priori the specific collector area that will provide a given solar heating fraction. In the iterative approach a range of collector areas is selected, the performance of each is evaluated, and finally the cost curve is generated. The process of optimization by this performance prediction technique is illustrated in flowchart form in Fig. 5.3. The first step in the calculation is to determine the annual heating or cooling load L, the levelized fuel cost $\bar{C}_f$, and the applicable capital recovery factor CRF. A collector area A_c is selected to begin the calculations, its cost and performance are determined, and the total heating cost for this system size is found from Eq. (5.7). This calculation

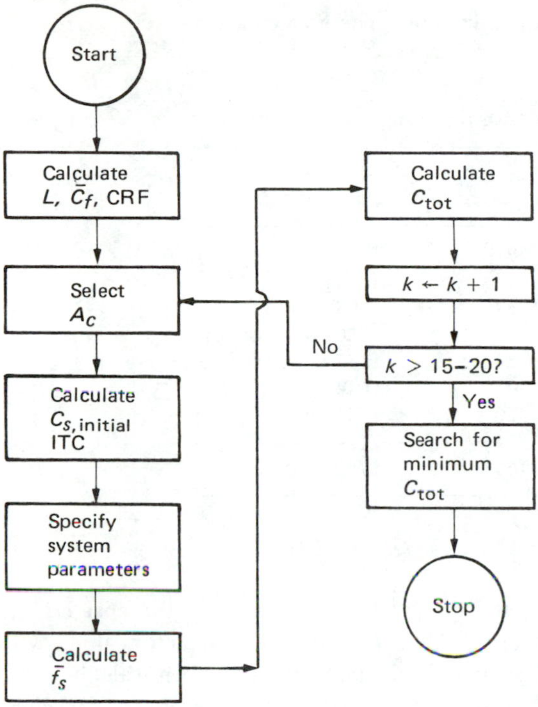

Figure 5.3 Solar system economic optimization flowchart showing iterative procedure required to identify the minimum-cost combination of solar and nonsolar energy.

sequence is repeated for a number of system sizes—typically 15 to 20—bracketing the region in which the minimum point is contained. The minimum can be identified by reviewing the calculations or, if they are done on a computer, an automatic routine for finding the minimum can be used. The number of collector sizes studied depends on the refinement required in the economic analysis.

The method in the flowchart is illustrated by the following example.

EXAMPLE

Find the optimum size of a passive solar heating system whose performance is tabulated below. Assume that the levelized cost of fuel is \$20/MMBtu and the passive system costs \$18/ft^2 and is purchased over a 20-year period at 10 percent interest. The annual heat load is 88.38 MMBtu.

SOLUTION

The capital recovery factor is 0.11746; therefore the passive system costs $0.11746 \times 18 = \$2.11/\text{ft}^2$ per year. This cost, multiplied by the collector area, gives the solar system cost per year. The solar system cost is then added to the fuel cost to find the total heating cost. These calculations are summarized in the table below.

Solar, Backup Fuel, and Total Heating Costs for Example Passive System

Area (ft^2)	Solar cost* ($/year)	$\bar{f}_S$	Fuel cost† ($/year)	Total cost ($/year)
53	112	0.1	1591	1703
112	236	0.2	1414	1650
176	371	0.3	1237	1608
246	519	0.4	1061	1580 ← least cost
333	703	0.5	884	1587
444	937	0.6	707	1644
593	1251	0.7	530	1781
842	1777	0.8	354	2131
1230	2595	0.9	177	2772

*Solar cost = $2.11 × area.
†Fuel cost = $(1 - \bar{f}_S) \times 88.38 \times 20$.

A more desirable approach to optimization is to specify the cost criteria contained in the levelized fuel cost and the capital recovery factor, along with the load and possibly certain system characteristics, and then calculate the optimum collector area directly. This approach has not been developed for active systems, but for passive systems a method called the *P* chart is available as described in Chap. 7. The *P* chart permits direct calculation of collector area from specified inputs as shown in the flowchart in Fig. 5.4. The direct approach is substantially simpler than the iterative approach used in most performance prediction methods. An additional complication of many estimation methods for active systems is that the calculation of $\bar{f}_S$ requires the performance to be found for each month of the year. Therefore the block in Fig. 5.3 labeled "calculate $\bar{f}_S$" actually represents the total of 12 monthly calculations. The direct calculation method avoids this problem. A method similar to the *P* chart will probably be developed for active systems in the early 1980s and will simplify the design of these systems as well.

The optimization curve shown schematically in Fig. 5.2 and the example optimization problem worked above exhibit a characteristic common to most solar system analyses. That is, the cost curve does not have a sharp minimum, and variations of several percent in the size of the solar system from the theoretical, absolute minimum point have very little effect on the life-cycle economics of the system being analyzed. It is important to consider this characteristic of solar optimization curves, since it may be possible to reduce the initial investment by undersizing the system and still have almost the same life-cycle benefits. Likewise, if the budget is limited or space for collector area restricted, it is possible to use reduced collector areas without much effect on cost saving. If a significantly smaller budget or available area for collector is

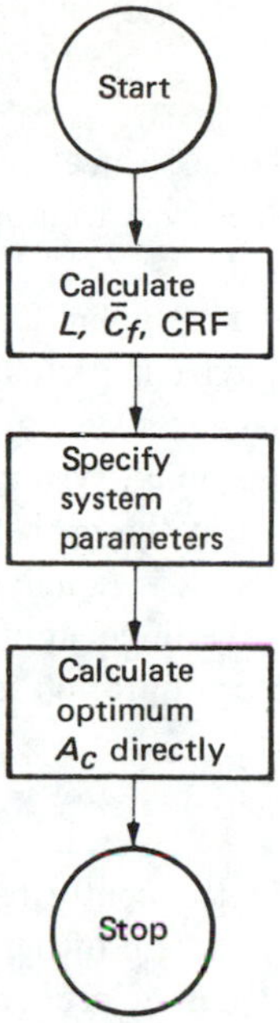

Figure 5.4 Simplified optimization procedure using direct calculation of optimal collector area, as opposed to the iterative procedure shown in Fig. 5.3.

restraining a specific project, the cost saving for the nonoptimal system can still be identified by using the approach of Fig. 5.2. The saving for any system size is measured by the line labeled "savings" in the figure. Although the savings may not be optimal for restricted systems, there will be an incentive to use solar energy to save a large fraction of the maximum amount possible.

The optimization method for sizing solar systems clearly is vastly different from using rules of thumb to determine optimal collector areas. The procedure above is the preferred method of sizing collectors since it includes the cost of the system, value of energy saved, load characteristics that reflect local climatic parameters, and regional effects on solar system performance embodied in the $\bar{f}_s$ calculation. None of these important effects are included in the rule-of-thumb sizing methods advocated by some designers.

Payback Period

Another index sometimes used in solar system economic analyses is the payoff period POP, or payback period. The payback period is defined as the time required to recoup the initial investment in a solar system by accumulated fuel savings. Various other definitions of payback period may be used to include discount rates and other factors. The simple payback period, including only initial cost, investment tax credits, and fuel savings, is given in equation form below.

$$\text{POP} = \frac{C_{s,\text{initial}} - \text{ITC}}{\bar{f}_s L C_{f,0}} \tag{5.9}$$

The initial fuel cost $C_{f,0}$ is used in the simple payback calculation, since the period of analysis N is not known a priori and therefore a levelized cost cannot be calculated.

The payback period has several deficiencies that make it a very limited economic index. It does not consider cash flows beyond the payback period, when the economic value of the solar savings will be largest. It penalizes systems with long operating lives, and it is not useful in determining the optimum system size, since it does not include the two terms of the cost equation in the proper additive role. The one useful feature of the payback period is that it is familiar to many types of system owners. It is a simple concept that can be understood by many people; however, its utility in solar system sizing and optimization is very limited.

Feasibility Studies

The output of an economic and optimization analysis will frequently be summarized in a written feasibility study for the prospective solar system owner. The purpose of a feasibility study is to determine early in the design process whether a solar heating system has the potential for significant energy savings during its lifetime. A more detailed feasibility study may be carried out during later design stages as costs are refined, but the preliminary feasibility study will be the basis for a solar go-no go decision.

Feasibility studies should list all thermal and economic assumptions, methods of calculation, data used, and expected accuracy of results. The single most important conclusion of the report is a statement about feasibility or infeasibility, as determined by supporting thermal and economic calculations and graphic displays of both.

To conduct a feasibility study, the following information is needed as a minimum:

Thermal information:
- Space heating and cooling loads and phasing
- Water heating load and phasing
- Water source and delivery temperature
- Night setback, if any
- Types of solar system to be considered
- Limits to component sizes
- Type of auxiliary system and fuel used
- Internal heat gain estimates

Economic information:
- System cost, fixed plus variable
- Discount rate
- Net fuel cost, including burner efficiency
- Fuel inflation rates

Budgetary limits
Length of period of analysis
Applicable tax credits

The information above, when used in the simplified economic approach and optimization methods described above, will produce a decision about the efficacy of solar that is accurate within 10 or 15 percent; this is sufficient for a decision regarding the feasibility of solar energy for a specific project.

DETAILED ECONOMIC ANALYSIS METHODS

The simplified economic analysis and optimization methods presented up to this point have included only the major cash flow terms for solar economic analyses. The costs considered were solar capital investment and interest charges—less tax credits—backup energy costs, and their inflation rate. Many other terms enter a detailed economic analysis, and in this section these terms are described in detail.

For residential economic analyses it is generally not necessary to include the detailed costs described in this section. In a residential context, it is acceptable in most cases to use the simplified calculation method described earlier. For commercial systems, which involve significantly greater investments and more stringent critēria for economic feasibility, the detailed method presented here should be used. Likewise, for most publicly funded projects all cash flow terms should be included.

The detailed economic method includes the following costs:

Initial capital cost increment less investment tax credit
Interest charges
Replacement of solar system components during economic lifetime
Operating energy costs
Property taxes
Maintenance
Insurance

These negative cash flows are offset to some extent by several positive cash flows, including

Salvage value at end of economic lifetime
Tax deductions for interest payments on solar loans
Depreciation
Operating energy tax deductions
Maintenance tax deductions
Insurance tax deductions
Replacement tax deductions

The last five items in this list are available only to profit-making organizations.

However, the income tax credit for interest payments is available to individual homeowners as well.

Quantitative Representation of Cash Flow Terms

The discipline of engineering economics enables the solar analyst to quantitatively calculate each of the terms listed above. These terms are presented in Table 5.6 as either negative cash flows, which represent payments by the system owner, or positive cash flows, which represent tax credits and other cash flows that tend to offset payments by the system owner. Nine terms are shown in Table 5.6, and for six of these terms a multiplying factor is used if the system is owned by a profit-making organization. The symbols used in Table 5.6 are defined as follows:

$C_{s,\text{tot}}$ = total initial solar investment, including the cost of all equipment, installation, any sales taxes, and any fees or ancillary costs (capital cost of solar components is described below)
$C_{s,\text{salv}}$ = solar system salvage value
$C_{s,\text{asses}}$ = assessed value of solar system
C_e = first year energy cost operate solar system per year
$\text{CRF}(i', N)$ = capital recovery factor for N years at interest rate i'
$i' = (i - j)/(1 + j)$ = real discount rate
$i'' = (i - j_e)/(1 + j_e)$ = inflation-adjusted discount rate for energy
i = interest or discount rate bounded below by the owner's rate of return forgone on the next best alternative investment and bounded above by the cost of borrowing; the rate i is strictly the marginal discount rate, i.e., the rate applied to the owner's next set of investment decisions, and it is frequently different from the average of discount rates on prior investments
M = maintenance cost per year
j = general inflation rate (measured by consumer price index)
i_m = market mortgage rate (real mortgage rate plus general inflation rate)
j_e = total energy inflation rate per year
k = index for years at which replacements or repairs are made
I = insurance costs per year
R_k = replacements made in year k
$i_m P_k$ = interest charge in year k

The principal in year k is given by

$$P_k = (C_{s,\text{tot}} - \text{ITC})\ [(1 + i_m)^{k-1} + \frac{(1 + i_m)^{k-1} - 1}{(1 + i_m)^{-N} - 1}$$

$\text{PWF}(i', N)$ = present-worth factor for year N based on interest rate i'
T_{inc} = income tax rate: state tax rate + federal tax rate − state tax rate × federal tax rate, where rates are based on last dollar earned

Table 5.6 Comprehensive Listing of Annual Cash Flow Terms for Solar Economic Analyses

Term description	Negative cash flow	Positive cash flow	Commercial multiplier*
Capital and interest	$(C_{s,\text{initial}} - \text{ITC})\,\text{CRF}(i', N)$	–	–
Salvage value	–	$C_{s,\text{salv}}\,\text{PWF}(i', N)\,\text{CRF}(i', N)$	$(1 - T_{\text{salv}})$
Replacements, year k	$\left[\sum_{k=1}^{N} R_k\,\text{PWF}(i', k)\right]\text{CRF}(i', N)$	–	$(1 - T_{\text{inc}})$
Operating energy	$C_e\,\dfrac{\text{CRF}(i', N)}{\text{CRF}(i'', N)}$	–	$(1 - T_{\text{inc}})$
Property taxes	$C_{s,\text{assess}}\,T_{\text{prop}}\,(1 - T_{\text{inc}})$	–	–
Maintenance	M	–	$(1 - T_{\text{inc}})$
Insurance	I	–	$(1 - T_{\text{inc}})$
Interest deduction	–	$T_{\text{inc}}\,i_m\left[\sum_{k=1}^{N} P_k\,\text{PWF}(i', k)\right] \times \text{CRF}(i', N)$	–
Depreciation (commercial systems only)	–	$T_{\text{inc}}\left[\sum_{k=1}^{N} D_k\,\text{PWF}(i', k)\right] \times \text{CRF}(i', N)$	1

*Multiply term in columns to left by this factor for commercial systems owned by for-profit firms.

[The summation term for interest deduction in Table 5.6 can be expressed in closed form as

$$i_m \sum \frac{P_k}{(1+i)^k} = \left\{ \frac{\mathrm{CRF}(i_m, N)}{\mathrm{CRF}(i, N)} + \frac{1}{1+i_m} \frac{i_m - \mathrm{CRF}(i_m, N)}{\mathrm{CRF}[(i - i_m)/(1+i_m), N]} \right\} \cdot (C_{s,\mathrm{tot}} - \mathrm{ITC})$$

If $i = i_m$:

$$i_m \sum \frac{P_k}{(1+i)^k} = \left\{ 1 + \frac{N}{1+i_m} [i_m - \mathrm{CRF}(i_m, N)] \right\} (C_{s,\mathrm{tot}} - \mathrm{ITC})]$$

T_{salv} = tax rate applicable to salvage value

ITC = investment tax credit for initial solar system cost

The expressions in Table 5.6 are based on the assumption that the interest rate for the solar loan and the discount rate for the system owner are the same. This assumption is made to reduce the complexity of the expressions so that they are commensurate with the level of economic information presented in this book. If the interest rate and discount rate are different, the terms in the table will require an additional factor [5].

The approach used for the detailed economic analysis is similar to that for the simplified analysis regarding the time scale used. Each term is expressed on a levelized, annual basis, i.e., as a series of uniform payments made annually throughout the period of economic analysis, taken to be N years long. Each cash flow term can be converted to a total present-worth basis for the system lifecycle by dividing by the capital recovery factor $\mathrm{CRF}(i', N)$.

The terms included in Table 5.6 are based on the real discount rate i'. The calculation may be carried out in current dollars rather than constant dollars by using the nominal discount rate, which includes inflation in each term. That is, i' would be replaced by i. If current dollars are used, however, each term must be based on the same discount and inflation rates for the results to be meaningful. The inflation rate in operating energy, for example, would be the total inflation rate including general and differential inflation, rather than the differential inflation rate alone, which is used with constant-dollar analysis.

A graphic exhibition of the cash flow terms, both positive and negative, that occur during the solar system lifetime can give the prospective system owner additional insight. Of course, the graphic representation and the analytical representation summarized in Table 5.6 will always lead to the same conclusion regarding system size. However, the graphic technique can show the order of magnitude of various cash flow terms.

Figure 5.5 shows cash flows during several years of the life cycle of a solar system. It includes all nine terms in Table 5.6, but it does not show the terms to scale, since their relative magnitude will vary from system to system. In addition to cash outflows, the value of fuel savings escalating at inflation rate j_e is shown in Fig. 5.5. For solar energy to be economically feasible, the sum of the positive cash flows must exceed the sum of the negative cash flows.

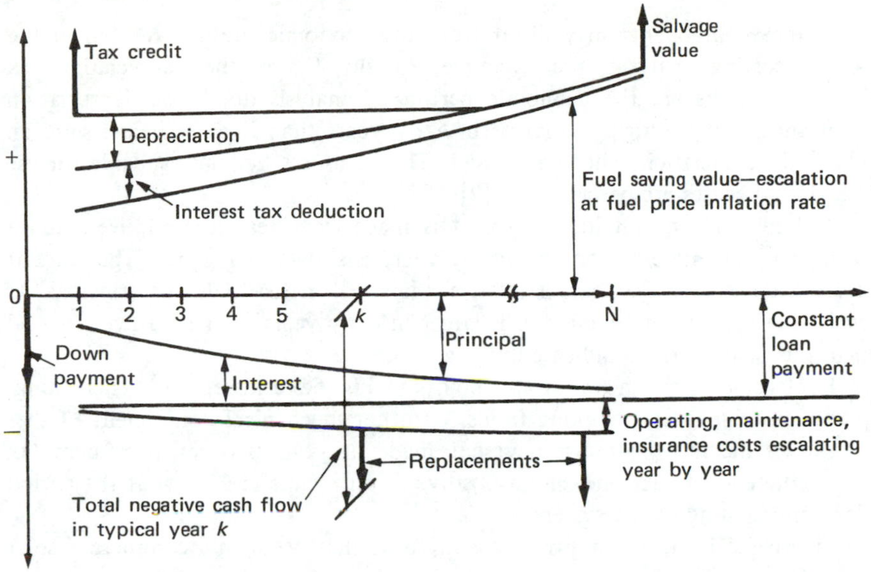

Figure 5.5 Year-by-year cash flow terms involved in solar system life-cycle costing.

The payments and savings are shown in Fig. 5.5 in nondiscounted dollars. For proper application of engineering economic principles, the cash flows must be reduced to their annualized present worth by factoring out the time value of money. This is done, as indicated in previous sections, by using the present-worth and capital recovery factor approach. During the early years of solar system investment, annual cash flows may be negative when viewed as a cumulative sum. That is, a net payment is required, since the fuel savings and other positive cash flows do not offset the negative terms associated with owning the solar system. In future years, as fuel savings inflate, the net cash flow will change sign and cumulative savings will be realized throughout the balance of the system lifetime.

Since negative cash flows do not produce net savings for a system owner, the owner may postpone investment in a solar system until the first year for projected positive cash flow. If the option of investing in a solar system in the future as opposed to the present is available, it is always advantageous to make the investment in the first year that shows positive savings. This time can be established by rigorous economic analysis, but it is clear on a little reflection that negative cash flows tend to offset the cumulative effect of positive cash flows occurring after the first year of positive cash flow. Therefore, to maximize the benefit realized by a solar system that will last N years, it is always best to make the purchase of that system coincide with the first year for net positive cash flow. Of course, the option of installing a solar system at some future time is often not available to a building owner; the investment is frequently made at the time the building is constructed, and a few years of negative cash flow may be the result.

Several other features of Fig. 5.5 are of interest. Depreciation lifetimes of solar system components may differ from the economic lifetime N used in the life-cycle costing analysis. For example, in the figure, the depreciation tax deduction ends before the economic period of analysis does. The depreciation amount shown in the figure is larger in early years than in later years, since an accelerated depreciation schedule is used. The various schedules available for use in the United States are described in [2].

Although a constant loan payment is made each year, the relative amount of principal and interest changes year by year, as shown in Fig. 5.5. The amount paid for interest in early years is relatively larger than that paid for principal. As a result, the interest tax deduction is larger in early years of the life-cycle period than in the final years, as indicated in Fig. 5.5.

Replacements are shown at two points in Fig. 5.5 and represent such things as pump motors, filters, working fluids, or other mechanical components. Other costs shown include the down payment made at year zero, an investment tax credit at the end of year one, and a positive salvage value cash flow at the period of decommissioning of the system.

Figure 5.5 includes the primary cash flows that occur in a commercial solar system. Certain applications may involve other terms, such as a distributed tax credit over several years or a nonuniform increase in the future value of fuel savings. Some analysts use several inflation rates for fuel and future years rather than a single value j_e as used here. This complication is not required for the economic analyses treated in this book, however.

Year-by-year cash flows shown in Fig. 5.5 and levelized in Table 5.6 will be calculated in the following example.

EXAMPLE

Calculate the terms in Table 5.6 and Fig. 5.5 and annual cash flows in constant and discounted dollars for a solar system costing \$10,000 less state and federal tax credits. Assume that the salvage value is \$1000 after the life-cycle period of 3 years. Insurance, maintenance, and property taxes total \$400 a year. A repair costing \$100 is made in year 2 and straight-line depreciation is used. The tax rate is 50 percent, the real discount rate i' is 5 percent, and the analysis is to be carried out for a 3-year period. The 3-year period is chosen only to minimize the calculations for this example; in practice, a lifetime of 20 or 25 years would be used.

SOLUTION

The owner is a for-profit business. Required discounted cash flow factors are

$\text{CRF}(i', N) = 0.367209$

$\text{PWF}(i', N) = 0.8638$

$\text{PWF}(i', 2) = 0.9070$

The effect of general inflation will be omitted to simplify the calculations. The annual cost terms shown in Table 5.6 are

Capital and interest, \$10,000 CRF(i', N)	−\$3672.09
Salvage value, \$1000 PWF(i', N) CRF(i', N) $(1 - T_{inc})$	+ 158.60
Replacements, \$100 PWF(i', 2) CRF(i', N) $(1 - T_{inc})$	− 16.65
Maintenance and insurance, property taxes, operating energy, \$400 $(1 - T_{inc})$	− 200.00
Interest tax deductions (see below)	+ 172.02
Depreciation, \$3000 $(1 - T_{inc})$	+ 1500.00
Uniform annual cash flow	−\$2058.12

Interest payments (5 percent of outstanding balance) are calculated on a year-by-year basis as summarized below:

Year	Payment (\$)	Interest payment (\$)	Principal payment (\$)	Discounted interest payment (\$)
1	3,672.09	500.00	3,172.09	476.19
2	3,672.09	341.40	3,330.69	309.66
3	3,672.09	174.80	3,497.22	151.06
		1,016.27	10,000.00	936.91

On a uniform annual basis, the discounted interest payment is \$936.91 × CRF(i', N) = \$344.04. For a 50 percent tax rate, the deduction is \$172.02.

Additional insight into cash flows can be developed by calculating each term in Fig. 5.5 first, as shown above, in constant dollars, then in discounted dollars.

Cash Flow Summary: Constant Dollars

Year	Solar cost (\$)	Salvage (\$)	Repairs (\$)	Operation and maintenance (\$)	Tax saving (\$)	Depreciation (\$)
1	− 3,672.09	–	–	−200	250.00	1,500
2	− 3,672.09	–	−50	−200	170.70	1,500
3	− 3,672.09	500	–	−200	87.44	1,500
	−11,016.27	500	−50	−600	508.14	4,500

The cash flows in the table above do not have the time value of money factored out. By applying the proper present-worth factor to each term, the following discounted cash flows are found.

Cash Flow and Total Present-Worth Summary in Discounted Dollars*

Year	Solar cost ($)	Salvage ($)	Repairs ($)	Operation and Maintenance ($)	Tax saving ($)	Depreciation ($)
1	− 3,497.22	–	–	–190.48	238.10	1,428.57
2	− 3,330.69	–	–45.35	–181.41	154.83	1,360.54
3	− 3,172.09	431.92	–	–172.77	75.53	1,295.76
	–10,000.00	431.92	–45.35	–544.66	468.45	4,048.87

*Discount and interest rates both 5 percent.

The algebraic total of all life totals from the table is −$5604.76. When this total is multiplied by CRF(i', N) to annualize it, the result is −$2058.12 per year, the same as the value calculated from the annual cost terms in Table 5.6. In practice, the year-by-year method is much too laborious and the Table 5.6 approach is preferred. The results will always be identical.

Internal Rate of Return Method

When solar energy systems are in competition with conventional methods of energy conservation, the internal rate of return (IRR) method can be used to rate the various options in order of maximum economic benefit. IRR studies are frequently carried out in the industrial environment but rarely in the residential market. The IRR method will be briefly discussed in this section; additional details may be found in [3].

The internal rate of return is defined as the discount rate i that, when used to calculate the terms in Table 5.6, will result in the sum of all nine terms being identically equal to the value of savings in fuel. That is, the algebraic total of all the cash flows shown in Fig. 5.5, when discounted with the IRR, will be exactly zero. Either total life-cycle costs or annual costs may be used. Since the internal rate of return enters the solar cost terms in Table 5.6 in both capital recovery factors and present-worth factors, it is usually impossible to solve for the IRR value directly. An iterative solution is nearly always required to find the value of i that reduces the total of all cash flows and energy savings to zero. The IRR method of finding the optimum investment requires more calculations than the annualized cost method described in the preceding paragraphs. The example below illustrates the IRR method.

EXAMPLE

A solar energy system on a large institutional facility is expected to save $125,000 per year over the next 10 years. (In this calculation we ignore inflation for simplicity.) The system's initial cost less tax credits is $750,000. What is the rate of return on the investment if the project is publicly owned and therefore tax exempt and if annual operating costs are expected to be \$3000 per year? Assume that replacement and salvage values, when discounted, offset each other. Depreciation is not considered since no taxes are involved in this analysis.

The equation to be solved for the internal rate of return is given below.

SOLUTION

$$125{,}000 = 750{,}000\ \text{CRF}(i, 10) + 3000$$

So

$$\text{CRF}(i, 10) = 0.1627$$

From Table 5.2

$$\text{CRF}(0.12, 10) = 0.17698$$

$$\text{CRF}(0.10, 10) = 0.16275$$

$$\text{CRF}(0.08, 10) = 0.14903$$

Therefore the internal rate of return is 10 percent. This value happens to be the rate established by the Office of Management and Budget for U.S. government energy feasibility analyses.

Accuracy of Economic Analyses

The differentiation between the simple and detailed methods of life-cycle costing presented in this chapter is correlated with the significance of the cash flow terms involved. To determine the confidence with which solar life-cycle costing studies are made, it is advisable to carry out a sensitivity study of the results of the costing analysis with respect to all the parameters. However, as indicated in Table 5.6, very many parameters enter life-cycle costing studies and an exhaustive examination of all would be too time-consuming.

The major cost factors are precisely those considered in the simplified method of economic analysis. Therefore the sensitivity of life-cycle economic results to the first-order or most important economic inputs can be established by examining the following components:

Solar system cost
Mortgage interest rate
Backup fuel cost
Backup fuel energy inflation rate

It is assumed that the investment tax credit is well defined and no variation in this parameter need be included, although it is part of the simplified economic method. Of the four costs listed above, the one known with the least confidence is the inflation rate in fuel prices. Backup energy requirements can be calculated within a few percent, as can effective discount rates. Solar system costs may involve somewhat greater uncertainty, and the effect of the initial cost of the system and the fuel inflation rate on the conclusions of life-cycle cost studies should be examined. The inflation rate in fuels is typically the most sensitive of all parameters in a life-cycle economic analysis. Unfortunately, it is the least well known, so a band of values should be examined. If a given solar system is feasible throughout the range examined, it can be recommended by the designer with confidence. However, if it is marginally feasible with the highest fuel inflation rate considered, it will be infeasible with reduced inflation rates, and the long-term feasibility of the solar application is probably questionable.

REFERENCES

1. J. L. Riggs, *Engineering Economics,* McGraw-Hill, New York, 1977.
2. F. Kreith and J. F. Kreider, *Principles of Solar Engineering,* Hemisphere, Washington, D.C., 1978.
3. J. F. Kreider, *Medium and High Temperature Solar Processes,* Academic, New York, 1979.
4. U.S. Department of Energy, *DOE Facilities Solar Design Handbook,* Rep. DOE/AD-0006/1, National Technical Information Service, Springfield, Va., 1978.
5. T. Sav, *Energy,* vol. 4, 1979, p. 415.
6. U.S. Department of Energy, *Project Experience Handbook,* Rep. DOE/CS-0045/D National Technical Information Service, Springfield, Va., 1978.

6 Solar Water Heating: Design and Operation

This chapter describes service water heating systems that have been used in many parts of the world for decades. Water heating is one of the simplest applications of solar heat in buildings and one of the least expensive. Therefore it is also the most common application. Solar water-heating systems have excellent life-cycle cost characteristics, since water-heating loads are rather uniform year-round compared to space-heating loads and the payback in fuel savings is therefore accelerated compared to that for all solar space-heating systems.

The generic classes of solar water-heating systems discussed cover the mainstream of solar system design. A great many types of water-heating systems have been conceived of by inventors and solar engineers; however, very few have reached the marketplace. The first section of this chapter describes briefly the calculation of water-heating loads; following sections describe the generic classes of water-heating systems. A detailed description of the operation of one type of water-heating system and rules for design and performance prediction and sizing are given in the final sections. In addition, a section describing solar swimming pool heating is included.

Figure 6.1 shows an active solar water-heating system installed on a condominium building in Boulder, Colorado. This system serves six units and provides approximately 65 percent of the annual water-heating load.

WATER-HEATING LOADS

Solar water-heating energy demands can be calculated precisely if the quantity of hot water required is known. The amount of energy needed to heat

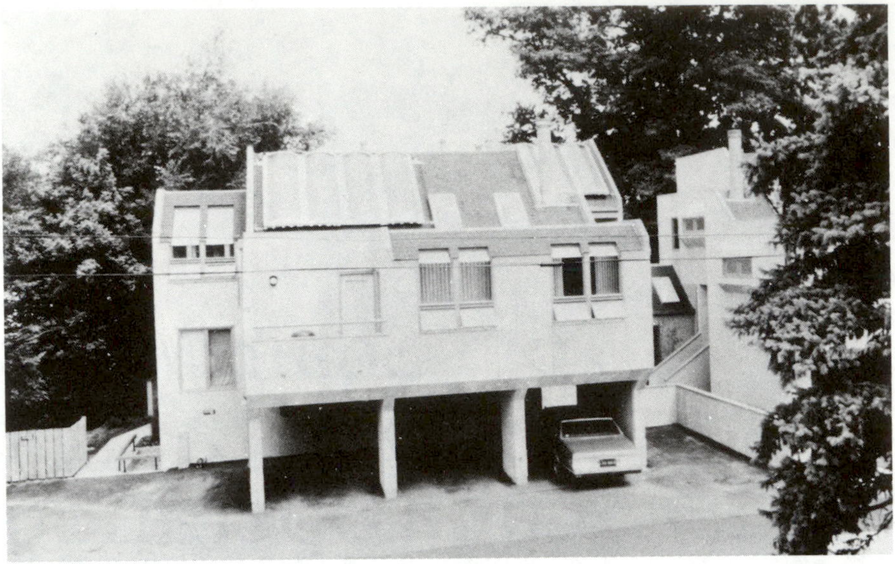

Figure 6.1 Example active solar water-heating system located on a six-unit condominium building in Boulder, Colorado. (*Courtesy of Jan F. Kreider Associates, Inc.; owner, F. Kreith.*)

water is the product of the volume of water, its density, its specific heat, and the required temperature increase. Expressed in equation form, the amount of heat required to heat water is given by

$$Q_{hw} = V(\rho c)(T_{\text{set}} - T_{\text{source}}) \tag{6.1}$$

in which Q_{hw} is the energy requirement per day, V is the volume of water required per day, ρ is the density, c is the specific heat [1.0 Btu/(lb·°F) or 4.18 kJ/(kg·°C)], T_{set} is the thermostat set point and desired delivery temperature, and T_{source} is the temperature of the inlet water from city water mains or a well. The source temperature may vary seasonally if the water is provided from a city water system, as shown in Table 6.1 for a number of cities in the United States.

A second thermal demand is present for solar water-heating systems. It is the amount of heat lost from the water-heating tank and the recirculation system, if one is used. (A recirculation system is used in large buildings to ensure that hot water is present continuously at all hot-water outlets.) This parasitic heat loss can consume 20 to 25 percent of the fuel used in a hot-water installation over the course of a year. The tank and piping loss is given by the thermal conductance of the insulation U_{hw} multiplied by its surface area A_{hw} and by the temperature difference between the water and the surroundings. In simplified form, it can be expressed by

$$Q_{\text{standby}} = U_{hw}A_{hw}(T_{\text{set}} - T_a)N_h \tag{6.2}$$

in which T_a is the ambient temperature in the vicinity of the water-heating and delivery piping if a recirculation loop is used and N_h is the number of hours of

Table 6.1 Average Monthly Cold-Water Source Temperature, °F

State	City	Jan.	Feb.	Mar.	Apr.	May	June	July	Aug.	Sept.	Oct.	Nov.	Dec.	Deep well temperature
Alabama	Montgomery	50	52	58	63	73	78	82	81	79	69	56	50	67
Alaska	Anchorage	-	—	—	—	—	—	—	—	—	—	—	—	—
Arizona	Phoenix	48	48	50	52	57	59	63	75	79	69	59	54	67
Arkansas	Little Rock	48	47	57	68	77	85	88	86	82	77	64	58	65
California	Los Angeles	50	50	54	63	68	73	74	76	75	69	61	55	65
	San Francisco	60	60	60	60	60	60	60	60	60	60	60	60	60
Colorado	Denver	39	40	43	49	55	60	63	64	63	56	45	37	50
Connecticut	Hartford	40	40	40	44	54	59	66	70	69	59	50	40	51
Delaware	Wilmington	50	48	50	52	68	73	78	79	73	69	60	55	57
District of Columbia	Washington	42	42	52	56	63	67	67	78	79	68	55	46	57
Florida	Tampa	75	75	76	80	85	87	85	85	83	80	77	75	75
Georgia	Atlanta	43	48	53	59	72	78	84	80	78	70	60	48	65
Hawaii	Honolulu	80	80	80	80	80	80	80	80	80	80	80	80	80
Idaho	Boise	52	52	52	52	52	52	52	52	52	52	52	52	52
Illinois	Chicago	32	32	34	42	51	57	65	67	62	57	45	35	52
Indiana	Indianapolis	50	50	50	53	68	73	80	82	77	70	60	50	55
Iowa	Des Moines	40	40	40	44	49	58	66	73	71	62	50	45	52
Kansas	Topeka	50	52	57	64	70	76	84	80	74	64	58	50	56
Kentucky	Louisville	40	40	45	49	69	77	82	82	77	70	60	50	58
Louisiana	New Orleans	55	55	60	66	77	86	89	90	90	80	70	60	70
Maine	Portland	35	35	40	45	50	56	64	66	64	58	47	40	47
Maryland	Baltimore	40	40	43	47	53	61	66	70	64	58	50	46	57
Massachusetts	Boston	32	36	39	52	58	71	74	67	60	56	48	45	50
Michigan	Detroit	35	35	38	41	56	64	75	74	68	63	55	45	47
Minnesota	Minneapolis	32	32	35	41	61	69	80	73	68	60	50	40	44
Mississippi	Jackson	60	60	65	70	75	80	82	74	74	69	65	61	68
Missouri	St. Louis	47	47	50	53	69	77	85	83	75	70	62	55	57
Montana	Great Falls	42	42	42	42	42	42	42	42	42	42	42	42	42
Nebraska	Lincoln	40	40	45	51	56	68	81	79	69	60	50	45	53

Table 6.1 Average Monthly Cold-Water Source Temperature, °F (*continued*)

State	City	Jan.	Feb.	Mar.	Apr.	May	June	July	Aug.	Sept.	Oct.	Nov.	Dec.	Deep well temperature
Nevada	Las Vegas	73	73	73	73	73	73	73	73	73	73	73	73	73
New Hampshire	Manchester	35	35	38	40	45	52	58	67	65	55	45	40	47
New Jersey	Newark	40	40	45	45	56	64	69	71	71	65	60	50	55
	Trenton	35	35	38	40	68	71	79	77	72	65	55	45	55
New Mexico	Albuquerque	72	72	72	72	72	72	72	72	72	72	72	72	62
New York	Albany	32	32	35	40	52	60	56	66	65	55	45	40	49
North Carolina	Greensboro	55	55	55	60	67	77	83	82	79	72	67	60	62
North Dakota	Bismarck	42	42	42	42	42	42	42	42	42	42	42	42	42
Ohio	Columbus	38	38	40	46	64	72	76	76	74	65	55	45	54
Oklahoma	Oklahoma City	45	45	50	55	68	73	77	77	72	65	55	50	62
Oregon	Portland	35	35	38	44	50	56	62	55	52	45	40	35	50
Pennsylvania	Philadelphia	35	35	38	40	68	71	79	77	72	60	50	40	53
	Pittsburgh	38	38	40	46	66	75	81	81	75	68	55	40	52
Puerto Rico	San Juan	80	80	80	80	80	80	80	80	80	80	80	80	80
Rhode Island	Providence	40	40	45	48	56	62	64	65	63	50	45	40	52
South Carolina	Charleston	55	58	60	65	75	81	83	88	80	75	70	65	67
South Dakota	Rapid City	55	55	55	55	55	55	55	55	55	55	55	55	47
Tennessee	Memphis	55	55	58	60	60	68	68	70	70	65	60	60	62
Texas	Dallas/Ft. Worth	56	49	57	70	75	81	79	83	81	72	56	46	67
	San Antonio	65	65	65	68	70	75	78	80	77	75	70	65	67
Utah	Salt Lake	35	37	38	41	43	47	53	52	48	43	38	37	52
Vermont	Burlington	32	32	38	45	50	58	63	66	68	60	50	40	47
Virginia	Norfolk	50	50	55	62	70	78	83	83	80	70	60	55	62
Washington	Seattle	39	37	43	45	48	57	60	68	66	57	48	43	52
West Virginia	Charleston	50	50	55	60	70	79	83	82	77	70	60	55	55
Wisconsin	Madison	35	35	40	45	50	53	52	53	52	50	48	40	47
Wyoming	Lander	45	45	45	45	45	45	45	45	45	45	45	45	45

Source: American Heliothermal Corp., Denver, Colorado.

use per day, usually 24. It may be necessary to express the right-hand side of Eq. (6.2) as two terms, one accounting for tank losses and the second for recirculation loop losses, if their ambient temperatures T_a are different. If the ambient temperatures are the same, the *UA*-product in Eq. (6.2) can be calculated to include both tank and pipe heat loss area and thermal conductance. The total water-heating demand is expressed by the sum of Eqs. (6.1) and (6.2).

Before considering an investment in solar energy, it is always worthwhile to reduce standby losses from water-heating systems. This is accomplished by reducing U_{hw} to the minimum practical value by properly insulating tanks and distribution systems, if used. However, if addition of an insulation blanket to an existing water-heating system is considered, care must be taken not to block off air entry to the burner or to insulate the control elements, which could result in overheating and malfunction of the water heater. Improper insulation may cause more trouble than the energy saved is worth.

Table 6.2 lists the average gallons-per-day requirement for service hot water in many types of buildings. The values in the table are averages that are useful for preliminary design. In the absence of more detailed data, they may be used for final design. However, if solar water heating is to be retrofitted to an existing system, hot-water consumption should be measured over a sufficient period to establish a long-term average rather than referring to Table 6.2 for sizing estimates. The temperature T_{set} for water-heating systems is limited in some places by state energy codes and may not be at the discretion of the building owner. Note that the volumes listed in Table 6.2 are based on a delivery temperature of 140°F. Reduction of T_{set} to a value below this will reduce standby losses but will increase the volume of water used. For systems in which high temperatures are required at only a few points, it is usually more cost effective to boost temperature locally than to have a central system set at an elevated temperature. Hot water provided by this type of system would have to be mixed with cold water for the majority of lower-temperature uses.

For recirculation systems in commercial buildings, the recirculation pump can be controlled by clock or by temperature sensor. Clocks are used to operate the pump during known periods of water usage. A temperature sensor located at the most distant point in the circulation loop can also be used to turn off the pump if the temperature set point is met. If the set point is met at the most distant point of the loop it will be met at every upstream point. The return line for recirculation loops should be insulated to reduce standby losses as well.

EXAMPLE

Calculate the daily energy requirement for water heating for the following situation: 1000 gallons of water are to be heated from 50 to 120°F per day. The water-heating system includes a 250-ft long recirculation loop insulated at R-4. The water heater has a 1000-gallon capacity, is insulated at R-4, and is 5 ft in diameter. Find the total daily energy requirement, using Eqs. (6.1) and (6.2).

SOLUTION

From Eq. (6.1), the water-heating portion of the 1000 gal/day load is

$$Q_{hw} = \left(\frac{1000}{7.46\ \text{gal/ft}^3}\right)(62.4 \times 1.0)(120 - 50) = 585{,}500\ \text{Btu/day}$$

For standby losses two terms are involved, one for the tank and one for the recirculation loop. In the absence of other information, assume that the entire 250-ft loop is at T_{set}. If 1-in pipe is used with $1\frac{1}{2}$-in insulation, the pipe heat loss area is

$$A_{pipe} = \pi\left(\frac{4}{12}\right) \times 250 = 262\ \text{ft}^2$$

For the tank area to be found, its height must be known. It can be calculated from the volume:

$$h = \frac{1000/7.46}{\pi/4(5^2)} = 6.8\ \text{ft}$$

The tank area (including both ends) is then

$$A_{tk} = \pi Dh + \frac{2\pi D^2}{4} = \pi\left(5 \times 6.8 + 2 \times \frac{25}{4}\right) = 146\ \text{ft}^2$$

Using Eq. (6.2) to find the standby loss for 24 h/day operation, we have (recall that $U = 1/R$):

$$Q_{standby} = \left(\frac{146}{4} + \frac{262}{4}\right)(120 - 70) \times 24 = 122{,}430\ \text{Btu/day}$$

The total energy demand is $Q_{hw} + Q_{standby}$:

$$Q_{hw} + Q_{standby} = 585{,}500 + 122{,}430 = 707{,}930\ \text{Btu/day}$$

of which 17.3 percent accounts for standby losses.

GENERIC TYPES OF SOLAR WATER-HEATING SYSTEMS

One method of classifying solar water-heating systems divides them into direct and indirect designs. Direct systems are those in which the potable water is heated directly in the solar collector. Indirect systems are those in which a separate heat transfer medium is used in the solar collector loop and potable water is heated indirectly by a heat exchanger. Variations of these two basic design classes are described in this section.

Direct Systems

The simplest type of direct system is shown in Fig. 6.2. It consists of solar collectors, a circulation pump, and a combined preheat tank and backup heating

Table 6.2 Average Hot-Water Demands for Various Building Types*

Type of building	Maximum hour	Maximum day	Average day
Men's dormitories	3.8 gal/student	22.0 gal/student	13.1 gal/student
Women's dormitories	5.0 gal/student	26.5 gal/student	12.3 gal/student
Motels: no. of units†			
20 or less	6.0 gal/unit	35.0 gal/unit	20.0 gal/unit
60	5.0 gal/unit	25.0 gal/unit	14.0 gal/unit
100 or more	4.0 gal/unit	15.0 gal/unit	10.0 gal/unit
Nursing homes	4.5 gal/bed	30.0 gal/bed	18.4 gal/bed
Office buildings	0.4 gal/person	2.0 gal/person	1.0 gal/person
Food service establishments:			
Type A–full-meal restaurants and cafeterias	1.5 gal/(max meals·h)	11.0 gal/(max meals·h)	2.4 gal/(avg. meals·day)‡
Type B–drive-ins, grilles, luncheonettes, sandwich and snack shops	0.7 gal/(max meals·h)	6.0 gal/(max meals·h)	0.7 gal/(avg. meals·day)‡
Apartment houses: no of apartments			
20 or less	12.0 gal/apt.	80.0 gal/apt.	42.0 gal/apt.
50	10.0 gal/apt.	73.0 gal/apt.	40.0 gal/apt.
75	8.5 gal/apt.	66.0 gal/apt.	38.0 gal/apt.
100	7.0 gal/apt.	60.0 gal/apt.	37.0 gal/apt.
200 or more	5.0 gal/apt.	50.0 gal/apt.	35.0 gal/apt.
Elementary schools	0.6 gal/student	1.5 gal/student	0.6 gal/student‡
Junior and senior high schools	1.0 gal/student	3.6 gal/student	1.8 gal/student‡

*Based on a delivery temperature of 140°F.
†Interpolate for intermediate values.
‡Per day of operation.
Source: Reprinted with permission from the 1980 Systems Volume, *ASHRAE Handbook & Product Directory*, American Society of Heating, Refrigerating, and Air-Conditioning Engineers, New York, 1980.

system. In this simplest system, potable water from a city or well source is circulated directly through the collectors from the tank. As water is used, it is drawn off from the top of the tank and fresh cold water is supplied to the bottom of the tank. It is important to use this flow regime so that the coldest possible water is introduced into the collector to permit proper operation and maximum efficiency. During sunlight periods, the pump P1 circulates water from the tank through the collectors at the rate of about one complete circulation of the tank per hour.

The combined preheat and backup tank contains the auxiliary energy source required to maintain outlet water temperature at the set point. This set point temperature is frequently higher than the temperature delivered by the solar collector under conditions of medium sunlight. It is therefore important that the backup heating system not heat water that will eventually be introduced into the collector. This is avoided, as shown in Fig. 6.2, by introducing solar-heated water *below the heating element*. It is essential that the heating

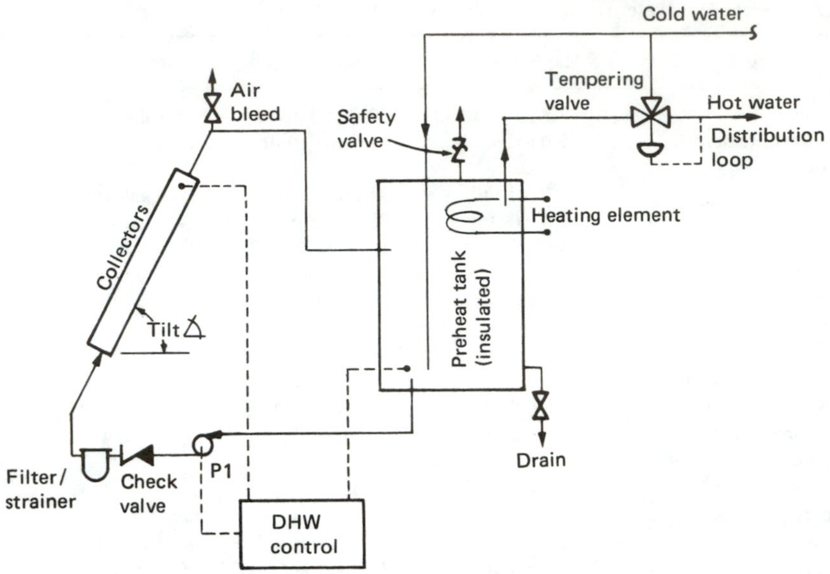

Figure 6.2 Single-tank direct solar water-heating system. Instrumentation and miscellaneous fittings are not shown.

element be placed above the water return line from the collector to permit efficient collector operation and to avoid rejecting auxiliary energy through the collectors during low-insolation periods. Since the backup heating element must be placed in the upper part of the tank, the backup energy source is restricted to electricity in most commercially available systems. One system currently under development by F. deWinter uses gas or fuel oil as the backup with a heat pipe to heat the upper part of the tank. However, this system will not be described since it is not commercially available or tested.

Figure 6.2 shows a collector loop check valve downstream of collector pump P1. A check valve is a small but essential part of the collector loop, since it prevents reverse flow in the loop during the night. Reverse flow would occur in the absence of a check valve since the fluid in the collectors becomes the coldest fluid in the entire water-heating system overnight. Cold fluid is heavy and tends to sink to the lowest point of the system. Since collectors are the highest point of the system, this cold fluid would flow down from the collectors to the mechanical space in which the storage tank is located. To replace the cold fluid in the collector, warm fluid would rise from the top of the preheat tank to the collector, and reverse circulation would occur. Of course, if the collectors are drained whenever pump P1 is off, a check valve is not required.

Direct systems require a solar collector capable of withstanding city water pressure. Some collector designs cannot be used since their construction is restricted to operating pressures of 30 to 40 lb/in^2. If the city water or well water is hard, solar collectors used for direct water heating may be subject to scaling and require occasional cleaning. Hardness always precipitates out solids at the warmest point in a water-heating system, which for solar systems is in the collectors. Local swimming pool manufacturers can advise on water hardness, or city water authorities may be consulted. Scale can sometimes be removed from solar collectors by flushing occasionally with swimming pool acid. An alternative, of course, is to run source water through a water softener before it is introduced into the collectors.

Direct systems have several advantages over indirect systems, which are described in the next section. The direct systems have relatively few components, a simpler design, and theoretically higher thermal performance owing to the absence of a heat exchanger between the collector and the water system. The reduced system complexity and smaller number of components should result in a lower initial cost and lower operating and maintenance costs over the life-cycle period of the system.

However, direct water-heating systems have several fundamental difficulties that militate against their use in many parts of the world. The principal difficulty is the need for protection against freezing in cold climates. During cold winter nights when the temperature drops below freezing for many hours, water contained in solar collectors will freeze. Expansion of the freezing water can destroy collectors, piping, and fluid conduits and must be avoided. Several methods of freeze protection have been used in direct systems. In relatively mild climates, collector pump P1 may be operated at night to circulate warm water from the tank through the collectors to prevent freezing. However, this introduces cold water into the top of the water-heating tank, which will eventually turn on the auxiliary heater and consume additional energy. Recall that the upper zone of the tank is always maintained at the set point temperature by the auxiliary heater.

A second method of freeze protection is to drain the collectors back into the storage tank at night. Draining is usually effected by gravity, since collectors are commonly located above the storage tank. Theoretically, the drain-down system has a number of advantages that indicate that it might be practical for use in water-heating systems as well as space-heating systems. Less heat is lost when a collector cools at night if the collector is not filled with water (since water left in the collector must be rewarmed before use the next day). The use of drain-down eliminates extra capital and maintenance costs for heat exchangers, pumps, and special collector fluids. As shown in the next section, controls are simplified and an expansion tank and pump can be deleted; several other small components are also not present in direct systems with drain-down. In addition, collector efficiency is higher if no heat exchanger is required; this results in a somewhat longer operating period for the collector during the day. Most antifreeze liquids used in indirect systems have a higher viscosity and lower specific heat than pure water, hence larger pumps are required and pumping costs are higher.

The theoretical advantages of drain-down systems are offset, for the most part, by practical difficulties. Failure of a drain-down system can have catastrophic effects. Collectors and/or piping may freeze, requiring replacement in the dead of winter or deactivation of the solar heating system until the weather moderates. For example, on December 7, 1978, 590 collectors in Phoenix, Arizona, froze during an unexpected period of cold when controllers malfunctioned. The fundamental difficulty with drain-down systems is that there are no absolutely fail-safe designs for collector drainage under all possible conditions when it might be required. For a liquid collection system to drain, every collector, every inlet and outlet header, and every connecting pipe run must be sloped. Plugging of a collector or water entrapment in orifices such as those used for flow balancing can mean incomplete drainage and result in collector freeze-up. The cost penalties associated with collector freeze-up are the primary deterrent to use of drain-down systems in direct water-heating applications. Also, annual power consumed by solenoid values can exceed that required by the collector pump.

Other difficulties with the drain-down approach are described in [1], which shows that of 27 possible mechanisms for collector freezing, 25 are associated with direct systems—20 with drain-down freeze protection approaches, and 5 with the warm fluid circulation approach described earlier.

Thus there is a significant risk involved in the use of drain-down systems. Many hundreds of these systems froze during the 1970s for a multitude of reasons. The risk in drain-down freeze protection can be minimized by careful design and by very careful installation. The final decision about whether to use a drain-down system should be based on a risk-benefit analysis that trades off the possible cost of system damage against the benefits of increased thermal performance and reduced system complexity. The decision of the majority of solar system designers in cold regions of the United States has been not to use the drain-down approach, since its reliability cannot be guaranteed for a long period and the potential cost of system malfunction can wipe out any possible fuel savings over the life cycle of the system. The drain-down freeze protection idea will not be covered in further detail in this book.

The use of potable water directly in solar collectors restricts the selection of materials available for these collectors to copper. If galvanized steel pipe or aluminum conduits were used, severe corrosion would certainly occur owing to the presence of oxygen. Also, the impeller on circulation pump P1 (Fig. 6.2) must be bronze or stainless steel if untreated water is used in the collector circulation loop.

A direct solar water-heating system using two separate tanks rather than a combined preheat and backup tank is shown in Fig. 6.3. The principal advantage of the two-tank system is the ability to use any form of backup heat in the auxiliary tank. In residential applications, the auxiliary tank is frequently a standard water heater, which automatically provides the proper outlet temperature from the water-heating system. The preheat tank is heated only by solar heat, contrary to the single-tank design. Where gas or fuel oil is less expensive than electricity, the backup fuel bill may be somewhat reduced in the two-tank system. Extra storage capacity is provided in the two-tank system, although this

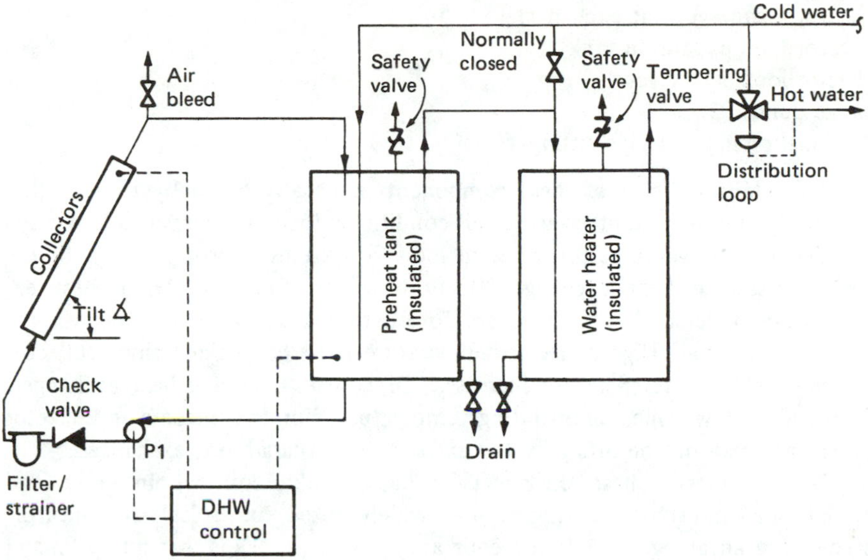

Figure 6.3 Double-tank direct solar water-heating system. Instrumentation and miscellaneous fittings are not shown.

does not increase overall thermal performance during the year by a great deal. Since inlet water to the water heater in the two-tank system is normally well above the cold-water inlet temperature, the recovery rate of the auxiliary tank will be higher than if city water were directly introduced into the tank.

Drawbacks to the two-tank system are twofold. First, increased cost is involved in the purchase and installation of the extra tank and associated piping. Additional floor space is required in the mechanical room for a second tank. Second, standby losses from a two-tank system are greater than from a single-tank system. Since the auxiliary system is maintained at the set point temperature, heat losses from this tank are substantially greater than from the upper, auxiliary zone of the single-tank system shown in Fig. 6.2. A third possible drawback of the two-tank system is that when excess solar heat is available, it cannot offset standby losses in the auxiliary tank. In the single-tank system, solar heat can make up part of the standby losses. For solar to provide this function in a two-tank system without introduction of a circulation pump between the tanks, the auxiliary tank should be located above the preheat tank and another pipe added to permit gravity flow from the preheat tank to the auxiliary tank when the temperature of the former is greater than that of the latter.

Indirect Systems

Indirect systems use a nonfreezing fluid in the solar collector loop for heat collection. Heat is then transferred from this fluid to potable water in a heat exchanger connected to the storage tank. Figure 6.4 shows an indirect one-tank system. It is seen that the following additional components are required:

Collector-to-storage heat exchanger
Collector loop expansion tank
Collector loop safety valve
Storage pump P2
Additional piping and pipe fittings

The additional cost of these components is usually offset by the fact that freeze protection is guaranteed under all conditions, including power failure.

The key to good performance in indirect systems is proper design of the collector-to-storage heat exchanger HX1 shown in Fig. 6.4. When heat exchangers are undersized, collectors are forced to operate at temperatures much higher than normal. High collector temperature is to be avoided since collector efficiency drops as temperature increases. Of course, oversized heat exchangers will provide a low collector operating temperature, but the increase in collector performance may not be offset by the extra cost of a large heat exchanger.

Several types of heat exchangers–jacketed tanks, coils within tanks, and external shell-and-tube exchangers–are widely used in solar water-heating systems. The advantage of internal coils and jacketed tanks is that pump P2 and its piping are not required. Jacketed tank and internal coil heat exchangers have relatively low effectiveness compared with external shell-and-tube heat exchangers of the type shown in Fig. 6.4. The fundamental difficulty arises from the reliance on free-convection circulation within the tank to remove heat from the

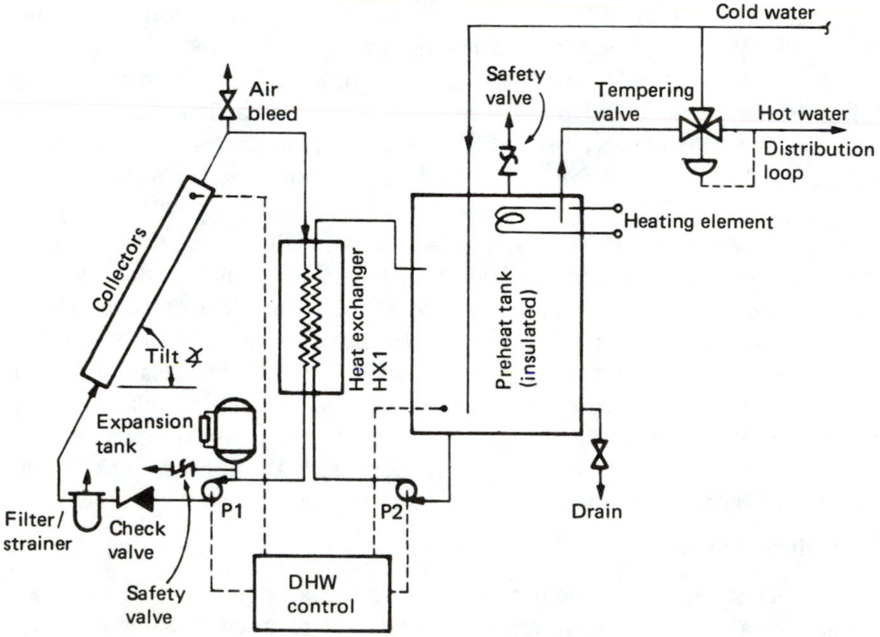

Figure 6.4 Single-tank indirect solar water-heating system. Instrumentation and miscellaneous fittings are not shown.

coil. Free convection is a relatively ineffective heat transfer mechanism compared with forced convection, which is used in shell-and-tube heat exchangers. To compensate for the reduced heat transfer coefficients in free convection, larger areas for heat transfer are required. As is the case in most solar design questions, the decision to use coil or jacketed heat exchangers or external shell-and-tube heat exchangers must be based on their relative economics and performance. Most commercially available free-convection heat exchangers for solar water heaters have effectiveness values below 0.4. External exchangers typically have values in the range 0.7 to 0.8. Heat exchangers penalties are therefore reduced for the latter. All three heat exchanger configurations are commercially available. (For a discussion of effectiveness calculation, see [1].)

Heat exchanger HX1 (Fig. 6.4) is required by several building codes to have a double-wall design. This requirement, although associated with reduced performance, is necessary to prevent nonpotable antifreeze solutions from entering the potable hot-water system, which can occur if there is corrosion of the wall dividing the two fluid streams. A double-wall heat exchanger will indicate a leak before corrosion of the second wall. Working fluids used for freeze protection include ethylene glycol, which has significant toxicity; in any glycol system a two-wall heat exchanger is mandatory. Another fluid used is propylene glycol, which is much less toxic. Propylene glycol of food grade is nontoxic and is present in many prepared foods. Proposed changes in plumbing codes will permit use of a single-wall heat exchanger in systems that use propylene glycol as the working fluid. Of course, glycols must be replaced periodically as inhibitors become depleted. If a single-wall heat exchanger is used in a system originally filled with propylene glycol, it is essential that the system be refilled with propylene glycol.

A closed solar collector loop requires an expansion tank to accommodate changes in fluid volume as collector and piping temperatures change throughout the day and over the operating lifetime of the system. The size of the tank is determined by the volume of fluid contained in the solar collector loop. The total volume can be calculated by finding the volume of fluid in the collectors, collector headers, piping, collector-to-storage heat exchanger, and the expansion tank itself. The typical expansion rate for glycol antifreeze solutions in water is approximately 10 to 12 percent of the initial volume of the fluid at ambient temperature. Collectors will operate between −20 and 200°F, and this large temperature excursion must be accounted for in expansion tank selection. Manufacturers' data can be used to find fluid volumes at various temperatures. The fluid expansion is related to the acceptance volume of the expansion tank. The total volume of the tank is typically three to four times the required acceptance volume.

The expansion tank is located at the inlet of the collector loop pump to provide a constant reference pressure at this point. The precise design pressure depends on the relative height of the collector above the pump. It is desirable to maintain a positive pressure of 3 to 4 lb/in^2 at the topmost point of the collector. This requirement plus the static pressure head from the top collector header to the pump determines the operating pressure of the expansion tank. The expansion tank can be equipped with a sight glass to check the fluid volume in the collector loop; the tank is also the location of the fluid loop safety valve, as shown in Fig. 6.4, since it is the reference pressure point for the loop.

The fluid collector loop shown in Fig. 6.4 contains a check valve for reasons given in the preceding section. In indirect systems, however, the check valve is not optional; it must be present to avoid reverse flow of the collector fluid. In addition to possible heat loss associated with reverse circulation, water in the tubes of the collector-to-storage heat exchanger can freeze during very cold periods if subfreezing antifreeze is permitted to flow back through the heat exchanger for prolonged periods at night. This has occurred with a great number of indirect water heating systems that were not equipped with a check valve. In very large commercial systems, it may be worthwhile to invest in a solenoid valve in place of the check valve. The solenoid valve would close whenever pump P1 does not operate.

Figure 6.5 is a diagram of an indirect two-tank system. This system is a combination of the direct two-tank system shown in Fig. 6.3 with the indirect collector loop shown in Fig. 6.4. The advantage of a two-tank system is the flexibility it allows in the choice of backup energy type, as described above. Its principal drawbacks are the extra cost and additional standby losses from the second tank.

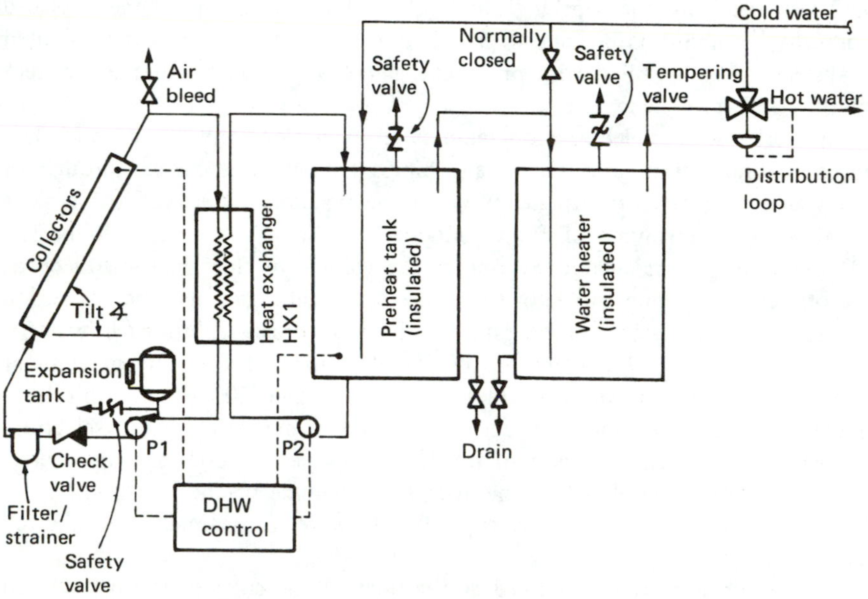

Figure 6.5 Double-tank indirect solar water-heating system. Instrumentation and miscellaneous fittings are not shown.

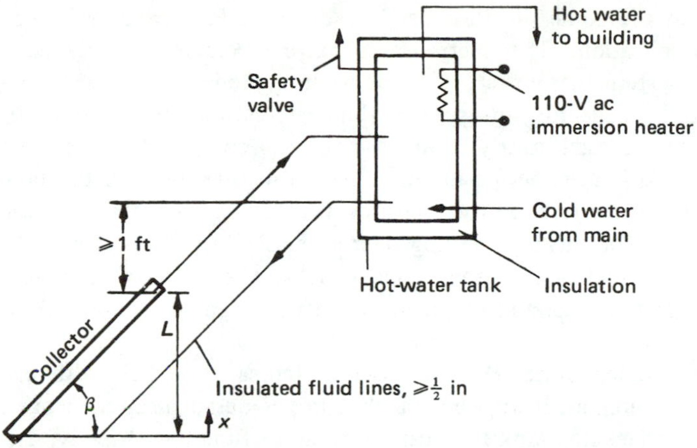

Figure 6.6 Passive thermosiphon single-tank direct system for solar water heating. Collector is positioned below the tank to avoid reverse circulation.

Passive Thermosiphon Water-Heating Systems

Figure 6.6 is a simplified diagram of a thermosiphon system for water heating. Thermosiphon systems are passive water-heating systems, requiring no pumps or controls. The motive force for fluid circulation is the density difference, caused by solar fluid heating, between the collector inlet and outlet over the vertical height L shown in the figure. This density difference depends on the amount of heat added to the collector fluid. The thermosiphon is the oldest type of water-heating system and has been used in many parts of the world. Thousands of these systems were in use in California and Florida between 1920 and 1950. As indicated later, the performance of thermosiphon systems is equal to and in some cases better than that of the active systems described in the two preceding sections.

Thermosiphon systems are fundamentally simpler to design and build than are active systems of either direct or indirect design. The only requirement for proper functioning is that relatively larger piping be used, since the pressure available to cause fluid flow is relatively small. Piping should have a minimum of tight elbows, and sweeping bends are preferred to reduce pressure drops. To prevent reverse flow at night in thermosiphon systems, the collector can be placed below the storage tank, as shown in Fig. 6.6. Reverse flow is completely eliminated if the top of the collector is at any point below the bottom fitting of the storage tank.

A storage tank above roof-mounted collectors is architecturally awkward. For thermosiphon applications where the storage tank cannot be placed above the collector but can be placed at the same height as the collector, a check valve is required to prevent backflow. Conventional types of check valves used in active systems are unsatisfactory for this purpose since relatively large pressure drops are required to overcome the flow barrier. A thermostatic check valve can be used in this application. The thermostatic valve operates similarly to a automotive cooling system thermostat; it opens when a predetermined temperature is reached on the collector surface, and fluid circulation begins. At night, when the temperature is significantly lower, the valve is completely closed and blocks reverse flow. Although thermosiphon systems can operate with the tank either above or at the same level as the collector, they do not operate when the tank is below the collector thereby limiting their applications. This is clearly the case since the warmest fluid in the loop, namely that in the collector, is located at the highest point in the loop, and no gravity circulation can occur under these conditions.

Indirect thermosiphon systems can be used as indicated in [2]. Although they are relatively uncommon, it appears that by proper design heat exchangers of the internal coil or jacketed tank type can be made to function properly. The key to indirect thermosiphon system design is minimizing the pressure drop in the heat exchanger. Although increased pressure drop relative to the direct system design cannot be avoided, the effect can be made small. In addition, thermosiphon systems tend to be self-compensating in the sense that reduced flow results in increased fluid residence time in the collector. Fluid is therefore hotter and less dense than it would be under higher flow conditions, and the increased pressure available will cause proper flow to occur even though there is an additional pressure drop in the heat exchanger. Computer model studies, summarized in [2], indicated that indirect thermosiphon systems can operate very effectively. System performance was shown to be essentially independent of collector loop flow resistance over a relatively wide range of resistance values.

Other Water-Heating System Designs

Direct and indirect pumped and thermosiphon systems represent the vast majority of commercially available water-heating systems. Two other concepts have been used on a restricted basis, however.

H. Tabor has used an approach for water heating that involves a single circulation of water through the collector per day instead of the circulation typical of most active systems—approximately one cycle per hour. This single-pass approach significantly reduces the pump power requirement and improves collector efficiency. The inlet temperature to the collector is always the water source temperature, which is typically at or below the ambient temperature most of the year. Therefore collector efficiency is high, since the heat losses from the collector are approximately zero. Relatively low flows in the collector can reduce heat transfer coefficients within the collector passage, and the heat removal factor F_R described in Chap. 3 will be somewhat lower. However, this is more than compensated for by the absence of heat losses from the collector. The single-pass design has been used in relatively few installations, but it merits serious consideration for its possible economic advantages.

Another idea for producing hot water from solar energy involves the use of a heat pump. In this approach, the collector is also the evaporator of the heat pump. A specially designed heat pump, with the capacity to operate at temperatures below freezing, extracts heat from the collector whether it is above or below freezing and adds it to the water-heating tank. Preliminary estimates indicate that the energy delivery per dollar of initial cost is quite high for these systems. They have not been widely used commercially because it is difficult to design a reliable compressor for operation at a relatively low temperature while maintaining cost-competitiveness with other types of systems described in this chapter.

SOLAR WATER-HEATING SYSTEM OPERATION

The active systems described in the previous section operate in essentially the same number of modes, using the same control point criteria. Therefore the operation of each system will not be detailed, but the most complex–namely the indirect, two-tank system–will be described.

Control System Operation

The controller shown in Figs. 6.2 through 6.5 measures two temperatures–the collector plate temperature and the storage tank bottom temperature–to determine whether solar collection is feasible. Collector temperature must be measured on the collector plate and not at some external point. In systems with external sensors, boiling has occurred in the collectors before a temperature signal was transmitted to the remote sensor. The best approach is to use a sensor bolted to the absorber plate by the collector manufacturer.

The storage temperature sensor should be located in the lower quarter of the storage zone of the tank and should be of the averaging type. In addition, it should be located away from the point of introduction of city water into the storage tank so that a false signal is not produced.

When the measured temperature of the collector panel arises 15 or 20°F above the measured storage tank temperature, pumps P1 and P2 in the collector and storage loops begin to operate. Heat is transferred from the collector loop fluid through the heat exchanger into the storage tank. Pump P2 in indirect systems operates whenever pump P1 operates. Flow continues until the temperature in the preheat tank approaches that of the collector. This can occur if the solar flux on the collector begins to drop off in the afternoon or if little hot water is being used. At the point where the collector and tank temperatures are within approximately 3°F, the controller shuts off pumps P1 and P2 until the start-up condition of 15 to 20°F is again achieved.

Domestic hot-water controllers are available from many manufacturers. In 1980 the cost of the controller with two sensors of the type described was in the range $50 to $75.

Other control elements are present in water-heating systems in addition to the pump controller. As shown in Figs. 6.2 through 6.5, a tempering valve is used at the outlet of the water-heating system. This is an essential component of any solar water-heating system. Its function is to keep the water delivered to a building below a maximum temperature, typically 120 to 140°F. It is possible for a solar collector to heat water well above this point. Very hot water is a hazard to water users since it can cause scalding.

Although it is rare for solar-heated water to reach very high temperatures, this can occur during periods of high sun with low solar use in spring or fall. Under these conditions, collectors will intercept the maximum amount of radiation on a clear day. Preheat tank temperature will continue to rise above 190°F for an efficient collector. Scalding can occur with either single- or double-tank systems if a tempering valve is not used. In a single-tank system, the very hot water is drawn off immediately from the combined preheat and auxiliary tank to the distribution system. In a two-tank system, the overheated water first passes into the second tank, but as water is drawn off from this tank the overheated water eventually makes its way into the outlet line. The tempering valve is located at this point to control the temperature to a maximum of 120 to 140°.

The discussion above concerns system function under low-usage conditions. Under no-usage conditions it may be necessary to reject heat from the solar collection system since no hot water is drawn off. Several methods of heat rejection have been used in solar water-heating systems. The first involves a pressure-temperature (PT) valve. This valve opens when either the temperature or the pressure at the top of the storage tank exceeds design set points—typically 190°F or 120 to 150 $\mathrm{lb/in^2}$. The temperature condition will usually occur before the pressure condition. When the PT valve opens, hot fluid is dumped from the preheat tank and replaced by cold fluid from the city water supply, which enters the bottom of the tank. The PT valve should be located at the top of the preheat tank, where the water is hottest.

PT valves are not designed for repeated operation as heat rejection devices. In systems where frequent heat rejection is expected owing to limited water usage, another method of heat dumping should be used. This involves sensing the collector or storage temperature with an on-off thermal switch. As the temperature rises above a prescribed valve, the switch snaps closed, thereby closing a circuit to a solenoid valve located near the top of the preheat tank. The solenoid valve opens and hot water is dumped from the top of the preheat tank. In turn, this water is replaced by cold city water, introduced in the bottom of the tank. In direct water-heating systems, the cold water immediately passes into the collector inlet, the collector temperature is reduced, the switch opens, and the solenoid valve closes. In indirect systems, the effect of cold water is introduced into the collector fluid loop by way of the counterflow heat exchanger HX1, shown, for example, in Fig. 6.5. In either system the effect is to quickly reduce the system temperature to its proper operating level.

A high-temperature control mode used in many systems in the 1970s involved simply shutting off collector pump P1 whenever the collector temperature exceeded a specified value. *It is not recommended that pump P1 be shut off during high-temperature periods if glycol fluids are used.* Serious problems can occur when this simplistic control mode is used. First, working fluid in the collector can rise above 400°F. Glycols decompose rapidly at this temperature, forming glycolic and then oxalic acid. Above 250°F glycol fluids degrade, and replacement of the antifreeze will be required if collector stagnation is permitted. (High-temperature heat transfer oils can be used; they will not degrade at 400°F but are very costly.) Second, collectors can be subjected to very high temperatures for extended periods. Although solar collectors are designed to withstand stagnation episodes, it is better not to expose them repeatedly to this severe situation over a 20- or 30-year period. Materials within the collector degrade at elevated temperatures and reliability is reduced.

A third problem arises with this method of heat rejection. If the system pressure is below that corresponding to the stagnation temperature of the collector, fluid may boil in the collectors. Of course, damage will not result, since fluid will be rejected through the safety valve. However, on cooling, the system may be placed under a vacuum and certain components such as heat exchangers or expansion tanks may collapse. A vacuum breaker can solve this problem, but the problem can be avoided entirely by using proper heat rejection methodology.

Control design caused more problems in water-heating systems built in the past decade than any other single component. There were several reasons for these problems, including use of poor quality solid-state devices in controllers, poor understanding of the various modes in which solar systems must operate, and delicate control designs not suitable for installation by electricians in the field. At present, reliable solar system controllers are available, and with proper understanding of system function no difficulty should be encountered in proper system design. Key considerations in control system selection are:

Collector sensor location
Storage sensor location
Proper on-off hysteresis characteristics
Reliable solid-state devices
Control located in space maintained above 50°F
Controller connected according to proper instructions
Control designed for all possible system operating modes, including heat collection, heat rejection, system start-up after power outage, and freeze protection

Figure 6.7 shows a modern water-heating controller module, including the control, pumps P1 and P2, heat exchanger HX1, an expansion tank, and the safety valve. This unit can be joined to the collectors via a check valve and storage tank by making four piping connections, thereby completing the installation of the solar water-heating system. This prewired, prepiped approach to domestic hot-water system design has lower installation costs and is more reliable than the early component-by-component installation approach. Modules similar to those shown in Fig. 6.7 are manufactured by several companies in the United States.

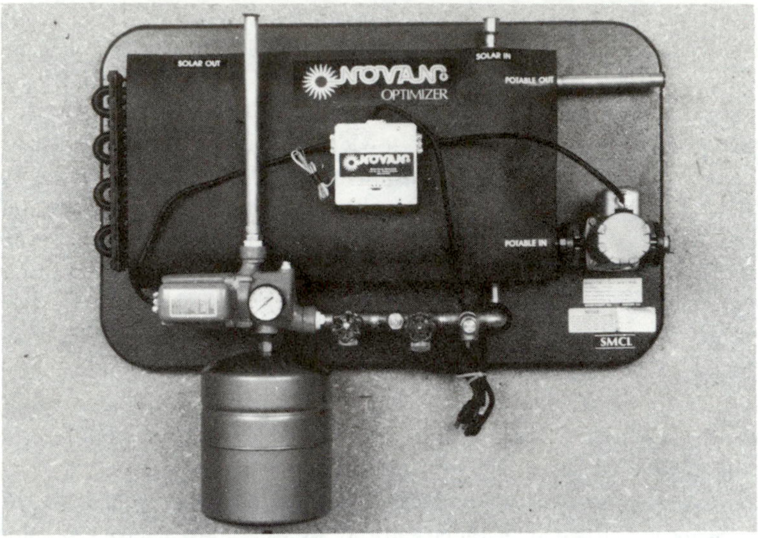

Figure 6.7 Example domestic hot-water control and hardware module including pumps, expansion tank, check valve, controls, and heat exchanger in a single unit. (*Courtesy of Novan Energy Systems.*)

Mechanical Components

The important mechanical components in solar water-heating systems include the following:

Collectors
Heat exchangers
Pumps
Tanks
Expansion tanks
Safety valves
Backup heating system
Pipes and pipe fittings

Solar collectors are described in detail in Chap. 3, and specific system design considerations are discussed in the next section. No special features of solar collectors are required in water-heating systems. In direct systems, of course, the collector must be selected to operate under city water pressure. In residential systems, flow balancing in water-heating systems is not a problem. In commercial systems with large arrays, flow balancing may need to be considered. This subject is covered under space heating in Chap. 8.

The collector-to-storage heat exchanger was described earlier. As noted then, it is important for economic reasons that the exchanger be neither too large nor too small. Heat exchangers in water-heating and space-heating systems are identical in design, and the reader is referred to the heat exchanger section in Chap. 8 for details of their selection.

One or two pumps are required in water-heating systems. The selection of a pump requires specification of head and flow rates, materials, and operating pressure. In addition, the net positive suction head must be specified. The pressure drop in the collector loop of indirect solar water-heating systems can be found by adding all the pressure drops in the loop. Sources of pressure drop in the collector loop include pipe fittings, check valve, filters, collectors, collector headers, heat exchangers, and an air separator, if present at the expansion tank. The design pressure drop based on the proper antifreeze fluid can be determined from manufacturers' literature at the required flow rate. The sum of these pressures is the head that the pump must develop. If a drain-back system is used, the pump must also be capable of lifting fluid from the storage tank to the topmost point of the collector piping array, normally located on the roof. This head requirement is typically larger than that to account for the frictional pressure drop in the piping loop.

The flow rate in a solar collector loop is described in the next section and is related to collector area. Rules given in the next section specify flow rates about 0.025 gal/min water equivalent per square foot of collector area. Pump flow is this number multiplied by the net collector area.

Storage pump P2 is usually sized for a flow rate approximately 50 percent greater than that of P1 to provide improved heat exchanger performance. The small pressure drop to be overcome in pump P2 is that in the heat exchanger tube side and the associated piping.

Pump material selection depends primarily on the operating temperature and fluid used. The use of glycol fluids at elevated temperatures may place restrictions on the type of pump seals that can be employed. This should be checked with the manufacturer. For systems in which no free oxygen is present, such as the collector loop, cast iron impellers may be used. However, when potable water is being pumped, as in direct water-heating systems or by pump P2 in indirect systems, bronze or stainless steel impellers are required to avoid corrosion. Pump materials should be compatible with other piping materials in the fluid loop to avoid galvanic corrosion. If material compatibility cannot be achieved, insulating unions should be used on the pump to reduce the rate of corrosion although they will not eliminate it.

Storage tanks for water-heating systems must be lined with an inert substance. Typically, glass-lined steel tanks are used in water-heating systems and are generally available. The tank pressure should be selected on the basis of the maximum city water pressure that will be encountered. Safety valves are selected to release pressure above the design operating pressure of the storage tanks. All storage tanks in solar systems should be insulated; the economic amount of insulation can be calculated by a method outlined in [1]. This method usually results in specification of R-25 to R-30 on the tank. Simply insulating the tank surface itself is not adequate, however. Thermal breaks must be provided between the tank and every point of contact with the environment, including tank supports, piping, and other possible exposure points. All areas of the tank should be insulated; piping running to and from the tank will, of course, be insulated, and there should be no breaks in this insulation at the tank. An alternative to insulating the tank is to house it in an insulated box. If the tank is not insulated by the manufacturer, the latter approach may be easier. Use of buried storage tanks involves many problems, including difficulty of insulation, shifting of the tank as earth settles, and waterproofing of insulation. These problems must be considered very carefully before deciding to use a buried tank; they are discussed in detail in [1].

Each closed loop in a solar water-heating system requires an expansion tank, as described earlier. In most applications, the water delivery loop itself need not be protected with an expansion tank, since drawing off water at any point in the distribution loop will immediately relieve any pressure buildup caused by fluid expansion. The delivery loop is considered an open loop and does not require an expansion tank. Expansion tanks in closed loops can be sized either by using the rule of thumb described above or by reference to manufacturers' design charts for the fluid under consideration.

Valves of various sorts are present in solar water-heating systems. PT and check valves have been described above. Tempering valves should also be used in any system. Although not shown in Figs. 6.2 through 6.5, gate valves are used at many points in fluid loops to isolate components. For example, four valves would be used at each pipe connection to heat exchanger HX1 to permit removal of the heat exchanger without complete drain-down of both liquid systems attached to it. Care should be taken to protect any portion of a pipe loop that can be isolated by valves if heat can be added to that component. For example, if both valves on the storage side of HX1 were closed and pump P1 were operating to collect solar heat, pressure buildup in the blocked shell side of the exchanger would require relief through a safety valve. Safety valves are fundamentally different from PT valves since they operate only on a pressure signal and not a temperature signal.

All water-heating systems require an auxiliary heat source for periods of solar outage. In most cases the auxiliary system is a standard nonsolar water-heating system connected downstream of the water-to-solar fluid loop. Alternatively, in residential single-tank systems it may be included within the preheat tank itself, as shown in Figs. 6.2 and 6.4. Auxiliary system design requires no specific additional consideration because of the use of solar energy. The backup system operates much as it would if no solar heating system were present. Backup energy sources, including steam, gas, fuel oil, electricity, and wood, are activated automatically if the temperature in the auxiliary tank falls below the set point specified in the system design. Auxiliary heating units are sized to carry the full peak water-heating and standby loads in case of complete solar unavailability in severe weather. The peak water-heating load can be estimated from Table 6.2. Peak usage rates indicated there are multiplied by the expected temperature rise in winter, when source water temperatures are lowest. To this quantity the expected standby loss is added to find the peak heat rating of the backup system. Chapter 8 gives additional information on component design for active liquid-based heating systems. Sizing of components in water-heating systems is described in the next section.

In the design of water-heating systems, material compatibility is important since water is a very corrosive working fluid. Metals in contact with each other in an aqueous environment can corrode very rapidly if they are widely separated in the electromotive series. Table 6.3 summarizes the generally acceptable use conditions for aluminum, copper, steel, stainless steel, galvanized steel, and brass. The table also lists conditions under which these metals should not be used in aqueous environments. The information in the table should be carefully employed in the design of water-heating systems. Similar information on materials compatibility requirements in open systems is presented in [3]. The corrosion potential is greater in open systems with free oxygen present than in closed systems.

Generic System Performance Comparison

In addition to reliability and cost considerations, the thermal performance of generic system types should be considered in the selection of one for a particular application. Few careful tests have been carried out side by side for the classes of systems described in this chapter. However, in one study the National Bureau of Standards (NBS) compared most of the common system types.

The NBS study [4] ranked the various types of systems in order of decreasing cost-effectiveness as follows:

Thermosiphon
Single-tank direct
Single-tank indirect
Double-tank direct
Double-tank indirect

In this study the same load was applied to all systems, which were operated side by side at NBS laboratories. The conclusions apply only for the specific location studied, but the differences among the various systems were significant. It is therefore expected that approximately the same trend would apply in other locations for the same types of systems. The conclusions of the study can be summarized as follows:

1 Thermosiphon systems are clearly the most cost effective because of their low initial cost, good thermal efficiency, and minimal parasitic energy consumption. Architectural problems may rule them out sometimes.

2 Direct systems outperformed indirect systems from a cost-effectiveness point of view. Direct systems have inherently higher heat transfer rates than indirect systems. This conclusion about cost-effectiveness does not include the effects of increased corrosion and scaling within the collector, however.

3 Single-tank systems outperform double-tank systems. According to [4], this conclusion might have been different if the double-tank system had been optimized more thoroughly with respect to tank insulation, collector area, and storage size. However, double-tank systems will always have greater heat losses than single-tank systems given the same tank R values.

Table 6.3 Acceptable and Unacceptable Use Conditions for Metals in Contact with Aqueous Working Fluids in Closed Solar Systems

Generally unacceptable use conditions	Generally acceptable use conditions
Aluminum	
1. When in direct contact with untreated tap water with pH <5 or >9. 2. When in direct contact with liquid containing copper, iron, or halide ions. 3. When specified data regarding the behavior of a particular alloy are not available, the velocity of aqueous liquids shall not exceed 4 ft/s.	1. When in direct contact with distilled or deionized water that contains appropriate corrosion inhibitors. 2. When in direct contact with stable anhydrous organic liquids.
Copper	
1. When in direct contact with an aqueous liquid having a velocity greater than 4 ft/s. 2. When in contact with chemicals that can form copper complexes such as ammonium compounds.	1. When in direct contact with untreated tap, distilled, or deionized water. 2. When in direct contact with stable anhydrous organic liquids. 3. When in direct contact with aqueous liquids that do not form complexes with copper.
Steel	
1. When in direct contact with liquid having a velocity greater than 6 ft/s. 2. When in direct contact with untreated tap, distilled, or deionized water with pH <5 or >12.	1. When in direct contact with untreated tap, distilled, or deionized water. 2. When in direct contact with stable anhydrous organic liquids. 3. When in direct contact with aqueous liquids of $5 < \text{pH} < 12$.
Stainless steel	
1. When the grade of stainless steel selected is not corrosion-resistant in the anticipated heat transfer liquid. 2. When in direct contact with a liquid that is in contact with corrosive fluxes.	1. When the grade of stainless steel selected is resistant to pitting, crevice corrosion, intergranular attack, and stress corrosion cracking in the anticipated use conditions. 2. When in direct contact with stable anhydrous organic liquids.
Galvanized steel	
1. When in direct contact with water with pH <7 or >12. 2. When in direct contact with an aqueous liquid with a temperature $>55°C$ (131°F)	1. When in contact with water of pH >7 but <12.

Table 6.3 Acceptable and Unacceptable Use Conditions for Metals in Contact with Aqueous Working Fluids in Closed Solar Systems (*continued*)

Generally unacceptable use conditions	Generally acceptable use conditions
Brass and other copper alloys	
Binary copper-zinc brass alloys (CDA 2XXX series) exhibit generally the same behavior as copper when exposed to the same conditions. However, the brass selected shall resist dezincification in the operating conditions anticipated. At zinc contents of 15 percent and greater, these alloys become increasingly susceptible to stress corrosion. Selection of brass with a zinc content below 15 percent is advised. There are a variety of other copper alloys available, notably copper-nickel alloys, which have been developed to provide improved corrosion performance in aqueous environments.	

Source: U.S. Department of Housing and Urban Development [3].

SOLAR WATER-HEATING SYSTEM SIZING

The preceding sections of this chapter summarized the operation of all common commercially available solar water-heating systems. As described in Chap. 5, the sizing of water-heating systems is an economic question based on a trade-off of solar energy cost versus backup energy cost. The law of diminishing returns affects solar water-heating systems as it does any other solar system. The system size is therefore optimized, in most cases, to provide something less than 100 percent of the annual heating requirement. Performance prediction for water-heating systems is described in the next section. It has been found, however, that the size of all components in the solar system can be related to collector area. Since performance is also related to collector area, the optimal sizing procedure can be based on this index of system size alone. Once the cost-optimum area is identified by the performance prediction procedure, the size of each component can be determined. Sizes of all components given in the next paragraphs are therefore based on collector area (ft_c^2).

Sizing Rules for Domestic Water-Heating Systems

The following rules have been developed from component tests in situ as well as detailed computer simulations. Conclusions of both types of studies agree with the following results.

Collector area: determined by economic analysis [approximately 1 $ft^2/(gal \cdot day)$] as described in the following section on water heating system performance prediction

Storage: 1.5 to 2 gal/ft_c^2

Water equivalent collector flow rate: 0.025 $gal/(min \cdot ft_c^2)$

Storage flow rate: 0.03 to 0.04 $gal/(min \cdot ft_c^2)$ (indirect systems)

Heat exchanger: 0.05 to 0.1 ft^2_{hx}/ft^2_c (indirect systems)
Collector tilt: latitude ±5°
Expansion tank volume: 12 percent of collector fluid loop (indirect systems)
Heat rejector capacity: equivalent to peak collection rate possible under clear-sky conditions at heat rejection specified temperature
Controller turnon Δt: 15 to 20°F
Controller turnoff Δt: 3 to 5°F
System operating pressure: to provide 3 lb/in^2 gauge at topmost collector manifold
Storage tank insulation: R-25 to R-30
Mixing valve set point: 120 to 140°F
Pipe diameter: to maintain fluid velocity below 6 ft/s and above 2 ft/s

Example System Design

Figure 6.8 is the one-line schematic diagram for a solar water-heating system installed on the Wyoming state capitol building in 1981. The collector area determined by an economic analysis for this system was 2750 ft^2. This area provided 84 percent of the nominal 3000 gal/day demand. A collector tilt angle of 45° was used to slightly favor the winter sun. Flow rates in pumps P1 and P2 were selected on the basis of rules in the previous section.

Heat rejection in the system shown is achieved by HRU-1, a weatherproof unit heater whose rating is based on collector energy delivery at an operating temperature of 200°F and a temperature of 90°F for the air entering the heat rejection unit. The control of the heat rejecter is part of the solar system controller.

Storage for the system is provided by a 4125-gallon lined steel tank connected in series to the nonsolar gas-fired water heater. A mixing-valve temperature of 140°F is used in accordance with the architectural program for this building.

The system in Fig. 6.8 also includes an automatic glycol makeup feature. If fluid leaks from the system, pressure in the closed collector loop drops and a pressure switch located at the collector loop expansion tank closes. The closed pressure switch activates the glycol fill pump, which pumps fluid from the glycol makeup tank into the collector loop until the design pressure is reached. At this point the pressure switch opens and the glycol fill pump ceases operation. Since relatively small amounts of fluid are moved by this pump against a relatively high pressure, a gear pump is specified.

A glycol overflow tank is also included in the design to accept any fluid purged through the 30 lb/in^2 safety valve. Fluid loss would occur if pump P1 ceased operation owing to power failure or controller malfunction. Since glycol is expensive, it should not be wasted. The system automatically refills as pressure in the loop drops. Because of its toxicity, ethylene glycol should not be dumped into sewer systems—the second reason for using a glycol overflow tank.

The collector loop expansion tank TK-1 is connected to the fluid circulation loop by a device called an air eliminator. This component removes air

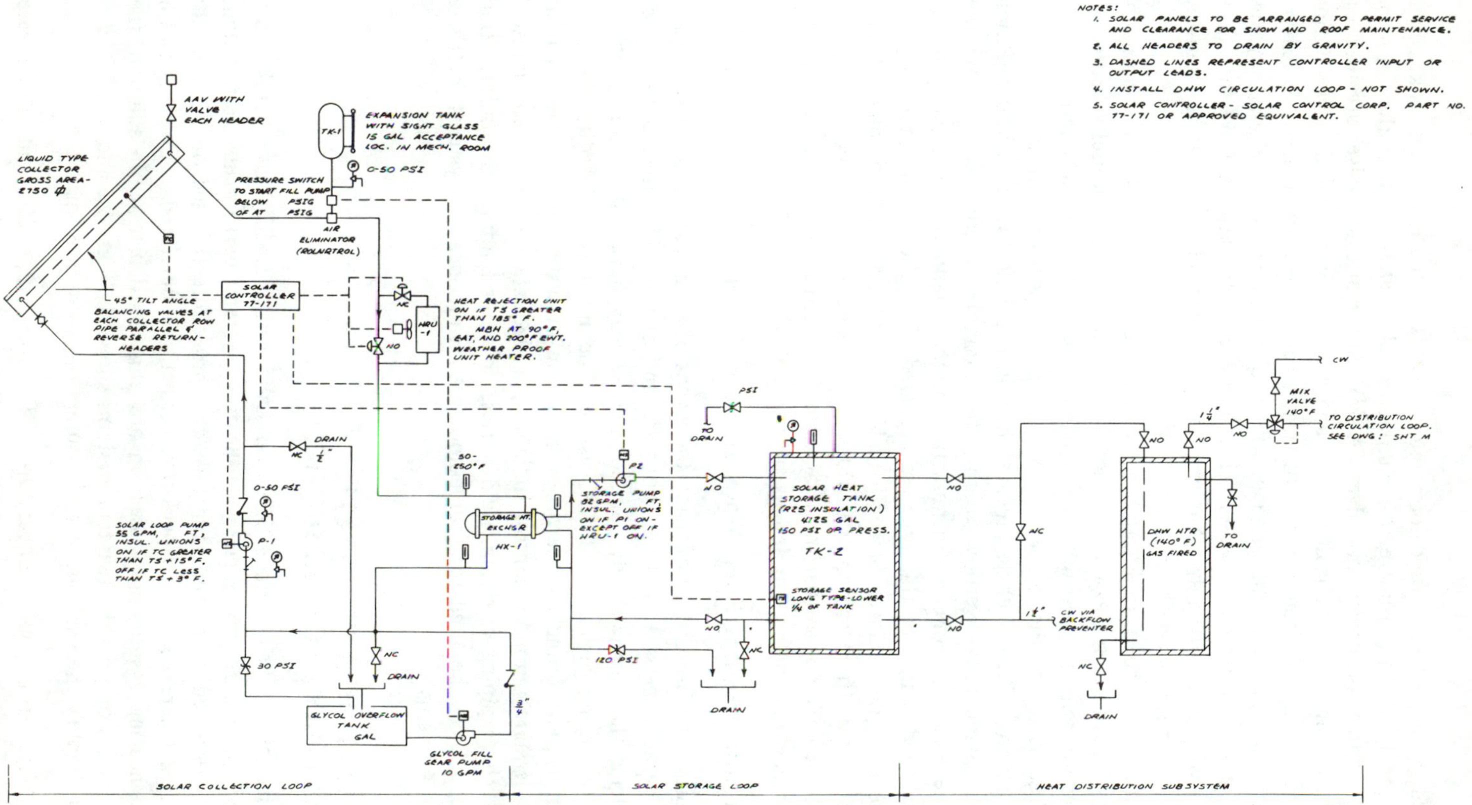

Figure 6.8 Example one-line solar system design-development drawing for Wyoming state north office building water-heating system. (*Courtesy of Jan F. Kreider Associates, Inc.*)

entrained in the collector fluid loop and diverts it to the expansion tank, where it is purged to the atmosphere. Air elimination also takes place at the collector headers through automatic air valves (AAV). Automatic air valves are used on large systems since it is difficult to open and close many manual air valves during the initial system fill. However, the automatic valves have the undesirable feature that they leak readily in the presence of glycol fluids, which have relatively low surface tensions compared to water. Therefore each automatic air valve should be equipped with a manual valve that can be closed after all air is purged from the system some weeks after the initial fill.

Gate valves shown in Fig. 6.8 are labeled normally open (NO) or normally closed (NC). Although theoretically not a requirement, this is a useful feature of solar flow diagrams for persons not familiar with how the system functions.

Close study of Fig. 6.8 will show that some specifications are incomplete on the drawing. These include pressure drop in the collector loop, heat rejection rate, and glycol tank overflow volume, for example. These quantities are not specifiable until the collector manufacturer is known. In the preliminary design stages, when drawings of this type are prepared, the collector manufacturer may not have been determined. Since there is little standardization among collector manufacturers with regard to size, flow, or pressure drop, it can be useful to preselect the collector so that the solar system design can be finished at the same time as other building system designs are completed.

Troubleshooting Solar Water-Heating Systems

During the installation of many thousands of solar water-heating systems in the last decade, the solar domestic hot-water industry has identified certain types of problems that are relatively common. These may occur either during system start-up or during operation of the system. The main types of difficulties are shown in Table 6.4 and include drops in system or water pressure, lack of fluid flow, inadequate pump or control functions, inadequate hot water, and unusual noises, such as piping vibration, fluid noises, or water hammer. The information in Table 6.4 can be used to troubleshoot solar water-heating systems for these common problems.

WATER–HEATING SYSTEM PERFORMANCE PREDICTION

Scientists at Los Alamos Scientific Laboratory (LASL) have developed a simplified procedure for solar water-heating system performance estimation, which is described here. Unfortunately, this procedure is based on 12 monthly calculations, whereas the other performance methods described in Chaps. 7 and 8 for space-heating systems are based on are annual calculation. In addition, the method applies only to the two-tank systems shown in Figs. 6.3 and 6.5.

To simplify the analysis, several assumptions have been made:

1 Water set point temperatures of 110, 130, 150, and 170°F are considered. For other temperatures interpolation is required.

Table 6.4 Troubleshooting Guide for Solar Water-heating Systems

Symptom	Source	Cause	Action
Drop in system pressure	Collectors	Leak in absorber	Repair leak. Replace wet insulation. Replace absorber. Replace collector.
(Note: After leaks have been fixed, remove all wet insulation and replace with new, dry insulation.)	Piping	Leak from soldered joint	Resweat joint.
		Leak from threaded joint	Tighten-replace Teflon tape, add pipe dope.
		Leak from clamped joint	Tighten hose clamp.
	Valves (gate, globe, etc.)	Leak from valve packing	Tighten packing nut, replace packing.
	Drains	Leak from drain	Close drain tightly, add cap.
	Air vents	Leak from needle valve	Inspect, clean mechanism, repair or replace.
		Leak from coin vent	Inspect, tighten.
	Pressure relief valve	Preset relief valve has expelled excess pressure	Install relief valve with higher setting. Recharge system to operating pressure. Lower pressure in system.
		Adjustable relief valve has expelled excess pressure	Set for higher temperatures, recharge system. Lower pressure in system.
	Expansion tank (diaphragm)	Bladder exceeded	Inspect for fluid in air chamber, replace.
	Expansion tank (standard)	Inadequate air space	Adjust water level.
	Heat exchanger	Leak from exchanger	Inspect, repair, or replace.
Drop in water pressure at both hot and cold outlets	Inlet water pressure	Well pump problems	Check wiring; check pressure tank and and controls; check water level in well; check pump motor and impellers; repair or replace.
		Frozen inlet pipe	Thaw frozen section. Keep water running until hard freeze is over.
		Strainer in house supply line clogged	Clean basket.
		Problem with community water supply	Notify local water department.
		Broken or compressed pipe between main and house	Replace pipe.
		Pressure-reducing valve	Check for plugged ports, and proper operation. Clean, repair, and replace.

Table 6.4 Troubleshooting Guide for Solar Water-heating Systems (*continued*)

Symptom	Source	Cause	Action
Drop in water pressure at hot outlets only. Cold outlet pressure OK	Obstruction in domestic hot-water system	Valve improperly closed	Open manual valves. Check controls for automatic valves; check automatic valve operation; repair or replace. Look at flow direction arrows on check valves and backflow preventer. Reinstall correctly. Check for jammed elements. Repair and replace.
		Strainers clogged	Clean basket.
		Bent or crushed pipe restricting flow	Repair or replace
Inadequate or no circulation of coolant	Collector	Passages clogged	Remove collector, backflush passages.
		Air locked	Install air vents at all high points.
	Pump	Undersized	Check specs, add larger motor and/or impeller, change pump if necessary.
		Air locked	If present, loosen vent on pump head.
		Impeller bound	If possible, turn impeller manually to free it; repair or replace.
		Installed backward	Check flow direction arrow and reverse.
		Installed incorrectly	Check manufacturer literature for proper operating position.
		Pump isolated	Make sure all shutoff valves are open.
	Air vents	Inadequate number	Install additional vents at high points.
		Improperly placed	Install at correct positions.
		Faulty, causing air lock	Operate needle valve manually; check float control; inspect, clean, repair, or replace.
	Shutoff valves	Closed	Open.
	Automative valves	Closed	Check controls, operation, and springs; repair or replace.

	Piping	Air locked	Install vents if necessary, check pitch.
		Too small	Increase size, install larger pump or pump motor or, install second pump.
	Check valve	Installed backward	Check flow direction arrow and reverse.
		Jammed	Repair or replace.
Pump functioning improperly (cycles excessively, or always active or inactive). Freeze protection modes not operating	Differential controller and control system	Sensors improperly located	Place sensors as specified by manufacturer.
		Sensor improperly secured	Ensure tight bond between sensor materials.
		Break in wire	Inspect, repair.
		Wired incorrectly	Check all wiring with drawings, inspect connections.
		Controller settings	Ensure proper settings.
		Sensors faulty	Test by heating one sensor, cooling other(s). Use ohmmeter to determine if sensor is producing proper resistance at various temperatures.
		Controller faulty	Check fuses. Check relays. Check for loose circuit boards; repair, replace.
	Power wiring	No power to controller	Inspect electrical connections and wiring; repair.
		No power to pump	Inspect electrical connections and wiring; repair.
Decrease in performance of solar portions of systems	Collectors	Outgassing	Clean surface; contact supplier.
		Condensation	Inspect and repair glazing seal. Inspect and repair pipe gaskets. Inspect weep holes for clogging and proper location. Replace desiccant.
		Absorber coating degradation	Recoat absorber. Contact installer.
		Dirt on glazing	If rain or dew offers inadequate washing, rinse cover (check with manufacturer for suggested methods).
	Piping	Night losses–reverse thermosiphon	Install check valve on collector supply.
		Insulation seams and joints deteriorated	Reglue, tape, or staple pipe insulation.

Table 6.4 Troubleshooting Guide for Solar Water-heating Systems (*continued*)

Symptom	Source	Cause	Action
Excessive vibration noise in system	Pipe hanger	Vibration from equipment transmitted to framing members	Locate area of noise; disconnect hangers from framing, replace with flexible-type or wrap hanger on outside of insulation.
	Pump	Impeller shaft misaligned	Inspect, repair, replace.
		Pump vibrates excessively	Clean and rebalance impeller. Install vibration isolators on both sides of pump (strainers and check valves between isolators increase mass).
Fluid noise	Piping	Entrapped air	Force purge system, install additional vents.
		Air purge installed backward	Check flow direction arrow and reverse.
		Gravity return line oversized, creates waterfall noise	Replace with smaller-diameter tubing.
	Air vents	Not venting	Check for tight cap, vent manually.
		Air locked	Force purge, vent pump if it has bleeder.
Water hammer	Piping	Pressure	Install shock suppressor.
		Loose support of pipes	Secure pipe supports to structure.
		Entrapped air	Install additional vents.
	Automatic valves	Open or close too fast	Install valves with slower action. Modify controls to slow valve action.
Squeaking noise	Piping	Pipe expansion in contact with building materials	Cut holes for pipes $\frac{1}{4}$ in larger than pipe O.D. Pack penetration with insulation.
Humming of mechanical equipment motors	Differential controller or control system	Controller is not fully deactivating component, allows slight continuous current to motors	Use differential controls troubleshooting procedure. Check solenoid at motor starter for complete break. Adjust or replace.

Insufficient hot water	Collector loop	Failure in fluid system circulators, controls	Use appropriate troubleshooting procedure.
	Tank	Too small	Install second tank.
		Conventional heater problems	Check backup source, wiring, gas, oil lines, burners, and set point. Check for deposits on heater.
		High storage losses	Insulate tank with additional layer.
	Piping	Reverse thermosiphon	Install check valve on collector supply.
	Mixing valve	Improperly adjusted	Check adjustment temperature indicator, set higher.
		Faulty	Replace.
Hot water in cold-water lines	Mixing valve	Hot water from top of tank expanding into cold feed of mixing valve	Increase distance so that top of tank to bottom of mixing valve is at least 16 in. Install check valve in cold-water line at mixing valve.

Source: *Solar Age*, May 1980, pp. 45–46; reprinted with permission of the author, R. Schwolsky.

2 Preheat tanks are sized on the basis of 1.8 gal per square foot of collector net area.

3 R-6 insulation is used on the preheat tank.

4 A double-glazed collector with flat-black absorber surface is used. (Performance for a single-glazed collector with a selective surface is nearly identical to that for the double-glazed collector.)

5 A standard usage profile is used.

The Method

The LASL method can be used in stepwise fashion:

1 Calculate monthly averages of *daily* collector plane insolation $\bar{I}_c$ (Chap. 2), or read from App. 6.

2 Calculate the average daily water-heating load L ($=Q_{hw} + Q_{\text{standby}}$) from Eqs. (6.1) and (6.2).

3 Specify a collector area A_c. To start the calculations, use 1.0 ft^2 for each gallon-per-day expected consumption.

4 Find the solar load ratio (SLR) from

$$\text{SLR} = \frac{A_c \bar{I}_c}{L} \tag{6.3}$$

5 Calculate the monthly solar fraction f_s from

$$f_s = \begin{cases} a\ \text{SLR} & 0 < \text{SLR} < 0.8 \\ 1.0 - be^{-c\ \text{SLR}} & 0.8 < \text{SLR} < 5.0 \end{cases} \tag{6.4}$$

Constants *a, b,* and *c* are contained in Table 6.5.

6 Finally, find the annual solar heat production Q_u by adding the monthly solar outputs $(f_s LN)_i$, where N_i is the number of days in month i:

$$Q_u = \sum_{i=1}^{12} (f_s LN)_i \tag{6.5}$$

7 If an optimization is to be carried out, use the methods given in Chap. 5 and vary A_c in step 3. Then repeat steps 4 to 6 as required.

Table 6.5 Domestic Hot-Water Solar Load Ratio Parameters for Eq. (6.4)

Desired water delivery temperature (°F)	Empirical constants		
	a	*b*	*c*
110	0.568	1.153	0.933
130	0.499	1.080	0.729
150	0.440	0.978	0.514
170	0.348	0.966	0.365

SOLAR SWIMMING POOL HEATING

Heating a swimming pool with solar energy is attractive because most pools are outdoors and are used in the summer, when solar energy is most available. Solar heating of swimming pools can therefore be economical compared to conventional heating. Although the volume of water to be heated is large, it is not necessary to use a large area of collector panels because the pool temperature does not need to be very high. Thus it is not surprising that most solar water-heating systems in operation today are used to heat swimming pools. Moreover, in some areas of the United States it is becoming illegal to use natural gas to heat pools, and this will offer additional opportunities for solar heating in the near future.

If a swimming pool has conventional heating, one can add the solar heating system to it and use existing pipes, pump, and filter, thus reducing the required retrofit investment. For a new swimming pool one can often save the cost of an expensive boiler installation.

Collector Sizing

The installation described here is suitable for residential swimming pool heating. Copper pipes are recommended because they will not be affected by chemicals in the water and other impurities associated with the operation of pools, but some plastics can also be used. To be drainable, the panels should be mounted above the pool water level, for example, on the roof of a house, an outbuilding, or a specially constructed frame.

DeWinter, in *How to Design and Build a Solar Swimming Pool Heater* [5], recommends a panel area (single-glazed collectors) equal to half the pool's surface area, but some solar panels designed for swimming pools have no glazing and therefore a larger area may be needed. The more efficient the collector panels, the smaller the area needed. If the area is too small, evaporation and convection losses from the pool prevent the attainment of reasonable temperatures. If it is too large, the panels may overheat and lose energy to the atmosphere, giving a diminished return on the investment. Panel area must therefore be selected on the basis of pool area in relation to collector efficiency. Kiely [6] gives the following rule for calculating the area of swimming pool collector. First, take the area of the pool surface (e.g., 300 ft^2) and divide by 2 (150 ft^2); if the unit collector area is 15 ft^2, ten panels are required. Next, calculate the volume of the pool by multiplying the area by the average depth ($300\ ft^2 \times 4.5\ ft = 1350\ ft^3$). Multiply this answer by 7.46 to find the number of gallons ($1350\ ft^3 \times 7.46\ gal/ft^3 = 10070$ gal). Finally, divide the gallonage by 1250 (i.e., one 15-ft^2 panel per 1250 gal), and you find that eight panels are required. Whichever number is greater is the number of panels to use; in this case, use 10 panels.

Installation

The panels should be placed so that they do not obstruct sunshine falling on the pool, which also helps raise the water temperature (Fig. 6.9). As the

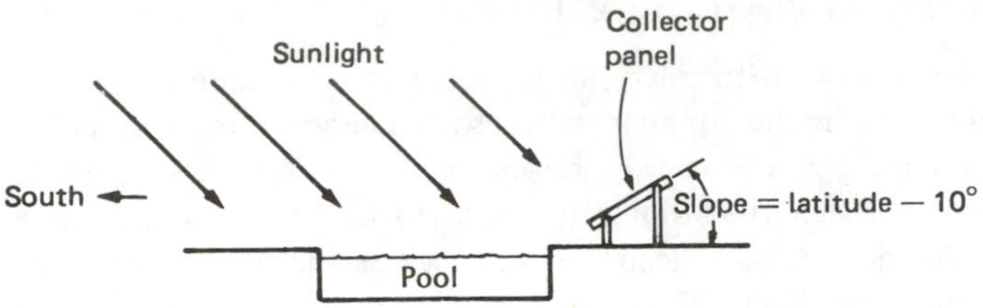

Figure 6.9 Collector panel orientation for a swimming pool.

panels heat a pool in the summer, a lower angle of inclination should be used than in a domestic system, where energy collection is required year-round. An angle of latitude less 10° is recommended for the United States.

If a frame is needed it can be made from wood, as shown in Fig. 6.9. Frame-mounted panels catch the wind and can be damaged in severe winds if they are not securely fastened. Wooden frames must be treated with protective coating. Aluminum angle and steel tube can also be used to make a suitable frame, but use of these materials increases the cost.

Panels should be connected in parallel, as shown in Fig. 6.10; if they are connected in series, the water may boil by the time it reaches the last panel in hot weather. Also, panels linked in series require a lower flow rate than panels linked in parallel. This increases the time for the pool water to circulate and reduces efficiency.

The system must be drainable to prevent water from freezing in the panels in cold weather. If the panel array is placed above the level of the pool, water can automatically drain if there are no loops in the piping. An automatic air vent is necessary to allow air into the panels to assist drainage. It should be placed on the top manifold at the position shown in Fig. 6.10.

Figure 6.11 shows a typical plumbing diagram with solar panels and an existing heater. The valves can be manipulated to direct flow through the existing heater, through both heater and solar panels, or through the solar panels only. The heater can be operated independently when there is little solar radiation, or

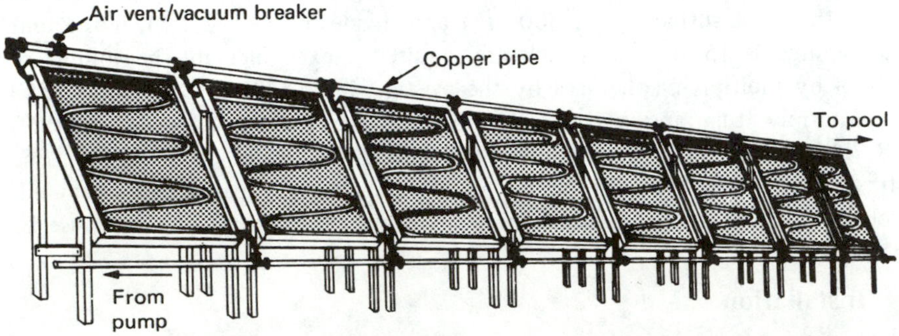

Figure 6.10 Collector panels mounted on frame and connected in parallel.

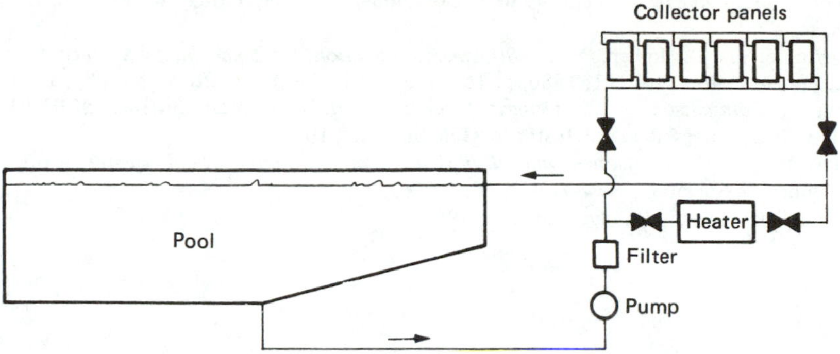

Figure 6.11 Plumbing diagram for retrofit solar swimming pool heating system using existing pump, heater, and filter.

it can be shut down when there is enough radiation for the collectors to heat the pool by themselves.

Flow Rates

The flow rate of water through the solar panels is not critical, but very high flow rates tend to be as inefficient as very low ones. If the water passes through the panels rapidly, it will lower the panel temperature and improve the collection efficiency, but at the same time the pumping power increases. Also evaporation losses may increase due to higher turbulence at the surface of the pool. A very low flow rate will result in higher collector temperatures and lower efficiency. A good rule is that the pool volume should be circulated every 8 hours; this usually requires a flow rate through each panel of 2 to 3 gal/min. A 10-panel installation connected in parallel would therefore have a flow rate of 20 to 30 gal/min. Control of circulation can be achieved by manual operation, thermostat, time switch, or differential controller.

A solar-heated pool should be covered at night to reduce heat loss. A transparent, slightly pressurized "bubble" cover will allow solar energy to enter the pool, help to raise the temperature, and prevent heat loss during the day as well as during the night.

A properly installed solar heating system for a pool saves money on heating bills and can extend the swimming season. Temperatures between 21 and 27°C (70 and 80°F) can easily be achieved during the summer in cold climates. In warmer climates, even higher temperatures can be attained.

REFERENCES

1. J. F. Kreider, *Solar Heating Design Process,* McGraw-Hill, New York, 1981.
2. A. Mertol, W. Place, T. Webster, and R. Greif, *Proc. 1980 American Section, Int. Solar Energy Society Annu. Meet., Phoenix, Ariz.*, vol. 3.1, 1980, p. 309.
3. U.S. Department of Housing and Urban Development, *Intermediate Minimum Property*

Standards for Solar Heating, No. 4930.2, Government Printing Office, Washington, D.C., 1977.
4. R. Farrington and D. Noreen, *Proc. 1980 American Section, Int. Solar Energy Society Annu. Meet., Phoenix, Ariz.,* vol. 3.1, 1980, p. 162; see also R. Farrington, L. M. Murphy, and D. Noreen, *A Comparison of Six Generic Solar Domestic Hot Water Systems,* SERI/RR-351-413, Solar Energy Research Institute, Golden, Colo., 1980.
5. F. deWinter, *How to Design and Build a Solar Swimming Pool Heater,* Copper Development Association, New York, N.Y., 1976.
6. C. Kiely, *Solar Energy,* Hamlyn, London, 1977.

7 Passive Solar Space-heating Systems

In this chapter the most common types of passive solar systems for space heating of commercial and residential buildings are described. The three types of systems most commonly used differ in the type of storage and the relationship of storage to the passive aperture. After each system and its components are described, the method of calculating the heating load for the system is given. This approach will also be used in Chap. 8 for active systems. In the final section calculation of the performance of passive systems is described in detail, using a new method called the *P* chart.

INTRODUCTION

Passive systems are fundamentally different from active space-heating systems in that most passive system components are part of the building itself. Therefore the design of passive systems takes place earlier in the architectural design process than does the design of active systems. Active systems can be added on to buildings when loads and energy requirements are identified; therefore their design is focused in the design development architectural phase. However, passive systems, with their intimate association with the building and its structure, are front-loaded in the design process. That is, most of the passive design work is included in the programming and schematic design phases, with somewhat reduced emphasis required in design development relative to that for active systems.

There is also a fundamental difference between residential and commercial passive systems. Most passive systems built in the 1970s were applied to relatively small residences and the methodology for residential passive design appears to be relatively mature. During the same period only a handful of commercial passive systems were built—that is, systems used in office buildings, institutional buildings, hospitals, and so on. The differences between residential and commercial buildings are basically twofold. First, it is much more difficult to distribute heat in a large commercial building when the heat is produced along a southern zone or in the roof area than it is in relatively small residences. Therefore temperature control by zones is more complex for the commercial application. Second, internal loads due to business machines, lighting, and people frequently dominate the energy performance of commercial buildings. In some applications very little heating is required. Therefore the heating season during which a passive system investment must be paid off in fuel saving may be relatively shorter in a commercial than in a residential building, which has substantially smaller internal loads per square foot of floor area. In this chapter the design of residential passive systems is emphasized.

DIRECT GAIN PASSIVE SYSTEMS

In this type of system the building space is directly heated by sunlight, which passes through relatively large south-facing windows. The first direct gain buildings were built in the 1940s and established the technical feasibility of the concept, although the early buildings had significant heat losses through the large south-facing glazing area during long winter nights. This difficulty has been remedied in modern buildings, as described below.

Solar energy passing through the transparent south wall is converted to heat through absorption by components within the building such as floors, walls, furniture, and, after reflection, ceilings. Direct gain systems are very effective in heat production during daytime hours; high temperatures may be reached during sunny days in spring and fall.

A requirement for direct gain passive systems that are to provide more than approximately one-third of the heat load is the use of thermal storage. If thermal storage is sized and located properly, daytime overheating during periods of reduced heating loads can be avoided and solar heat can be stored for later use during darkness or overcast periods.

Figure 7.1 shows two applications of the direct gain concept. In Fig. 7.1*a*, direct gain is used to heat a space immediately to the north of the large south glazing. Heating of spaces in zones to the north is somewhat more difficult with the direct gain approach unless a heat circulation system is included. These hybrid systems are described shortly. In the example in Fig. 7.1*a*, thermal storage is provided by a relatively massive floor slab, which is insulated to prevent heat loss to the ground beneath. The insulation used must be capable of supporting significant loads, and only one or two types of insulation appear to be suitable for this purpose.

Figure 7.1*b* shows a two-level direct gain application including both a large south glazing and a clerestory to produce heat in the upper level of the building.

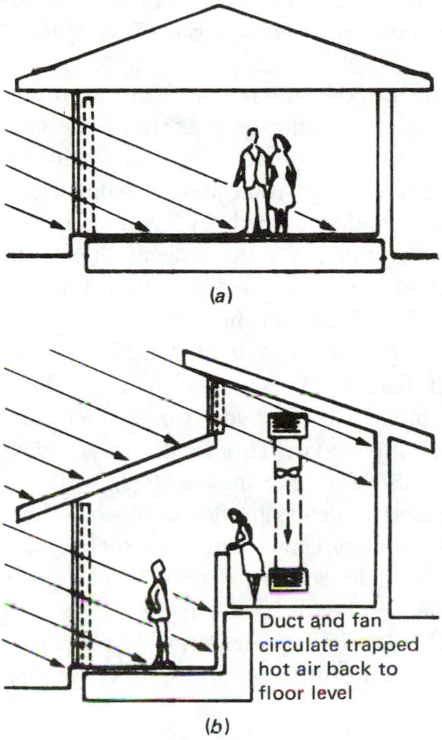

Figure 7.1 Schematic diagrams of direct gain passive heating systems. (*a*) Simple direct gain system to heat space immediately adjacent to a south-facing glazing. Movable insulation is shown by a dashed line, and heat is stored in the insulated floor slab. (*b*) Direct gain system with an additional clerestory to heat the north space and an air mover to circulate hot air from the upper level. The clerestory also provides natural daylight.

Because heat produced above the floor is not useful for heating spaces where people are, a simple air circulation system, shown schematically in Fig. 7.1*b*, is required. The air circulation system helps to prevent stratification, which can occur in passively heated spaces, particularly spaces that are relatively tall—more than one story in height. The clerestory approach is an excellent way to heat north zones of buildings by direct conversion of sunlight to heat.

Direct gain spaces are very effective in producing heat during daylight, but they are also very effective in losing heat at night. In winter, the number of hours of night is approximately twice the number of hours of daylight. Therefore there will be significant heat loss through large south-facing glazings when no solar gains occur. One method of eliminating heat loss in direct gain systems is to install insulation, either outside or immediately behind the south-facing aperture. The south-facing aperture is almost invariably constructed of double-pane or triple-pane glass. The added nighttime insulation can be either a movable arrangement of rigid insulation panels finished to serve as an interior surface of the building or a roll-up type of flexible insulation. The latter can be rolled up near the ceiling during the day and does not consume expensive floor space in the building.

The thermal resistance or R value of movable insulation should be at least 5 and preferably closer to 10, subject to an economic analysis of the cost-effectiveness of the increased R value. The R value of double-glazed windows is approximately 2. Therefore addition of movable insulation at the recommended R-8 level can reduce heat losses through the window space by a factor of 5.

In addition to clerestories or south glazings, a third type of direct gain system has been used in some applications—the skylight. Skylights, unfortunately, do not have optimum orientation for the best heat gain in winter and, in fact, present significant cooling difficulties when placed on roofs of typical residential pitch. Since the sun is much higher in the sky in summer than in winter, skylights placed on roofs with a shallow pitch will intercept two to three times as much sunlight in summer as they do in winter. This results in an added cooling load in summer and probably very little net heating benefit in winter. If skylights must be used for direct gain passive systems, a reflective opaque shield should be provided to completely block the skylight aperture in summer and reject sunlight directly back through the skylight without permitting heat to be produced. The investment in skylights must be relatively low if this approach to passive heating is to be economically viable. The better orientation of clerestories and vertical south glazings makes them significantly more cost effective than skylights for most applications.

Design Information for Direct Gain Systems

In a residential context, for direct gain to heat an adjacent space, the depth of that space should be no more than $2\frac{1}{2}$ times the height of the south-facing window. This rule is based on the geometry of the sun during winter [1]. The directly illuminated floor also provides thermal storage. South-facing vertical apertures must be confined to azimuth angle ranges between southeast and southwest, with approximately due south providing the optimum performance on an overall annual basis.

The area of the double-glazed aperture for direct gain heating is approximately one-third to one-fourth of the floor area of the immediately adjacent space. Night insulation of some sort is required, as described above. If movable insulation is unacceptable for a particular application, infrared heat mirrors, which were first marketed in 1980, may be adequate. The overall U value for a double-glazed window with an intermediate heat mirror is approximately 0.2, corresponding to an R value of 5. This is slightly below the recommendation above, but reduced maintenance costs and operating complexity may offset the added heat loss value.

One troublesome feature of direct gain systems is that they almost invariably produce objectionable glare and overheating on a sunny day. This is particularly a problem in commercial office buildings, where there are already high internal gains in the daytime. The designer of passive systems must coordinate material selections with the tenant finish designer to minimize glare problems. In choosing materials that will be exposed to sunlight in direct gain spaces one must consider the possibility of high solar intensity levels and the tendency of many colors to fade in such an environment. In addition, plastic materials may be degraded by extended exposure to ultraviolet rays. Nearly all fibers react to a greater or lesser extent with sunlight. The rate of deterioration depends on the fiber content, yarn, and fabric construction, the finish applied to the textile, and the type of dyeing or printing process employed. In addition to fading, fabric exposed to sunlight may also deteriorate in physical strength. With prolonged exposure to sunlight, fabric can even disintegrate and later be destroyed during an ordinary cleaning process. The selection of fabrics for upholstery, carpets, and draperies is normally within the purview of the interior decorator; however, inputs from the solar designer should be considered if large direct gain apertures are involved. Durability of fabrics in the presence of sunlight can be increased by the use of ultraviolet screens and inhibitors. Artificial fabrics are less prone to solar degradation than are natural ones such as wool.

In addition to material properties, carpets are of interest from a thermal point of view if they act as insulators. In this case, the coupling between solar heat and any storage mass in the floor will be nonexistent and the extra cost of a massive floor will be wasted. The insulating characteristics of textiles derive from dead air trapped among fibers. The amount of entrapped air depends on the diameter, shape, and length of the fiber, the amount of twist, and the fabric construction. Fabrics that provide maximum insulation, which should be avoided, have irregularly shaped fibers with low twist. Pile surfaces or open weaves with expanded foam mats beneath provide so much insulation that almost no heat can be stored in the floor. In summary, the use of carpets on floors intended to provide direct gain storage should be examined carefully. An attractive carpet selected for interior use will probably present a greater insulating barrier than can be tolerated if thermal storage is to operate properly.

Selection of drapery materials used to cover direct gain windows should also be considered by the solar designer. If draperies are to serve as solar barriers, the reflectance of the drapery liner should be considered. If its appearance is acceptable, even an aluminized backing can be used. Other designs include multilayer reflective insulation materials. The thermal durability of these materials seems to be good and they may be considered for use as movable insulation. Both direct gain and indirect gain systems employ movable insulation, and it will be described in more detail below.

Since direct gain solar systems are characterized by large day-to-night temperature swings, inclusion of storage is mandatory. Without storage, temperatures may rise to 90°F on a winter day and fall to 50°F or so at night if the backup system is not energized. To be most effective, thermal storage should be directly illuminated. Therefore floors and walls immediately north of the glazing aperture are best. Dark surface colors with absorptances above 75 percent improve the effectiveness of storage. Nonilluminated, nonstorage surfaces should be light in color to reflect sunlight onto the darker storage elements elsewhere in the space.

The amount of storage recommended for direct gain systems [2] is as follows. Interior walls and floors of solid masonry should be about 4 in thick. If water storage is used, approximately 2 to 3 ft^3 of water is required in an interior wall for each square foot of direct gain glazing. The larger the amount of storage, the smaller the day-to-night temperature excursion. Of course, there is a limit to the cost-effectiveness of storage, and excess storage will not pay for itself in improved system performance, although comfort levels may be marginally better. Insulation used in direct gain system storage should be exterior to all mass storage components.

Direct Gain Aperture Design Considerations

Direct gain apertures are typically vertical and south-facing and consist of one or two glass panes, the latter being by far the more common. Selection of glass to be used should be based on the physical load on the glass as well as its solar transmittance and absorptance and thermal properties. It is most desirable, within the constraints of cost-effectiveness, to have the highest possible transmission and the lowest possible absorption. This can be accomplished by using low-iron glass, although this material is frequently not available in the large sheets required for direct gain glazings.

Another important consideration in the selection of glazing material is structural strength. Large glazing surfaces present tempting targets for vandals and tempered glass is therefore the material of choice. Further, this type of glass is also essential for the inner glazing of double-glazed passive apertures, which can reach temperatures in excess of 200°F in summer if overhangs are not used on the building. This elevated temperature will be reached if movable insulation is put in place during the day to keep heat from entering the space inside the direct gain glazing. In this case, untempered glass is certain to fracture eventually due to the thermal stresses that occur at shadow lines between the illuminated portion of the glazing and the shaded areas. The designer should also ensure that the edges of all glass panes have been smoothly polished prior to tempering. Nicks in the edges of tempered glass act as stress raisers, and fracture of the glass is certain to result. Thermal stresses on either glazing pane are most likely to cause breakage during part-load periods of the year, when the glazing is partially illuminated and partially in the shadow of the overhang, as shown in Fig. 7.1.

Glazing must be well sealed to the building to avoid infiltration losses. The large perimeters of glass panels used in direct gain systems present an opportunity for significant air infiltration. The sealant material should be selected with somewhat more attention than is needed for conventional fenestration. Seals will be subject to a combination of elevated temperatures and high levels of ultraviolet radiation, which can result in relatively rapid chemical reaction of the sealant material and its deterioration. If operable vents are included in the direct gain glazing for summer heat rejection (see below), the seals on the movable fenestration should also be selected for wide ranges of operating temperature during the course of the year.

Adjacent panes of glass are frequently sealed with an overlay cap strip for appearance as well as for physical support of the sealant material. Like other components of the glazing systems, these cap strips will be subject to significant expansion and contraction during the thermal cycles that occur over the years. This factor must be included in the design; for example, clearance must be provided at the end of each strip for expansion in summer, and holes drilled to mount the cap strip with screws should be somewhat larger than the typical clearance hole for the screws. In some applications where standard holes were used, the expansion and contraction actually sheared the heads from the cap strip mounting screws.

Location of windows is another important feature of passive direct gain design. To provide daylight, it is desirable to have the windows as high in the space as possible. However, for heating it is desirable to have the windows as low as possible to minimize stratification. Therefore the relative magnitude of direct gain aperture versus skylight aperture should be considered by the daylighting designer and the passive heating designer together. Daylighting is not the subject of this book, but it should be integrated with direct gain design as suggested above. The design of daylighting systems is an ancient art that has been lost in the twentieth century. The principles of daylighting, including rules for sizing and locating apertures to achieve a given footcandle level, are being rediscovered, but simple guidelines for use at the schematic or design development level do not yet exist.

One of the major difficulties of daylighting design is predicting the amount of energy that can be displaced by the use of daylighting systems under various solar angle and solar intensity conditions. At present, daylighting design appears to be strictly a science practiced by a few experts using very expensive computer models. Some daylighting consultants suggest that a scale model be used in actual sunlight conditions to determine the efficacy of daylighting apertures. Since models are typically built by architectural designers during the design process, with additional attention to interior finishes, the same models could be used for preliminary daylighting studies, although the results obtained in this way will not necessarily be highly accurate.

This discussion of direct gain aperture glazings also applies to the next type of passive solar heating system described—the indirect gain system.

INDIRECT GAIN PASSIVE SYSTEMS

Indirect gain systems convert sunlight to heat at an intermediate location and then release the heat to the space by means of natural radiation and convection processes. Figure 7.2 shows two types of indirect gain systems. Figure 7.2*a* represents what is known in architecture as the Trombe design, although it was originally developed in the nineteenth century by E. S. Morse [3], who did

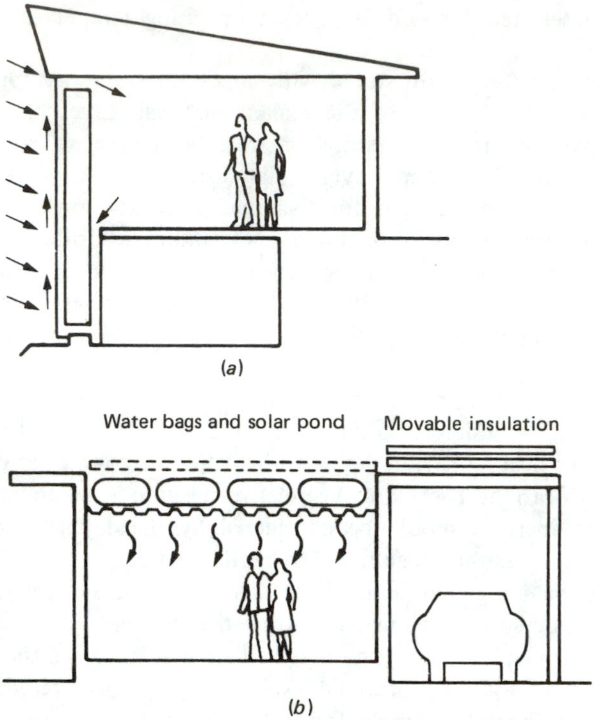

Figure 7.2 Indirect gain passive systems. (*a*) Thermal storage wall (TSW) system, using thermal circulation airflow for delivery during the day. (*b*) Horizontal roof pond in indirect system; movable insulation is shown by dashed lines.

not include storage directly at the absorbing surface. The concept of indirectly heating a space by means of intermediate conversion of sunlight to heat at a passive collector surface has also been used for centuries in the Middle East.

The system shown schematically in Fig. 7.2*a* includes an air circulation loop for space heating. When an air circulation loop is used, heat is transferred to the space by airflow in addition to radiation and convection from the inner surface of the storage wall. Therefore heat is transferred not only through the wall but from the outer surface of the wall directly to the space. Airflow takes place by virtue of the density difference between the air in the living space and the warmer air in the gap between the south glazing and the storage wall. Typically 50 percent of the solar heat produced at the south wall is transferred into the air circulation loop on a sunny day. This heat is used mostly for daytime heating. The remaining 50 percent slowly diffuses through the thermal storage wall (TSW) for night heating.

Addition of air circulation vents to the basic indirect gain system is useful in many locations as a balancing device between daytime and nighttime load periods. Heat produced and delivered by the airstream is useful during the day since it is immediately delivered to the space, whereas heat diffusing through the storage wall is available 8 to 12 hours after it is produced at the south surface of the wall. The selection of wall materials is important in determining the rate of heat migration through the wall at night. Likewise, wall thickness is important if the proper phase lag between heat production and delivery is to be achieved.

The second indirect gain system, shown in Fig. 7.2*b*, is based on an idea long used in the American Southwest for building conditioning. In this so-called roof pond system, a layer of water about 1 ft deep is placed over a metal roof painted a dark color. As sunlight passes through the water it is partially absorbed within the water, with the balance being converted at the black roof surface (or the black lower surface of the water container). The heat produced can be stored in the pond until needed or can be conducted and radiated to the space below. To eliminate heat loss from a roof pond during the night, movable insulation is placed above it.

The discussion of solar angles in Chap. 2 indicated that a horizontal surface is far from optimum for a solar heating system since the sun is very low in the sky in winter, when heat is required. Therefore a horizontal roof pond is not oriented at the best angle for solar collection in northern latitudes, where space-heating loads are significant. In addition, in freezing climates water is not a suitable storage medium, and some type of antifreeze protection would be required at significantly increased cost. The structure to support the large storage mass is also rather expensive. Since the emphasis in this chapter is on passive space-heating systems that are practical in many areas of the country, the roof pond system will not be considered further because of these drawbacks.

Design of Storage for Indirect Systems

In the last section we described the design of glazing for direct gain systems. Precisely the same considerations are involved in the design of glazing for indirect gain systems. The difference between the two systems is the intervention of storage between the glazing and the space to be heated. Design considerations for this storage mass are the subject of this section. The most common materials used for storage in passive systems are masonry elements or water in dark-colored containers. Storage requirements per unit collector area for indirect gain systems are greater than for active systems by approximately a factor of 2. As discussed in Chap. 8, the amount of thermal storage required for active systems is roughly 15 $\text{Btu}/(^\circ\text{F}\cdot\text{ft}_c^2)$. Therefore for passive systems the amount of storage is approximately 30 to 40 $\text{Btu}/(^\circ\text{F}\cdot\text{ft}_c^2)$. A larger amount is needed for passive systems since the flow of heat into storage is accomplished with smaller driving forces than are present in active systems.

Storage in indirect gain systems is located in intimate contact with the collector surfaces and in many cases is a part of the collector itself. For example, in the TSW concept shown in Fig. 7.2*a*, the sun-facing side of storage is the collector surface. Likewise, if water is used instead of masonry, the water container's dark outer surface serves as the collector.

In addition to thermal requirements, the color of passive storage elements is dictated by appearance. A relatively dark color, of course, is essential for high solar absorptance, but the color need not be black. A typical black paint applied to a thermal storage wall may have a solar absorptance of 90 percent, but a more attractive color may reduce this value by only 10 percent or so. In the interest of architectural compatibility, slightly reduced thermal performance would be acceptable to most building owners. Table 7.1 lists the absorptances of most materials considered for use in passive storage walls. A value above 60 percent should be used if thermal performance is to be maintained.

Table 7.1 Solar Absorptance of Passive Wall Materials*

Material	Absorptance
Optical flat black paint	0.98
Flat black paint	0.95
Black lacquer	0.92
Dark gray paint	0.91
Black concrete	0.91
Dark blue lacquer	0.91
Black oil paint	0.90
Stafford blue bricks	0.89
Dark olive drab paint	0.89
Dark brown paint	0.88
Dark blue-gray paint	0.88
Azure blue or dark green lacquer	0.88
Brown concrete	0.85
Medium brown paint	0.84
Medium light brown paint	0.80
Brown or green lacquer	0.79
Medium rust paint	0.78
Light gray oil paint	0.75
Red oil paint	0.74
Red bricks	0.70
Uncolored concrete	0.65
Moderately light buff bricks	0.60
Medium dull green paint	0.59
Medium orange paint	0.58
Medium yellow paint	0.57
Medium blue paint	0.51
Medium kelly green paint	0.51
Light green paint	0.47†
White semigloss paint	0.30†

*These values are meant to serve as a guide only. They will be affected by variations in texture, tone, overcoats, pigments, binders, etc.

†Materials with absorptance values below 0.50 are of little use for passive wall outer surfaces.

Sources: U.S. Department of Energy [2]; G. G. Gubareff et al. [4]; S. Moore, Los Alamos Scientific Laboratory, unpublished data.

Table 7.2 Thermal Properties of Various Passive Storage Wall Materials

			Material					
Property	Symbol	Units	Gravel and sand concrete	Limestone rock	Brick	Wood (pine)	Dry sand	Adobe
Density	ρ	lb/ft^3	144	153	112	31	95	120
Specific heat	c	Btu/(lb·°F)	0.19	0.22	0.22	0.67	0.19	0.20
Thermal conductivity	k	(Btu·ft)/(ft^2·°F·h)	1.05	0.54	0.40	0.097	0.19	0.332

Source: From *Solar Energy Handbook* by Jan F. Kreider and Frank Kreith [5].

Storage Size

The thickness of the TSW used in indirect gain systems can be calculated from the specified thermal capacitance noted above. To do this, the densities and specific heats of the materials must be known. These are given for common materials in Table 7.2. Wood is also included in the table to show that its storage capacity is much less than that of materials more commonly used for storage. Thermal conductivities are also shown in Table 7.2. A number of computer studies have shown that wall thickness should be in the range specified below, but it is not a critical variable from the point of view of thermal performance. For masonry, an 8-in-thick wall will be near the performance maximum. Of course, the cost of wall added beyond 8 in must be compared with any increase in thermal performance derived by increasing the amount of storage. If increased storage is not paid for in increased thermal performance, its addition is questionable.

Another determinant of wall thickness in TSW systems is comfort. Although an 8-in and a 10-in wall may provide almost identical heating energy per year, the 10-in wall will have smaller temperature swings. Thinner walls may feel too warm on sunny, mild days and too cool on overcast, cold days. Therefore the hard-to-quantify design consideration of comfort may push the design toward thicker walls than would be dictated solely on the basis of thermal performance and energy saving economics.

The designer must also specify the finish on interior and exterior surfaces of the TSW. It is not essential that the surface of masonry walls be perfectly flat if a texture is desired for improved appearance. Since the airflow between the glazing and the TSW is laminar, the pressure drop in this cavity is not sensitive to the microcharacteristics of the surface, as is the case in high-velocity turbulent flows. Therefore the wall can be given texture with little effect on performance.

The interior surface of the wall may also be textured as desired by the interior designer. However the interior finish of the TSW must not present a thermal barrier to heat transfer from storage into the heated space. At most, only paint should be applied; no insulation, paneling, draperies, or large pictures should be attached to the inner surface of the storage wall. Further, no large furniture should be placed where it would obstruct natural air circulation or radiation from the wall inner surface to the space. Of course, baseboard heaters and heating system vent outlets should not be placed near the thermal storage wall.

Thermal storage for indirect gain systems will present a large physical load to be supported by the building. The design of the structure and foundation for this support is the responsibility of the structural engineer, who must be informed in advance of the anticipated mass location and weight of storage. This information is required prior to foundation design. Of course, soil properties determine the physical size of the foundation. The extra cost of increased foundations must be considered in the economic feasibility analysis of passive solar heating systems.

Wall Vent Size

In most locations in the United States, TSW systems can benefit from the inclusion of thermal circulation vents as described above. The sizing of these vents is determined by the solar load fraction, as described in [2, 6]. For relatively low solar fractions the vents must be relatively larger to permit rapid daytime heating of the space. For relatively larger solar fractions the vents should be smaller to force more heat into storage and to control daytime overheating. Figure 7.3 shows the recommended size of TSW vents for the typical range of solar heating fractions encountered. The solar heating fraction itself is based on economic optimization, as described in Chap. 5. When the optimum passive solar fraction is known, the vent size can be read directly from Fig. 7.3.

Thermal circulation vents require backdraft dampers for control of

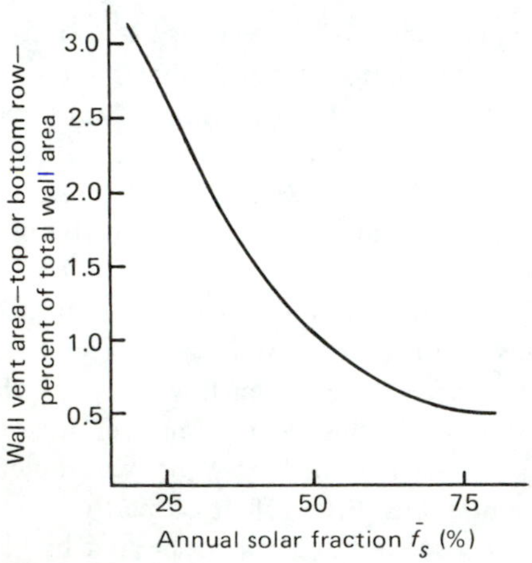

Figure 7.3 Thermal storage wall vent size related to annual solar heating fraction.

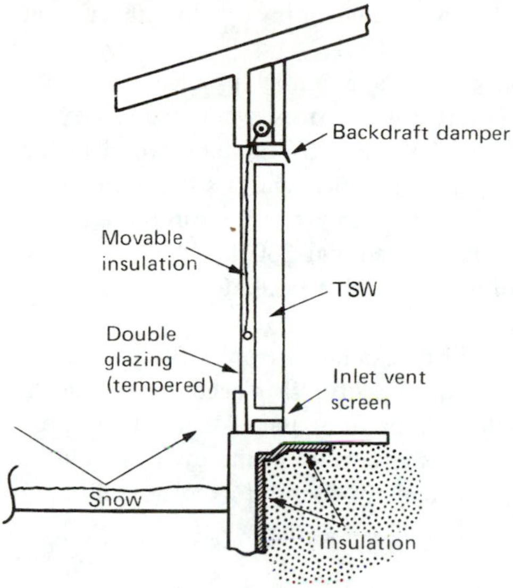

Figure 7.4 TSW cross section, showing thermal circulation vents, backdraft damper, inlet screen, and movable insulation (not to scale).

potential backflow at night or during overcast periods. Backdraft dampers are best placed at the inner surface of the upper vent, as shown in Fig. 7.4. Here the upper vent is blocked by the damper when any tendency for backflow occurs. Of course, the same effect could be achieved by use of a damper at the inner surface of the lower vent, but repairing a damper in this location would be very difficult.

The inlet to the lower thermal circulation vent should be several inches above the floor to reduce dust entry and to make sure that baseboard trim or other obstructions do not block the lower vent. A coarse screen should be placed across the lower vent opening to prevent any materials from being placed in the duct or the wall cavity. A screen with approximately a $\frac{1}{4}$-in mesh is suitable for this purpose.

The space between the sun-facing surface of a TSW and the inner glazing of passive apertures is normally 3 to 4 in wide. This dimension is not critical and can be adjusted to accommodate specific details of movable insulation in windows. However, the window-to-wall gap should not be made excessively large, since a thermal circulation loop within a large gap could be formed, reducing airflow to the heated space.

The design of indirect gain passive heating systems should consider the potential for fire. The designer will try to optimize the movement of air within the wall cavity and other circulation paths in the building so that stratification is minimized and passive heat delivery is simultaneously maximized. However, the passive airflow paths become excellent paths for fire propagation as well, since the same physical principles control both. Flames can propagate with unbelievable speed along the same path that air would normally flow through in passive systems.

Spontaneous combustion of wood can be a potential problem in passive solar buildings. This is particularly true in direct gain spaces, where wood may be exposed to sunlight for very long periods. Also, in TSW systems wood in the wall-to-glazing cavity may become quite hot in summer, and a self-igniting form of carbon called pyrophoric carbon may be formed. Pyrophoric carbon has been known to ignite spontaneously at temperatures around 250°F, which could be achieved on a sunny day in spring or fall when movable insulation is in its closed position to block solar gain.

In addition, the insulation itself should be examined from a fire prevention point of view. Insulation located in the wall cavity, if placed in the down position in summer, can become very hot. Therefore, the likelihood of spontaneous combustion of the insulation as well as any wood in the cavity should be considered. An overhang or shutter is particularly useful for solar control in summer to alleviate the problem of overheating in the cavity.

Passive System Insulation

Performance of both direct and indirect gain passive systems benefits from the use of insulation to retain solar heat within the building during nighttime and overcast periods. Without insulation the performance of these passive systems is severely decreased, and in many cases they become uneconomical owing to large heat losses in winter. Therefore most passive systems incorporate some type of heat loss control. The following discussion of insulation applies to both direct and indirect gain passive systems.

A number of designs for movable insulation have been marketed. Insulation for use in a TSW system must be removable from the TSW cavity. This can be accomplished by using either flexible roll-up insulation of the type shown in Fig. 7.5 or an insulated, overhead garage door of some kind. Although the garage door idea is attractive in terms of mass production potential, most designs available in 1980 did not include sufficient insulation material–a problem that is easily solved in principle. An insulation value between R-5 and R-10 should be specified and ultimately justified by cost-effectiveness.

As noted in the section on direct gain systems, heat mirrors placed between double glazing or on one surface of the glazing can serve as nighttime insulation. Heat mirrors are selective reflectors that cannot be seen by the naked eye. However, in the infrared region where reradiation of heat from building interiors occurs, the heat mirror is opaque and reflects all long-wavelength radiation back to the building interior. Heat mirrors have been used for many years in Europe and are becoming available in the United States; their cost is approximately $\$3/ft^2$ above that for a double-glazed window.

Passive apertures can also be insulated outside rather than inside the wall glazings. This less common approach has the advantage that the insulation is more accessible if servicing is required. One of the practical difficulties is the mechanism for moving the insulation itself. If insulation positioning is not

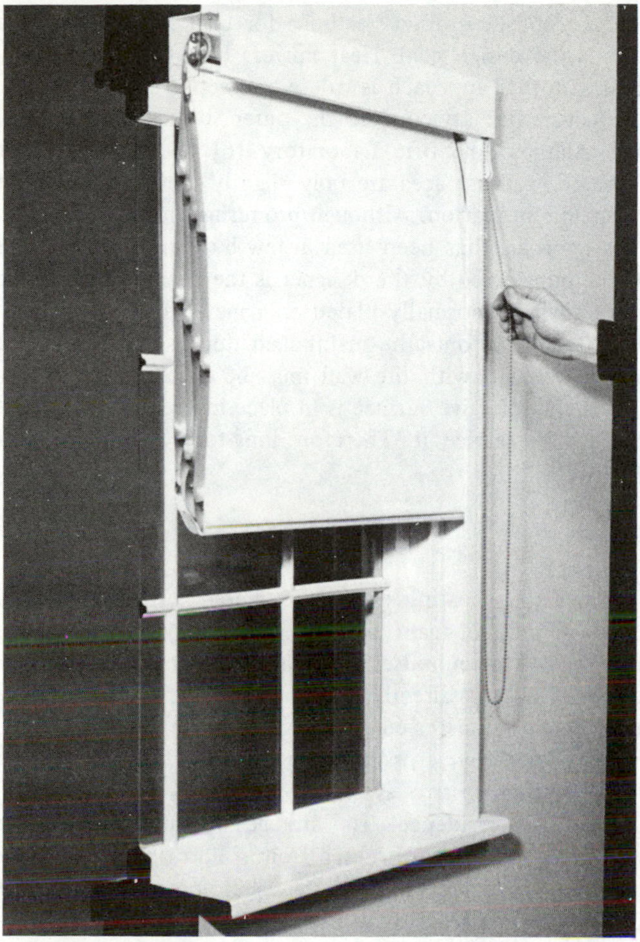

Figure 7.5 Example of movable insulation. (*Courtesy of Insulating Shade Co.*)

automated, large accumulations of snow may prevent its being closed after a snowfall. Likewise, insulation should be in the closed position during dust or wind storms to minimize damage. The long-term durability of exterior glazing insulation is not known.

Servicing of movable insulation in a TSW cavity should be considered carefully by the designer. Adequate space must be provided for access to the insulation retraction mechanism, as well as to any electric motors and switches involved in the insulation control. As sketched in Fig. 7.4, an opening can be provided above the upper TSW vent to permit removal of the insulation roller. Of course, the insulation and its roller are the full width of the thermal storage wall, so the removable access panel must also be a full width, not interrupted by any support.

Movable insulation is the only moving part in most passive heating systems.

It is therefore of interest to consider other methods for heat loss control if a completely passive system is the design goal. Heat mirrors have been mentioned above as one such method. Another approach is to use a selective surface, similar to that used in active collectors but attached to the outer surface of the TSW. Computer studies at Los Alamos Scientific Laboratory [6] showed that heat losses from selectively treated wall surfaces are only slightly greater than those from a wall with R-9 nighttime insulation. Although promising and feasible from a technical viewpoint, this approach has been used in few buildings to date. The fundamental difficulty to be considered by the designer is the proper and reliable attachment of the selective surface (normally plated on a metal foil) to the wall. Physical integrity of the attachment must be maintained during the lifetime of the building, and its thermal contact with the wall must be excellent for proper heat transfer to occur. Once the selective surface is in place in the TSW cavity, it will be impossible to replace or service it. Therefore long-term durability is of paramount importance.

GREENHOUSES

The third type of passive solar heating system considered is the greenhouse. It combines the features of indirect gain and direct gain systems with an intervening thermal storage wall. A greenhouse is shown schematically in Fig. 7.6. The greenhouse space itself is heated directly by sunlight and functions as the direct gain zone. However, heat is transferred to the space to be heated through the thermal storage wall, shown by cross-hatching in Fig. 7.6. Active air movement may be added to this passive system to improve air distribution between the greenhouse and the residence. The storage wall common to the greenhouse and the residence is usually constructed from solid masonry or water containers. Movable insulation helps prevent excessive heat loss at night but is rarely used. However, if the greenhouse is to be used for horticulture, insulation is required to avoid fréezing of plants.

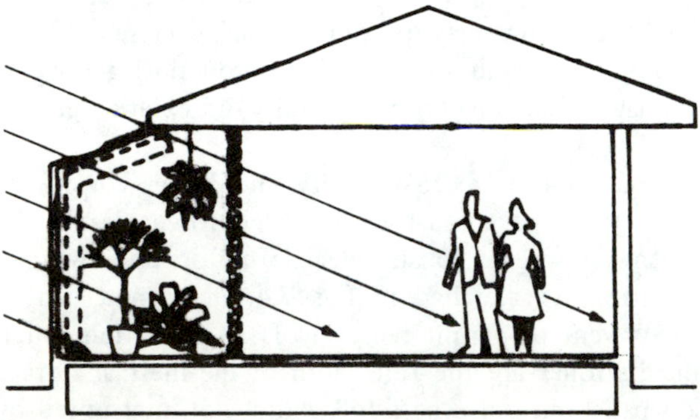

Figure 7.6 Greenhouse used for passive heating through a TSW, shown by cross-hatching. Movable insulation is shown by dashed lines.

Greenhouses are basically enlarged thermal storage wall cavities of the type used in a TSW system and placed to the south of the storage element. They have relatively larger amounts of glazing per square foot of wall area than do TSW systems. Therefore heat losses will be relatively greater and energy delivered to the building space proportionately less. Computer studies [2] suggest that approximately 50 percent more greenhouse aperture area is required to heat the same space for the same percent of the time as a TSW.

The amount of storage used in a greenhouse is roughly the same as for the TSW system described earlier. However, if the greenhouse is not of the lean-to type but is detached, the amount of storage needed may be larger. Detailed sizing rules do not yet exist, and the amount of storage in a freestanding configuration may have to be determined by experiment. Storage is added to the building until the day-to-night temperature swing is within the range that can be tolerated by the plant species in the greenhouse.

The design of greenhouses cannot be carried out as precisely as that of direct and indirect gain systems. Few greenhouses have been accurately monitored or studied by computer simulation. In the absence of detailed design rules, the best approach seems to be use of design guidelines for systems described earlier in this chapter.

Solar gain controls must be provided in summer to prevent overheating, and a method of heat rejection, usually employing high vents or fans, must be included in any greenhouse design. Solar gains in winter are larger on south faces than on east or west surfaces; therefore the south wall should be largest and the east and west walls smaller for best performance. Opaque end walls are often used. East and west walls are also sources of unwanted solar gain in summer.

OTHER PASSIVE SYSTEM DESIGN INFORMATION

Hybrid Systems

The various types of passive systems described in the preceding sections are frequently used in combination. Also, active features may be applied to a passive system to improve control and heat distribution. These various combined system options are called hybrid systems. Design guidelines for hybrid systems are not well developed, although a few rules of thumb have appeared. For example, in combining direct gain and TSW passive systems it is recommended [6] that approximately one-third of the aperture be direct gain and the remaining two-thirds indirect gain.

Combination of active and passive systems has rarely been treated in the solar design literature, and such systems may have to be designed on a case-by-case basis. Active features included in some passive systems built to date have consisted of air movers to reducing stratification, as indicated schematically in Fig. 7.1*b*. For spaces higher than one story, stratification may be a problem even in indirect gain systems, and a fan and heat recovery system to remove hot air from the highest zone of the space may be useful. Heat removed from upper areas can be ducted to the inlet of a forced-air heating system for redistribution or can simply be directed to the floor by a low-velocity fan and duct system. Both approaches have been used.

Another class of hybrid active and passive systems consists of full active solar heating systems used on passively heated buildings. The passive system is used for base load and reduces the load to be met by the active system. In most good solar building designs, as many passive features are included as feasible to reduce the net load. Since the size of an active system is directly proportional to the load to be met, any reductions in the load have a major beneficial impact on the cost of the active system. The best mix of active and passive features must be determined on an individual design basis, since guidelines and general rules have not been established for this class of hybrid systems. The methods of economic optimization described in Chap. 5 can be used to determine the best mix of active, passive, and energy conservation features in a building. Interaction of these three systems is relatively complex, and it appears that the use of computer models is essential to determine the best mix of these three building energy features.

Passive Controls and Heat Rejection

Controls of passive systems are relatively simple compared to those involved in active systems, described in the next chapter. The only active control device in a passive system is that which controls movable insulation used for night heat loss control. The control can be either automatic or manual. If it is automatic, a simple sun sensor that places the movable insulation in its down position if sunlight is below a certain level is all that is required. A typical threshold insolation value is 50 to 75 $Btu/(h \cdot ft^2)$. The threshold level should be adjustable in the controller to permit fine tuning of the system after installation. The sun sensor is placed in an unshaded region of the south aperture. The control actuator for this system is a small motor, which is used to roll and unroll the movable insulation or operate whatever mechanism is used for this purpose. Movable insulation must seal repeatably to the south-facing glazing to avoid reverse circulation between the glazing and the insulation. Any airflow in this area will seriously diminish the effectiveness of movable insulation.

In hybrid systems using active air movers for improved temperature control, a method of fan control is required. For example, if a fan is used to extract heat from a TSW cavity and deliver it to north zone of a building, it can be actuated by space-heating thermostats. If a call for heat exists in a north zone, the fan can be actuated and a motorized damper opened to permit airflow from the cavity to the north zone. A secondary thermostat placed within the TSW cavity can make the decision as to whether heat is available for use in the north zone. If heat is not available, this thermostat will remain open and the motorized damper and fan will not operate. Only standard components are required for this simple control system. Passive controls present in passive heating systems include backdraft dampers for blocking reverse circulation in TSW systems at night. Likewise, most passive TSW and direct gain systems use overhangs to reduce solar gain into south-facing apertures in summer. This is part of the heat rejection system that should be included in any passive system.

Rejection of heat during periods of the year when heating loads are low is a feature of passive systems that must be included in any proper design. The principal method of heat control for vertical surfaces is the use of shutters, awnings, or overhangs, as shown in Fig. 7.7. Properly designed overhangs can completely cut off the beam component of radiation to a direct gain or TSW aperture in summer. The sizing of the overhang is determined by the period for which shading is desired. As indicated in Chap. 2, the sunpath diagram provides a simple and accurate method for sizing overhangs. The sunpath diagram in Fig. 7.7*b* shows that the 60° profile angle corresponds to blockage of beam radiation from mid-April through the end of August. A profile angle of 60° is used only for example; the actual angle can be adjusted to suit local climatic and building load requirements. For more details on overhang design using hour-by-hour and month-by-month calculations, see [7].

Inclusion of well-designed overhangs and vertical fins is essential to ensure the proper function of passive heating systems. Without an overhang or awning of some type, the amount of heat gain through south-facing passive walls can be substantial in summer. In one early project built without an overhang, the TSW was in the range 85 to 90°F throughout the summer, with the result that comfort levels within the space were very poor and increased air conditioning was required. On another project in a similar location with properly designed overhangs and movable insulation, the interior space was consistently 15° below the outdoor ambient temperature and no air conditioning was used.

The example shown in Fig. 7.7 can be refined to improve solar gain control. The heavy line in Fig. 7.7*b* representing the overhang applies strictly for a very long overhang. If the overhang is of finite width, the extreme right- and left-hand edges of the shadow map would not exist and early morning and late afternoon solar gain would be experienced. To ensure shading during these periods with a finite width overhang, vertical fins may be added to either side of the window. The construction of the fin effect on the sunpath diagram in Fig. 7.7*b* is shown by dashed lines, which form a triangular zone to the bottom right and left of the solid line that applies for the overhang only. If the overhang shown in Fig. 7.7*a* were equipped with vertical fins having a 60° azimuth angle cutoff, the shadow map would be as shown by the cross-hatching in Fig. 7.7*b*. Fins do have a small negative effect, since they reduce winter gains through the south-facing aperture early in the morning and late in the afternoon. However, this is of little consequence, since early morning and late afternoon winter sunlight contains very little usable heat.

As is the case for both solar gain control designs, the size of overhangs should be determined by economic analysis. Overhangs and fins are expensive, and cost should be considered as well as the benefits of reduced heat gain achieved by their use.

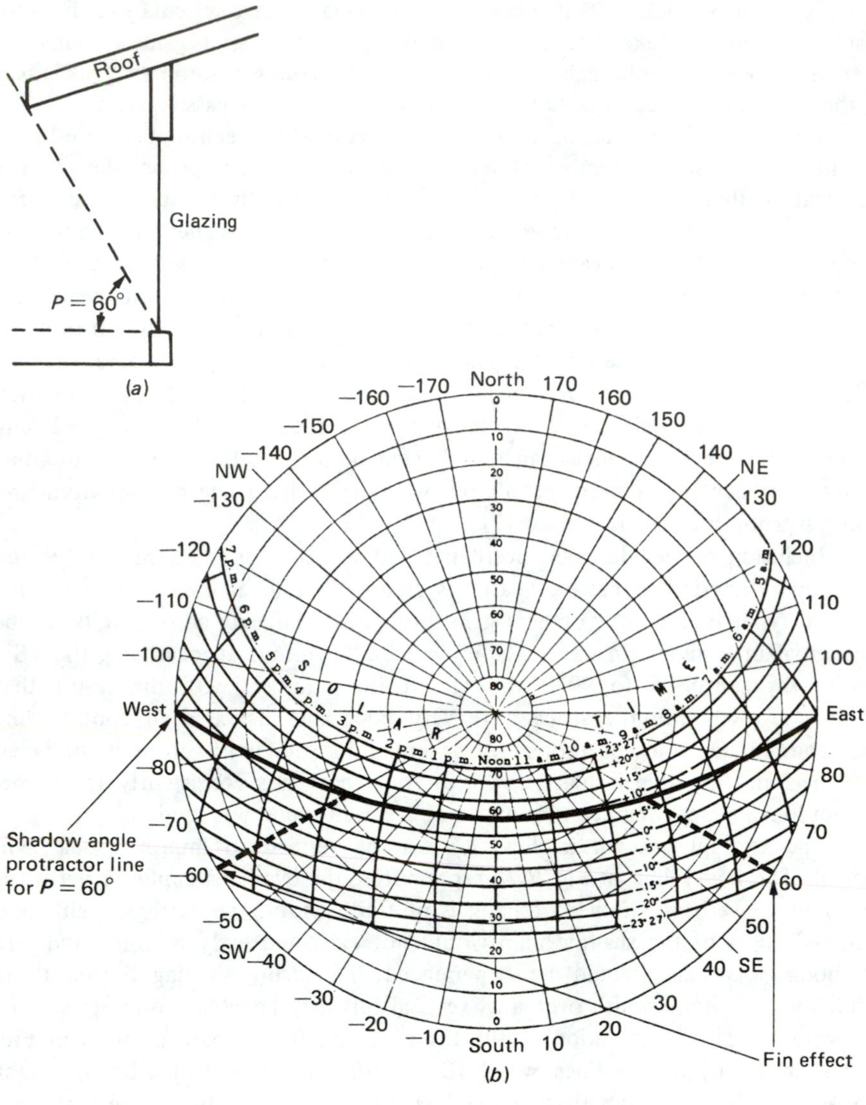

Figure 7.7 Example roof overhang design, using a sunpath diagram. (*a*) Required profile angle shown is 60°. (*b*) A 60° overhang is indicated by a solid line on the sunpath diagram; addition of 60° azimuth vertical fins to right and left edges of glazing is indicated by dashed lines in the sunpath diagram.

Overheating is particularly troublesome in greenhouses since they contain some upward-facing glazing elements. With high summer sun angles, these glazing components intercept large amounts of solar radiation when it is least needed. Therefore any greenhouse design must include an effective method for providing high rates of air change for heat rejection.

SPACE-HEATING LOADS

Passive solar systems whose function has been described in this chapter are used to offset heat losses from buildings. The precise sizing of passive apertures and storage is based on the amount of heat to be provided. Therefore the first step in designing a passive system is calculation of the heating requirement of the building to be served. This section briefly describes the origin of heat losses from buildings and a simple method of calculation. Standard procedures for detailed heat load calculations are discussed in the ASHRAE *Handbook of Fundamentals* [8].

Heat is lost from buildings by several mechanisms. The first loss mechanism is that associated with transmission through walls, windows, doors, and roofs of buildings. This heat loss is usually taken to be proportional to the temperature difference between the interior of the building and the environment. The constant of proportionality is the U value of the wall or window multiplied by its area. Thermal conductivities for walls, windows, and other building elements are tabulated in Apps. 7 and 8. Table 7.3 gives the analytical expressions for calculating transmission heat losses as well as other losses described below.

The second source of heat loss from a building is displacement of warm air in the building by cool air from the environment. This air exchange may occur by design or may be unintended; when by design it is called ventilation and when unintended, infiltration. Calculation of ventilation heat loads is quite

Table 7.3 Unit Heat Base Loss [Btu/(h·°F) or W/°C] Equations for Space Heating*

Type of loss	Expression	Parameters used
Transmission	$\Sigma_i U_i A_i$	U_i = wall, window, door, etc. U value, Btu/(h·ft²·°F), W/(m²·°C)† A_i = area, ft², m²
Air change		
Ventilation	$\rho(\dot{V})c_p \times 60$	ρ = density, lb/ft³, kg/m³ ‡ $\dot{V}$ = volumetric flow rate, ft³/h, m³/h c_p = air specific heat, Btu/(lb·°F), kJ/(kg·°C)
Infiltration	$\rho(nV)c_p$	n = air changes per hour V = building volume, ft³
Floor slab		
Exposed	$F_e P_e$	F_e = linear loss coefficient [8, 9], Btu/(h·ft·°F), W/(m·°C) P_e = edge perimeter, ft, m
Buried	U_f^*	U_f^* = buried slab coefficient [8, 9], Btu/(h·ft²), W/m²; varies with ground water temperature but not air temperature

*Multiply the given *unit* heat load by the design temperature difference to find the *peak* heat load.

†See Apps. 7 and 8.

‡The density can be calculated from $\rho = 39.8 \exp(-H/27000)/(460 + T)$ where T is the temperature (°F) and H is the altitude above sea level (ft).

straightforward if air mover flow rates are known. The volume of outside air introduced is multiplied by the density, the specific heat, and the temperature rise required to heat outdoor air to the temperature of the indoor air. Recall that the density of air depends on altitude and temperature and should be calculated from the ideal gas law or the equation in Table 7.3 footnotes.

Infiltration rates are difficult to estimate since they depend on construction variables as well as climate variables. Two methods of calculating infiltration losses are described in [8]; they are known as the air change method and the crack method. The latter uses the physical length of the crack and a construction index to estimate heat loss. The former uses rules for air change rates based on construction quality and wind speed.

The third source of heat loss is floors. Heat loss through floors depends on the nature of their construction. If a slab on grade is used with exposed slab edges, the first expression for slab loss in Table 7.3 is used. If the slab is insulated or a basement is used, a constant rate of heat flow through the floor will occur since the outer surface of the slab is exposed to earth temperature, which is relatively constant, to first order, during the heating season.

Refer to Table 7.3 for the equations to be used to calculate heat loads in Btu's per hour or kilowatts for the three heat loss mechanisms described above. Heat loads can be expressed in different ways depending on the use to which they are to be put. In nonsolar calculations, heat load values are used to size heating systems for the statistically worst-possible set of climatic conditions to be expected in winter in a given location. This heat load is called the *peak* heat load and is expressed in kilowatts or Btu's per hour. The peak heat load, however, is not of interest in the design of solar systems, since they are rarely intended to carry the peak heat load owing to economic considerations described in Chap. 5.

For solar systems, the heat load value is more useful when expressed in Btu's per hour per degree Fahrenheit or kilowatts per degree Celsius. This is directly related to the peak heat load and can be calculated from it by dividing by the design, peak temperature difference. The unit heat load expressed in Btu's per hour per degree Fahrenheit is a characteristic of a given building and is independent of the location of the building (except for second-order wind effects on infiltration). This unit load can be converted to the form used in most solar designs by multiplying by 24. The units then become Btu's per day per degree or Btu's per heating degree-day.

The heating degree-day is the standard measure of heating demand used throughout the United States and is tabulated for many hundreds of locations. Appendix 6 contains these statistics for more than 200 locations. Most heating degree-day values in the United States have historically been based on 65°F. That is, it was assumed that no heating was required for outdoor temperatures above 65°F, even though the interior temperature of the building may have been 70°F. The difference between 65 and 70°F was made up by internal heat sources and solar gains. Prior to 1975 this design temperature basis for new buildings was satisfactory. However, with increasing insulation levels and reduced thermostat settings, the 65°F base must be modified. To modify degree-days for other bases on a month-by-month basis, a method described in [9] should be used. Simply applying an empirical constant to monthly degree-days is inadequate for revising the 65°F base.

Monthly degree-days are not required for the design procedures presented in this book, and a simple method for correcting *annual* degree-day totals for bases other than 65°F developed by Harris [10] can be used. The annual degree-days to any base DD_b accounting for variations in the interior temperature T_i, internal heat sources q_i, and building envelope characteristics embodied in UA are given by

$$DD_b = \left[1 - k_d\left(65 - T_i + \frac{q_i}{UA}\right)\right] DD_{65} \tag{7.1}$$

English units are used and DD_{65} is the annual total of standard heating degree-days to base 65°F. The parameter k_d is given by

$$k_d = 6.398\, DD_{65}^{-0.577} \tag{7.2}$$

Finally, the annual heating requirement can be calculated by multiplying the number of annual total degree-days to the appropriate base by the unit heat load described above. The unit heat load should be expressed in Btu's per heating degree-day when used for this purpose. The average year for which long-term degree-day statistics are tabulated may rarely be encountered, however, and variations from predictions based on the degree-day method may be ±20 percent.

The degree-day method is accurate only for skin-dominated, ventilation-dominated, or residential buildings. For commercial buildings with large internal heat sources and substantial internal mass or many internal zones, the degree-day method is inadequate and detailed load calculation by computer or other techniques may be required. Intermediate between the degree-day method and the computer methods is the bin-hour method described in [11]. In addition, the degree-day method will not precisely calculate the annual heating load if night setback is used, since night setback involves a variation in interior temperature T_i in Eq. (7.1). If the distribution of day and night heating degree-days is not known, it is not possible to use Eq. (7.1) to adjust nighttime values downward on the basis of reduced T_i levels.

The cost of heating a building with conventional fuels to which solar will ultimately be compared can be estimated from the Btu's per year calculation described above. The annual load is multiplied by the cost of energy ($/MMBtu) to arrive at the annual cost of providing space heating to a building. Of course, the cost of fuel must be expressed on a net basis, accounting for inefficiencies in combustion processes if fossil fuels are used. Likewise, for heat pump systems the cost of backup energy is the cost of electricity divided by the seasonal performance factor or long-term coefficient of performance (COP) of the heat pump. The efficiency for electric backup systems can usually be assumed to be unity. If the cost of providing heat to the building thus calculated is greater than the cost of heating the same building to the same temperature with an active or passive heating system, then an economic incentive exists to consider the use of solar energy for space heating. This comparative methodology using life-cycle costing terms is described in Chap. 5. It will be applied to passive systems in the next section.

EXAMPLE

Find the annual energy requirement for heating a residence in Sheridan, Wyoming, that has a unit heat load UA of 450 Btu/(h·°F) including infiltration. Internal heat sources averaged over day and night are 5500 Btu/h and the inside temperature is 68°F.

SOLUTION

The Harris method can be used to adjust degree-days (7708 for 65°F base, App. 6) for both reduced T_i and internal gain q_i. From Eq. (7.2),

$$k_d = 6.398(7708)^{-0.577} = 0.0366$$

and from Eq. (7.1),

$$DD_b = \left[1 - 0.0366\left(65 - 68 + \frac{5500}{450}\right)\right] 7708 = 5106$$

The adjusted degree-days are 5106. The annual load L is

$$L = (24UA)DD_b = 24 \times 450 \times 5106 = 55 \text{ MMBtu/year}$$

Energy Conservation in Solar Buildings

Any building that is to incorporate a significant amount of heating provided by an expensive energy source such as active or passive solar heating should be designed with care to minimize heating energy consumption by all practical cost-effective energy conservation methods. The payback period of most standard energy conservation features is less than that of a solar heating system. Therefore energy conservation features should be included in the building design from the outset. Details of energy conservation in buildings are covered in a number of books (e.g., [11]) and are not the subject of this book. The subject of energy conservation in buildings is broad and the designer should be aware of all codes and measures for reducing energy use.

The amount of energy conservation included in a building is limited by the cost of energy conservation features. Adding 6 in of insulation to a wall that already incorporates 12 in is not as cost-effective as providing double glazing in place of single glazing or infiltration control in the absence of any. Priorities for energy conservation features must be established by economic analyses. Methods described in Chap. 5 can be adapted for this purpose. In addition to heat loss control associated with the three heat loss loads described in the previous section, other building and site features can be considered for energy conservation. For example, use of thermal mass in a building can reduce peak heat loads.

It was shown at the National Bureau of Standards that this feature does not reduce total annual requirements significantly. With widespread use of demand rates in electrically heated buildings, reduction of peak heating requirements is important.

Orientation of buildings is an important variable when determining heat losses. In areas with significant heating energy requirements, an orientation favoring the south with a higher aspect ratio will result in reduced heating requirements in winter owing to improved solar gains and direct gain passive system performance. Microclimatic variables related to orientation should also be considered, especially the direction of prevailing winds in winter and the hours of maximum sunlight if diurnal, bimodal, or skewed solar intensity patterns exist.

Benefits provided by the many energy conservation features available can be analyzed in a few cases by the simple degree-day method described earlier. However, more sophisticated methods are frequently necessary. Computer models of buildings are very useful in determining economic trade-offs among energy conservation features. A well-known computer program for this purpose is DOE2, developed by various laboratories under contract to the U.S. Department of Energy. This model is driven by hourly weather and solar radiation data and calculates peak, monthly, and annual heating and cooling loads for an arbitrary number of zones of arbitrary occupancy schedule in any building. This model is written in the Fortran computer language and is available to the public for a nominal charge or through various computer time-sharing services. Some familiarity must be developed with input formats before this or any other building model can be used on a routine basis.

In addition to deciding among various energy conservation opportunities, it is necessary to determine the relative investment in aggregate energy conservation features and solar systems. This is done by using economic methodologies and performance prediction methods, and a simplified overview of this approach is given in [12]. Often the comparison of solar and energy conservation economics is not a standard feature of the building design process. If it is to be included, additional fees will be required. Since energy conservation features are so varied, it is not possible to calculate the best mix of solar and conservation features by means of simple rules of thumb or experience. It would appear at this time that the only rational approach is use of computer simulation, although this approach adds significant expense to the design process.

In the 1970s much attention was focused on energy consumption in buildings and codes were promulgated to reduce it. However, there are limits to energy conservation when viewed macroscopically. One side effect of reduced energy consumption can be the compromising of comfort levels in an office building, for example, to save on energy payments. If the comfort level of an office is reduced beyond the point at which workers can operate efficiently, the cost to the firm of reduced performance far outweighs the relatively small benefits that accrue from reduced heating or cooling expenditures. For example, a worker paid \$20,000 a year may use 100 ft^2 of office space. Therefore the worker costs the employer \$200/(ft^2·year) in direct labor. A typical building may consume 100,000 Btu/(ft^2·year) or approximately 0.1 MMBtu. Heating energy costs are approximately \$8–10/MMBtu; therefore, the cost of energy is approximately \$1/(ft^2·year). If energy conservation features are employed to reduce this energy requirement by one-half, the saving will be approximately 50¢/(ft^2·year). However, if these conservation features reduce the comfort and hence the efficiency of the office workers by only 0.25 percent, the entire effect of energy conservation on the profit-and-loss statement of the firm will be obliterated. Energy conservation features that affect comfort include reduced thermostat settings in winter, reduced ventilation rates, increased thermostat settings in summer, and elevated humidity levels in summer.

PASSIVE SYSTEM PERFORMANCE PREDICTION

The final decision about passive system sizing depends on the relative cost of solar and nonsolar heating energies, as described in Chap. 5. The optimum passive system size is associated with the minimum point of the total cost curve shown in Fig. 5.2. The first step in the construction of the cost curve is calculating the expected performance of the solar heating system. Performance prediction for active or passive solar systems is a very complicated matter and is not discussed in this book. However, the conclusions of many performance studies can be correlated in a relatively simple form that can be used in a prescriptive fashion without giving the background of the method in detail. For passive solar systems, a method developed by Arney et al. [13] will be described. This is the *P*-chart method, and for its use only the annual building heat load must be known. The unit heat load expressed in Btu's per degree-day is multiplied by annual degree-days, accounting for internal heat sources, to determine the total annual heat load L (see the example in the preceding section for this procedure).

The *P*-chart will specify the solar fraction and optimum size of a passive solar system of the direct gain, masonry storage wall, or water storage wall types. It does not require iterative calculation of solar performance for many system sizes as indicated in the flowchart in Fig. 5.3; instead, the cost-optimal system area can be calculated directly (Fig. 5.4). The following five steps should be followed to find the cost-optimal collector area:

1 Determine the building unit heat load (excluding the passive aperture area) in Btu's per degree-day. For a well-insulated residential building it will be 5 to 7 Btu's per square foot of building per degree-day (the P chart is useful only in the English system of units).

2 From App. 6, find the total annual degree-days for the site nearest the location to be studied. Multiply the unit load from step 1 by the degree-day total [adjusted with Eq. (7.1) if required] and divide by 1 million to find the annual heat load L in millions of Btu's per year.

3 Select one of the three types of passive systems for application to your building and read constants A and B from the P-chart table (App. 9) for your location.

4 Evaluate economic parameters including the levelized cost of backup energy $\bar{C}_{bu}$, \$/MMBtu (Chap. 5); the capital recovery factor CRF(i, N), where i is the discount rate and N is the period of analysis; and the extra variable cost C_s of the passive features expressed in dollars per square foot of aperture area. The extra cost of a passive system is the cost in excess of that for an identical nonsolar building.

5 Calculate the optimum collector area from

$$A_{c,\text{opt}} = L\left[\frac{A\bar{C}_{bu}}{\text{CRF}(i, N) \times C_s} - \frac{1}{B}\right] \tag{7.3}$$

If desired, the annual solar fraction $\bar{f}_s$ can now be calculated from

$$\bar{f}_s = A \ln\left(1 + B\frac{A_c}{L}\right) \tag{7.4}$$

Instead of using Eq. (7.3) to find the optimum area, the P-chart nomograph shown in Fig. 7.8 can be used directly as follows. Start at the upper right quadrant at axis 1 with the value of CRF(i, N) $\times$ C_s. Move vertically up to intersect the curved line with the value of $\bar{C}_{bu}$, the levelized backup energy cost. Move horizontally left to intersect a line labeled with the appropriate value of P-chart constant A. Then drop vertically downward to the lower left quadrant of the figure.

Now start at axis 2 in the lower right quadrant with the suitable value of P-chart constant B. Move up to the curved line and then horizontally left to intersect the line constructed downward from the upper left quadrant. The point where the two construction lines intersect is the design ratio of $A_{c,\text{opt}}$ to annual load L.

Finally, to find the optimal collector size, multiply the ratio by the value of L calculated in step 2.

The P-chart method is illustrated in the following example.

EXAMPLE

A water storage wall system with night insulation is to be used on a residence in Eagle, Colorado, with a heat load of 10,000 Btu's per degree-day. If the backup fuel is liquefied petroleum gas (LPG) at \$5/MMBtu inflating at 17 percent per year, what is the optimum system size if the passive extra cost C_s is \$15/ft^2. Use a 20-year period of analysis N and 9 percent interest rate i.

SOLUTION

For this location $A = 0.701$ and $B = 0.216$. To base 65°F, 8426 degree-days are measured on the average. Hence $L = 10{,}000 \times 8426 \times 10^{-6} = 84.26$ MMBtu/year.

The levelized cost of energy $\bar{C}_f$ is found from Eq. (5.4). To find it, i' is needed:

$$i' = \frac{0.09 - 0.16}{1 + 0.16} = -0.0603$$

and

$$\text{CRF}(i', N) = 0.0244$$

$$\text{CRF}(i, N) = 0.1095$$

Then

$$\bar{C}_{bu} = \$5 \times \frac{0.1095}{0.0244} = \$22.45 \text{ MMBtu}$$

Using Eq. (7.3), the optimal area is

$$A_{c,\text{opt}} = 84.26\left(\frac{0.701 \times 22.45}{0.1095 \times 15} - \frac{1}{0.216}\right) = 417 \text{ ft}^2$$

This area will provide the following solar fraction:

$$\bar{f}_s = 0.701 \ln \left(1 + 0.216 \frac{418}{84.26}\right) = 0.51$$

Thus 51 percent of the annual load is provided by the cost-optimal system using 417 ft^2 net of aperture.

The *P* chart is based on the solar load ratio (SLR) passive prediction method developed at Los Alamos Scientific Laboratory [14]. The *P*-chart prediction of passive system performance is within $1\frac{1}{2}$ percent (root-mean-square) of the prediction of the SLR method for all cities listed in App. 9. The method applies strictly to the standard passive system assumed in the development of the SLR approach. Included in the standard system are 45 Btu/($^\circ$F$\cdot$ft$_c^2$) thermal storage, thermal circulation vents for masonry wall storage amounting to 3 percent of the wall area, double glazing, R-9 night insulation (if used), 30 percent solar reflection from collector foreground, vertical south-facing glass, wall absorptance for sunlight of 90 percent, and no overhang shading of passive apertures during the heating season. Other assumptions about the thermal properties of the wall are given in [14]. These specifications above describe a passive system with standard values of most parameters. If the designer wishes to examine the effects of variations in important parameters, the monthly SLR method must be used. The SLR method is described in a step-by-step fashion in [7].

The *P* chart applies only to the standard system described above and can be used only with English units, although the optimal system area in square feet can be converted to metric units by dividing by 10.76. Quantities used in Eq. (7.3) for the *P*-chart nomograph to identify the optimum system include backup energy costs, inflation rates, and solar system extra costs. Backup energy costs

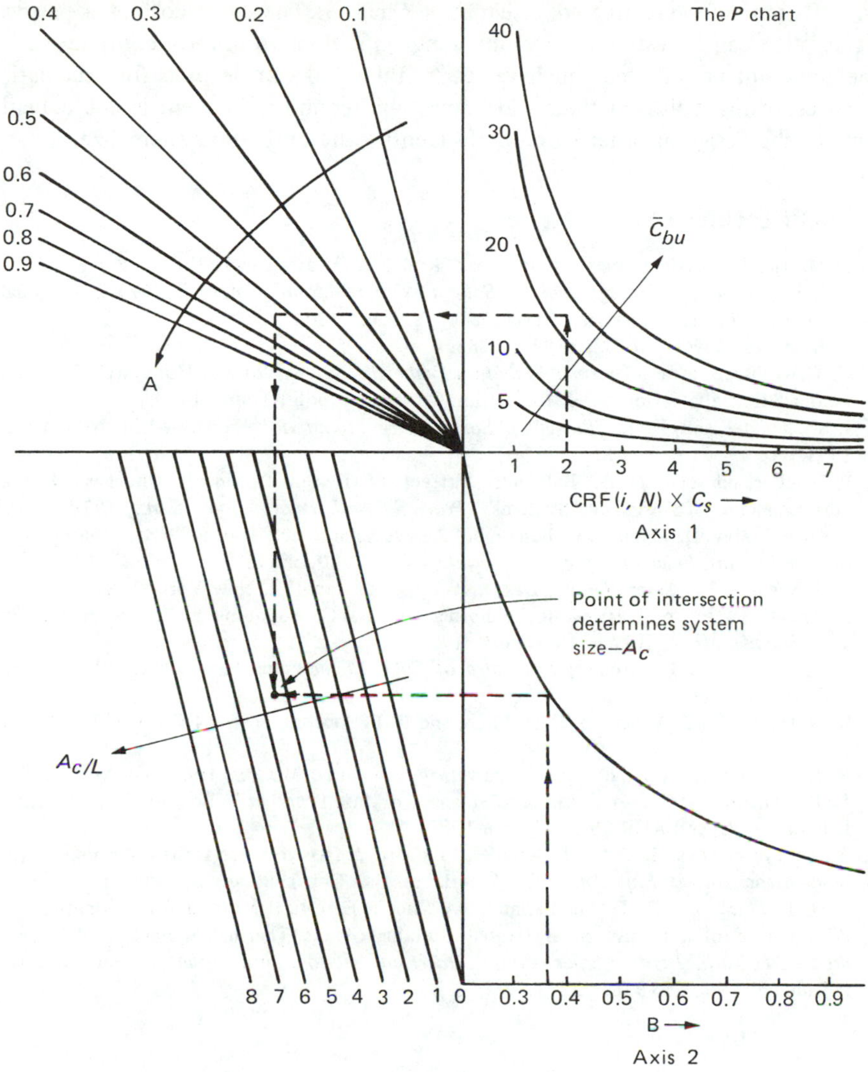

Figure 7.8 *P*-chart nomograph for optimal passive system sizing. (*Courtesy of Solar Energy Design Corp. of America.*)

Table 7.4 Costs (1980 Dollars) for Generic Passive Systems

System type	Net extra cost per square foot of collector glazing ($)
Thermal storage wall	8–18
Direct gain	0–11
Attached sun space or greenhouse	5–15

and inflation rates are treated in detail in Chap. 5. The extra cost of a passive solar system can be estimated by obtaining quotations from local contractors. If time does not permit, the rough values in Table 7.4 can be used for schematic design estimates, although their wide range will require refinement in subsequent stages of the design in order to properly identify the optimum system size.

REFERENCES

1. E. Mazria, *The Passive Solar Energy Book,* Rodale, Emmaus, Pa., 1979.
2. U.S. Department of Energy, *Passive Solar Design Handbook,* DOE/CS-0127/2, National Technical Information Service, Springfield, Va. Jan. 1980.
3. E. S. Morse, *Science*, vol. 2, 1883, p. 283.
4. G. G. Gubareff et al., *Thermal Radiation Properties Survey,* 2d ed., Honeywell Research Center, Minneapolis-Honeywell Regulation Co., Minneapolis, Minn., 1960.
5. J. F. Kreider and F. Kreith (eds.), *Solar Energy Handbook,* McGraw-Hill, New York, 1981.
6. R. McFarland and J. D. Balcomb, "Effect of Design Parameter Changes on the Performance of TSW Passive Systems," *Proc. 3d Natl. Passive Solar Conf.,* 1979, p. 56; see also R. McFarland and J. D. Balcomb, "Passive Solar Space Heating," *Proc. 1980 American Section, Int. Solar Energy Society Conf.*, vol. 2, 1980, p. 696.
7. J. F. Kreider, *The Solar Heating Design Process,* McGraw-Hill, New York, 1981.
8. American Society of Heating, Refrigerating, and Air-Conditioning Engineers, *Handbook of Fundamentals,* ASHRAE, New York, 1977.
9. F. Kreith and J. F. Kreider, *Principles of Solar Engineering,* Hemisphere, Washington, D.C. 1978.
10. W. S. Harris, G. Y. Anderson, C. H. Fitch, and D. F. Spurling, *ASHRAE J.*, vol. 7, 1965, p. 50.
11. F. Dubin and G. Long, *Energy Conservation Standards,* McGraw-Hill, New York, 1978.
12. J. D. Balcomb, "Conservation and Solar: Working Together," Los Alamos Scientific Laboratory Paper LAUR-79-3195, Jan. 1980.
13. W. Arney, P. Seward, and J. F. Kreider, "P-chart: A Passive Solar Design Methodology," *Proc. Assoc. Energy Eng. Annu. Conf.,* Atlanta, Ga., Oct. 1980, ed. A. Thuman.
14. J. D. Balcomb and R. D. MacFarland, "A Simple Empirical Method for Estimating the Performance of a Passive Solar Heated Building of the Thermal Storage Wall Type," *Proc. 2d Natl. Passive Solar Conf. American Section, Int. Solar Energy Society,* Philadelphia, vol. 2, 1978, p. 377.

8 Mechanical Solar Space-heating Systems

INTRODUCTION

In this chapter, the operation and design of mechanical solar space-heating systems are described. Mechanical systems—sometimes called active systems—use pumps and fans to move energy from one point in the system to another. This is in contradistinction to passive systems, described in Chap. 7, which use a minimum amount of mechanical assistance to carry out their heating function.

Space heating is one of the few mature solar technologies, and reliable design methods exist. Many tens of thousands of space-heating systems have been built in the past decade, and much of the experience with these early systems has indicated which system designs work and which do not. The emphasis in this chapter is on space-heating systems, either air- or liquid-based, that are in the mainstream of solar system design in the United States. Many solar heating concepts have been developed by inventors since the 1930s; however, very few have been shown to be reliable and economical. It is these systems that are emphasized in this chapter.

The first liquid-based solar space-heating systems were simply enlarged water-heating systems. That is, water-heating collectors were used on a larger scale to provide space heating. Along with larger collector arrays, enlarged storage tanks, pumps, and heat exchangers were used. In the early systems it was not clear whether the storage should be sized for relatively short periods, that is, 1 or 2 days, or very long periods, on the order of months. Current design approaches for residential systems most often use short-term storage. For district heating systems, however, seasonal storage may be appropriate, although very few systems of this type have been built and the economics of this application are not yet clarified.

In the modern solar heating era, beginning in the early 1970s, the most common systems were liquid-based ones employing collectors with tubes attached to metal plates in the flat-plate configuration described in Chap. 3. In the mid-1970s air systems began to be used for both large and small applications, and the significant advantages of these systems have accelerated their adoption for both residential and commercial buildings. Both air and liquid systems are treated in detail in this chapter.

Figure 8.1 shows an example modern solar space-heating system. This system is located in Boulder, Colorado, and provides approximately three-fourths of the heating for the University of Colorado credit unit. It is of the liquid-based type and uses approximately 2200 ft^2 of collector for heating the 15,000-ft^2 office building. The piping diagram for this system is shown in Fig. 8.2. The details of the system will be described in subsequent sections of this chapter. At this point, it is sufficient to note that the principal system components are collectors, heat exchangers, pumps, storage tank, energy delivery system, and controls. These components are also present in air-based systems, and both air and liquid systems function almost identically from a systems perspective. The system shown in Fig. 8.1 has been in operation for several years, and the average heating

Figure 8.1 Active mechanical solar heating system used in the University of Colorado credit union in Boulder, Colorado. The 2200-ft^2 system provides approximately three-quarters of the heating load of 15,000-ft^2 building. (*Courtesy of Jan F. Kreider Associates, Inc.*)

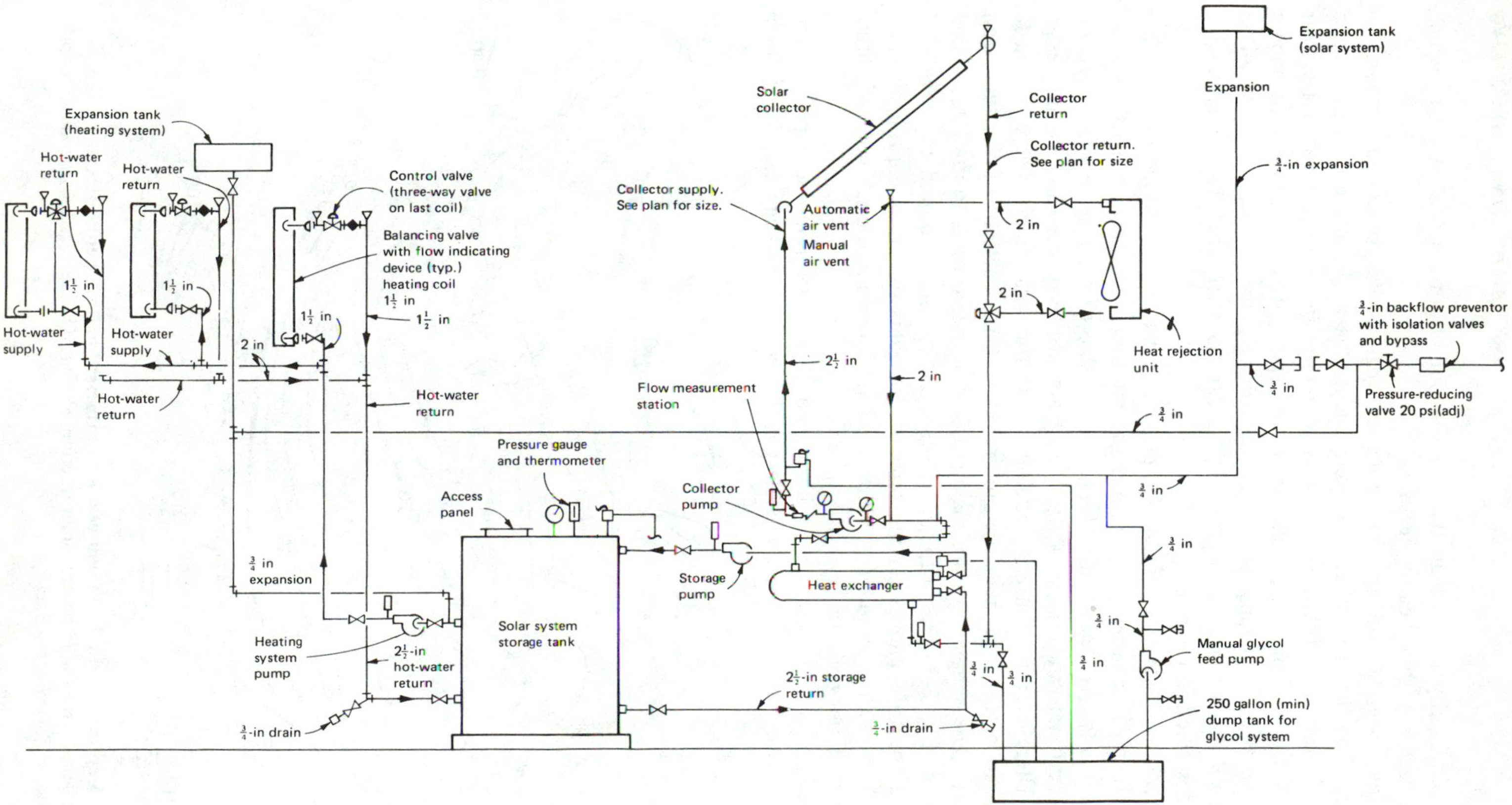

Figure 8.2 Solar system diagram for the building shown in Fig. 8.1. Compcnents include collectors, storage pumps, heat exchangers, and backup system. (*Courtesy of Lee and Associates, Architects.*)

bill for this facility is less than that for the building manager's residence, which does not have any solar heating features.

It has been found in the design of many solar space-heating systems worldwide that the sizes of all the components listed above are directly related to the collector area. Therefore the most common index of size for active solar systems is the net or active collector area expressed in square feet or square meters. The sizes of pumps, pipes, heat exchangers, storage, and so on can all be related to collector area in a manner similar to that used for water-heating systems described in Chap. 6.

Early in the solar heating design process it is necessary to have an estimate of collector area in order to determine the approximate initial cost of the solar heating systems. As described in Chap. 5, the size of solar systems is based on economic criteria, and many of these data are not available early in the design process. Therefore an arbitrary decision about the solar fraction is often made and the collector area required to provide this fraction of heating load determined. The simplest way of making this determination is to use information presented in Fig. 8.3. This map shows the collector area required to provide 75 percent of the total annual heating demand. Numbers shown on contour lines on the map express the ratio of building load in English units (Btu's per heating degree-day) to net collector area in square feet.

The use of the map can be illustrated as follows. Suppose a preliminary building heating load of 15,000 Btu's per degree-day has been calculated for a

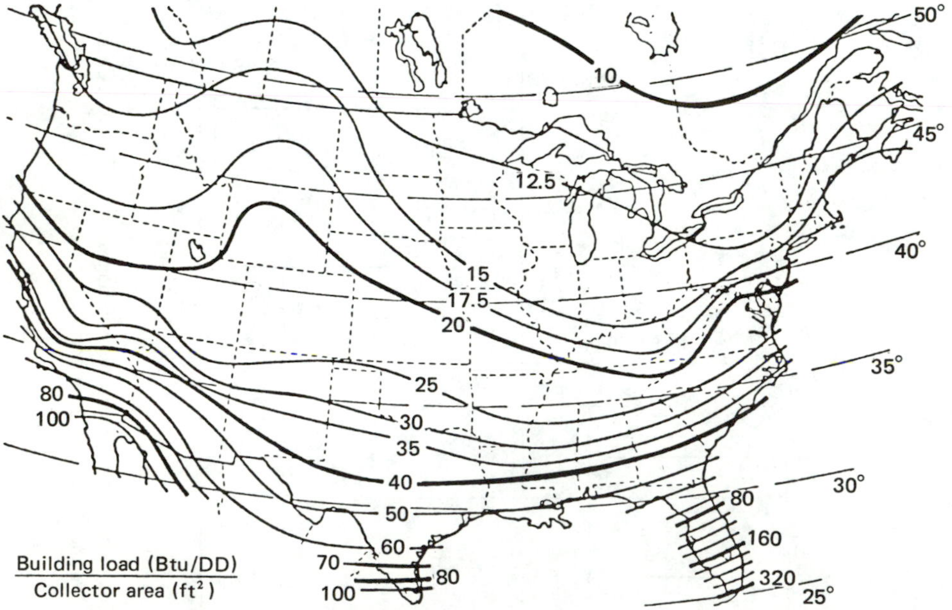

Figure 8.3 Map indicating ratio of unit heat load expressed in Btu's per heating degree-day to required collector area for 75 percent solar fraction. (*From [1]; courtesy of American Section of International Solar Energy Society.*)

building located in central Missouri. According to the map in Fig. 8.3, the ratio of heating load to collector area is 20. Therefore the collector area required to provide 75 percent of the heating load is obtained by dividing 15,000 Btu's per degree-day by 20; that is, 750 ft^2 of net collector area will suffice to provide approximately 75 percent of the heating load. Figure 8.3 applies for both air and liquid systems, so the decision about which system to use need not be made at this early phase of the design. The specification of 75 percent solar load fraction is, of course, arbitrary, and the subsequent steps of the design will use additional economic criteria to determine the cost-optimal solar heating fraction for the specific building. The map should be used only in the early schematic design phase to indicate the approximate collector area required. Costs per square foot for active systems are given in Chap. 5 and the estimated initial investment is simply the product of the collector area determined from the map and the per-square-foot cost from Table 5.5.

In the remainder of this chapter we describe the function of both liquid and air systems and the components of each system. Although detailed design guidelines are not given here, the important features of all active system components are described. For detailed step-by-step design procedures the reader is referred to [2]. The next sections describe the operation of liquid-based and air-based heating systems. A separate section is devoted to controls for both air and liquid systems, since these components, although small and inexpensive, have been troublesome from a design and operational point of view. The final section of this chapter describes a performance prediction method used for sizing of active air- and liquid-based solar systems.

LIQUID-BASED MECHANICAL SOLAR HEATING SYSTEMS

Figure 8.4 is a schematic diagram of an active space-heating system. All major components are shown and their interrelations are specified. The system includes a domestic water preheating system of the two-tank variety described in Chap. 6. The reader is referred to Chap. 6 for specifications of domestic water-heating systems and trade-offs between single- and double-tank systems. Design of the added water-heating system in Fig. 8.4 is not covered in this chapter since it has been presented in Chap. 6.

In this section the operation of the system will be described qualitatively; the performance of the major components is covered in subsequent sections. Operation of the system shown in Fig. 8.4 can be understood by referring in turn to the several fluid loops that make up the system. In this particular system, five loops are present: collector, storage, delivery, water preheat, and hot-water delivery. The first three loops will be described in this section.

Heat Collection

The collection loop in most liquid-based systems contains a nonfreezing fluid for freeze protection. As described in Chap. 6, other methods of freeze

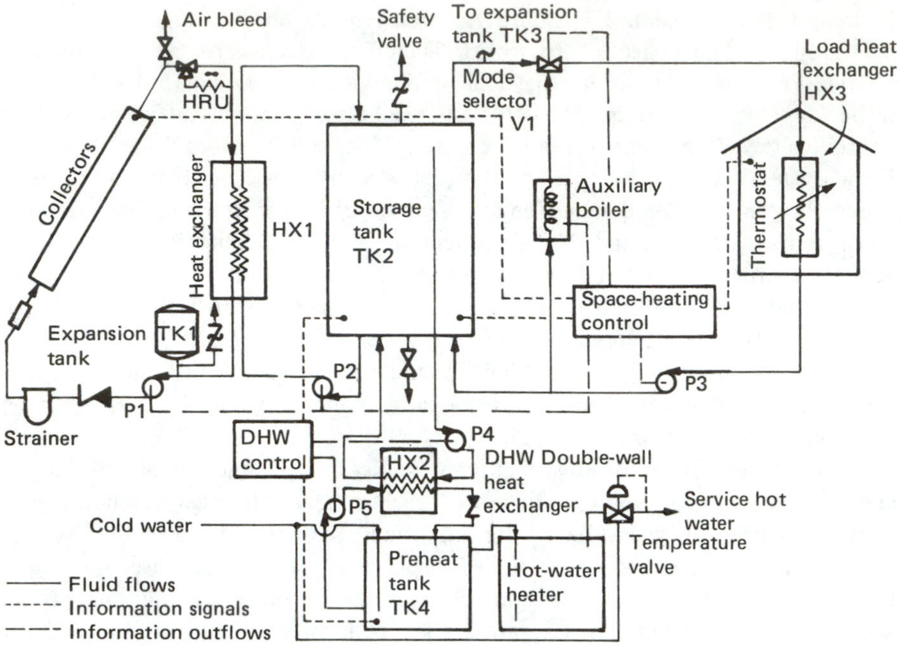

Figure 8.4 Schematic diagram of liquid-based space-heating solar system including domestic water preheat subsystem.

protection are available but are much less reliable from a risk-benefit viewpoint than the indirect freeze protection approach. This is the most common method of freeze protection used in modern liquid-based heating systems in areas with significant heating loads. Collector loop pump P1 causes fluid to circulate through the loop. Flow rates (described later) are selected such that the temperature rise in bright sun across the collector is approximately 15°F. The heat added to the fluid in the collector is removed in heat exchanger HX1 and delivered to storage.

The collector loop is a closed loop and therefore requires an expansion tank and safety valve arrangement. The expansion tank TK1 accommodates changes in fluid volume as a result of temperature changes in the collector loop. Over the course of a year, collector loops experience temperatures ranging from −20 to 220°F. A safety valve is attached to the expansion tank to relieve excess pressure buildup, which could occur in the event of pump or power failure.

A check valve is located downstream of pump P1. This is necessary to prevent reverse circulation at night, when collectors on the roof become colder than the fluid in the remainder of the loop inside the building. The natural tendency for gravity flow to occur is blocked by the check valve, and a heat leak from storage is thereby prevented. In addition to heat leakage from storage that occurs with reverse flow in the collector loop, it would be possible under very severe conditions for heat exchanger HX1 to freeze, since very cold liquid would pass through the heat exchanger and be exposed to storage tank water. Therefore this small component, the check valve, is very important for system reliability.

Strainers or filters are also important in collector loops to prevent fluid conduit blockage in the system. Many collectors have relatively small fluid conduits, and any particles of dirt remaining from the construction process or from chemical reactions in the glycol loop must be avoided. Although they are not shown in Fig. 8.4, strainers and filters should have pressure measurement points; an increase in pressure drop across a filter indicates that the filter requires cleaning.

After passing through the collectors, the fluid is at its highest temperature. At elevated temperatures aqueous working fluids tend to reject dissolved air. It is therefore necessary to have air bleeds in closed loops as shown in Fig. 8.4. The air bleed can be either automatic or manual. For large systems, automatic air bleeds are most commonly used owing to their functional simplicity during filling of the system. If manual bleeds are used in a large system, it is very difficult to reliably remove air from all collector outlet manifolds. However, for residential applications using only one or two manifolds, the manual air bleed approach may be acceptable.

Heat Storage

Heat exchanger HX1 is the point of interaction between the collection loop and the storage loop. The storage loop is relatively simple in operation, consisting of only a circulation pump P2, the heat exchanger, and associated piping. Whenever pump P1 is on, P2 is also on, and heat is transferred to storage. In liquid systems some stratification of the storage tank is achievable if fluid from the bottom of the tank is introduced into the heat exchanger and hotter returning fluid from the heat exchanger is introduced into the top of the storage tank. Flow rates in pumps P1 and P2 are of the same order of magnitude; however, P2 may have increased flow to improve the effectiveness of heat exchanger HX1. These matters are described shortly.

Storage tank TK2 is well insulated, with thermal breaks between it and the environment to minimize heat loss. A safety valve is also required here as in any closed loop to which heat can be added. Storage tanks should be equipped with a drain that can be used to change the storage fluid. Although water is the nominal storage fluid, it is usually treated with inhibitors to prevent corrosion; occasionally, these inhibitors may require replenishment.

Heat Delivery

The third loop shown in Fig. 8.4 is the energy delivery loop operated by pump P3. This pump is operated whenever a load requirement occurs in the space, as indicated by information received by the space-heating controller. Specifically, if the space thermostat calls for heat, pump P3 will operate. Mode selector valve V1 will be in a position to attempt to extract heat from storage by removing storage fluid from the top of the tank and passing it through the load heat exchanger HX3. The load exchanger may consist of baseboard elements specifically designed for solar applications or a finned-tube coil located in the forced-air duct. Other common terminal load devices may also be used if properly sized for reduced solar fluid temperatures.

If sufficient heat is contained in the storage tank, the space will be heated by solar heat and the thermostat contacts will eventually open. Upon opening, pump P3 ceases to operate. However, if inadequate heat is contained in storage to carry the space-heating load, the temperature sensed by the thermostat will continue to drop. Thermostats of a dual-point design are used for most space-heating applications. As temperature continues to drop by 1 or 2°F in the space, the second contact on the thermostat will close, indicating that sufficient solar capacity does not exist. The controller will then make the decision to operate mode selector valve V1 to pass fluid through the auxiliary boiler instead of the solar storage tank. At the same time the boiler or other backup system, such as a heat pump or electric resistance element, is activated by the space-heating controller. The auxiliary system is designed to have sufficient capacity to carry the space-heating load under the worst possible climatic conditions. As heat is added to the space, the space temperature rises and both contacts of the thermostat ultimately open, indicating that no further call for heat exists. Pump P3 ceases to operate and the auxiliary system is deactivated.

Heat Rejection

The preceding description of system operation pertains to the normal operating modes of the liquid-based system, namely

Solar heat collection
Heat delivery to storage
Heat extraction from storage
Auxiliary system heating

An additional operating mode is frequently required during periods of the year when heating loads are low. Occasionally, the solar collector may collect more energy than is required and storage tank TK2 may become fully charged. In this case, the solar system controller makes the decision to reject heat through heat rejection unit HRU. This unit can be either a cooling tower on large installations or a weatherproof unit heater on smaller systems. The heat rejection system is activated whenever the storage tank temperature approaches its upper design limit. At this point, P1 will continue to operate but P2 will be turned off. The heat rejector three-way valve will divert solar fluid through the heat rejector and the heat rejector fan will begin to operate, reducing the temperature of the collector loop fluid to the design point for heat rejection conditions. The heat rejector unit is sized to reject all heat collected during the peak heat collection period of the year. Design of the heat rejector is therefore based on peak collection conditions, not on average conditions. If average conditions are used for design, the heat rejector will not be capable of dissipating sufficient heat on sunny, mild days in spring or fall. As will be shown later, the collector tilt angle for active systems is such that very good performance of collectors is ensured in winter, but a by-product of this orientation is also excellent collection in spring and fall, when large heat loads do not occur.

The method of heat rejection shown in Fig. 8.4 is not the only method possible. Hot water can be dumped from storage tank TK2 and be replaced with cold water, for example, or the collector and storage pumps P1 and P2 can be operated at night to reject heat from the storage tank. This method is technically feasible but requires operating two pumps for many hours to dissipate storage heat. Collectors are designed to be effective heat collectors and not effective heat rejectors. Therefore in most cases the use of a heat rejection unit is indicated.

Heating Loads

Calculation of heating loads on passively heated residential and commercial buildings has been described in Chap. 7. The same approach is used for actively heated solar buildings. Therefore the reader is referred to Chap. 7 to review the procedure for heat load calculations. Details beyond those given in Chap. 7 are presented in the ASHRAE *Handbook of Fundamentals* [3].

In the following sections we describe each major component of the system shown in Fig. 8.4 except for controls, which are described later in this chapter.

Solar Collectors

The performance of solar collectors and their design has been described in Chap. 3. This section will summarize a number of practical details in the design of solar collectors which should be noted by the designer in making a selection for a specific project.

Collector Materials

Solar collector materials must be selected with care since solar collectors operate over a very wide temperature range, from well below zero to approximately 400°F during stagnation periods. Certain materials are immediately ruled out as a result of these stringent requirements. For example, wood should not be used in a solar collector. In addition to outgassing problems, wood has been shown to self-ignite after several months of stagnation. For example, on May 25, 1980, a flat-plate solar collector with a wood housing self-ignited in Boulder, Colorado, after rigid foam insulation behind the absorber plate deteriorated, exposing the wood housing to very high temperatures. After prolonged exposure to high temperature, wood changes chemically to pyrophoric carbon, which self-ignites at temperatures around 250°F; such temperatures are relatively common for even a moderately efficient collector under bright sun and no-flow conditions.

Likewise, plastics and low-temperature insulations should not be used in collectors. A problem with many early collector designs was outgassing of plastics or other nonmetals in the collector. Outgassing is accelerated under high-temperature conditions and the outgassed products invariably condense on the inner surface of the collector glazing, thereby reducing optical efficiency. Of course, certain plastics are specifically designed for high-temperature applications, and their use causes no problems in the solar environment. A critical component of many collectors is the glazing-to-housing seal. Ethylene propylene diamine monomer (EPDM) seems to be one of the best materials for this application and does not have any outgassing problems. Dust seals are also used in solar collectors to keep dirt and moisture from entering near the duct or piping attachment points.

Plastics have been used for glazings on solar collectors, but in most cases their reliability has been marginal. Some glazings made of plastic have been relatively thin, and after exposure to wind for prolonged periods of time they have experienced fatigue and eventually fractured. In addition, prolonged exposure to ultraviolet radiation in a high-temperature environment can decrease the transmission of a glazing. If plastic is used, it must be selected with great care. Glass is the most common material since its performance in the solar environment is excellent and has been well documented for many decades.

Collector Fluid Flow Considerations

In addition to studying materials used in collectors, the designer should inspect the collector for proper flow balancing provisions. Uniform flow of fluids through solar collectors is important for good performance. In liquid collectors a series of pipes or other flow passages approximately $\frac{1}{4}$ to $\frac{3}{8}$ in in diameter are connected together at the top and bottom of the absorber plate in a manifold. To ensure equal flow in each of these risers, the pressure drop in the manifold connecting them must be much smaller than that in the riser itself. A typical ratio of manifold pressure drop to riser pressure drop is 0.1. If this method of flow balancing has been used by the manufacturer, automatic flow distribution within the collector is ensured. The same consideration applies when connecting many collectors in series to the collector inlet and outlet manifolds. Flow balancing for these conditions is described later.

The pressure drop across a solar collector should not be made very small. Some collector manufacturers consider this to be a selling point because it results in low pump energy requirements. However, arrays made up of collectors with a very small pressure drop are very difficult to balance, since a finite pressure drop must be present across the collector to ensure uniform flow in a large array connected in parallel. Collector pressure drops below 6 or 8 in of water will require a separate orifice to be inserted to ensure adequate pressure drop for system flow balancing. This added expense more than offsets any small saving in pump operating cost.

A further characteristic of collector piping should be reviewed by the designer. Internally manifolded collectors that can be connected with only a pipe union have a significant potential for cost saving during installation. If a separate piping connection is required for each collector, increased labor and material costs are certain to result. Also, insulation of collector headers can be eliminated if the internal header approach is used, since the collector insulation serves as piping insulation. Manufacturers should be consulted about the maximum number of collectors that can be interconnected while preserving flow balance.

Even if only six or eight panels can be connected internally, significant savings in piping and insulation costs will result.

Collector Loads and Mounting

Installation of solar collectors on a building requires a structure for collector mounting. Collectors can be either attached to a sloped roof or mounted on a separate structure above a flat roof. In either case, the attachments and structure should be designed to carry the maximum dead load, which includes the weight of the collector, operating fluid, and any collector structure. Live loads include wind, snow, and hail and can be determined from the historical climatic conditions at the location. Wind loads on collectors are manifested as either upward or downward forces on the structure. The drag coefficient on surfaces tilted at various angles is shown in Fig. 8.5, which can be used to estimate the force present under design wind conditions. Snow and hail loads are tabulated in [4]. Liquid-based collectors typically weigh less than 11 lb/ft^2. Lighter designs are available that weigh half as much.

Many methods can be used to mount solar collectors on sloped or flat roofs. Figure 8.6 shows one approach for mounting on a flat roof. This design uses pitch pocket penetration through the roof membrane and attachments to structural members beneath the roof. Pitch pockets must be designed with great care, and no more than one vertical support should be used per pocket. This approach will require continuous maintenance and replenishment of pitch over the years if a leak-proof roof is to be preserved. Other approaches for flat-roof mounting include curb mounting, in which collectors are attached to concrete curbs that are integral parts of the roof. The leak problem is thereby avoided, but there is an increased cost in roof construction.

On sloping roofs, collectors can be attached directly to the subroof and structural members beneath it. The collector can then be flashed onto the roof as an integral part of the roof. However, if collector replacement is ever required, significant extra costs for labor and roof replacement will be imposed. An alternative approach is to locate collectors above a roof on spacers; thus the collector is not part of the roof, and a complete roof beneath the collector is

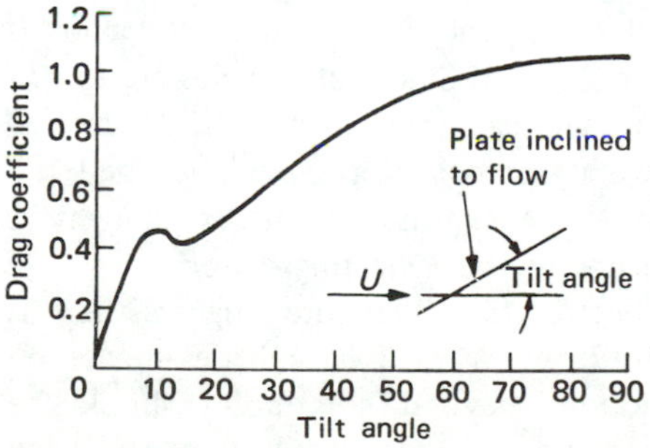

Figure 8.5 Flat-plate collector drag coefficient plotted against tilt angle. Total wind load is equal to the drag coefficient multiplied by the collector area and wind pressure $\frac{1}{2}\rho U^2$ (ρ is the density). (*From [5]*.)

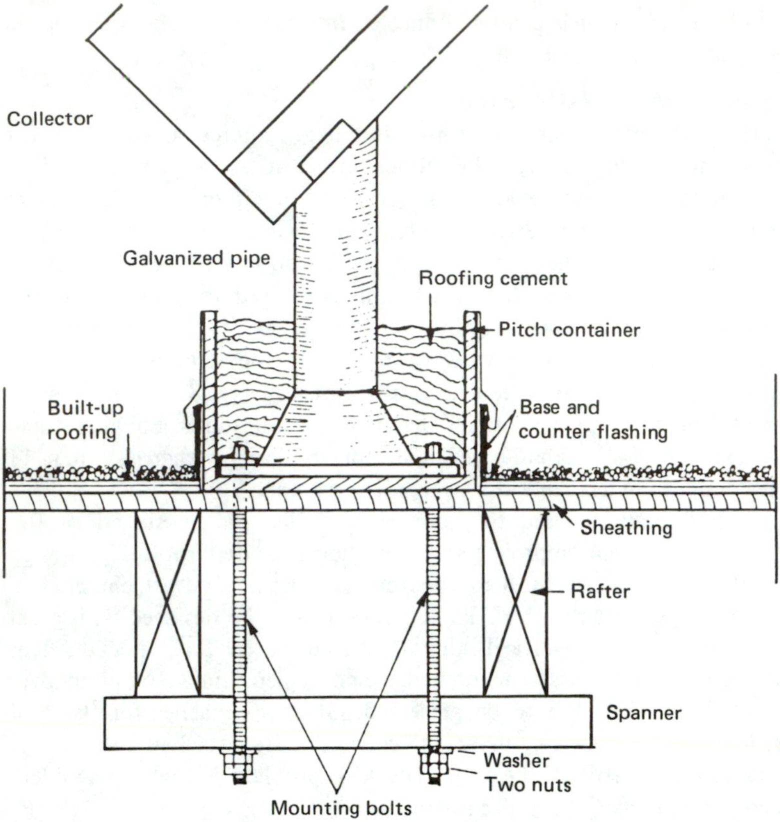

Figure 8.6 Example collector mounting detail in the pitch pocket approach for a flat, wooden, residential roof.

required. This approach permits ready replacement of collectors, although roof replacement is somewhat more difficult, requiring collector removal.

Collector Tilt

The angle from the horizontal selected for space-heating solar collectors is based on peak solar heat requirements in midwinter. Since solar angles are low in winter, the collectors are tilted up by a significant angle from the horizontal plane. A rule that seems to work well is that the collector should be tilted 15° plus the local latitude. At this tilt angle the capture of winter sunlight is enhanced and good performance results during the peak heating season.

The 15° rule may result in collector tilts that are higher than is architecturally acceptable. Reducing the tilt by 5° will have little impact on year-round performance, although reductions beyond this point should be avoided if possible. Of course, cost considerations are also involved, as less steep roofs cost less to fabricate. If the collector is to be mounted on the sloped roof of a building, an economic study should be carried out comparing the cost of a steeper roof with the benefit accruing from improved solar system performance in winter.

For best performance in most parts of the world, solar collectors used for space heating should face due south. A variation of ±15° from due south has little effect on annual energy delivery, however. For orientations more than 15° from due south, it may not be cost effective to place the collectors at the best angle. In this case, an off-south collector can be used if its area is increased to compensate for reduced energy capture. A study should be carried out by the designer to determine the best tilt angle if off-south orientation is to be used. The 15° rule applies only to south-facing collectors. For off-south collectors, somewhat reduced tilt may be advantageous to improve energy capture during the winter.

Collector Fluid Flow

For most liquid collectors, the rate of flow per square foot of collector is in the range 0.02 to 0.03 gal/min. This is based on the law of diminishing returns, which applies to collector fluid flows. At very high flows, collector performance is slightly improved but at the cost of increased pumping power. At very low flows, fluid residence time in the collector is too long and excess temperature rise occurs. At higher outlet temperatures, the collector is operating less efficiently than it does at the proper flow rate.

To determine the total flow rate for sizing of pump P1 in Fig. 8.4, the net collector area is multiplied by the nominal flow rate per square foot given above. If system size and design fees so indicate, a cost-effectiveness study of flow rate can be made by the system designer. Such a study compares the economic value of increased performance, increased pumping power, and increased pump capital investment to determine the optimum flow. Details of this approach are described in [2].

Thermal Storage for Liquid-based Systems

Water in pressurized tanks is invariably used for liquid-based system storage. The storage described in this book is called short-term storage and is useful for 1 or 2 days of solar outage. Long-term storage does not appear viable in most residential and medium-scale systems, although in district heating systems it may be effective.

The design of storage is relatively simple, requiring only specification of the amount of storage, the pressure and material requirements for the tank, and the amount of insulation to be used. As is the case for most components of a solar system, the law of diminishing returns is instrumental in specifying the amount of storage required. For most liquid systems, the storage ranges between 1.5 and 2 gal/ft^2 net of collector. With smaller amounts, the storage overheats and heat must be rejected. With larger sizes, the increased capacity is rarely used and does not pay for itself since its load factor is small.

Steel tanks are frequently used, and concrete and aluminum have been employed occasionally. Steel tanks are readily available for many sizes and pressures useful in solar heating systems. Concrete tanks cannot be pressurized and are difficult to seal reliably.

Closed storage tanks are desirable to reduce evaporation, which causes increased heat loss, increased humidity within the structure, and increased water hardness buildup in the storage tank. Although these problems are technically surmountable, it is best to avoid them by using closed systems. The delivery loop of most liquid-based solar systems operates at approximately 15 lb/in^2 and high-pressure tanks are not required.

Piping Connections to Storage Tanks

Proper piping connections to storage tanks are essential for proper system function and storage utilization. Figure 8.7 shows schematically the proper method for connecting piping. A cross flow of both heated collector fluid and load fluid is indicated. This ensures that the entire tank is used for heat storage and that heat is provided to the load whenever available.

Collector fluid is introduced into the top of the tank through an optional diffuser, as shown in Fig. 8.7. Return to the collector (really the collector heat exchanger) occurs from the diagonally opposite port at the bottom of the tank. This ensures that the collector heat exchanger loop is exposed to the lowest temperature possible for best collector performance.

Heat extracted from storage for delivery to the load is removed at the top of the tank, where stratification will ensure availability of the hottest possible fluid. Cooler fluid returned from the load is introduced into the diagonally opposite corner of the tank through an optional diffuser. The cross flow in this loop also ensures maximum storage utilization.

An important by-product of the cross-flow arrangement for storage piping is a reliable controller sensor signal. The storage sensor, as described in a later section, is crucial for proper operation of the solar controller. If cross flow is not used in the storage tank, the sensor may be bypassed by fluid in motion and may sense a temperature having no relation to that in the fluid loops.

Storage Insulation

The exterior of a solar storage tank must be well insulated to retain heat. Losses from the storage tank to the building are not controlled and may occur

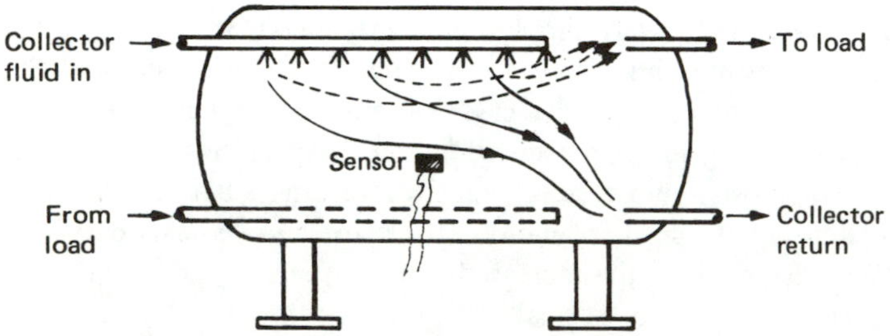

Figure 8.7 Schematic representation of flows in liquid storage tank. Solid lines show collector heat exchanger flow paths; dashed lines show fluid flow to load.

when no heat load is present. Insulation of storage tanks was a problem in some early solar heating installations.

Several types of insulation can be used, including foam installed at the plant, batts installed on site, or a separate insulated housing for the tank itself. All these approaches are feasible and the selection is based on the particular project.

The amount of insulation used is an economic trade-off between the cost of added insulation and the benefit of greater retention of solar heat. Equation (8.1) can be used to calculate the optimum thickness of insulation on a solar storage tank:

$$t_{opt} = \frac{\bar{C}_f \, \Delta T \, \Delta t \, k}{\mathrm{CRF}(i, N) b} - R_0 k \qquad (8.1)$$

where t_{opt} = optimum insulation thickness (ft)

$\bar{C}_f$ = levelized cost of fuel (see Chap. 5) ($/Btu)

ΔT = average temperature difference across insulation (°F)

Δt = number of hours per year during which heat loss occurs, i.e., length of heating season (h)

k = insulation thermal conductivity in English engineering units

CRF = capital recovery factor

b = cost of insulation (dollars per foot of thickness per square foot of area; see App. 7)

R_0 = thermal resistance for heat loss in the absence of insulation

Storage Tank Location

Storage can be located either inside or outside the structure. In most applications storage has been located within the building, where freeze protection is not a problem, access for repair is relatively easy, and piping connections to the balance of the system are straightforward. Outdoor storage may be required if space is unavailable indoors or if the cost of space is prohibitive. Buried storage involves several problems, including difficulty of servicing, possible accelerated heat loss to underground aquifers, and location in a corrosive environment. Storage mounted aboveground is technically feasible, but freeze protection and high insulation levels are necessary to prevent freeze-up and excess heat loss during very cold periods of winter.

The weight of storage also influences its location. Liquid storage is quite heavy, and if it is located within a building extra provisions for the floor slab are necessary, including added thickness. For outdoor locations, special foundations may be required to support the relatively large weight. Supports must also be insulated to reduce heat loss if a metal structure is used. The structural engineer should be consulted for sizing of storage supports, floor slabs, and foundations to support them.

Heat Exchangers

Heat exchangers of two types are used in liquid-based systems, as shown in Fig. 8.4. The first is a liquid-to-liquid heat exchanger between the collector loop and storage, used to isolate the expensive freeze protection fluid from water storage. The second is the load heat exchanger, HX3 in Fig. 8.4, which is used to deliver heat to the space to be heated. In this section both types of heat exchangers are described, although the technical details are not covered. The reader should refer to [6] for additional information on heat exchanger design.

Liquid-to-liquid Heat Exchangers

In counterflow shell-and-tube heat exchangers, the hot fluid is contained in the shell of the heat exchanger and flows in the opposite direction to the fluid in the tubes. The tube fluid in solar heating applications is the storage tank fluid.

A measure of the size of shell-and-tube heat exchangers is the heat transfer area multiplied by the overall heat transfer coefficient U. This product, UA, when multiplied by an appropriate average temperature difference between the two fluid streams, can be used to calculate the heat transfer under a specific set of conditions. However, solar collector loops operate under a wide range of conditions and a different approach is used in solar applications.

An alternative measure of heat exchanger size is the heat exchanger *effectiveness*. This is defined as the actual rate of heat transfer divided by the maximum possible rate of heat transfer between the two fluid streams, that is, the rate limited only by the second law of thermodynamics. The maximum heat transfer rate q_{max} is given by Eq. (8.2):

$$q_{max} = (\dot{m}c_p)_{min}(T_{hot,in} - T_{cold,in}) \tag{8.2}$$

where $(\dot{m}c_p)_{min}$ is the minimum of the two fluid capacitance rates in the heat exchanger and the temperature difference is the maximum occurring across the heat exchanger. The minimum capacitance rate is used in the expression for q_{max} since heat transferred by one stream must be absorbable by the other. The maximum capacitance rate cannot be used since the minimum capacitance stream could not absorb the amount of heat so calculated. The actual heat transfer rate in a heat exchanger can be expressed as shown in Eq. (8.3):

$$q = E_{hx}q_{max} = E_{hx}(\dot{m}c_p)_{min}(T_{hot,in} - T_{cold,in}) \tag{8.3}$$

where E_{hx} is the effectiveness. This expression follows directly from the definition of heat exchanger effectiveness. Equation (8.3) is convenient for solar design since the inlet temperatures of the hot and cold fluids may take on any value and the heat rate can be calculated directly. The effectiveness E_{hx} is only a weak function of temperature and for preliminary solar designs can be taken to be constant.

Unfortunately, the heat exchanger industry does not use the effectiveness approach for sizing of heat exchangers. Therefore the solar designer must revert to the standard mean temperature difference approach when specifying an exchanger to be used for a given application. The example below illustrates the method of converting from the effectiveness to the log mean temperature difference (LMTD) used in the heat exchanger industry.

EXAMPLE

Convert a 70 percent effectiveness rating to fluid stream temperatures and flows required by a heat exchanger manufacturer for a 500-ft² solar system in Denver. Also calculate the LMTD.

SOLUTION

The maximum heat rate Q for HX1 must be known. This will occur on a sunny day in fall or spring, when solar incidence angles are small near noon. Under these conditions, assume that the collector is found to operate at 53 percent efficiency. Therefore the heat collected Q_u is

$$Q_u = 0.53 \times 310 \text{ Btu/(h}\cdot\text{ft}^2) \times 500 \text{ ft}^2 = 82{,}150 \text{ Btu/h}$$

Clear-sky solar radiation data—in this case 310 Btu/(h·ft²)—for various latitudes are given in App. 5.

The next step is to pick an average storage temperature, which becomes the tube-side inlet temperature. Here we will use 130°F. Recall that the heat rate Q_u is

$$Q_u = E_{hx}(\dot{m}c)_{\min}(T_{\text{hot,in}} - T_{\text{cold,in}})$$

For a liquid system, the minimum capacitance rate is through the collector, since storage-side flow is usually 50 to 100 percent above shell-side flow to improve the effectiveness. Using the 0.02 gal/min rule for collector flow (50 percent glycol) and glycol properties,

$$(\dot{m}c_p)_{\min} = [(0.02 \times 500) \times 0.9] \times 500 \frac{\text{Btu/(h}\cdot{}^\circ\text{F)}}{\text{gal/min}} = 4500 \text{ Btu/(h}\cdot{}^\circ\text{F)}$$

From Eq. (4.6) the hot-fluid inlet temperature is

$$T_{\text{hot,in}} = \frac{Q}{E_{hx}(\dot{m}c)_{\min}} + T_{\text{cold,in}} = 156.1^\circ\text{F}$$

The final information for the manufacturer is the tube-side flow rate. Using the 50 percent rule above,

$$\text{Tube flow} = 1.5 \times (0.02 \times 500) = 15 \text{ gal/min}$$

In summary, the thermal data for the manufacturer are

Heat rate: 82,150 Btu/h
Shell entering water temperature (EWT): 156.1°F
Tube EWT: 130.0°F
Shell fluid and flow: 50% glycol, 10 gal/min
Tube fluid and flow: water, 15 gal/min
Type: counterflow

To find the LMTD, the shell and tube outlet temperatures must be found. Using a heat balance on the shell side

$$Q = (\dot{m}c)_{hot}(T_{hot,in} - T_{hot,out})$$

Solving for $T_{hot,out}$

$$T_{hot,out} = T_{hot,in} - \frac{Q}{(\dot{m}c)_{hot}} = 156.1 - \frac{82,150}{4500} = 137.8°F$$

Likewise, for the tube side

$$T_{cold,out} = T_{cold,in} + \frac{Q}{(\dot{m}c)_{cold}} = 130.0 + \frac{82,150}{15 \times 500} = 140.9°F$$

The LMTD is

$$\text{LMTD} = \frac{\Delta T_1 - \Delta T_2}{\ln (\Delta T_1/\Delta T_2)}$$

where $\Delta T_1 = 156.1 - 140.9°F = 15.2°F$

$\Delta T_2 = 137.8 - 130°F = 7.8°F$

$\text{LMTD} = (15.2 - 7.8)/\ln (15.2/7.8) = 11.1°F$

The size of the heat exchanger used between collector and storage is determined by economic considerations. Since heat exchangers require a finite temperature difference between two fluid streams, they impose a performance penalty on a solar collection system. That is, the solar collector will be forced to operate at a temperature several degrees higher to deliver fluid to storage than it would if no heat exchanger were present. The greater the temperature difference across the heat exchanger, the warmer and less efficient the solar collector. A very large heat exchanger could be used to alleviate this problem. However, oversized heat exchangers are very expensive and are not cost effective. At the other end of the scale, very small and inexpensive exchangers cause very high collector temperatures, and the reduced solar system performance will have prohibitive economic consequences. The precise size of the heat exchanger to be used depends on its cost, the value of the fossil fuel saved, and other economic factors. The details of heat exchanger sizing are beyond the scope of this book, but many system designs have indicated that relatively large effectiveness values, in the vicinity of 0.7 to 0.8, are justified. This may be larger than the size suggested by heat exchanger vendors owing to their lack of familiarity with the solar application. However, solar energy is expensive, and a relatively large investment in heat exchangers can be justified.

In addition to specifying the thermal performance of a heat exchanger, as illustrated in the example above, the designer should specify materials that will be compatible with the fluids used and the piping materials used elsewhere in the fluid loops. The operating pressure of the heat exchanger (both shell and tube sides) should be specified, as well as the size of the piping connections and the level of insulation to be used. As is the case with all mechanical system components, the heat exchanger requires careful insulation to minimize parasitic heat loss.

Air-to-liquid Heat Exchangers

Transferring the solar heat contained in a liquid working fluid to air- or space heating requires an air-to-liquid heat exchanger. The most typical design in active solar heating systems utilizes a finned-tube arrangement, with water flowing through a series of tubes connected by thin metal fins. Air is caused to flow across the pipes and fins, resulting in transfer of heat to the air.

Finned-tube coils are standard, off-the-shelf equipment. In a solar heating context, this heat exchanger is most readily specified in terms of effectiveness, and conversion to heat exchanger industrial nomenclature will be required as in the previous example. A suggested rule for sizing the load heat exchanger is indicated in Eq. (8.4):

$$E_{\text{load}}(\dot{m}c_p)_{\text{air}} \sim 2(UA)_{\text{bldg}} \tag{8.4}$$

That is, the load heat exchanger effectiveness E_{load} multiplied by the air capacitance rate across the exchanger should be approximately twice the building heating load expressed in Btu's per hour per degree Fahrenheit or watts per degree Celsius. The constant 2 is established, as usual, by an economic trade-off, and values between 1 and 3 will bracket the most cost-effective design for most forced-air systems.

Optionally, baseboard units may be sized to provide the terminal load device function. Baseboard units are typically designed for operation at 180°F, well above temperatures available from most solar space-heating systems in winter. The solution to this problem is to add baseboard length beyond that required for a nonsolar system. Baseboard systems can operate satisfactorily for solar heating applications if extra capacity is added. One manufacturer's baseboard has a performance of approximately 6 Btu/(h·°F) per foot of baseboard. The length of baseboard required can immediately be determined from Eq. (8.4). For example, if the heating load on a building is calculated to be 1500 Btu/(h·°F), the length required is this number multiplied by 2 and divided by 6, or 500 ft of baseboard. This is approximately twice the length of baseboard normally used for a nonsolar application.

Pumps, Pipes, and Pipe Fittings

Specification of pumps P1, P2, and P3 in Fig. 8.4 requires that the loop pressure drop and flow rate be known. In this section the method of calculating pressure drops in piping and other components is described.

Flow rates in various system loops were discussed earlier. The pressure drop that P1 must overcome results from pressure drops in each of the components of the collector fluid loop. This loop has been isolated in Fig. 8.8, and it is seen that pressure drops occur in the following components of the loop:

Check valve	Storage heat exchanger
Pipe tees, coupling elbows	Piping
Collector manifold	Miscellaneous valves
Collector risers	Air separator

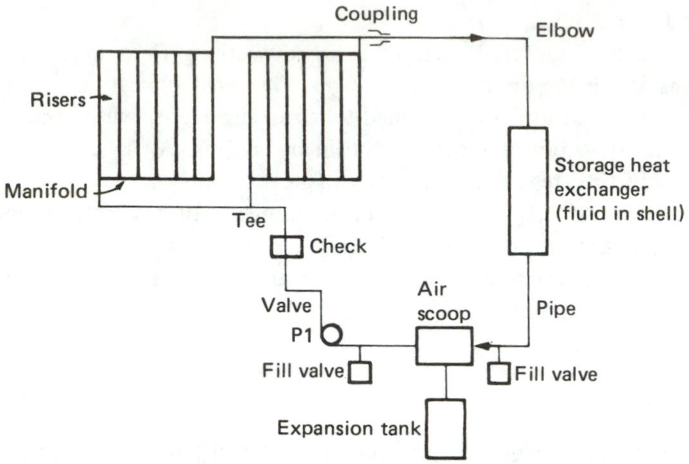

Figure 8.8 Schematic diagram indicating sources of pressure drop in collector loop for liquid systems.

The pressure drop in each component can be determined at the specified flow rate from the manufacturer's literature. This pressure drop and the flow rate for pump P1 are used to specify a pump from the manufacturer's literature.

In addition to pressure drop and flow rates, pump material, operating pressure, and pipe fitting size should be specified. For fluid loops in which no oxygen is present, such as the collector loop, cast iron fittings are suitable. However, in domestic water-heating systems, where fresh water is continously supplied, brass or stainless steel pump impellers are needed to avoid corrosion.

Figure 8.9 and 8.10 can be used to calculate pressure drops across standard pipe fittings and 100-ft lengths of standard pipe sizes. Figure 8.9 is a nomograph that is used by selecting the type of fitting to be analyzed and its size and constructing a line between them. The point of intersection of the line on the center axis specifies the equivalent length of straight pipe. This equivalent length is added to the known physical length of pipe in the loop to determine the pressure drop. Figure 8.10 is used for this purpose. The pressure drop per 100 ft of pipe can be read from the specified pipe diameter and known flow rate. A wide range of pipe diameters can be used to carry fluid at a given flow rate, and further information is required to specify or select the pipe size. It has been found in the piping industry that flow rates in excess of 6 ft/s in copper or steel pipes are uneconomical because of erosion and the pumping power needed. Therefore, for preliminary pipe size selection, Fig. 8.10 can be used with a pipe velocity of 6 ft/s. The known volumetric flow rate and this velocity then specify a pipe diameter, which can be used to find the pressure drop in the collector or any other fluid loop.

Pump Horsepower

To determine the amount of energy required by various pumps in solar systems, the horsepower must be known. When the horsepower is multiplied by

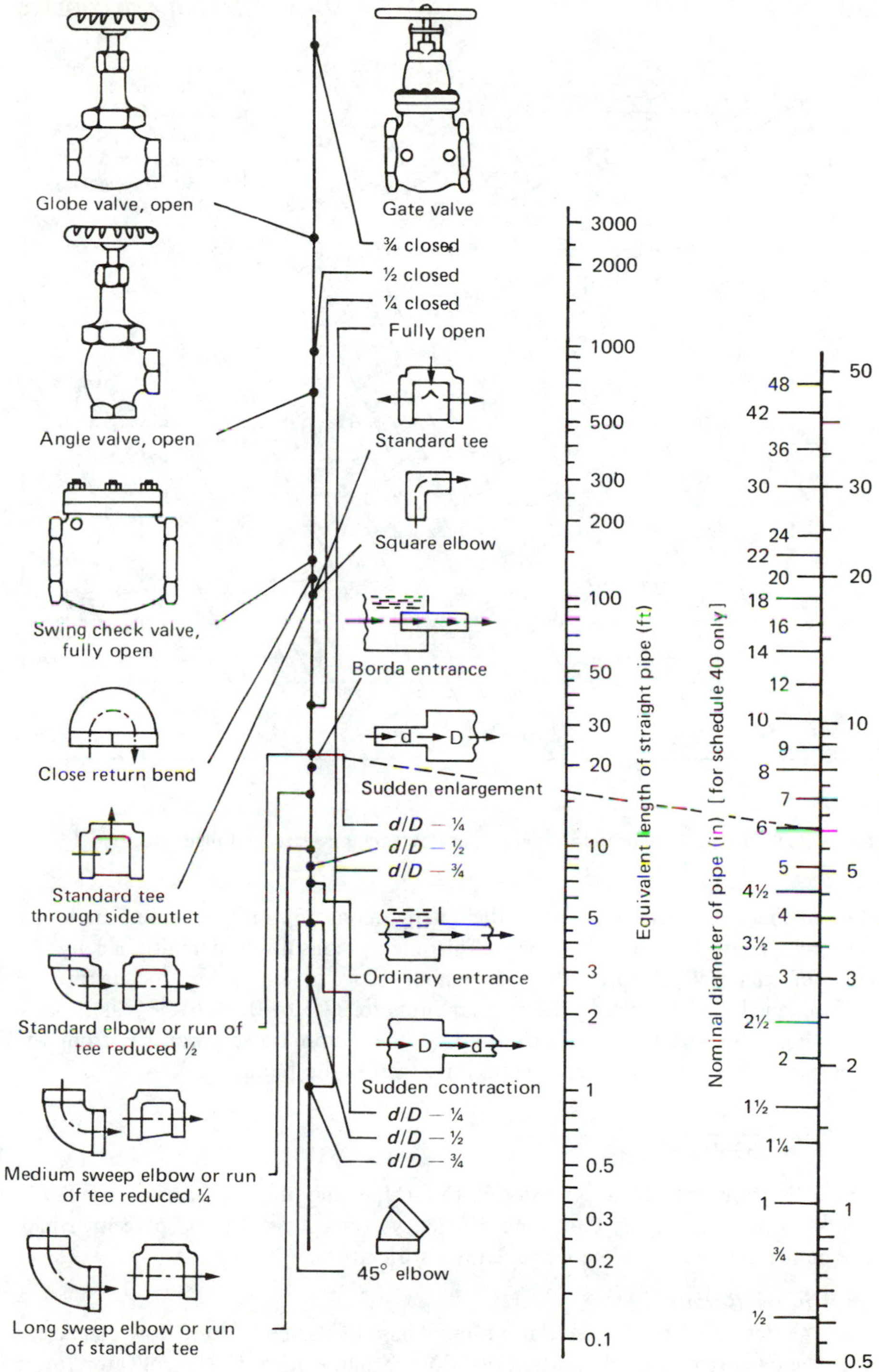

Figure 8.9 Nomograph for calculating equivalent length of pipe in standard pipe fittings and valves.

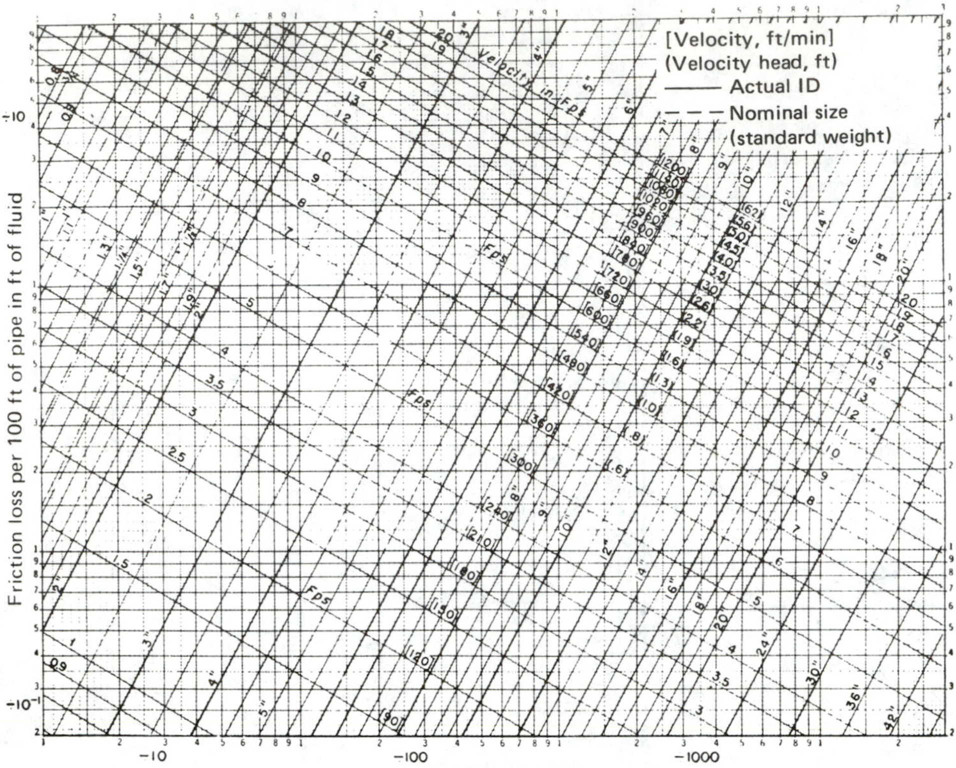

Figure 8.10 Pipe pressure drop as a function of flow rate, pipe size, and pipe velocity.

the number of operating hours, the energy consumption expressed in horsepower-hours or kilowatts is known. This energy consumption multiplied by the cost of electricity represents the annual cost of operating the pump. In well-designed liquid systems, the cost amounts to 0.03 or 0.04¢/(ft^2·year).

Equation (8.5) can be used to find the pump horsepower HP from the known flow rates and required pressure rise across the pump:

$$\text{HP} = \dot{Q} \times \frac{\Delta p}{3960 \times E_p} \tag{8.5}$$

where the flow rate $\dot{Q}$ is expressed in gallons per minute, the pressure rise Δp in feet of water, and E_p is the pump efficiency, typically 50 to 60 percent, which can be determined from the manufacturer's literature.

Flow Balancing

As described in the section on liquid-based solar collectors, flow balance in a collector array is as important as flow balance in a single collector. Most collectors used in large systems are connected in parallel, since this results in a reduced pressure drop and reduced horsepower requirement. The difficulty with parallel arrangements, however, is ensuring that each collector in the array receives precisely the same amount of fluid. There are two methods for achieving this flow balance. One is by brute force, using balancing cocks in each collector fluid branch. However, these devices are relatively expensive and require a balancing contractor to fine-tune the system to ensure proper operation. In addition, the small orifices of these devices are prone to clogging in solar applications, where relatively low flows are used.

The alternative method is to design the fluid loop from the outset for proper balance. This can be done by sizing the collector inlet and outlet manifolds so that the pressure drop in the manifold is approximately one-tenth of the total array pressure drop. That is, 90 percent of the pressure drop occurs across the collector and 10 percent in the manifolds. Collector pressure drop is controlled by collector design. As noted earlier, it is important that there be a finite pressure drop across solar collectors. Otherwise, the 90 percent rule is achievable only if extremely large headers with very small pressure drops are used.

The pressure drop in a parallel array header is difficult to calculate, since fluid additions and withdrawals from the header occur at every collector connection point. However, the pressure drop can be estimated to engineering accuracy from Eq. (8.6):

$$\Delta p_{\text{header}} = \frac{K\dot{m}^2}{6} \; [j(2j + 1)(j + 1)] \tag{8.6}$$

This equation applies to $j + 1$ identical collectors with the same flow rate $\dot{m}$ per collector. The dimensional constant K in Eq. (8.6) is defined in Eq. (8.7):

$$K = \frac{8\bar{f}L}{\pi^2 \rho D^5} \tag{8.7}$$

where ρ = fluid density
D = pipe inside diameter
L = distance between collector connection points, expressed in consistent units
$\bar{f}$ = friction factor in the header evaluated at half the total array flow $(j + 1)\dot{m}/2$

The friction factor can be calculated from information contained in Fig. 8.10, as shown in Eq. (8.8):

$$f = \frac{\Delta p}{(L/D)(\frac{1}{2}\,\rho V^2)} \tag{8.8}$$

That is, f is the pressure drop Δp divided by the product of the length-to-diameter ratio and the velocity head $\frac{1}{2}\,\rho V^2$, where V, $\Delta p/L$, and D can all be found from Fig. 8.10. The value of f from Eq. (8.8) is inserted into Eq. (8.7) to find K, which is used in Eq. (8.6) to define the header pressure drop. This is the pressure drop that should be approximately one-tenth of that for each collector at the design flow rate $\dot{m}$.

It is seen that the header pressure drop is very sensitive to pipe diameter, depending on the inverse fifth power of diameter. Therefore small pressure drops in the header required to meet the 90 percent rule are usually readily achieved by increases in header diameter. The final selection of diameter is an iterative process, since the friction factor depends on diameter as well as pressure drop in the header. However, two or three iterations are usually sufficient to find the proper diameter to meet the 90 percent rule.

Thermal Design of Piping Loops

Two thermal considerations are of importance in the design of piping loops. The first is the rate of parasitic heat loss from the surface of the pipe and the second the accommodation required for expansion and contraction in piping loops.

Heat loss from pipes is calculated from well-known equations that can be found in any heat transfer text. Equation (8.9) expresses the heat transfer through cylindrical insulation placed around a pipe:

$$q_{\text{pipe}} = \frac{2\pi kL}{\ln (D + 2t)/D} (T_f - T_a) \tag{8.9}$$

Insulation usually provides the principal resistance to heat transfer, and use of Eq. (8.9) is adequate for heat loss calculations. Losses from piping are part of the heating load on both water- and space-heating systems and should be included as part of the load to be met. In Eq. (8.9), k is the insulation conductivity, L the pipe length, D the pipe outside diameter, t the insulation thickness, T_f the fluid temperature, and T_a the surrounding ambient temperature; q is expressed in units of Btu's per hour or watts. Values of insulation conductivity can be read from Apps. 7 and 8 for common types of pipe insulation.

The amount of insulation used for piping depends on the relative cost-effectiveness of the insulation and the value of the heat lost. A number of publications are available for determining the proper size of insulation to use on piping. Alternatively, recommendations in energy conservation codes such as ASHRAE 90-75 can be used to specify insulation thickness. This standard contains tables of insulation thickness for different operating temperatures and pipe diameters; however, the thicknesses listed are probably less than the economic optimum.

Accounting for expansion and contraction in piping networks is a particularly troublesome problem in the design of large arrays with very long pipe runs. Expansion loops, with their associated added cost, or expansion joints may be required. Equation (8.10) can be used to calculate the change in length ΔL for a length L_0 of pipe undergoing a change in temperature ΔT:

$$\Delta L = \bar{\alpha} L_0 \, \Delta T \tag{8.10}$$

The coefficient of linear expansion α is given in Table 8.1 for several common materials. It is seen that aluminum expands approximately twice as much as steel

Table 8.1 Coefficient of Linear Expansion for Common Materials

Material	$\bar{\alpha}(°C^{-1}) \times 10^6$	$\bar{\alpha}(°F^{-1}) \times 10^6$
Aluminum	23	13
Brass	19	11
Copper	17	9
Glass	9	5
Steel (carbon)	15	8.4

or copper, the more common materials for solar piping loops. If proper accommodation for expansion and contraction is not provided, enormous forces can be produced within piping arrays and can physically distort both the collector and its mounting subsystem. Of course, expansion and contraction occur in storage tanks, solar collector absorber plates, and all metal components that undergo changes in temperature. Most collector manufacturers have designed collectors to accommodate expansion over a 400°F temperature range. Likewise, storage support design usually includes provision for expansion and contraction.

A final consideration in the design of piping systems is related to galvanic corrosion. Dissimilar metals react chemically when they are connected electrically. If the metals are copper and aluminum, this causes aluminum to pass into solution. If this happens over a prolonged period, the aluminum surface will ultimately be dissolved away and a leak will form. Design of piping loops for compatibility in this regard is not a unique solar consideration, and plumbing handbooks will delineate solutions to the problem. With a mixture of metals the problem can be partially fixed by using dielectric unions between dissimilar metals, but since water is an effective electrolyte, a dielectric union only delays the galvanic corrosion problem. The best solution appears to be to use only one metal.

Secondary Components of Liquid Systems

In this section the selection of fluids, heat rejection units, backup systems, and instrumentation for liquid-based solar systems is briefly described. Many of these matters have been discussed in Chap. 6 for water-heating systems.

Fluids

The working fluid in all loops other than the collector loop in liquid-based systems is treated water. Water is treated if it is necessary to retard corrosion. However, if fluid loops are tightly closed, oxygen initially in the circulating water will be quickly consumed and treatment is probably not necessary.

For the collector loop, selection of the working fluid requires information about the most severe thermal environment to which the collector will be exposed. This minimum temperature can be used to select the fluid. The fluid used most commonly in collector loops is an aqueous solution of ethylene or propylene glycol. Propylene glycol is preferred by most designers because of its low toxicity. The thermal characteristics and costs of both glycols are roughly the same. Figure 8.11 can be used to determine the percent-by-weight glycol requirement for a given freezing point. For example, for protection at −20°F, a 48 percent solution of propylene glycol in water would be required. Of course, the graph also indicates that a 91 percent solution would provide the same freeze protection, but the added cost and increased viscosity of the fluid at this concentration indicate that the lower concentration should be used.

Fluids other than glycols have been used for freeze protection. Organic heat transfer fluids used for many years in industry are available, as are silicone oils. These fluids have very low surface tensions, are expensive, and are prone to leak where aqueous solutions will not leak. These fluids have the advantage that periodic replacement, which is necessary for glycols, is no longer a requirement. However, it is not clear at present that the extra cost of the fluid and the extra care required in system fabrication are ultimately paid off in added reliability over the life cycle of the system. The economic consequences of a spill involving these more expensive fluids are significant.

The fluid that is used is a factor in the selection of fluid circulation pumps. Since nonaqueous fluids are more viscous than water, increased pump capacity is required. Organic fluids also have a lower specific heat than water, so that a higher volumetric flow rate is necessary to achieve the same heat addition with the same temperature rise in the collector fluid. Increased flow and reduced specific heat both require increased pump size. In selecting the pump model and

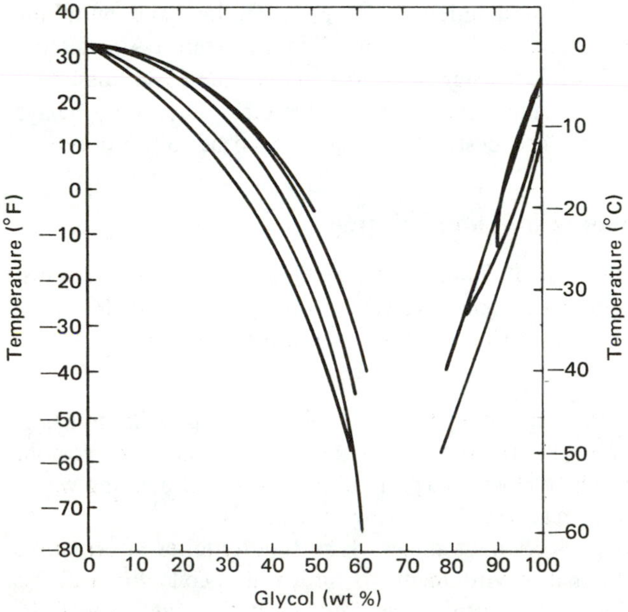

Figure 8.11 Freezing point of common glycols as a function of glycol weight percent. (*Courtesy of Union Carbide Corp.*)

Table 8.2 Important Properties of Liquid Heat Transfer Media*

Property	Water	50% Ethylene glycol/water	50% Propylene glycol/water	Silicone fluid	Aromatics	Paraffinic oil
Freezing point, °F (°C)	32 (0)	−33 (−36)	−28 (−33)	−58 (−50)	−100 to −25 (−73 to −32)	–
Boiling point, °F (°C) (at atmospheric pressure)	212 (100)	230 (110)	–	None	300 to 400 (149 to 204)	700 (371)
Fluid stability	Requires pH or inhibitor monitoring	Requires pH or inhibitor monitoring	Requires pH or inhibitor monitoring	Good	Good	Good
Flash point,† °F (°C)	None	None	600 (315)	600 (315)	145 to 300 (63 to 149)	455 (235)
Specific heat at 73°F, Btu/(lb·°F)	1.0	0.80	0.85	0.34 to 0.48	0.36 to 0.42	0.46
Viscosity, cSt at 77°F	0.9	21	5	50 to 50,000	1 to 100	–
Toxicity	Depends on inhibitor used	Depends on inhibitor used	Depends on inhibitor used	Low	Moderate	–

*These data are extracted from manufacturers' literature to illustrate the properties of a few types of liquids that have been used as transfer fluids.

†It is important to identify the conditions of tests for measuring flash point. Since the manufacturer's literature does not always specify the test, these values may not be directly comparable.

Source: U.S. Department of Housing and Urban Development.

rating, this additional consideration in the collector loop must be taken into account. Water is used in the storage and delivery loops, and no adjustment for special working fluids need be made.

Table 8.2 summarizes important properties of common heat transfer fluids used in liquid-based systems. The key thermodynamic parameters are specific heat and viscosity, but toxicity is also important. For propylene glycol solutions, the requirement for double-wall heat exchangers may be relaxed in plumbing codes, and a significant performance and economic benefit is therefore associated with use of this fluid. Systems with ethylene glycol antifreezes must have double-wall heat exchanger protection between the glycol loop and any potable water loop, as described in Chap. 6.

Expansion Tanks

All closed fluid loops require expansion tanks to accommodate volume changes that occur over the relatively wide temperature excursions present in solar system applications. The sizing of expansion tanks depends on the expected temperature change, the loop pressure, and the type of fluid used. Expansion tank vendors can assist the solar designer in selecting the proper tank. The acceptance volume of the expansion tank needed in the collector loop is typically approximately 10 percent of the volume of confined loop fluid. The acceptance volume of an expansion tank is approximately one-third the total physical volume of the tank. The most reliable type of expansion tank design uses a rubber diaphragm pressurized by an inert gas charge.

To reduce parasitic heat losses, an expansion tank should be placed below the pipe to which it is connected. In this configuration it need not be insulated. An alternative to use of a closed expansion tank is an open expansion tank placed above the highest point in the collector fluid loop. Expansion is automatically achieved by permitting the level to rise and fall in this tank as the fluid loop temperature varies. Use of open tanks may present reliability and location problems, however.

Heat Rejection Units

The heat rejection unit HRU shown in Fig. 8.4 is used to dissipate collected solar heat during periods of the year when heat requirements are low. An active heat rejection system is required for most reliable operation. Some early system designs simply turned off the solar collector when storage temperature reached a prescribed upper limit. However, this approach is not recommended as it leads to several difficulties. First, glycol working fluids degrade around 260°F. On a sunny day a reasonably efficient collector may stagnate under no-flow conditions at 400°F. Deterioration of glycol into glycolic acid and oxalic acid is rapid and corrosion of metal piping and pipe fittings is certain to result.

The second difficulty with the simple pump turnoff approach is the possibility of fire. As noted above, wood used as part of the collector housing can spontaneously ignite if permitted to stagnate at elevated temperatures for a period of 1 or 2 months. Since the stagnation deterioration is cumulative over the lifetime of the collector, at some time all collectors in which the turnoff approach is used may cause a fire hazard if wood is present. The third difficulty with the turnoff approach is the effect on collector materials of long exposure to elevated temperatures. Although collectors are designed by most manufacturers to tolerate stagnation for extended periods, it is prudent to minimize their exposure to this condition, since deterioration will ultimately take place.

Active heat rejection of the type shown in Fig. 8.4 is a reliable alternative to simply turning off collector pump P1. The heat rejection unit is a liquid-to-air heat exchanger sized for the maximum expected hourly collection of solar radiation. This number is calculated from the known peak insolation in spring or fall, when most heat rejection problems take place, and the known collector efficiency under spring or fall conditions. The product of efficiency, insolation, and collector area is the number of Btu's per hour to be rejected. The entering fluid temperature required to specify the HRU size is determined from the design maximum tolerable temperature of storage. The temperature of air entering the

HRU is based on the highest expected temperature during the seasons when heat rejection will be required. Typically, this will be a peak temperature in September.

Heat rejection units on smaller systems can be weatherproof unit heaters, which are manufactured by several commercial firms; for larger systems a cooling tower may be required. An alternative to these two approaches would be to dump hot water from solar storage and replace it with cool water. This approach can be used if water is not in short supply, but is generally deemed an undesirable heat rejection method.

Backup Heating Systems

For most solar heating systems, a 100 percent backup system is required. That is, the same system is present for space heating as if no solar system were used. If this is the design approach, the backup system is sized according to conventional peak heat load calculations, as described in [3]. The specification of the backup system is independent of the existence of the solar system and need not concern us in this book.

It has been shown for some systems in some locations that the solar system can result in a reduction of peak heating loads, since some solar heat is available to offset peak heat losses. Whether this is the case for a specific project can be answered only by detailed hourly computer simulations. The conservative approach is to provide the 100 percent backup capability.

Freeze Protection

Freeze protection for liquid-based systems has been described in Chap. 6 in connection with solar water-heating systems. The reader is referred to that chapter for a discussion of various freeze protection approaches.

Instrumentation

Solar systems should be instrumented at a minimum level to ensure proper functioning. Pumps may operate and air movers move air but heat transfer may not take place. The addition of a few pressure gauges and thermometers can ensure that a solar system is, in fact, delivering energy. Thermometers should be inserted in each fluid loop where heat addition or heat withdrawal takes place. For example, in Fig. 8.4 a pair of thermometers would be added at the inlet and outlet connections for both streams of the heat exchanger as well as at the load heat exchanger. An additional thermometer could be inserted in thermal storage to monitor its average temperature.

Flowmeters can be used to evaluate flow rates in each fluid loop; however, pressure gauges are less expensive, more reliable, and serve the same purpose. A pair of pressure gauges inserted at the inlet and outlet of each pump can be used to determine whether the pump is providing the proper flow rate. The manufacturer's pump curve can be used to determine whether the pressure rise is that associated with the design flow rate. Pressure rise readings at each pump should be recorded when the system is new. At inspections in future years, pressure at these same locations can be recorded to determine whether flow rate levels have been maintained at their design point. If not, filter cleaning or loop flushing of some sort is indicated.

The third type of instrument that may be useful in troubleshooting or crudely monitoring system performance is a solar radiation meter. Photocell-based pyranometers can be purchased for approximately $50 and will suffice for determining proper system function. Of course, a much more accurate meter is needed to monitor performance on a continuous basis. Performance monitoring, however, is not the subject of this section.

AIR-BASED MECHANICAL SOLAR HEATING SYSTEMS

Air suggests itself as a solar working fluid for space-heating systems since the majority of such nonsolar systems in the United States employ forced air. Use of air instead of a liquid can result in system simplification, since a liquid-to-air heat exchanger is not required. Also, air is noncorrosive, does not freeze or boil, and does not require heat rejection during low heating load periods. The only fundamental difficulty with air-based systems is the poor heat transfer between air passing through the collector and the collector absorber surface. By proper system design this inherent inefficiency of air collectors can be overcome and air-based systems can provide approximately the same energy output per year as well-designed liquid systems. In this section, the operation of air systems is described with respect to each operational mode.

System Configuration

The basic components of an air system are shown in Fig. 8.12. These components include collectors, storage, a fan, control dampers, backup system, and water preheating subsystem with heat exchanger. The air system does not require either a collector-to-storage or a storage-to-load heat exchanger, both of which are present in liquid systems. In other respects, however, the operation of both systems is the same, although the air-based system can be operated in the direct collector-to-load mode, which is not possible in liquid systems.

Space heating directly from the collector can be visualized by referring to Fig. 8.12. Return air entering through a backdraft damper (similar to a check valve) and filter is introduced into the collector, where heat is added. Heated air passes through damper D1, the blower, and damper D2 on its way to the load. After the load is satisfied the control system so indicates and damper D2 is placed in the storage mode. During storage charging, air from the fan outlet passes through damper D2 and into the upper zone of pebble-bed storage. As the air passes through pebble-bed storage, heat is extracted, and the air exits the storage at approximately room temperature for return to the collector inlet. Use of pebble-bed storage is essential to ensure that the collector has the coolest possible inlet temperature. It is this single system consideration that permits an air-based system to provide energy as effectively as a liquid-based system. Phase-change storage or liquid storage reduces performance, since a 70°F inlet

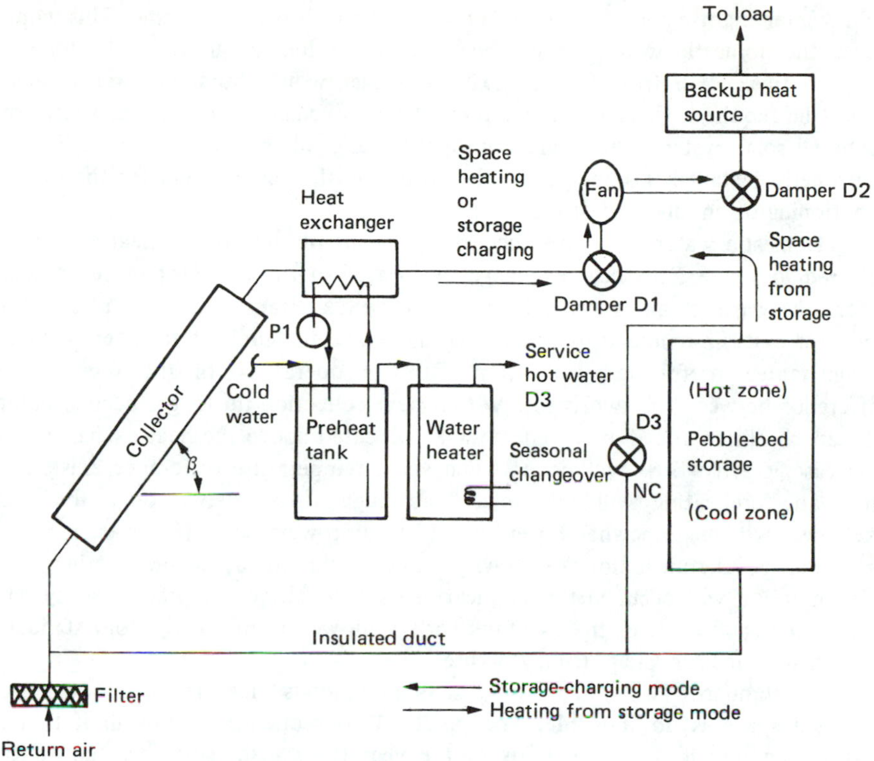

Figure 8.12 Schematic diagram of air-based solar heating system including domestic water preheat subsystem. The series backup location may be inappropriate for some forms of backup heating, as described in the text.

temperature to the collector cannot be guaranteed with these two storage types.

The final operational mode is used during periods of solar outage when a space-heating demand exists. In this case, air flows through the return duct and in the opposite direction to pebble-bed storage. This is achieved by repositioning damper D1 so that the fan inlet is connected to the hot zone from storage. After leaving the fan, storage-heated air passes through damper D2 to the load. The counterflow or reverse flow of heat through the storage bed during heat extraction from storage guarantees that warm air is provided to the space. If air were to flow in the charging direction through storage, heated air would simply be moved from one zone of storage to the other and would not provide useful space heating.

During the storage-to-load mode, it is absolutely essential that damper D1 seal completely. If it did not, leakage would occur through the solar collector, which is at approximately ambient temperature during periods of solar outage. Mixing ambient-temperature air with warm air from storage depletes the heating effect and can cause the backup system to operate even though heat is contained in storage. A second difficulty would arise if damper D1 were to leak. On very cold winter nights, collector outlet air would be well below freezing. This would cause the domestic water preheat heat exchanger located in the collector exit duct to freeze. On freezing, the heat exchanger would burst and replacement would be required. Freezing of this preheat heat exchanger has occurred in many air-based solar systems that were improperly designed, inadequately specified, or improperly installed. Damper D1 is therefore a critical component for the proper functioning of an air-based space-heating system.

Domestic water preheating occurs by means of the preheat heat exchanger referred to above. This heat exchanger is located in the collector outlet stream, where the warmest air is available to maximize heat transfer rates. Whenever the collector exit air temperature is sufficiently above the preheat tank temperature to guarantee useful heat collection, P1 will operate until the temperature difference between the two is so low that heat collection no longer occurs. Pump P1 can be eliminated if the preheat tank is located above the heat exchanger. In this case, gravity flow will occur whenever a temperature difference exists and automatic preheating will take place, although flow rates through the heat exchanger will be somewhat lower than if a pump were used. However, as noted in Chap. 6, thermosiphon or gravity flows tend to be self-correcting, and self-correction will occur in this application as well. The water preheat subsystem shown in Fig. 8.12 is of the two-tank variety; however, any of the four standard systems described in Chap. 6 may be used.

In standard air system designs, simultaneous heating of storage and occupied space is not possible. The practical consequence of this limitation is that during low heat load periods of the year the rate of solar heat collection may be significantly greater than the rate of heat loss from the space. Unless proper control design is used, space temperatures may rapidly rise before a thermostat senses that the load has been satisfied. In some air systems this thermal overshoot has caused the air-conditioning system to come on to return room temperatures to their design condition. Careful placement of thermostats and control system design should be able to avoid this problem.

Nonheating Season Operation

During summer, the air-based heating system will not be operated since the presence of a large, hot storage volume within the space is not desired. However, the collector may serve the function of water preheating in summer if a seasonal changeover damper is used, as shown in Fig. 8.12. During summer, the manual damper would be positioned to bypass storage as indicated. The collector fan would operate only when the pump P1 controller decided that collection is worthwhile for water heating. Since a relatively large collector would be serving a relatively small load in summer, only an hour or two of collector operation should be necessary. Some system designers have unnecessarily operated the air system for the entire day when only a few hours of useful energy collection occurred. A summer mode must therefore be programmed into the controller to change the operational criteria for the fan between the summer no-load situation and the winter heating situation.

Heat rejection from air-based collectors manufactured by several commercial firms has simply been by the stagnation method. That is, flow ceases through the collector, the collector rises in temperature to its equilibrium point, and heat is reradiated and convected from the very hot absorber surface. If the collector is designed for this type of heat rejection, this approach is adequate, although the thermal stress so produced may eventually cause deterioration of organic materials in the collector. Another method of heat rejection for summer is to operate the fan periodically and exhaust air from the collector outlet stream. Continuous operation of the fan should be avoided to limit electrical energy consumption.

The seasonal changeover damper D3 used to bypass storage during water preheating or heat rejection must be of high quality and have a very low leakage rate. Hot air leaking past D3 would enter the space and cause an increased air-conditioning load in summer. The seasonal changeover damper should be repositioned in fall before there is a significant heating requirement so that the pebble-bed storage will be fully charged at the outset of the heating season. An alternative to the summer operation described above is use of a second small fan in place of the full-sized system fan for summer operation. Reduced airflow through the collector will result in higher outlet temperatures, increased heat transfer at the water preheat heat exchanger, and reduced electricity consumption for fan power.

Backup System Operation

One series backup configuration is shown in Fig. 8.12. This location for backup is suitable if a heat pump is used. However, in residential applications the backup system is frequently a fossil fuel-fired furnace. In this situation it is not suggested that a series backup be used, since all solar-heated air passing through the furnace will have associated stack losses, which do not contribute to effective space heating. Therefore in the residential situation it is suggested that the backup heat source be placed in parallel with storage, similar to the liquid-based system shown in Fig. 8.4. Stack losses are thereby completely avoided and proper auxiliary heating ensured.

A second difficulty with the series configuration, when furnaces are used, is related to fan motor design. In residential gas- or oil-fired furnaces, blower motors are designed to operate at ambient temperatures below 100°F. If solar-heated air passes through a furnace whose fan is used as the solar blower, the fan will be exposed to temperatures above 100°F in spring and fall. In this event the blower motor will overheat and cease to operate when the over-temperature circuit breaker opens. The parallel auxiliary configuration also solves this difficulty. Parallel location requires additional dampers, however, and if fossil fuel backup in the form of a forced-air furnace is not used, the series configuration is adequate as shown in Fig. 8.12.

Air-based Collectors

Air collectors are described in detail in Chap. 3 and the reader is referred to that chapter for technical details. In addition, the section in this chapter on liquid systems has described many of the practical details of collector installation and construction, which also apply to air-based systems.

Connections between air collectors are different than those between liquid collectors, and they will be briefly described. Two methods are in common use commercially. In the first a rectangular or circular duct connects each collector array. This approach is based on common air ducting design in the HVAC industry. However, solar collectors are relatively small individually compared to the overall array used for space heating, and very many connections are required. Proper insulation of individual ducts and prevention of leaks can be expensive. Several manufacturers have improved the design by using internal ducting and having no external connections. Collectors are placed adjacent to each other, and openings for airflow between them are automatically sealed by gaskets compressed during the collector installation operation. For example, Solaron Corp. of Denver, Colorado, offers an air collector design that can be connected without any ducting in arrays 400 ft^2 in area. The reduced installation cost, improved reliability, and much reduced opportunity for leaks are obvious.

Pebble-Bed Storage

The only practical storage medium for air-based systems at present is the pebble bed. This type of storage is quite inexpensive, generally available, and conducive to the stratification required for proper air system function. It will be recalled that the collector inlet temperature must be as low as possible to guarantee proper air system performance. The lowest useful temperature is room temperature, so the collector inlet temperature should be approximately 65 to 70°F. Since heat transfer from pebble to pebble in a pebble bed is very slow, this type of storage unit guarantees that the collector inlet air will be at the temperature of the cool zone of storage. The cool zone of storage is always near room temperature during most of the heating season, since the storage-to-load mode exposes this end of storage to room air. Design of pebble-bed storage units has been described in Chap. 4, and only a few practical details are included in this section.

Among the most important features of a pebble bed are uniformity and cleanliness of the rocks used. The type of rock used is not of great consequence, since the specific heats and densities of most commonly available rocks are similar. Pressure drop will be somewhat reduced if polished, rounded river gravel is used; however, this is not a requirement. But it is necessary that the pebbles be of the same size and be very clean before installation in the pebble bed.

The rock bed itself is contained in a wood or concrete housing. In the residential environment either material is commonly used and both are relatively inexpensive. A concrete housing is probably preferred since it can be poured as part of the basement and will not undergo dimensional changes with age. Wood has the problem of warping and changing dimensionally with time, as warm rock tends to drive moisture from the wood during long periods of exposure. The rock bed should be insulated properly either on the interior or exterior. Interior insulation is preferred in concrete storage to act as a thermal break between the warm storage region and the colder floor slab. In some jurisdictions, building codes require that a nonflammable lining constitute the innermost surface of the pebble bed. This can be achieved easily by use of gypsum board.

Storage container tops have often been a source of leaks; therefore a careful design and detailed specification of this relatively simple component should be prepared. The top should be gasketed to the rest of the housing and attached positively with bolts or screws.

Proper air distribution in a rock bed is relatively easy to guarantee if adequate plenums on top and bottom are used. Figure 8.13 shows a typical pebble-bed design with approximately 8-in plenums at both top and bottom. The exact size of the plenum can be calculated by requiring that the pressure drop from one end of the plenum to the other be less than 10 percent of the pressure drop through the bed. The bed pressure drop can be calculated from Fig. 4.10, and the pressure drop in the plenum can be considered to be the entering velocity head given by Eq. (8.11):

$$\Delta p_{\text{plenum}} = 6.2 \times 10^{-8}\ \text{FPM}^2 \tag{8.11}$$

where FPM is the plenum inlet velocity in ft/min and the pressure drop Δp_{plenum} is expressed in inches of water gauge. If the frictional pressure drop through the rock bed is 9 or 10 times this velocity head, excellent flow balance through the entire bed is achieved. This consideration, along with economics, will dictate the path length of the rock bed. In residential applications, a roughly cubical configuration has been widely used.

Another important practical detail to consider in pebble-bed design is the

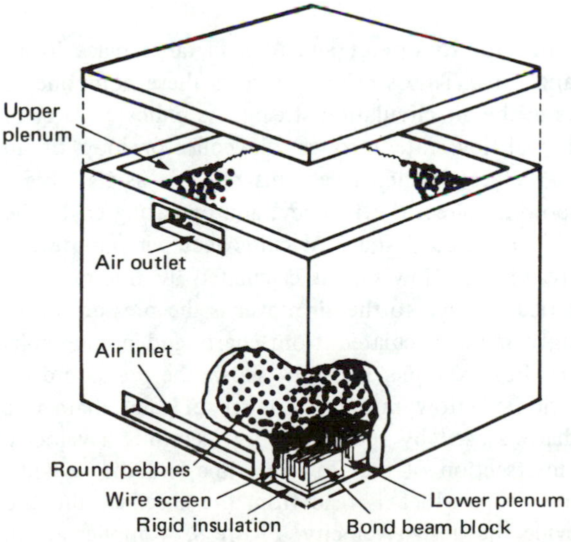

Figure 8.13 Cutaway drawing of example pebble-bed storage unit, showing inlet and outlet plenums and block for pebble support within the lower plenum. (*Courtesy of Solaron Corp.*)

location of the storage temperature sensor required for fan control. This sensor should be just above the lower plenum, that is, in the cold end of storage. This sensor is described in detail in the following section on controls. A simple method of protecting the storage sensor is to position it after the box has been filled with gravel. This can be done by inserting a piece of conduit from the outside of the storage box to the desired location. After filling, it is a simple matter to insert the thermistor in the conduit, slide it to the end of the conduit, and finally seal the end of the pipe to avoid air leakage from storage.

Pebble beds for large commercial buildings require somewhat more careful design than those for residential applications since the volume of rock required is very large. It may be difficult to ensure proper flow balancing in one very large storage bed. The approach of some designers has been to subdivide storage into several zones and feed each zone part of the collector outlet air. Proper flow balance can be ensured in each substorage volume without needing the very large plenum dimensions that would be required if a single large storage unit were used.

As indicated earlier, the best flow direction for rock beds is vertical, not horizontal. Horizontal flow has two problems. First, stratification is difficult to achieve with two-dimensional flow, which will occur when horizontal flow is attempted. A second difficulty has been observed in some installations where the horizontal pebble bed was not completely filled. In these cases, air has bypassed the pebble bed and moved over the top of the bed from one plenum to the other, so that no storage at all occurred. This subtle problem may arise some time after filling of the storage bed, as the gravel settles and a bypass path develops over the top of the bed.

Airflow System

Transport of heat from collector to storage and to load takes place in air systems via ducts, dampers, and fans. This section describes these components and other ancillary components in the air circulation stream. As indicated later in a section on system sizing, the airflow rate through the collector loop of an air-based system is based on a well-established rule. This rule has as its basis a trade-off between pump horsepower, collector efficiency, and operating cost. The rule is that the flow rate should be 2 to 3 standard cubic feet per minute per square foot of net collector area. If the flow rate is calculated by this method, the remaining item to be specified relative to the air mover is the pressure drop.

Pressure drops in ducting can be calculated from charts and nomographs prepared by equipment vendors. For example, Fig. 8.14 shows the pressure drop for 100 ft of duct as a function of flow rate for a range of duct diameters between $1\frac{1}{2}$ and 20 in. The chart is used by specifying, for example, a velocity and a known flow rate. The intersection of the flow rate line and the velocity line indicates both the pressure drop, which is read from the abscissa, and the duct diameter required to provide the design velocity. Figure 8.14 applies at sea level.

Figure 8.14 applies for straight runs of duct. The method for including

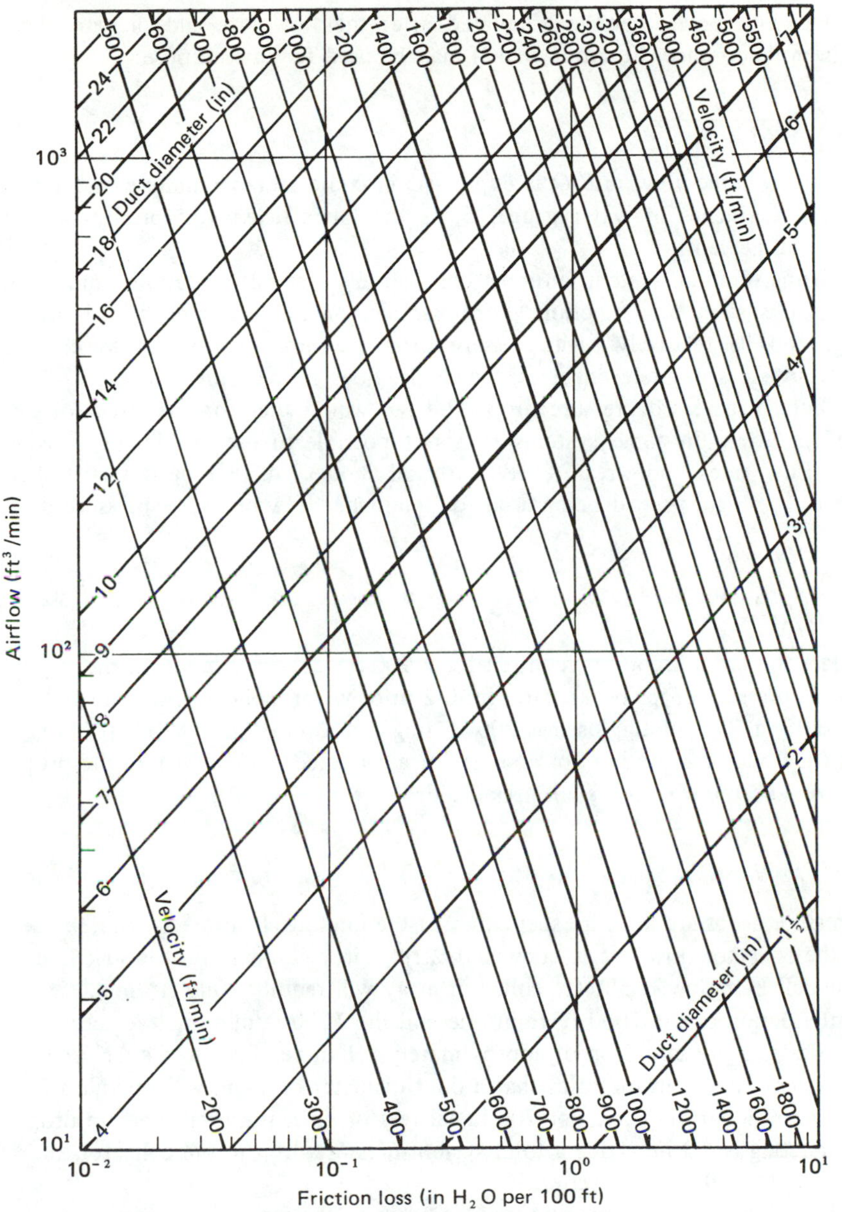

Figure 8.14 Duct pressure drop for airflow as a function of volumetric flow rate, duct diameter, and flow velocity.

bends, dampers, turning vanes, and the like is analogous to that used for pipe fittings, as shown in Fig. 8.9. The equivalent length of all fittings is determined and is added to the physical length of duct before Fig. 8.14 is used. Complete tables of equivalent length factors for many types of air duct fittings are given in [6].

Fan horsepower can be calculated after the flow rate and pressure drop have been determined. Equation (8.12) may be used for this purpose:

$$HP = \dot{Q} \times \frac{\Delta p}{6356 \times E_f} \tag{8.12}$$

where the volume flow rate $\dot{Q}$ is expressed in cubic feet per minute and Δp is expressed in inches of water gauge; E_f is the fan efficiency, typically on the order of 50 percent.

Sizing of an air system fan must consider all three basic operation modes in which the system can function. For example, in the collector-to-load mode, storage and its relatively high pressure drop are not involved. However, the collector-to-storage mode involves the same flow rate as the collector-to-load mode but an increased pressure drop. The fan would therefore be sized for this worse condition. In some systems it is not possible to use one fan to provide proper flow rates and pressure rises for all modes. A two-speed fan is then required, the higher speed being used for the higher flow or higher pressure drop mode.

EXAMPLE

Calculate the fan horsepower required for a 900-ft^2 collector in which the following pressure drops occur: storage, 0.2 in of water; collector, 0.1 in; filter, 0.05 in. The ducting used is 200 ft long and a velocity of 1000 ft/min is specified. Determine the fan horsepower for a 50 percent efficient fan, the proper duct size, and the overall system pressure drop.

SOLUTION

All pressure drops are known except that in the ducting. Figure 8.14 can be used since the collector flow rate is known. If 2 ft^3/min per square foot is specified for the collector flow, a 500-ft^2 collector array will require 1000 ft^3/min. The intersection of the 1000 ft^3/min line and the 1000 ft/min duct velocity line indicates a pressure drop of approximately 0.1 in per 100 ft of duct. The same intersection point indicates that a duct diameter of 14 in will provide the 1000 ft/min design velocity. For 200 lineal feet of pipe, the total pressure drop in the ducting is 0.2 in, giving a total system pressure drop in the collector-to-storage mode of 0.55 in of water.

Equation (8.12) can now be used to evaluate the fan horsepower:

$$HP = 1000 \times \frac{0.55}{6356 \times 0.5} = 0.17$$

Duct Insulation

The second parasitic loss associated with ducts is heat loss. The sizing of insulation for ducts is based on a trade-off between insulation thickness and the

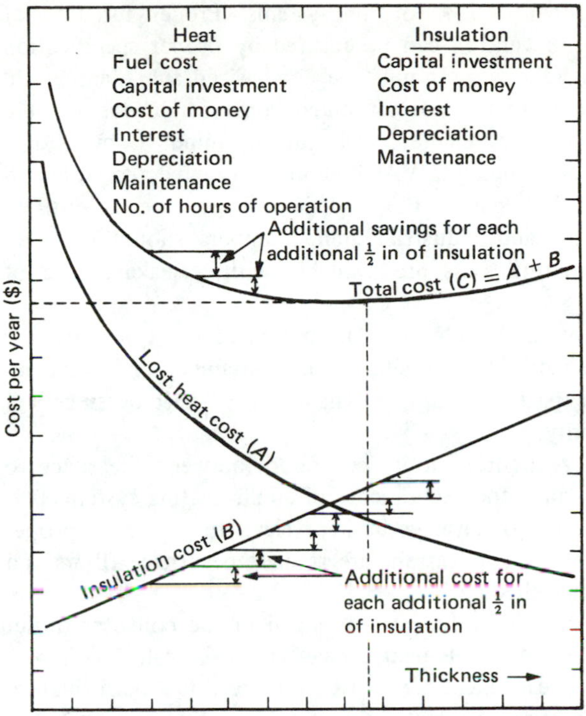

Figure 8.15 Example insulation optimization cost curve. The total cost consists of the heat loss value and the insulation cost. The minimum point of upper curve gives the optimum insulation thickness. This approach may be used for storage tank insulation, pipe insulation, and building insulation.

value of heat lost, as indicated in Fig. 8.15. This type of graph applies to insulation for storage tanks and pipes, as well as insulation and other energy conservation features of buildings. The minimum cost point is the required thickness of insulation for minimum life-cycle cost. For ducts in solar systems, insulation thicknesses in the range 1 to 3 in have most often been used. The ducts must be insulated whether they are inside or outside the building. The insulation optimum thickness can be calculated directly from Eq. (8.1) or a graph similar to that in Fig. 8.15 can be prepared. Insulation on ducting is always cost effective since solar heat is expensive and must not be dissipated.

Although ducts require insulation, fans should not be insulated. Adequate cooling air must be available for air movers and the motor must not be installed in the hot-air stream. This was inadvertently done in some early systems and intermittent operation of the motor resulted since the circuit breaker would open for high-temperature protection.

Dampers, Filters, and Heat Exchangers

Airflow control in air-based systems is achieved by means of dampers, as described in the first part of this section. Two dampers—D1 and D3 in Fig.

8.12—are of particularly critical design for air systems. Proper function of dampers with respect to leak prevention can be ensured by proper specification of damper seal material and adhesives. It must be emphasized that dampers in solar systems are exposed to relatively hot air under some conditions and the materials for the damper must be selected with this in mind. Conventional heating, ventilating, and air-conditioning (HVAC) dampers are expected to leak 5 to 30 percent of the air that passes them. This leakage rate is entirely unacceptable for solar systems and industrial grade dampers should be used. Butterfly valves with self-inflating seats are available with a leakage rate of approximately 1 percent.

Dampers are actuated by 24-V solenoids or motors of the type used for many years in the HVAC industry. Proper adjustment of damper drive linkages will ensure minimum leakage past the damper. High-quality shaft bearings will also result in improved reliability.

As indicated in Fig. 8.12, a filter should be present between the space to be heated and both collector and storage branches of an air-heating system. It is important that storage and collector airways be kept clean to guarantee proper heat transfer. Filters in residential applications are similar to furnace filters and should be replaced on the same schedule.

The domestic hot water heat exchanger contained in the collector outlet duct is of standard design. The size of this heat exchanger can be calculated from economic trade-offs between heat exchanger cost and projected performance. However, the calculations are laborious, and most manufacturers recommend a one- or two-row coil with a face area equal to the total collector outlet duct area. The pressure drop across this heat exchanger can be read from the manufacturer's literature, and recommended flow rates for pump P1 are also suggested by the coil manufacturer. An additional specification required for the preheat coil is fin spacing, and the manufacturer's recommendations can also be followed with regard to this parameter. The designer should select the fin spacing to minimize the pressure drop across the coil while maintaining adequate heat transfer. Typical heat transfer effectiveness values for this heat exchanger should be in the range 0.5 to 0.6. The air temperature drop across the coil will range from 3 to 5° under typical conditions.

Nominal Component Sizes

Collector area serves as the index of size for air-based systems as it does for liquid systems. Experience has shown that other components can be directly related to collector area. As indicated earlier, the airflow rate through standard air collectors should be in the range 2 to 3 ft^3/min per square foot of collector. Storage volumes are typically relatively short-term and vary between $\frac{1}{2}$ and $\frac{3}{4}$ ft^3 of pebble bed per square foot of collector. Collector tilt is optimal at latitude plus 15° for space-heating systems, although a variation of ±5° has little effect on life-cycle economics.

Design velocities in airflow ducts should be in the range 800 to 1000 ft/min. The method for determining duct insulation thickness was described earlier, and the results of this calculation will indicate an insulation thickness of 1 to 3 in. A heat rejection unit is not required for air systems and no provision for fluid expansion need be included. In the preceding section an approximate method for sizing the domestic hot water heat exchanger was described.

Airflows for space heating are specified in the standard way. The airflow required depends on the outlet air temperature and the air temperature rise needed to offset heat losses. This is a standard HVAC calculation described in [3].

MECHANICAL SOLAR HEATING SYSTEM CONTROLS

The control system is a key to proper heating system operation whether the system is air-based or liquid-based. Many difficulties in properly designed and installed systems have been traced to malfunctions in solar controllers. This section describes the important considerations in controller function and design and illustrates common problem areas. Historically, the controller has not been designed by the system designer but by the controller manufacturer. This is one of the causes of difficulty in the operation of some solar systems. The solar system controller is also relatively simple and inexpensive, and has occasionally been ignored in the design for this reason.

The control system should be designed after the mechanical configuration and all operating modes of the heating system have been finalized. Of course, during the design process the type of control that may ultimately be selected should be considered. A controller consists of three components: the sensors, which determine the state of the system; the control unit, which makes decisions based on preprogrammed criteria and sensor inputs; and the actuators, which carry out decisions made by the control unit. In solar heatng systems, actuators include diverting valves, dampers, pumps, heat rejection units, and solenoid valves.

Figure 8.16 shows an example control system for a liquid-based space-heating system similar to that depicted in Fig. 8.4. This system is similar in many ways to an air-based system and will be used to describe the function in each operating mode. Four modes are available for this system (a domestic water-heating subsystem is not shown; refer to Chap. 6): energy collection and delivery to storage, heat delivery to load, backup heat delivery to load, and summer heat rejection through a heat rejection unit. The first mode is controlled by a differential temperature measurement. Collector and storage temperatures are measured and the difference taken electronically. When this difference exceeds a preprogrammed level, pumps P1 and P2 are actuated by the controller and heat collection begins. For a liquid system the ΔT required for actuation is 15 to 20°F. As the day progresses, and periods of decreased solar intensity occur owing to cloud passage or sunset, the collector and storage pumps are turned off again on the basis of differential temperature measurements. However, the ΔT for turnoff is only 3°F. The difference between the ΔT's is required to avoid

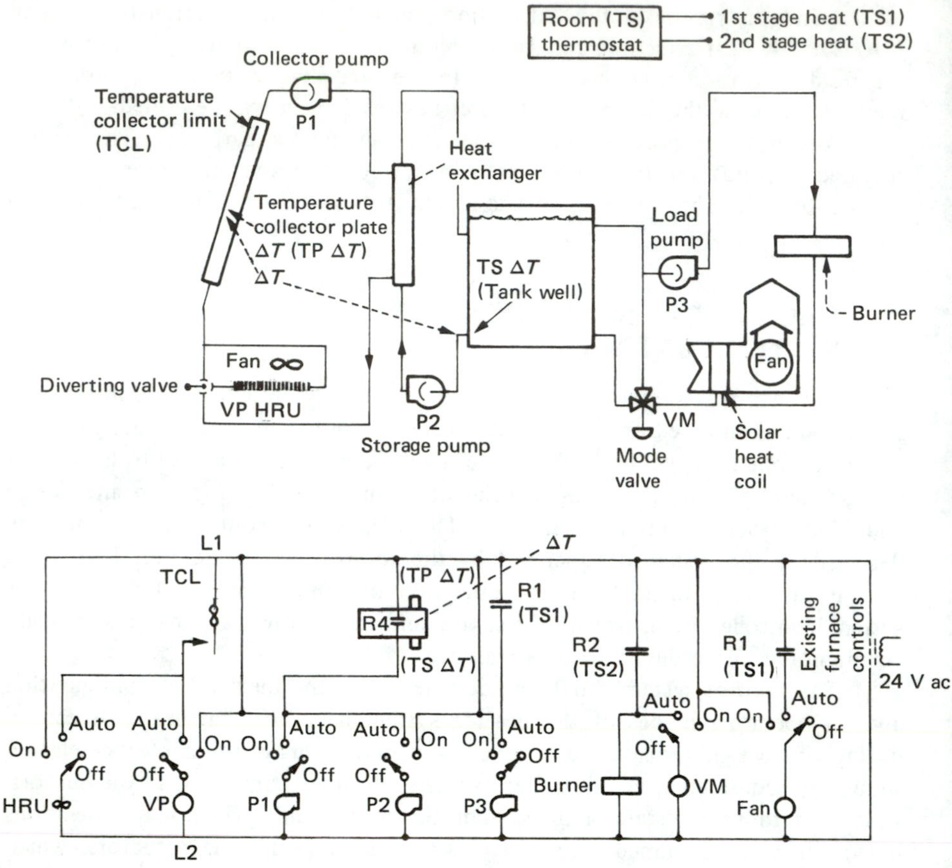

Figure 8.16 Simplified schematic diagram of control for liquid-based solar heating system with parallel backup.

excessive pump cycling during periods of initial pump turnon. Ratios of turnon to turnoff temperature differences for liquid systems should be approximately 5 to 7; for air systems a ratio of 1.5 to 3 is used with a turnon temperature condition of approximately 40°F.

The differential temperature controller also operates heat rejection unit HRU in Fig. 8.16. The decision to reject heat is based on storage temperature TS. For a storage temperature above the design maximum, pump P2 is turned off and diverting valve VP is operated. This causes collector fluid to pass through the heat rejection unit before returning to the collector. The controller also actuates HRU fans to dissipate heat. The design of heat rejectors has been described in the section on liquid-based systems.

Heat rejection can be carried out independently of the differential controller, however. In this method an on-off thermal switch, mounted on the collector plate, closes at a given maximum collector temperature (instead of storage temperature as described above), thereby operating the diverting valve and HRU fan. In this design, both collector and storage pumps are able to operate. Either heat rejection method has been used reliably for liquid systems. Of course, air systems do not require any heat rejection design.

The controller designer must provide for pump P1 operation during heat rejection periods. Normally, P1 and P2 are operated when there is a 20° temperature difference between collector and storage. At high temperatures this ΔT may never exist since collector efficiency is low, and an additional feature in the controller may be required to operate P1. This is particularly simple in the second method of HRU control described immediately above.

Heat delivery from storage uses standard HVAC controls for residential or commercial systems. The design philosophy is to attempt to provide the heat load first from solar storage if possible. If inadequate capacity exists in storage, the backup system is activated. In the residential context, a simple two-stage thermostat can be used to achieve this control. As the first contact closes, indicating a call for heat, pump P3 is activated to provide heat to the forced-air coil. If solar heat is adequate, heat added to the space will raise the temperature above its set point TS1 and pump P3 will be turned off.

However, if inadequate solar capacity exists, room temperature will continue to drop and the second-stage contact on the thermostat will close. This will activate the mode selector valve and burner to provide backup heat. Pump P3, which was activated by the first contact, will continue to operate. Heat will be added by the auxiliary system until contacts TS1 and TS2 both open. In larger buildings with multiple heating zones, the delivery loop control system is more complex. However, the same philosophy applies, with an attempt being made first to use solar heat sources and second to use auxiliary heat. [For air systems with heat pump backup two stages of backup heat are used: (1) the heat pump and (2) resistance strip heaters. Standard heat pump controls will activate the strip heaters, or a third contact on the room thermostat can be used for this purpose.]

Design of control systems will vary with application, but the underlying philosophy should be to make the controller fit the system. In other words, the control system should be designed considering all possible operation modes and their interaction with the building. Second, as simple a design as possible should be used. Inclusion of several optional modes, many of which are used infrequently, will reduce reliability, increase cost, and save little energy. A simple system does not imply a simple design exercise, however.

Solar Control Sensors

Decisions made in air- and liquid-based solar heating systems are based on temperature measurements, as indicated in the preceding section. Temperatures in solar systems are measured by use of semiconductor devices called thermistors. Alternatively, resistance temperature detectors (RTDs) can be used, although they are more expensive than thermistors. Thermocouples are not used since signal-to-noise ratios are poor.

Location of sensors in heating systems is of crucial importance. In the system shown in Fig. 8.16, three sensors are required, for collector temperature, storage temperature, and room temperature. The collector temperature must be sensed accurately, hence the sensor must be bonded mechanically and thermally to the outlet end of the absorber plate. This is normally done by the collector manufacturer so that warranties will not be voided by the controller subcontractor disassembling the collector to install the sensor. Collector sensors must not be located on pipes near liquid collectors or attached to the collector housing. Temperatures measured in these locations bear no significant relation to the important temperature, namely the absorber plate temperature. The method of bonding the sensor to the absorber plate should be able to accommodate high temperatures, in excess of 400°F, that may be experienced at the sensor point. Sensors located within a collector are difficult to access for future repair. It may be prudent to install several sensors within the collector during system installation. Then if one sensor fails, replacement is not required; only a change in electrical connections will be needed. For collectors in which the absorber plate is not accessible, such as evacuated tubes, the sensor should be located as close as possible to the collector outlet and a trickle flow of fluid provided through the collector to give this remotely located sensor a proper signal.

A thermistor sensor is also used to measure storage temperature. The temperature of interest is that near the bottom of the storage tank to take advantage of any stratification present. The storage sensor should therefore be located in the lower one-quarter of a liquid storage tank or of the pebble-bed storage for air system. In liquid storage, the sensor should be located away from all tank connections so that a true average temperature, not a local return fluid temperature, is measured. Storage sensor replacement in liquid systems is not difficult if storage is not buried; however, the tank may require draining. In rock-bed storage systems, removal of the storage medium from the container is not practical, and a method of replacing the sensor without removing the pebbles is needed. As suggested earlier, a piece of conduit inserted from the outside of the collector box can provide a very simple method of sensor replacement for air-based systems.

Storage sensors should be of the averaging type. That is, they should be relatively large in size, measuring temperatures over 1 ft or more within the storage container. A small sensor that measures a point temperature can be made to work properly if located carefully. However, a more reliable signal is available from the averaging-type sensor. Averaging sensors are commercially available from many controller manufacturers.

The third temperature to be measured to control heating systems is the room temperature. As indicated earlier, a dual-point thermostat will serve this purpose in residences. The thermostat should include an anticipator to guarantee proper sequencing of backup and storage heat sources. Thermostat location should adhere to good HVAC practice; it should be out of sunlight and away from any heated airflows or drafts. Dual-point thermostats are also commercially available.

The control designer must consider possible effects of power outages.

During the day in liquid systems the collector could boil, with resulting chemical degradation of glycol antifreezes. For air systems, daytime power failures are of little consequence since most air collectors are designed for stagnation. Power failures at night will have catastrophic results only in systems that use the drain-down approach for freeze protection. A simple method of protecting liquid collectors against power failure is to install a solenoid valve, powered closed, in the collector fluid loop. On loss of power this valve will open and the collector loop will drain into a receiving tank located in the mechanical space. Restart of the system after a daytime power failure is best accomplished by a manual start-up procedure. If an automatic approach is used, a decision to start must be based on collector temperature. During stagnation, collector temperatures in excess of 400°F will occur on a sunny day, and drained glycol solutions must not be exposed to this elevated temperature during refill. A separate collector refill temperature sensor can be used to prevent operation of the refill subsystem until the collectors have cooled after sunset. Another standard method of power failure protection, of course, is use of an auxiliary on-site electric power generator. In commercial systems this facility may be available for other purposes and pumps located in parallel with P1 and P2 in Fig. 8.4, but operated by the auxiliary power system, may be used. These pumps would not be required to provide the design collector flow rate but only sufficient flow to ensure that boiling not occur within the collector.

Night Setback

Night setback of building heating thermostats is used in some buildings to conserve energy during periods of no occupancy. Whether setback is used depends on the amount of thermal mass in the building and the relative magnitude of ventilation loads present during occupancy periods. If setback has been implemented by the heating system designer, its effect on solar system operation should be included. If the night setback is the standard 10 to 15°F below the daytime temperature setting, the morning reset of thermostat temperature will cause the backup system to come on every morning whether or not solar heat is available in storage. This is a result of the dual-point thermostat set points being separated by only 1 or 2°F in most applications. As a result, a 15° reset of temperature will cause backup operation.

Figure 8.17 shows that operation of the backup system can cause a relatively rapid increase of room temperature to the design daytime setting, since the backup system frequently operates at a higher terminal temperature than the solar system on an average day. Since solar systems operate at a reduced temperature their recovery rate is lower and provisions should be made for this in the control system. A time delay feature of some sort is therefore required. The amount of delay should be based on the ability of the solar system to reheat the building on a typical winter day followed by a typical winter night. Of course, the slope of the dashed recovery line shown in Fig. 8.17 for a solar system will vary each day, depending on the storage temperature history of the preceding night and the amount of solar radiation collected the preceding day. A

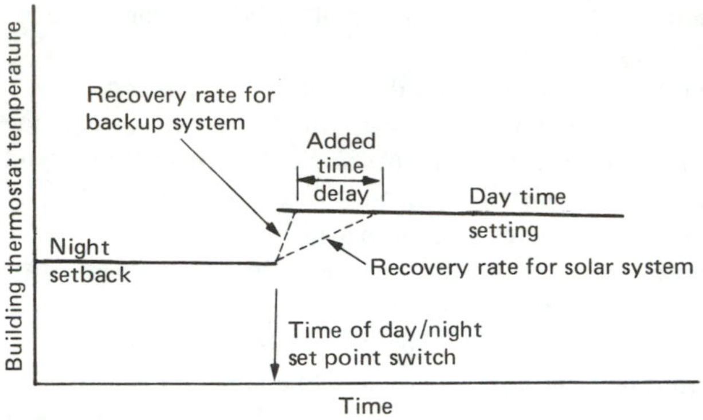

Figure 8.17 Characteristic room temperature recovery curves for solar and nonsolar heating systems in winter after night setback.

simple and reliable time delay feature can be included by use of a timer to lock out the auxiliary system energy source, whether it is electricity, gas, or fuel oil.

Hybrid Active and Passive Systems

In large commercial passive systems, an active air circulation subsystem is frequently present. The passive system itself requires no control, but the hybrid active component does require control.

The decision to call for airflow from the passive storage wall cavity can be based on a simultaneous call for heat in a zone of the building and adequate temperature in the thermal storage wall cavity. If the criteria are met, the dampers and fans can be actuated to deliver solar heat to a remote zone. Passive systems in small residences frequently do not have an active circulation mode requiring control.

SIMPLIFIED METHOD FOR SIZING ACTIVE SOLAR HEATING SYSTEMS

Background

Proper sizing of a solar space-heating system is important because it determines the initial investment as well as future heating bills. Often computers or hand-held programmable calculators are used to size solar heating systems. But since many people are not familiar with such computational equipment, a simplified method for sizing solar heating systems was developed at the Solar Energy Research Institute [7]. This method makes sizing possible without the aid of computational equipment. The method is based on the *f* chart [6], which is a sophisticated computer program used extensively for estimating collector size, long-term performance and costs, and the percentage of total required heat (called the solar fraction) provided by a solar heating system.

The simplified sizing method requires only basic information about the house, to be supplied on a worksheet as illustrated below. This, together with three easy-to-use tables and simple arithmetic calculations, sizes the system. The method determines three different collector sizes for heating the home and supplying domestic hot water, an estimate of the solar fraction for each size, and an estimate of how long it will take to recover the investment plus interest.

The method is based on the concept that there is no single "correct" or "optimum" solar system size for a particular house. Instead, there is a wide range of system sizes that will be economically advantageous and perform well, but will differ in costs and amount of heat provided. This method will help compare different types of systems (air or liquid) for three different size ranges.

The method does not advocate use of a detailed, long-term economic analysis (also known as life-cycle costing) as a major criterion in the sizing and decision-making process, because uncertainty about future fuel costs makes the prediction of the value of a solar system over a long period of time uncertain. With this method, you determine only the first-year savings for three system sizes. However, as an indication of what might happen in the future, an economic indicator was developed specifically for this method. This indicator, called RETURN, represents the number of cents saved on fuel bills in the first year per dollar invested in the system. The larger the RETURN, the sooner the system will pay for itself. However, as you will see after completing the worksheet, large RETURNS usually occur with small systems because small systems are less expensive initially and require less time than large systems to pay for themselves. But small systems also save less fuel than large systems.

Therefore a trade-off must be made by the owner. Remember that all long-term economic estimates contain uncertainties, and the RETURN value in this method is no exception.

Assumptions

The amount of useful heat provided by a solar heating system depends on many factors. These factors can be put into four broad categories: system design, installation quality, weather conditions, and occupant use patterns. In these four categories, the most unpredictable factor is generally the weather, but the other categories also contain uncertainties. Therefore some assumptions must be made about factors within these categories when sizing a solar heating system. The simplified solar sizing method has assumptions built into it concerning the four categories. The results obtained in the worksheet depend, therefore, on how closely the real system approximates the following assumptions:

1 *System design*: The system is properly designed and functions well. Select a qualified and experienced solar designer to help select the best system and properly design the layout of the system's components.

2 *Installation quality*: All components of the system are properly installed and adjusted.

3 *Weather conditions*: Predicting weather conditions for any one winter is impossible, but an average climate is predictable over several years. This method

uses weather data averaged over the past 20 years to predict probable climates for several locations. These climates are then used to predict system performance.

4 *Occupant use patterns*: The assumed life style is typically American, which in this context means that the occupants of the house are accustomed to receiving hot water at the turn of the faucet, expect a constant room temperature, and generally do not want to make large sacrifices in life style in exchange for owning a solar heating system. In addition, home insulation is probably the extent of energy conservation.

Variations from these assumptions will affect the results of this method. For example, family members may become more energy conscious after installing a solar heating system and may make several energy conservation improvements to the house, as well as altering their own energy use patterns. As a result, the decreased demand on the solar heating system will lead to an increase in solar system performance beyond that initially indicated on the worksheet. The reverse is also possible: a significant increase in demand on the solar heating system will result in lower solar system performance than originally predicted. If variations from the assumptions are minor, this method will provide a good prediction of system performance for the different types and sizes of systems described in this workbook.

Worksheet Instructions

A blank worksheet is provided in this section to compare different collector types, orientations, tilt angles, and glazings. A completed sample worksheet is also given and discussed in detail below.

Section A: Background Information

The information in this section must be as accurate as possible because it is used in the next section to estimate system performance. The line numbers below correspond to those on the worksheet.

Line 1: Enter the total square feet of heated space in the house, including the basement and attic if they are used as heated living spaces. With this sizing method, the amount of heat required is estimated according to size. (Remember, it is of benefit to take energy conservation measures whether or not you install a solar energy system.)

In the sample worksheet, the floor area of the building is 1500 ft^2.

Line 2: Enter the expected payment for both heating and hot water in the next 12 months without solar heat. This amount will not be used in designing the system; an estimate to the nearest $50 is sufficient. The local utility company should be able to assist with this estimate. Estimates of utility bill reductions for energy conservation measures can usually be obtained from an insulation contractor or, in the case of a new home, from the architect or builder.

In the sample worksheet, the owner expects to pay $365 in the next 12 months for space and water heating combined.

Line 3: To complete this line and the remainder of the worksheet, select a

Simplified Solar Sizing Worksheet
Sample: Boulder, Colorado

SECTION A: BACKGROUND INFORMATION

House size Enter floor area of house (ft^2)	(1)	1,500 ft^2
Fuel cost Enter current annual cost for heating and hot water combined	(2)	$ 365
Collector factor Copy entry from App. 10 that best describes collector type, orientation, etc.	(3)	8.5

SECTION B: COLLECTOR SIZING

		Small	Medium	Large
Location factor Copy entries from App. 10 for location	(4)	4.4	6.2	7.9
Solar fraction Multiply: Line 3 × Line 4	(5)	37	53	67
Collector-to-floor area ratio (Do nothing on this line)	(6)	0.12	0.20	0.30
Collector area Multiply: Line 1 × Line 6	(7)	180 ft^2	300 ft^2	450 ft^2
Gross cost Enter cost of each size system	(8)	$6,300	$9,900	$14,400
Tax credit (calculation) Calculate: Line 8 × 0.40	(9)	$2,520	$3,960	$ 5,760
Federal tax credit (actual) Enter Line 9 or $4000, whichever is less	(10)	$2,520	$3,960	$ 4,000
State tax credit, if any Enter state tax credit, if any	(11)	$1,890	$2,970	$ 3,000
Net cost Subtract: Line 8 – Line 10 – Line 11	(12)	$1,890	$2,970	$ 7,400

RETURN is introduced so that you can roughly compare the economics of different system sizes over a period of years. Calculations have indicated that a RETURN in the range of 3 to 10 will offer the best trade-off between prompt payback and high long-term savings. The best compromise appears to be a system with a RETURN closest to 6 (using a fuel inflation rate of 12 percent).*

RETURN Calculate: Line 2 × Line 5/Line 12	(13)	7.0	6.5	3.3

SECTION C: MORTGAGE PAYMENTS†

Term of mortgage Enter the term of your mortgage in years	(14)	20 yr
Interest rate Enter the interest rate of your mortgage as a percentage	(15)	10%
Compound interest Enter the corresponding number from App. 10	(16)	104

		Small	Medium	Large
Down payment Enter the down payment	(17)	$ 0	$ 500	$ 2,000
Amount financed Subtract: Line 8 – Line 17	(18)	$6,300	$9,400	$12,400
Monthly mortgage payment Divide: Line 18 by Line 16	(19)	$ 60.58	$ 90.38	$ 119.23

*You can expect these results from a system with a RETURN of 6: (1) Savings (in fuel bills) will accrue until the system has paid for itself in 9 years following installation; and (2) savings will pay for the system *plus* 8 percent interest on investment in 13 years following installation. Everything after that is pure savings.

†This section does not include savings from any state or federal credits presently being offered for the installation of solar energy systems.

SECTION A: BACKGROUND INFORMATION

House size Enter floor area of house (ft^2)	(1)	ft^2
Fuel cost Enter current annual cost for heating and hot water combined	(2)	$
Collector factor Copy entry from App. 10 that best describes collector type, orientation, etc.	(3)	

SECTION B: COLLECTOR SIZING

		Small	Medium	Large
Location factor Copy entries from App. 10 for location	(4)			
Solar fraction Multiply: Line 3 × Line 4	(5)			
Collector-to-floor area ratio (Do nothing on this line)	(6)	0.12	0.20	0.30
Collector area Multiply: Line 1 × Line 6	(7)	ft^2	ft^2	ft^2
Gross cost Enter cost of each size system	(8)	$	$	$
Tax credit (calculation) Calculate: Line 8 × 0.40	(9)	$	$	$
Federal tax credit (actual) Enter Line 9 or $4000, whichever is less	(10)	$	$	$
State tax credit, if any Enter state tax credit, if any	(11)	$	$	$
Net cost Subtract: Line 8 – Line 10 – Line 11	(12)	$	$	$

RETURN is introduced so that you can roughly compare the economics of different system sizes over a period of years. Calculations have indicated that a RETURN in the range of 3 to 10 will offer the best trade-off between prompt payback and high long-term savings. The best compromise appears to be a system with a RETURN closest to 6 (using a fuel inflation rate of 12 percent).*

RETURN Calculate: Line 2 × Line 5/Line 12	(13)			

SECTION C: MORTGAGE PAYMENTS†

Term of mortgage Enter the term of your mortgage in years	(14)	yr
Interest rate Enter the interest rate of your mortgage as a percentage	(15)	%
Compound interest Enter the corresponding number from App. 10	(16)	

		Small	Medium	Large
Down payment Enter the down payment	(17)	$	$	$
Amount financed Subtract: Line 8 – Line 17	(18)	$	$	$
Monthly mortgage payment Divide: Line 18 by Line 16	(19)	$	$	$

*You can expect these results from a system with a RETURN of 6: (1) Savings (in fuel bills) will accrue until the system has paid for itself in 9 years following installation; and (2) savings will pay for the system *plus* 8 percent interest on investment in 13 years following installation. Everything after that is pure savings.

†This section does not include savings from any state or federal credits presently being offered for the installation of solar energy systems.

particular collector model (air or liquid), determine the orientation of the collector (south, southeast, etc.), and decide on the collector tilt that is best for the particular situation. Appendix 10 lists several typical combinations. Select one and enter the number value (COLLECTOR FACTOR) on line 3. It is possible to try different combinations, using extra worksheets.

On the sample worksheet, the COLLECTOR FACTOR entered on line 3 corresponds to a liquid collector with a selective surface and a due-south orientation, mounted on a standard $\frac{4}{12}$ pitch roof.

Section B: Collector Sizing

In this section determine three sizes of solar collectors–small, medium, and large–and their expected performance. Each step must be completed for each of the three columns. Comparisons can then be made to determine what system size is best.

Line 4: Find in App. 10 the system location and enter the LOCATION FACTORS for small, medium, and large systems in the corresponding columns for line 4. If the location is not listed, enter the LOCATION FACTORS for the climate most similar to that where the system will be installed.

For the sample worksheet, the LOCATION FACTORS that correspond to Boulder, Colorado, are 4.4, 6.2, and 7.9, respectively, and are entered on line 4.

Line 5: Multiply the COLLECTOR FACTOR entered on line 3 by each LOCATION FACTOR entered on line 4 to get the solar fraction for each size range. The solar fraction is an estimate of the percentage of the total heating requirement that will be met by solar energy. In some cases it is also the approximate percentage by which your fuel bills will be reduced.

The sample worksheet shows that a small solar system will provide 37 percent, a medium system 53 percent, and a large system 67 percent of the total heating requirements. These fractions are also close estimates of how much the annual fuel bill will be reduced for the sample situation.

Line 6: This line requires no user input. The numbers entered represent the ratio of collector area to floor area as determined through extensive calculations.

Line 7: Multiply the number entered on line 1 by the numbers entered on line 6 and enter the answers on line 7. These numbers represent the collector area.

Line 8: Since the cost of solar systems varies widely, the dealer who will install the system must estimate the total cost for each size on line 7. This estimate should include installation costs but not applicable tax credits.

Lines 9 and 10: The federal government currently provides a tax credit of up to 40 percent of the cost of a solar energy system (with a maximum credit of $4000). These two lines show how much is deductible from federal income taxes for each size system in the first year after system installation.

Line 11: In addition to the federal government, some states offer tax credits. If applicable, enter this amount on line 11. Since these credits vary from state to state, no specific procedure can be given. Obtain details from the state department of revenue, state energy office, state solar energy association, or a solar consultant or contractor. (See App. 11 for organizations that can provide additional information about state tax credits.)

In the sample situation, the state of Colorado currently offers a tax credit of up to 30 percent (maximum, $3000) of the cost of a solar energy system including installation. This amounts to a deduction from state income taxes of $1890 for the small system, $2970 for the medium one, and $3000 for the large system in the first year after installation.

Line 12: Enter on this line the number obtained by subtracting line 10 from line 8, and then line 11 from the answer obtained. This figure represents the NET COST—the amount actually spent on the solar system.

For example, the NET COST for a medium-size system in the sample situation would be approximately $2970.

Line 13: Multiply line 2 by line 5, divide the answer by line 12, and enter the final answer on line 13. This number is the RETURN indicator, as explained above and on the worksheet.

In the sample worksheet, for every dollar invested in the medium system, 6.5 cents will be saved on the fuel bill in the first year. Thus $193.05 will be saved in the first year for the medium system (line 12 X line 13 = 19,305 cents = $193.05). Since a RETURN of 6.5 is closest to the desired figure of 6 (see worksheet remarks about RETURN), the medium system appears to be the best one in the sample situation as a compromise between prompt payback and higher long-term savings.

Section C: Mortgage

This section is optional and is included only for estimation of the monthly mortgage payment for the solar system.

Lines 14 and 15: Enter the mortgage terms. If the solar system is being built together with a new home, use the terms of the new home mortgage.

The sample worksheet shows a 20-year mortgage at a 10 percent interest rate.

Line 16: Find in App. 10, and enter on line 16, the number that corresponds to your mortgage terms.

The correct corresponding number for a 20-year mortgage at a 10% interest rate is 104.

Line 17: Enter the DOWN PAYMENT. If the mortgage covers the complete house, enter only the additional DOWN PAYMENT to be made because the solar system is included. For example, if a DOWN PAYMENT of $20,000 is being made whether the house is purchased with or without a $10,000 solar system, then the DOWN PAYMENT does not change and 0 should be entered on line 17.

In the sample, an additional DOWN PAYMENT of $2000 will be made if a large system is installed. However, if a small system is installed, no additional DOWN PAYMENT will be made.

Line 18: Subtract line 17 from line 8 and enter the answer on line 18. Note: the AMOUNT FINANCED will exceed the NET COST because tax credits (lines

10 and 11) have not, at the time of financing, been applied to the GROSS COST (line 8) of the system. This is because tax credits will be taken in the form of credits on federal and state income taxes later.

In the sample situation, a zero DOWN PAYMENT for a small system (line 17) results in the GROSS COST of $6300 (line 8) being used as the AMOUNT FINANCED (line 18). At the time of financing, the owner does not have the $4410 ($2520 plus $1890) that he or she will later be able to deduct from federal and state income taxes, so it must be borrowed. If the owner were to apply the savings from both federal and state income taxes ($4410) toward the AMOUNT FINANCED ($6300), the amount owed would be closer to the NET COST of the system ($1890).

Line 19: To approximate the monthly payment, divide line 18 by line 16 and enter the answer on line 19. In the case of a new home mortgage, this is the amount by which monthly mortgage payments will be increased because of the solar system.

In the sample situation, if a small system is installed the mortgage payment will be increased by $60.58 a month.

REFERENCES

1. J. D. Balcomb, "A Simplified Method for Calculating Required Solar Collector Array Size for Space Heating," *Proc. Int. Solar Energy Society Congr.*, vol. 4, 1976, p. 281.
2. J. F. Kreider, *The Solar Heating Design Process*, McGraw-Hill, New York, 1981.
3. American Society of Heating, Refrigerating, and Air-Conditioning Engineers, *Handbook of Fundamentals*, ASHRAE, New York, 1977.
4. U.S. Department of Housing and Urban Development, *Intermediate Minimum Property Standards for Solar Heating*, No. 4930.2, Government Printing Office, Washington, D.C., 1977.
5. Honeywell, Inc., *Design and Test Report for Transportable Solar Laboratory Program*, PB-240 609, National Technical Information Service, Springfield, Va., 1974.
6. F. Kreith and J. F. Kreider, *Principles of Solar Engineering*, Hemisphere, Washington, D.C., 1978.
7. P. Bendt and R. Soto, *Home Owner's Solar Sizing Workbook*, Solar Energy Research Institute, Golden, Colo., 1980.

9 Solar Cooling

The real cycle you're working on is a cycle called yourself.

ROBERT PIRSIG

Solar cooling of buildings represents a potentially significant application of solar energy for building air conditioning in most sunny regions of the United States. Solar cooling technology is not as advanced as solar heating technology, but research may be able to close the gap between the two by 1985. Several viable solar air-conditioning schemes are described in this chapter and methods for tentative system design are presented in detail.

OVERVIEW OF SOLAR COOLING SYSTEMS

A wide spectrum of physical, chemical, and electrical processes can be utilized to produce cooling effects for air conditioning of buildings. All of these processes can be integrated in either a closed or an open cycle. A closed cycle requires two separate process loops, one for refrigeration and the other for transfer of heat from the load. Since the two loops are closed and coupled with heat exchangers, one can optimize each loop so that the overall system can operate with maximum efficiency, minimum size, or minimum cost for a given application. Open cycles, on the other hand, combine the process and heat transfer loops, thus eliminating the interfacing heat exchanger. The price paid for eliminating this component is that compromises must be made to accommodate the different requirements for the two process loops involved.

Table 9.1 compares representative refrigeration processes and their application in both open- and closed-cycle configurations. Most current refrigeration and

Table 9.1 Spectrum of Refrigeration Possibilities

Process	Open cycle	Closed cycle
Mechanical compression	Air cycle	Rankine cycle
Absorption	Desiccant	Absorption
Adsorption	Desiccant	Adsorption
Thermoelectric*	Thermionic emission	Peltier effect

*Thermoelectric processes are not used commercially and will therefore not be treated here.

air-conditioning systems operate on the Rankine cycle, which is a closed mechanical vapor compression cycle.

The open-cycle absorption desiccant process in Table 9.1 represents an attractive alternative to conventional vapor compression air-conditioning systems when it is coupled with solar energy, using a recyclable air desiccation process with a material such as silica gel. Such a system can use inexpensive inorganic and inert materials that do not present corrosion or environmental problems, can tolerate some air leakage, could be easily serviced, and can operate over a wide range in solar input.

COOLING REQUIREMENTS

The cooling load of a building is the rate at which heat must be removed to maintain the air in the building at a given temperature. It is usually calculated on the basis of the peak load expected during the cooling season. The cooling load of a building depends primarily on

1 Design inside and outside dry-bulb temperatures
2 Design inside and outside relative humidities
3 Solar radiation heat load
4 Wind speed

A method of cooling-load calculation is presented in detail in the *ASHRAE Handbook of Fundamentals* [1].

The steps in calculating the cooling loads of a building are

1 Specify the building characteristics:
   Wall area, type of construction, and surface characteristics
   Roof area, type of construction, and surface characteristics
   Window area, setback, and glass type
   Building location and orientation
2 Specify outside and inside wet- and dry-bulb temperatures
3 Specify solar heat load and wind speed
4 Calculate building cooling loads due to:
   Heat transfer through windows
   Heat transfer through walls
   Heat transfer through roof

Sensible heat gains due to infiltration and exfiltration
Latent heat gains (water vapor)
Internal heat sources such as people, lights, etc.

Equations (9.1) through (9.7) may be used to calculate the various cooling loads for a building; cooling loads due to lights, occupants, etc. may be estimated from the *ASHRAE Handbook of Fundamentals*. For unshaded or partially shaded windows, the load is

$$Q_{\text{wi}} = A_{\text{wi}} \left[F_{\text{sh}} \bar{\tau}_{b,\text{wi}} I_{h,b} \frac{\cos i}{\sin \alpha} + \bar{\tau}_{d,\text{wi}} I_{h,d} + \bar{\tau}_{r,\text{wi}} I_r + U_{\text{wi}} (T_{\text{out}} - T_{\text{in}}) \right] \quad (9.1)$$

For shaded windows the load (neglecting diffuse sky radiation) is

$$Q_{\text{wi,sh}} = A_{\text{wi,sh}} U_{\text{wi}} (T_{\text{out}} - T_{\text{in}}) \quad (9.2)$$

For unshaded walls the load is

$$Q_{\text{wa}} = A_{\text{wa}} \left[\bar{\alpha}_{s,\text{wa}} \left(I_r + I_{h,d} + I_{h,b} \frac{\cos i}{\sin \alpha} \right) + U_{\text{wa}} (T_{\text{out}} - T_{\text{in}}) \right] \quad (9.3)$$

For shaded walls the load (neglecting diffuse sky radiation) is

$$Q_{\text{wa,sh}} = A_{\text{wa,sh}} \left[U_{\text{wa}} (T_{\text{out}} - T_{\text{in}}) \right] \quad (9.4)$$

For the roof the load is

$$Q_{\text{rf}} = A_{\text{rf}} \left[\bar{\alpha}_{s,\text{rf}} \left(I_{h,d} + I_{h,b} \frac{\cos i}{\sin \alpha} \right) + U_{\text{rf}} (T_{\text{out}} - T_{\text{in}}) \right] \quad (9.5)$$

For sensible heat infiltration and exfiltration the load is

$$Q_i = \dot{m}_a (h_{a,\text{out}} - h_{a,\text{in}}) \quad (9.6)$$

For moisture infiltration and exfiltration the load is

$$Q_w = \dot{m}_a (W_{\text{out}} - W_{\text{in}}) \lambda_w \quad (9.7)$$

where Q_{wi} = heat flow through unshaded windows of area A_{wi}, Btu/h
$Q_{\text{wi,sh}}$ = heat flow through shaded windows of area $A_{\text{wi,sh}}$, Btu/h
Q_{wa} = heat flow through unshaded walls of area A_{wa}, Btu/h
$Q_{\text{wa,sh}}$ = heat flow through shaded walls of area $A_{\text{wa,sh}}$, Btu/h
Q_{rf} = heat flow through roof of area A_{rf}, Btu/h
Q_i = heat load due to infiltration/exfiltration, Btu/h
Q_w = latent heat load, Btu/h
$I_{h,b}$ = beam component of insolation on horizontal surface, Btu/(ft^2·h)
$I_{h,d}$ = diffuse component of insolation on horizontal surface, Btu/(ft^2·h)
I_r = ground-reflected component of insolation, Btu/(ft^2·h) (direct plus diffuse)
$W_{\text{out}}, W_{\text{in}}$ = outside and inside humidity ratios, lb H_2O/(lb dry air)
$U_{\text{wi}}, U_{\text{wa}}, U_{\text{rf}}$ = overall heat transfer coefficients for windows, walls, and roof, including radiation, Btu/(ft^2·h·°F)

$\dot{m}_a$ = net infiltration and exfiltration of dry air, lb/h
T_{out} = outside dry-bulb temperature, °F
T_{in} = indoor dry-bulb temperature, °F
F_{sh} = shading factor (1.0 = unshaded, 0.0 = fully shaded)
$\bar{\alpha}_{s,wa}$ = wall solar absorptance
$\bar{\alpha}_{s,rf}$ = roof solar absorptance
i = solar incidence angle on walls, windows, and roof, deg
$h_{a,out}, h_{a,in}$ = outside and inside air enthalpy, Btu/lb
α = solar altitude angle, deg
λ_w = latent heat of water vapor, Btu/lb
$\bar{\tau}_{b,wi}$ = window transmittance for beam (direct) insolation
$\bar{\tau}_{d,wi}$ = window transmittance for diffuse insolation
$\bar{\tau}_{r,wi}$ = window transmittance for ground-reflected insolation

EXAMPLE

Determine the cooling load for a building in Phoenix, Arizona, with the specifications tabulated below:

Factor	Description or specification
Building characteristics:	
Roof:	
Type of roof	Flat, shaded
Area $A_{rf,sh}$, ft²	1700
Walls (painted white):	
Size, north and south, ft	8 × 60 (two)
Size, east and west, ft	8 × 40 (two)
Area A_{wa}, north and south walls, ft²	$480 - A_{wi} = 480 - 40 = 440$ (two)
Area A_{wa}, east and west walls, ft²	$320 - A_{wi} = 320 - 40 = 280$ (two)
Absorptance $\bar{\alpha}_{s,wa}$ of white paint	0.12
Windows:	
Size, north and south, ft	4 × 5 (two)
Size, east and west, ft	4 × 5 (two)
Shading factor F_{sh}	0.20
Insolation transmittance	$\bar{\tau}_{b,wi} = 0.60$; $\bar{\tau}_{d,wi} = 0.81$; $\bar{\tau}_{r,wi} = 0.60$
Location and latitude	Phoenix, Ariz.; 33°N
Date	August 1
Time and local solar hour angle H_s	Noon; $H_s = 0$
Solar declination δ_s, deg	18°14′
Wall surface tilt from horizontal β	90°
Temperature, outside and inside, °F	$T_{out} = 100$; $T_{in} = 75$
Insolation I, Btu/(ft²·h)	$I_{h,b} = 185$; $I_{h,d} = 80$; $I_r = 70$
U factor for walls, windows, and roof	$U_{wa} = 0.19$; $U_{wi} = 1.09$; $U_{rf} = 0.061$
Infiltration, lb dry air/h	Neglect
Exfiltration, lb dry air/h	Neglect
Internal loads	Neglect
Latent heat load Q_w, %	30% of wall sensible heat load*

*Approximate rule of thumb for Phoenix.

SOLUTION

To determine the cooling load for the building just described, calculate the following factors in the order listed:

1 Incidence angle for the south wall i

$$\cos i = \cos \delta_s \cos (L - \beta) + \sin \delta_s \sin (L - \beta) = 0.257$$

2 Solar altitude α

$$\sin \alpha = \sin \delta_s \sin L + \cos \delta_s \cos L \cos H_s = \cos (L - \delta_s) = \cos 15^\circ = 0.96$$

3 South window load [from Eq. (9.1)]

$$Q_{wi} = 40\left\{(0.2 \times 0.6)\left(185\,\frac{0.257}{0.96}\right) + (0.81 \times 80) + (0.60 \times 70) + [1.09(100^\circ - 75^\circ)]\right\} = 5600 \text{ Btu/h}$$

4 Shaded window load [from Eq. (9.2)]

$$Q_{wi,sh} = (3 \times 40)[1.09(100^\circ - 75^\circ)] = 3270 \text{ Btu/h}$$

5 South wall load [from Eq. (9.3)]

$$Q_{wa} = (480 - 40)\left\{0.12\left[70 + 80 + \left(185\,\frac{0.257}{0.96}\right)\right] + 0.19(100^\circ - 75^\circ)\right\} = 12{,}610 \text{ Btu/h}$$

6 Shaded wall load [from Eq. (9.4)]

$$Q_{wa,sh} = [(480 + 320 + 320) - (3 \times 40)]\,[0.19(100^\circ - 75^\circ)] = 4750 \text{ Btu/h}$$

7 Roof load [from Eq. (9.5)]

$$Q_{rf} = 1700[\bar{\alpha}_{s,rf} \times 0 + 0.061(100^\circ - 75^\circ)] = 2600 \text{ Btu/h}$$

8 Latent heat load (30% of sensible wall load)

$$Q_{wa} = 0.3[(480 + 480 + 320 + 320) - (4 \times 40)]\,[0.19(100^\circ - 75^\circ)] = 2050 \text{ Btu/h}$$

9 Infiltration/exfiltration load

$$Q_i = 0$$

10 Total cooling load for building described in example

$$Q_{tot} = Q_{wi} + Q_{wi,sh} + Q_{wa} + Q_{wa,sh} + Q_{rf} + Q_w + Q_i$$

$$Q_{tot} = 30{,}880 \text{ Btu/h} \sim 2.5 \text{ tons air conditioning}$$

This example is simplified for illustrative purposes. Heat loads must be calculated each hour of a design day to determine the maximum load. The maximum cooling load usually occurs between 3 and 4 p.m.

DESIGN OF A COLLECTOR FOR A SOLAR-COOLED BUILDING

Solar energy can be economical for building temperature control in areas where the local climate requires both air conditioning and heating, because use of solar energy for heating and air conditioning increases the system usage factor. The combined system shown in Fig. 9.1 is installed in the Solar Energy Applications Laboratory, Colorado State University, Fort Collins, Colo. The same solar collector and storage system can be used for combined, year-round heating and cooling operation. Löf and Tybout performed an optimization study of combined systems for the same geographic locations as their earlier solar heating study [2]. The general conclusions pertinent to collector design for combined systems in climates with hot summers are

1 Optimal collector tilt β is equal to latitude L (except in the south, where $\beta = L - 10°$).

2 Optimal collector area is always greater than that needed for heating only.

The Löf-Tybout study modeled a flat-plate collector and lithium bromide (LiBr) absorption air conditioner with a price based on assumed mass production costs for both. The results are only a guide to solar air-conditioning design because results of work and experience with solar air conditioning to date have not culminated in a reliable, inexpensive solar air-conditioning unit, and no system can be recommended for wide use at this time.

STORAGE OF ENERGY AT HIGH AND LOW TEMPERATURES

For a detailed description of high-temperature storage of energy the reader may refer to Chap. 4. In an air-conditioning system it is also possible to cool a tank of water or other storage medium if excess cooling capacity is available during operational hours. One of the attractions of this approach is that the temperature difference between cold storage (say 45°F at the coldest) and building temperature (~70°F) is less than the temperature difference between hot storage and room temperature. As a result, less cooling effect is lost from cold storage than from hot storage. Moreover, cold storage is preferable to hot storage because losses from hot storage in a building add to the summer air-conditioning load.

In addition, 1 Btu of cold storage is 1 Btu of cooling; 1 Btu of hot storage is equivalent to less than 1 Btu of cooling since the air conditioner is not 100 percent efficient. These advantages of cold storage are offset by the additional cost and upkeep required to install and maintain a second storage unit and the associated plumbing and ducting.

The decision to use hot, cold, or combined storage depends on the building location. In Phoenix one should design for cold storage since the investment in hot storage is uneconomical in a climate where heating loads are small (only

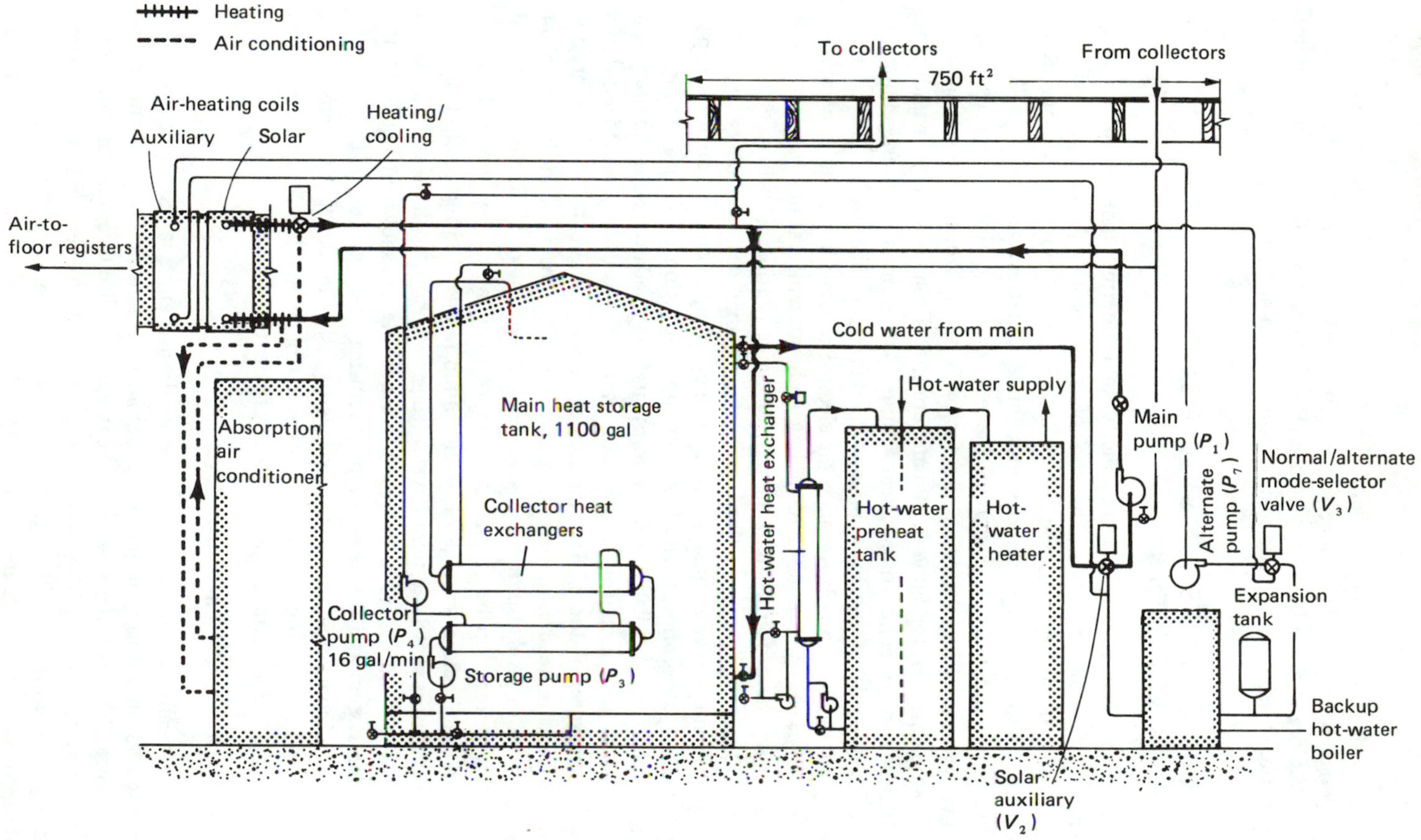

Figure 9.1 Combined solar heating and cooling system at the Solar Energy Applications Laboratory, Colorado State University, Fort Collins. (*Courtesy of G. O. G. Löf [3].*)

1500 degree-days per year). In Bemidji, Minnesota, on the other hand, one should design for hot storage since the cooling load is small. Recent test results from the Colorado State University house (Fig. 9.1) indicate that a small amount of cold buffer storage can improve solar cooling-system performance.

VAPOR COMPRESSION SYSTEMS AND HEAT PUMPS

As mentioned earlier, the periodicity and intermittence of solar energy incident on a collector require the use of a conventional backup (auxiliary) system. The size and cost of the total system depend not only on the Btu's collected but also on storage facilities; any concept that can reduce the collection and storage requirements can improve the economics of solar energy use. One attractive way to achieve a fuller use factor is to utilize a solar-assisted heat pump–air conditioner for all-year climate control. From an energy standpoint, the efficiency of a conventional heat pump system in a northern climate is reduced as a result of the low ambient temperatures; because of the reduced efficiency of the heat pump, reliance on electric resistance heating is necessary. However, solar energy can be used at low ambient temperatures to reduce and often eliminate the need for supplemental resistance heating in such systems.

A *heat pump* may be defined as a system that absorbs energy at a low temperature and delivers energy at a higher temperature through a vapor compression or absorption cycle. However, no absorption cycle heat pumps are presently available. Available heat pumps use a standard vapor compression refrigeration cycle operated in reverse: the evaporator is usually placed outdoors and the condenser indoors. The mechanical energy input to the compressor during vapor compression raises the internal energy absorbed at low ambient temperatures to a level useful for space heating. The quantity of heat extracted from such a system can be several times more than the energy required by the compressor. The basic advantage of a heat pump is this "heating multiplication."

In a solar-assisted heat pump system designed both to heat and to cool a building, two general modes of operation are possible. As shown in Fig. 9.2*a*, the solar heating system and the heat pump system may be separate. In Fig. 9.2*b* the two cycles are connected, and a solar collector assists in reducing the amount of heat that must be supplied by the compressor when heating the building. Obviously, the second arrangement is more efficient.

Solar-powered Heat Pump Cycle

Theoretically, a heat pump can be used with any thermodynamic cooling cycle, such as an absorption, jet compression, or mechanical vapor compression cycle. Figure 9.3 illustrates the basic heat flows in a conceptual solar-powered heat pump system in which the sun's energy directly provides shaft work instead of acting as a booster, as in Fig. 9.2. The amount of heat removed from the environment during the heating mode is Q_4. A heat balance of the cycle in Fig. 9.3 gives

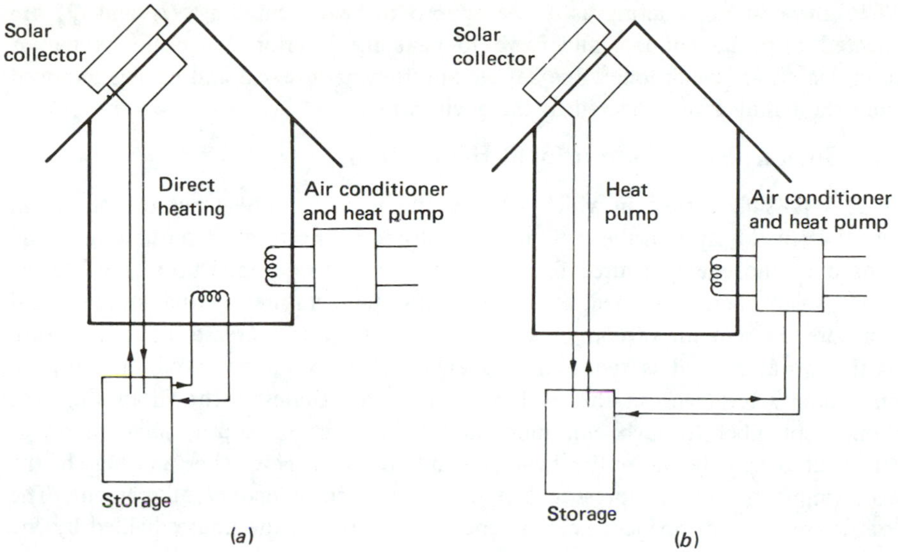

Figure 9.2 (*a*) Direct solar heating system; (*b*) solar-assisted heat pump.

$$Q_1 + Q_4 = Q_2 + Q_3 \tag{9.8}$$

where Q_1 = solar heat input

Q_2 = heat rejected in the power cycle

Q_3 = heat rejected in the cooling cycle

The efficiency $\eta_{C,hp}$ of the system when it acts as a heat pump is given by

$$\eta_{C,hp} = \frac{Q_2 + Q_3}{Q_1} \tag{9.9}$$

The efficiency in cooling, i.e., the heat transferred from a building divided by the energy required to drive the system, is called the *coefficient of performance* (COP) and is given as

$$\text{COP} = \frac{\text{cooling effect}}{\text{heat input}} = \frac{Q_4}{Q_1} \tag{9.10}$$

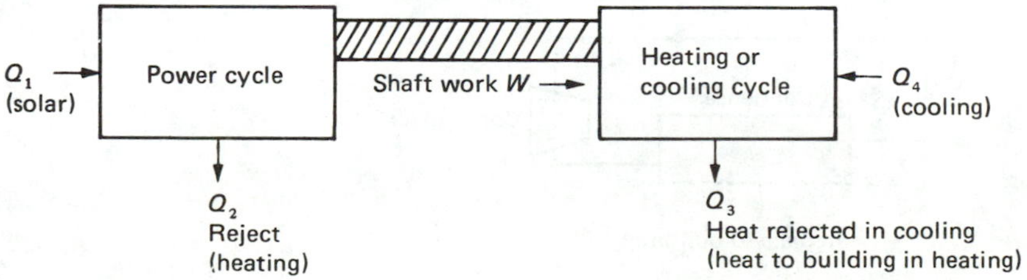

Figure 9.3 Basic solar-powered heat pump system.

When the system is acting as a heat pump to heat a building, Q_2 and Q_3 are rejected into the building and serve to heat the interior. When the system is acting as an air conditioner, the refrigerant flow is reversed and heat is removed from the building and rejected to the environment.

Solar-assisted Conventional Heat Pump

A possible system in which solar energy augments a heat pump is shown in Fig. 9.4. In this system thermal energy is stored in the form of sensible heat and is used as the energy source for a liquid-to-air heat pump. During the heating cycle, solar energy is used to increase the temperature of the storage, and therefore the amount of compression work necessary to increase the temperature of the working fluid is reduced. The energy that could be saved in a typical single-family residence in the northeastern United States is shown in Fig. 9.5. From September through November and March through April, solar energy is sufficient to supply the entire heat demand. Between November and March, the work input to the compressor is reduced by approximately 50 percent. The coefficient of performance, i.e., the energy delivered to the house divided by the electrical energy used to drive the system, shows an appreciable reduction in energy power consumption for the solar-assisted system. Less power is consumed because the average yearly coefficient of performance for a solar-activated system

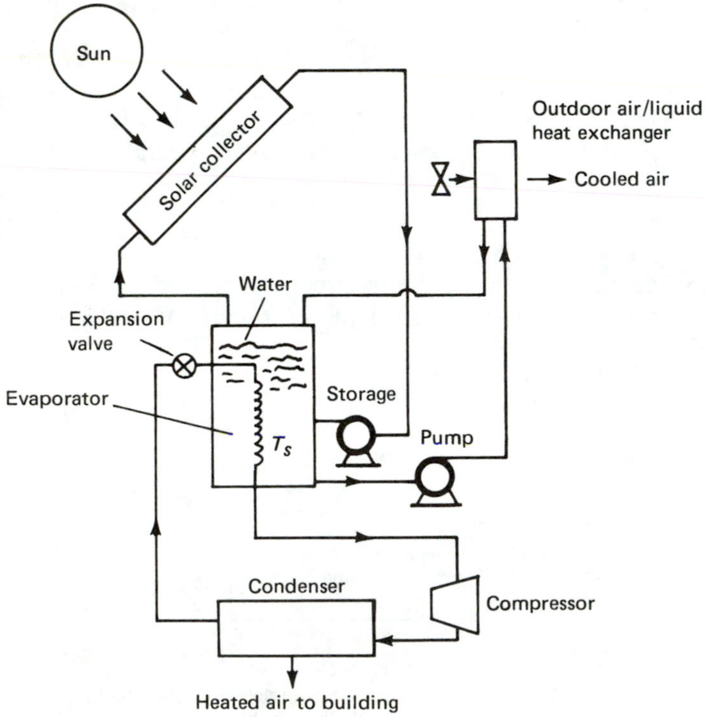

Figure 9.4 Solar-assisted heat pump system concept using a liquid-to-air heat pump cycle and showing the heating mode.

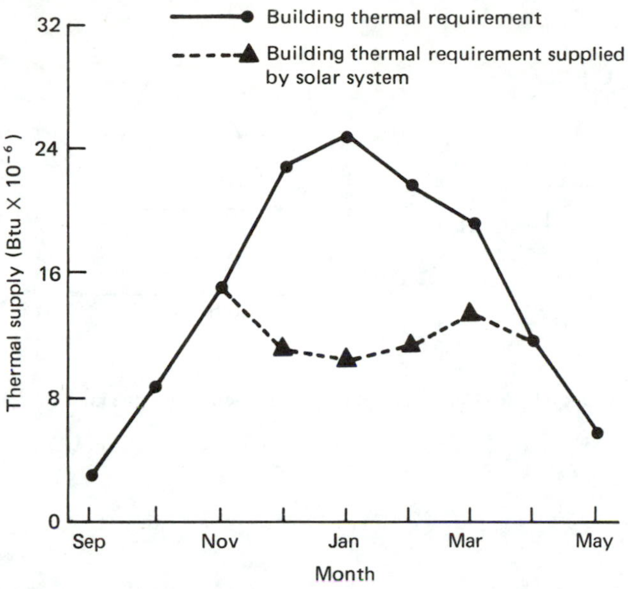

Figure 9.5 Thermal supply by heating month for solar-assisted heat pump concept.

is on the order of 3, in contrast to the yearly coefficient of performance for a conventional heat pump, which is approximately 2. Greater improvements could be expected if concentrator solar collectors were used. The performance of solar-assisted heat pumps at elevated evaporating temperatures is discussed by Kush [4].

A solar-assisted heat pump (without the fan loop) of the type shown in Fig. 9.4 has been operating successfully in a house constructed in 1974 in Colorado Springs. The north campus of the Community College of Denver, the third largest solar heating system in the world–300,000 ft^2 of heat floor area and 36,000 ft^2 of collector–uses the same concept.

Rankine Power Cycle

Many power cycles could be considered for a solar-assisted heat pump. One practical, recommended system is based on the Rankine cycle, illustrated in Fig. 9.6. The components are shown schematically on the left-hand side, and on the right-hand side a pressure-enthalpy diagram for the working fluid is presented.

The working fluid starts at state point 1 (P_1) and is pumped to the pressure in the boiler, state point 2 (P_2), where heat is added until the working fluid arrives at state point 3 (P_3) in a superheated form. The working fluid is then expanded at constant entropy to state point 4 (P_4), and heat is rejected at constant pressure in a condenser between points 4 and 1.

The efficiency of a Rankine cycle may be estimated by the so-called *Carnot efficiency* of the cycle, which is the highest efficiency that can be attained between the maximum and minimum temperatures available. The Carnot

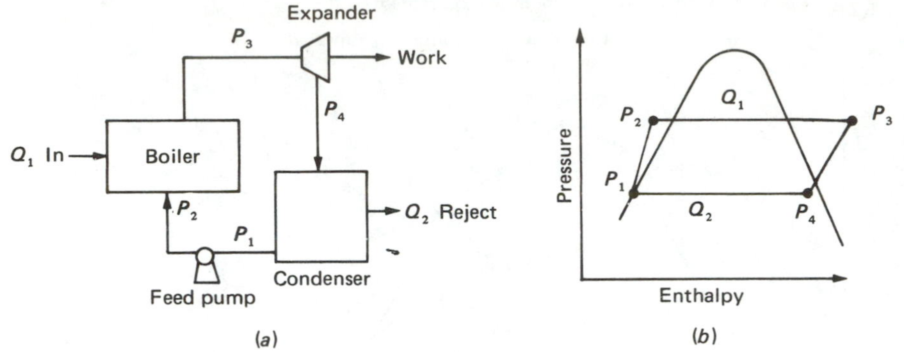

Figure 9.6 Simple Rankine engine cycle. (*a*) Components; (*b*) pressure-enthalpy diagram.

efficiency η_C is given by

$$\eta_C = \frac{T_3 - T_1}{T_3} = 1 - \frac{T_1}{T_3} \tag{9.11}$$

where T_3 is the maximum temperature attainable and T_1 is the lowest, or sink, temperature. Equation (9.11) shows the advantages of a concentrating collector, which could increase temperature T_3. It is not easy to change T_1, since it is generally dictated by the outside temperature for a power cycle. Figure 9.7 shows the effect of sink temperature and maximum temperature on the Carnot efficiency.

Any real power cycle can achieve only a fraction of the Carnot efficiency owing to the thermodynamic irreversibility of heat additions and other cycle

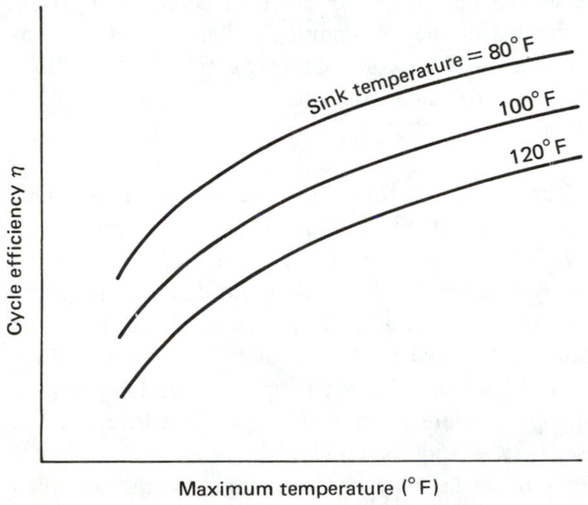

Figure 9.7 Ideal Carnot-cycle efficiencies as a function of sink temperature and maximum cycle temperature.

losses. The efficiency for a real cycle, which can be obtained by examination of Fig. 9.6, is given by

$$\eta = \frac{h_3 - h_4}{h_3 - h_1} \tag{9.12}$$

where h is the enthalpy at the condition indicated by the subscript. The work required by the feed pump is usually small and can be neglected.

Vapor Compression Cooling Cycle

Figure 9.8 shows a vapor compression cooling system that is most commonly used for cooling and air conditioning. The only major alternative at present is an absorption cycle cooling system, which will be discussed subsequently.

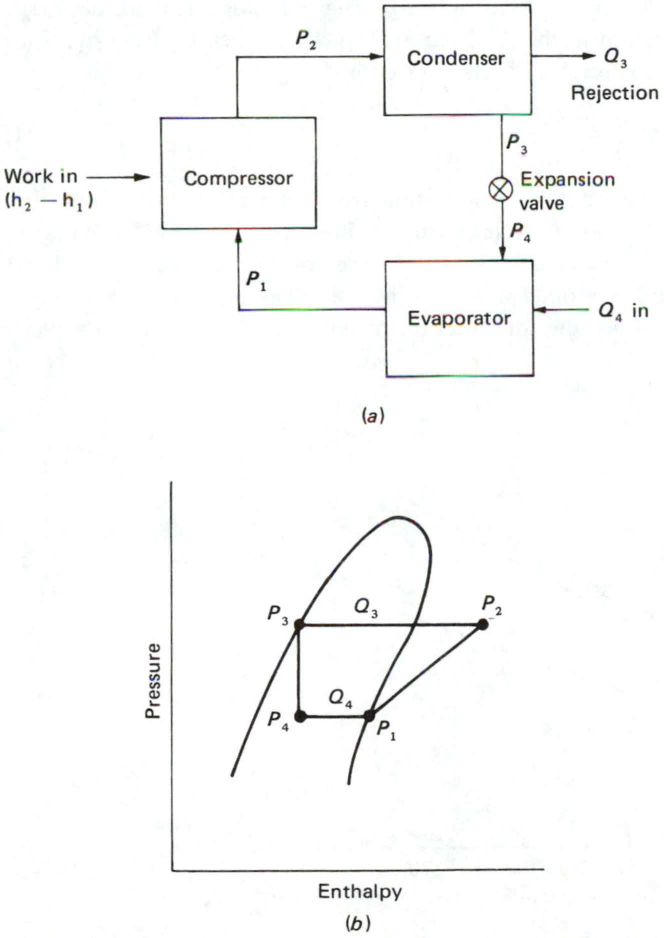

Figure 9.8 Vapor compression cooling cycle. (*a*) Components; (*b*) pressure-enthalpy diagram.

The Carnot COP for the vapor compression cycle is given by

$$COP_C = \frac{T_2}{T_1 - T_2} \tag{9.13}$$

where T_2 = lowest temperature of the working fluid at which heat from the interior of the house is absorbed

T_1 = upper temperature of the working fluid in the condenser, at which heat is rejected to the atmosphere

Figure 9.9 shows some typical numbers for Carnot cooling cycles. It can be seen that the COP is greater than unity; i.e., more cooling is accomplished than the shaft work required to drive the system.

Figure 9.8 shows the hardware for the vapor compression cycle schematically as well as a pressure-enthalpy diagram for the working fluid. It can be seen that the heat absorbed (equal to Q_4 in Fig. 9.3) is given by $h_1 - h_4$ and the heat rejected (Q_3 in Fig. 9.3) by $h_2 - h_3$. The compression work necessary to increase the temperature to the level required by h_2 is given by $h_2 - h_1$. Thus the COP of the vapor compression cycle is given by

$$COP = \frac{\text{cooling effect}}{\text{work input}} = \frac{h_1 - h_4}{h_2 - h_1} \tag{9.14}$$

In summary, the heat pump–air conditioner is a climate control system that employs a vapor compression cycle both to heat and to cool a building. Typically, the heat sink as well as the heat source for such a system is ambient air, although the ground or ground water can be used. The heat pump is simply a heat transformer, which, by the introduction of mechanical work into the cycle,

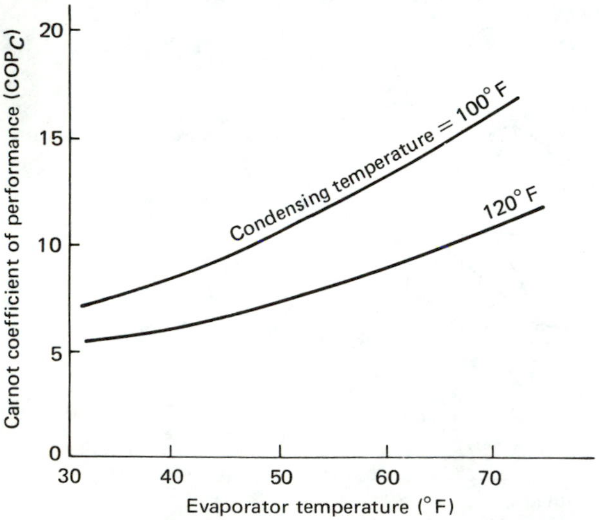

Figure 9.9 Ideal Carnot refrigeration cycle coefficient of performance as a function of evaporator and condenser temperatures.

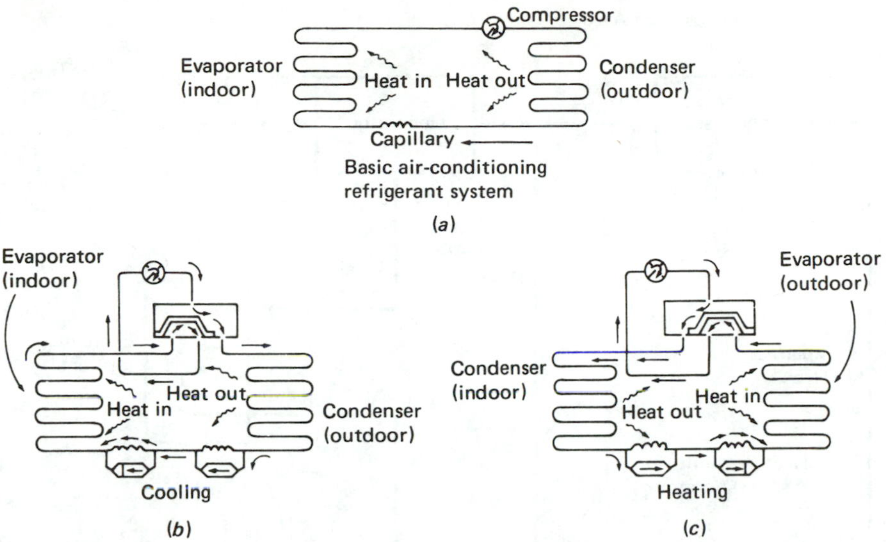

Figure 9.10 Vapor-compression cycle in (*a*) basic air-conditioning mode, (*b*) heat pump cooling mode, and (*c*) heat pump heating mode.

increases the temperature of the working fluid to a level where it can be used as a source for heating the home.

The heat pump has the same components that are used in standard vapor compression-air conditioning systems, as shown in Fig. 9.10. The only additions in a heat pump are a four-way valve to permit cycle reversal when the heating mode changes to the cooling mode, and an expansion device designed specifically for the particular operational (heating or cooling) refrigerant flow condition. The basic cycle can be designed to operate in many configurations, as discussed in standard texts on air conditioning.

ABSORPTION AIR CONDITIONING

Absorption air conditioning is the only air-conditioning system compatible with the upper collection temperature limits imposed by currently available flat-plate collectors. Home-sized absorption air-conditioning units are more expensive than vapor compression air-conditioning units, but to date only absorption air conditioning has been operated successfully in a full-scale installation.

At present, two types of absorption air-conditioning systems are marketed in the United States: the lithium bromide-water ($LiBr$–H_2O) system and the ammonia-water (NH_3-H_2O) system. The absorption air-conditioning system is shown in Fig. 9.11. Absorption air conditioning differs from vapor compression air conditioning only in the positive pressure gradient stage (right of the dashed line in Fig. 9.11). In absorption air-conditioning systems, pressurization is

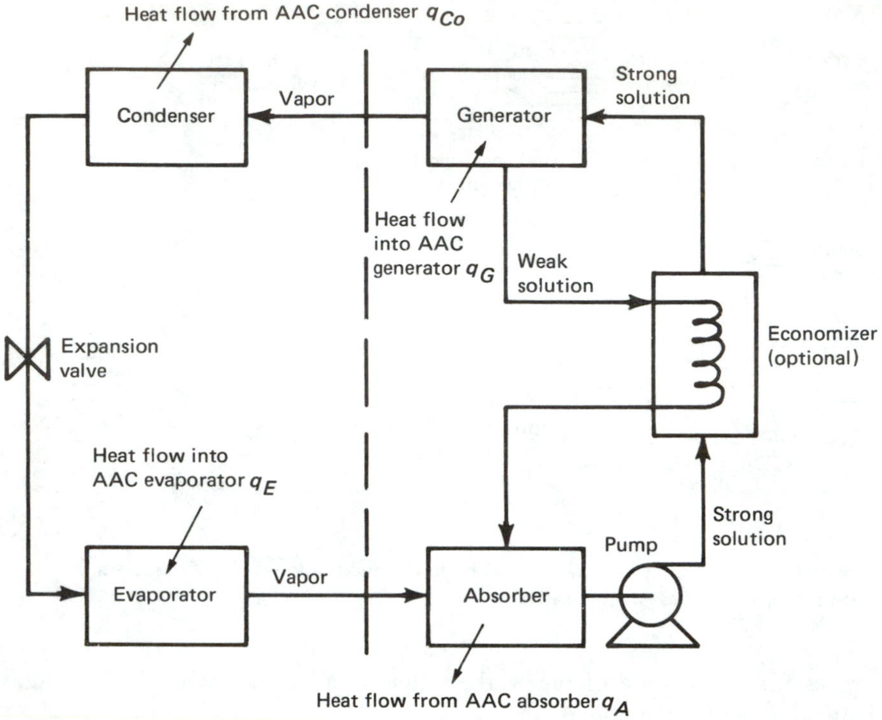

Figure 9.11 Absorption air conditioner (AAC) heat and fluid flow diagram with economizer.

accomplished by first dissolving the refrigerant in a liquid (the absorbent) in the absorber section, then pumping the solution to a high pressure with an ordinary liquid pump. The low-boiling refrigerant is then driven from solution by the addition of heat in the generator. By this means the refrigerant vapor is compressed without the large input of high-grade shaft work that the vapor compression air conditioning demands. The remainder of the system consists of a condenser, expansion valve, and evaporator, identical in function to those used in a vapor compression air-conditioning system.

Of the two common absorption air-conditioning systems, LiBr–H_2O is simpler since a rectifying column is not needed: in the NH_3–H_2O system a rectifying column ensures that no water vapor, mixed with NH_3, enters the evaporator, where it could freeze. In the LiBr–H_2O system water vapor is the refrigerant. In addition, the NH_3–H_2O system requires higher generator temperatures (250 to 300°F) than a flat-plate solar collector can provide without special techniques. The LiBr–H_2O system operates satisfactorily at a generator temperature of 190 to 200°F, achievable by a flat-plate collector; also, the LiBr–H_2O system has a larger COP than the NH_3–H_2O system. The disadvantage of LiBr–H_2O systems is that evaporators cannot operate at temperatures much below 40°F since the refrigerant is water vapor.

The effective performance of an absorption cycle depends on the two materials that comprise the refrigerant-absorbent pair. Desirable characteristics for the refrigerant–absorbent pair are

1 Absence of a solid-phase absorbent

2 Refrigerant more volatile than the absorbent so that it can be separated from the absorbent easily in the generator

3 Absorbent that has a small affinity for the refrigerant

4 High degree of stability for long-term operations

5 Refrigerant that has a large latent heat so that the circulation rate can be kept at the minimum

6 Low corrosion rate and nontoxicity for safety reasons

The main disadvantage of the LiBr–H_2O pair is the possible problem with crystallization in the generator.

Absorption air conditioners are manufactured by many of the large air-conditioning manufacturers in the United States: Carrier, Trane, York, Singer, Arkla-Servel, etc. These units are NH_3–H_2O systems. Only two manufacturers make a residential-sized (3 to 5 ton) LiBr–H_2O unit. In the past, Arkla-Servel manufactured a smaller residential LiBr–H_2O unit. The line was discontinued some years ago but has been revived because of the recent interest in solar-assisted air-conditioning systems. The present trade name of the Arkla-Servel system is Sol-aire.*

Performance

The COP of an absorption air conditioner can be calculated from Fig. 9.11.

$$\text{COP} = \frac{\text{cooling effect}}{\text{heat input}}$$

$$\text{COP} = \frac{q_{\text{refrig}}}{q_{\text{sup}}} = \frac{q_E}{q_G} \tag{9.15}$$

The pump work has been neglected since it is quite small; in some 3- and 5-ton units the pump can even be eliminated entirely and a percolation principle employed instead.

The COP values for absorption air conditioning range from 0.5 for a small, single-stage unit to 0.85 for a double-stage, steam-fired unit. These values are about 15 percent of the COP values that can be achieved by a vapor compression air conditioner. It is incorrect to compare the COP of an absorption air conditioner with that of a vapor compression air conditioner, however, because the efficiency of electric power generation or transmission is not included in the vapor compression air-conditioning COP. If the COP of the mechanical system is multiplied by the thermal efficiency of the power plant and the efficiency of the transmission network, it can be shown that the vapor compression air conditioner has little or no thermal performance advantage over the absorption air-conditioning system.

*Registered by Arkla Industries, Evansville, Indiana.

EXAMPLE

A water–lithium bromide absorption-refrigeration system such as that shown in Fig. 9.12 is to be analyzed for the following requirements:

1 The machine is to provide 100 tons of refrigeration with an evaporator temperature of 40°F, an absorber outlet temperature of 90°F, and a condenser temperature of 110°F.

2 The approach at the low-temperature end of the liquid heat exchanger is to be 10°F.

3 The generator is heated by a flat-plate solar collector capable of providing a temperature level of 192°F for evaporation of the refrigerant.

Determine the COP, absorbent and refrigerant flow rates, and heat input required for a 100-ton unit.

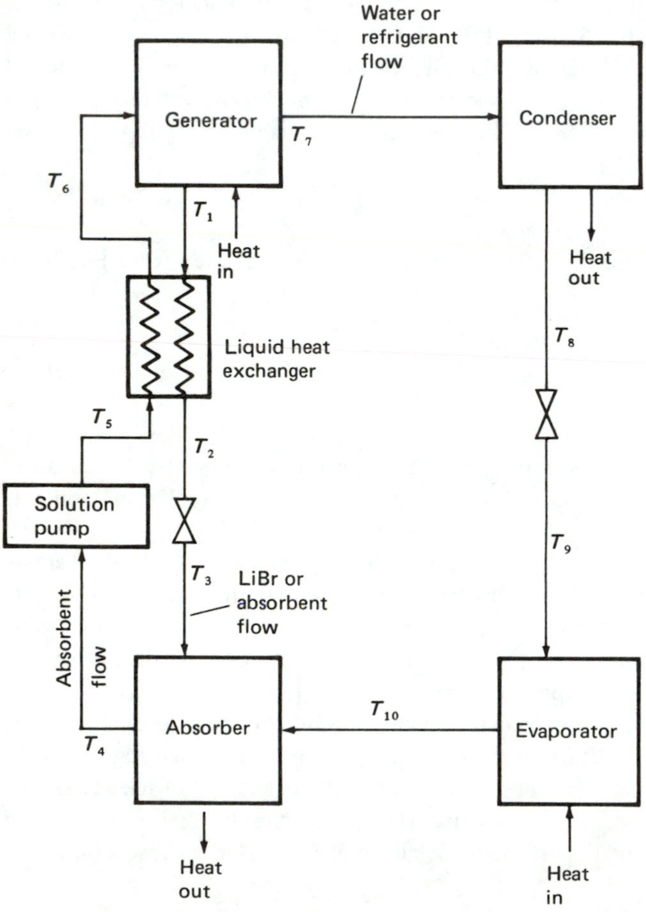

Figure 9.12 Lithium bromide–water absorption-refrigeration cycle (see Table 9.2).

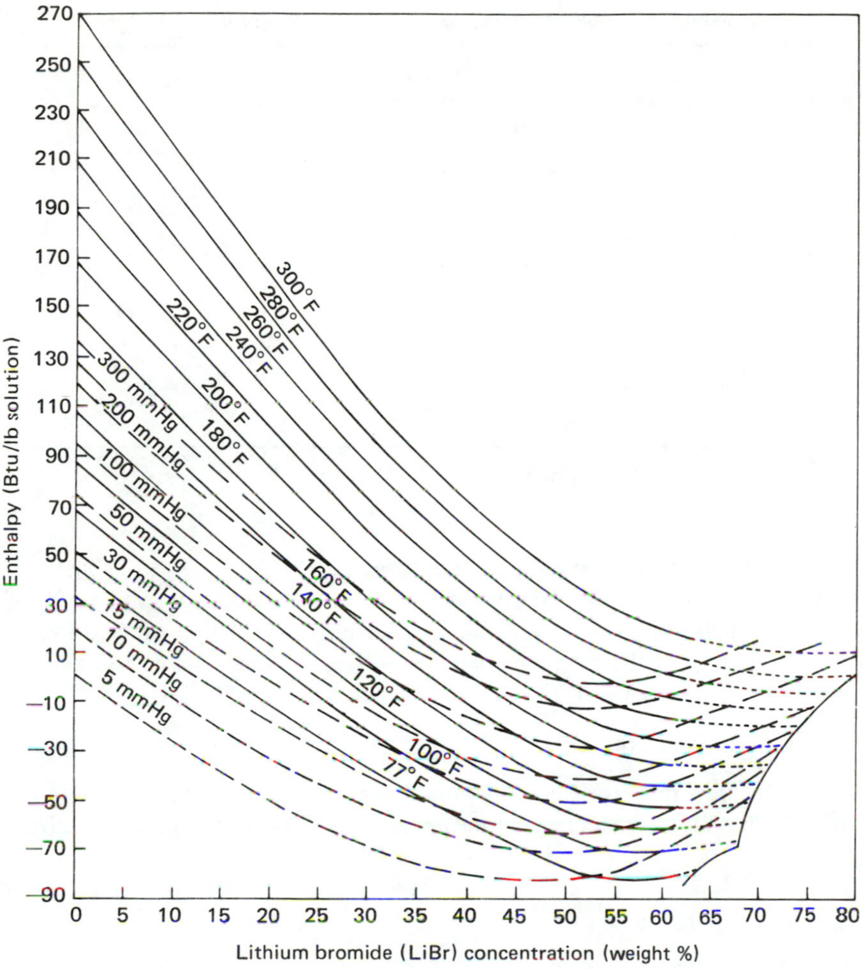

Figure 9.13 Enthalpy-concentration diagram for lithium bromide–water combination.

SOLUTION

Analytical evaluation of the LiBr-H_2O cycle requires that several simplifying factors be assumed:

1 At points in the cycle for which temperatures are specified, the refrigerant and absorbent phases are in equilibrium.

2 With the exception of pressure reductions across the expansion device between points 8 and 9 in Fig. 9.12, pressure reductions in the lines and heat exchangers are neglected.

3 The temperature difference at the inlet to the liquid heat exchanger is 10°F.

4 Pressures at the evaporator and condenser are equal to the vapor pressure of the refrigerant, i.e., water, as found in steam tables.

5 Enthalpies for LiBr-H_2O mixtures are given in Fig. 9.13.

Table 9.2 Thermodynamic Properties of Refrigerant and Absorbent for Fig. 9.12

Condition no. in Fig. 9.12	Temperature (°F)	Pressure (mmHg)	LiBr weight fraction (%)	Flow (lb/lb H_2O)	Enthalpy (Btu/lb)
T_1	192	66.0	0.61	11.2	−30
T_2	100	66.0	0.61	11.2	−70
T_3	100	6.3	0.61	11.2	−70
T_4	90	6.3	0.56	12.2	−75
T_5	90	66.0	0.56	12.2	−75
T_6	163	66.0	0.56	12.2	−38.2
T_7	192	66.0	0	1.0	1147
T_8	110	66.0*	0	1.0	78
T_9	40	6.3*	0	1.0	78
T_{10}	40	6.3	0	1.0	1079

*These values were derived from J. H. Keenan and F. G. Keyes [5].

As the first step in solving the problem, set up a table similar to Table 9.2; enter values of pressure in the appropriate table columns, enthalpy, and weight fraction for which sufficient information is available. For example, at point 8 the temperature is 110°F; the vapor pressure of steam corresponding to this pressure in the condenser is 1.28 lb/in^2 absolute, or 66 mmHg.

Mass Balance Equations

Relative flow rates for the absorbent (LiBr) and the refrigerant (H_2O) are obtained from material balances. A total material balance on the generator gives

$$\dot{m}_6 = \dot{m}_1 + \dot{m}_7$$

while an LiBr balance gives

$$\dot{m}_6 X_r = \dot{m}_1 X_{ab}$$

where X_{ab} = concentration of LiBr in absorbent, lb/lb solution
X_r = concentration of refrigerant in LiBr, lb/lb solution

Substituting $(\dot{m}_1 + \dot{m}_7)$ for $\dot{m}_6$ gives

$$\dot{m}_1 X_r + \dot{m}_7 X_r = \dot{m}_1 X_{ab}$$

Since the fluid entering the condenser is pure refrigerant, i.e., water, $\dot{m}_7$ is the same as the flow rate of the refrigerant $\dot{m}_r$:

$$\frac{\dot{m}_1}{\dot{m}_7} = \frac{X_r}{X_{ab} - X_r} = \frac{m_{ab}}{\dot{m}_r}$$

where $\dot{m}_{ab}$ = flow rate of absorbent, lb/h
$\dot{m}_r$ = flow rate of refrigerant, lb/h

Substituting for X_r and X_{ab} from the table gives the ratio of the absorbent and refrigerant flow rates:

$$\frac{\dot{m}_{ab}}{\dot{m}_r} = \frac{0.56}{0.61 - 0.56} = 11.2$$

The ratio of the refrigerant-absorbent solution flow rate $\dot{m}_s$ to the refrigerant-solution flow rate $\dot{m}_r$ is

$$\frac{\dot{m}_s}{\dot{m}_r} = \frac{\dot{m}_{ab} + \dot{m}_r}{\dot{m}_r} = 11.2 + 1 = 12.2$$

Energy Balance Equations

The enthalpy of the refrigerant–absorbent solution leaving the liquid heat exchanger at point 6 is obtained from an overall energy balance on the unit, or

$$\dot{m}_s h_5 + \dot{m}_{ab} h_1 = \dot{m}_{ab} h_2 + \dot{m}_s h_6$$

Hence

$$h_6 = h_5 + \left[\frac{\dot{m}_{ab}}{\dot{m}_s}(h_1 - h_2)\right]$$

$$= -75 + \frac{11.2}{12.2}[-30 - (-70)] = -38.2 \text{ Btu/lb solution}$$

The temperature corresponding to this value of enthalpy and a pressure of 66 mmHg is found from Fig. 9.13 to be 163°F.

The flow rate of refrigerant required to produce the desired 100 tons of refrigeration (equivalent to 1,200,000 Btu/h) is obtained from an energy balance about the evaporator:

$$q_{\text{refrig}} = \dot{m}_r(h_9 - h_{10})$$

where q_{refrig} is the cooling effect supplied the refrigeration unit and

$$\dot{m}_r = \frac{1{,}200{,}000}{1079 - 78} = 1200 \text{ lb/h}$$

The flow rate of the absorbent is

$$\dot{m}_{ab} = \frac{\dot{m}_{ab}}{\dot{m}_r}\dot{m}_r = 11.2 \times 1200 = 13{,}400 \text{ lb/h}$$

while the flow rate of the solution is

$$\dot{m}_s = \dot{m}_{ab} + \dot{m}_r = 13{,}400 + 1200 = 14{,}600 \text{ lb/h}$$

The rate at which heat must be supplied to the generator q_{sup} is obtained from the heat balance

$$q_{\text{sup}} = \dot{m}_r h_7 + \dot{m}_{ab} h_1 - \dot{m}_{ab} h_6 = [(1200 \times 1147) + (13{,}400 \times -30)]$$

$$- (14{,}600 \times -38.2) = 1{,}540{,}000 \text{ Btu/h}$$

This requirement, which determines the size of the solar collector, probably represents the maximum heat load the refrigeration unit must supply during the hottest part of the day.

The coefficient of performance COP is

$$\text{COP} = \frac{q_{\text{refrig}}}{q_{\text{sup}}} = \frac{1{,}200{,}000}{1{,}540{,}000} = 0.78$$

The rate of heat transfer in the other three heat exchanger units–the liquid heat exchanger, the water condenser, and the generator–is obtained from heat balances. For the liquid heat exchanger this gives

$$q_{1-2} = \dot{m}_{ab}(h_1 - h_2) = 13{,}400[(-30) - (-70)] = 540{,}000 \text{ Btu/h}$$

where q_{1-2} is heat transferred from the absorbent stream to the refrigerant-absorbent stream. For the water condenser the rate of heat transfer q_{7-8} to the environment is

$$q_{7-8} = \dot{m}_r(h_7 - h_8) = 1200(1147 - 78) = 1{,}280{,}000 \text{ Btu/h}$$

The rate of heat removal from the absorber can be calculated from an overall heat balance on this system:

$$q_A = q_{7-8} - q_{\text{sup}} - q_{\text{refrig}} = 1{,}280{,}000 - 1{,}540{,}000 - 1{,}200{,}000$$
$$= -1{,}460{,}000 \text{ Btu/h}$$

Explicit procedures for the mechanical and thermal design as well as sizing of the heat exchangers are presented in standard heat transfer texts. In large commercial units it may be possible to use higher concentrations of LiBr, operate at a higher absorber temperature, and thus save on heat exchanger cost. In a solar-driven unit this approach would require a concentrator-type absorber, because flat-plate solar collectors cannot achieve a sufficiently high temperature to raise the temperature in the absorber of an absorption air conditioner much above 190°F. Other design improvements in LiBr absorption chillers for solar applications are discussed in Grossman et al. [6].

COMPARISON OF MECHANICAL AND ABSORPTION REFRIGERATION SYSTEMS

Absorption refrigeration systems operate on cycles in which the primary fluid, a gaseous refrigerant, which has been vaporized in an evaporator, is absorbed by a secondary fluid called the absorbent. An absorption refrigeration cycle can be viewed thermodynamically as a combination of a heat engine cycle and a vapor compression refrigeration cycle, which are also the two components of a mechanical refrigeration system. Simplified diagrams for the two methods of providing refrigeration are shown in Figs. 9.14 and 9.15, respectively. A comparison of these two schematic diagrams indicates similarities between the main components in the absorption cycle and those in a heat engine cycle driving a mechanical refrigeration cycle.

In both cycles, heat from a high-temperature source is transferred in a heat exchanger to obtain a relatively high-pressure vapor. In the absorption cycle, the

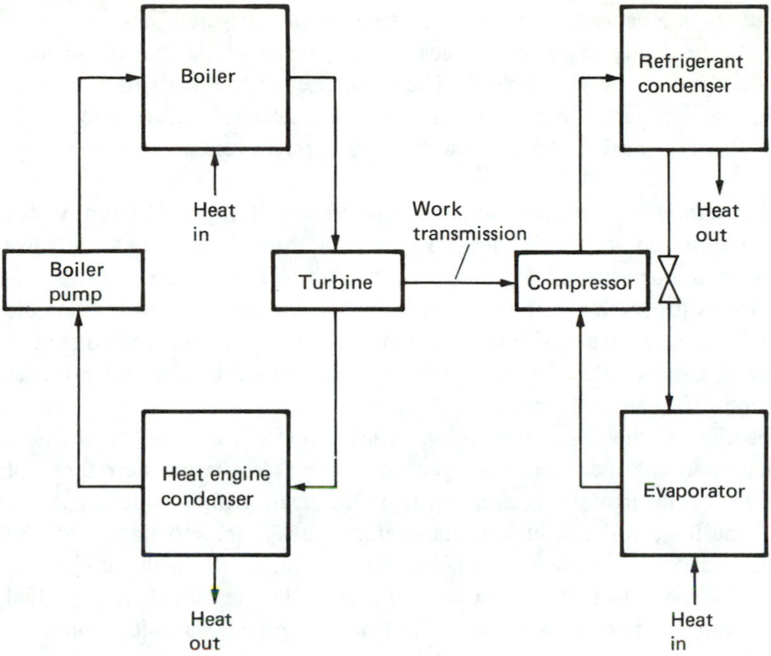

Figure 9.14 Basic absorption refrigeration cycle without economizer.

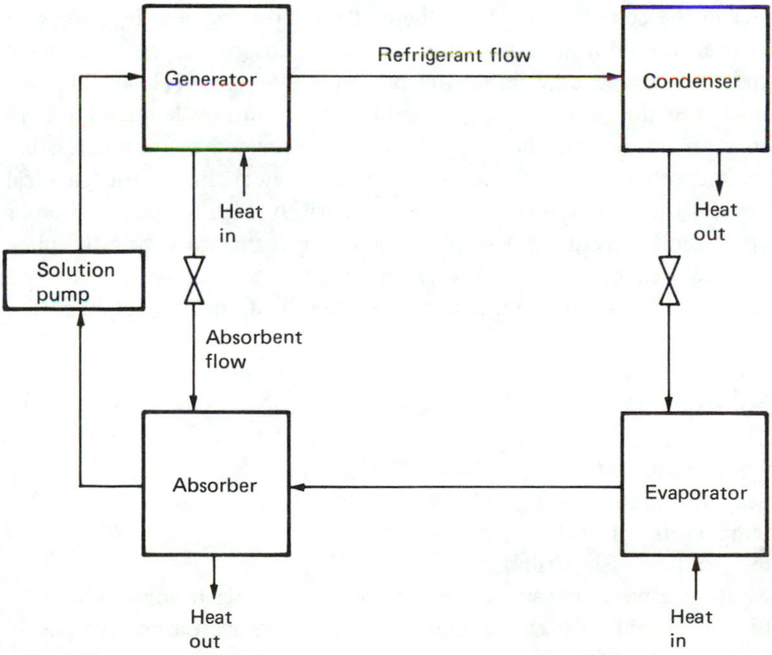

Figure 9.15 Combination of heat engine cycle and mechanical refrigeration cycle.

heat input occurs in a generator, from which streams of refrigerant and absorbent emanate. In the heat engine cycle, the heat input occurs at the boiler, where a vapor is produced that drives a turbine. The condenser in the absorption cycle is equivalent to the refrigerant condenser in the mechanical refrigeration cycle. In both heat exchanges, heat is transferred from the refrigerant at relatively high pressures.

In both methods of refrigeration, the high-pressure refrigerant (from which heat has been removed in the condenser) is passed through an expansion valve that reduces the temperature and pressure of the refrigerant before it enters the evaporator. In both methods, heat is transferred to the refrigerant in the evaporator, where, as a result of this heat transfer, the refrigerant is vaporized at relatively low pressures. It is the evaporator that absorbs the heat and provides the refrigeration effect in both methods.

The absorber in the absorption refrigeration cycle corresponds to the heat engine condenser in the mechanical refrigeration cycle. Heat is transferred out of the absorber in the absorption cycle, and out of the heat engine condenser in the combination method, to an intermediate-temperature sink to facilitate the conversion of relatively low-pressure vapor to the liquid state. In the absorption method, the absorbent is mixed with the refrigerant in the absorber. The final similarity between the two systems is the solution pump and the boiler pump. In both systems a small amount of work is necessary to increase the pressure of the liquid before it enters the boiler or generator of the cycle.

The turbine, which extracts heat energy from the high-temperature vapor from the boiler in the combination cycle, thereby transforming heat into work to drive a compressor, does not have a counterpart in the absorption method. In the absorption method, the energy input occurs in the form of heat into the generator, and hence the generator can operate at temperatures less than 200°F. In some absorption cycles, the heat supply can be provided by flat-plate solar collectors. On the other hand, in the combination heat engine and mechanical refrigeration cycle, the turbine drive requires a relatively high-temperature vapor for efficient operation; it is difficult to obtain good performance with a flat-plate solar collector, and a concentrator type is required.

The relation between work and heat for an ideal heat engine operating on a Carnot cycle is

$$W = q_g \frac{T_h - T_{hs}}{T_h} \tag{9.16}$$

where W = work output rate
q_g = heat input rate
T_h = temperature of heat source
T_{hs} = temperature of heat sink

The relation between the work required and the refrigeration load for an ideal mechanical refrigeration machine operating on the reverse Carnot cycle is

$$-W = q_e \frac{T_{hs} - T_1}{T_1} \tag{9.17}$$

where $-W$ = work input rate
q_e = rate of refrigeration
T_1 = temperature of refrigeration load
T_{hs} = temperature of heat sink

The COP for the combination of this engine cycle and the mechanical refrigeration machine is given by

$$\mathrm{COP} = \frac{q_e}{q_g} = \frac{T_1(T_h - T_{hs})}{T_h(T_{hs} - T_1)} \tag{9.18}$$

Equation (9.18) applies to the ideal absorption refrigeration process, as shown earlier, as well as to the combination heat engine and mechanical refrigeration cycle. In Fig. 9.16 the COP is plotted as a function of sink temperature for various heat source temperatures for a Carnot cycle with a refrigeration load at 40°F.

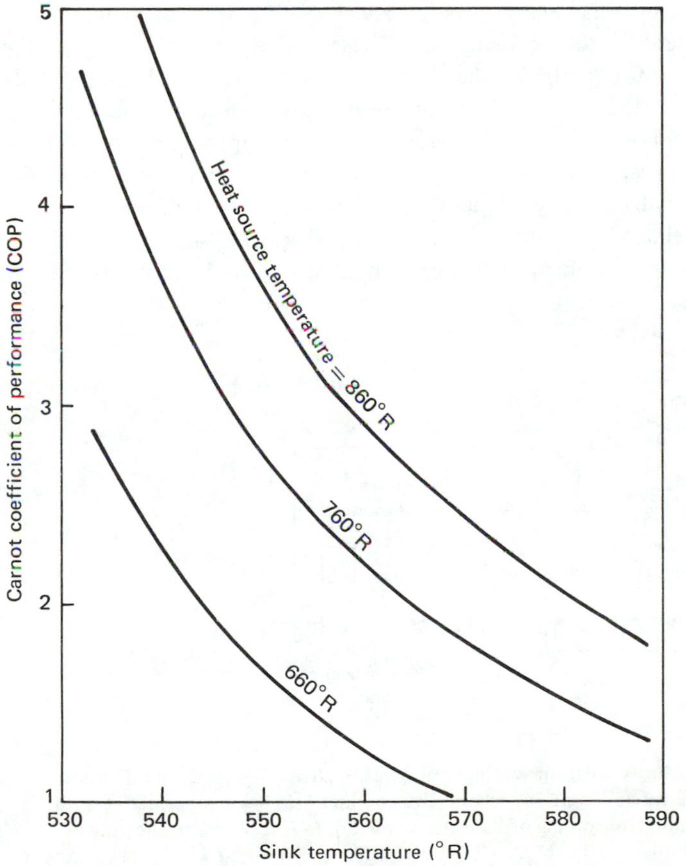

Figure 9.16 Ideal coefficient of performance for absorption-refrigeration cycles with evaporator (load) temperature T_1 of 500°R.

DESICCANT COOLING SYSTEMS

As shown in Table 9.1, cooling systems using desiccant materials that are regenerated by a solar-heated working fluid can be designed for open- and closed-cycle operation [7]. This section describes an example of each of these two approaches with solid desiccant material as the sorbent.

Open-Cycle Solar Desiccant Cooling

An open-cycle absorption system is shown in Fig. 9.17. The major, unique component in this type of system is the dehumidifier bed, which removes moisture from the process airstream. These systems usually can operate in one of two modes: ventilation or recirculation. In the ventilation system configuration, as shown in Fig. 9.17, a supply stream of ambient air is first adiabatically dehumidified, then sensibly cooled and evaporatively cooled, and finally introduced into the conditioned space. In the regenerating stream, air removed from the conditioned space is evaporatively cooled, then heated as it cools the supply stream, heated again by solar energy, cooled and humidified in the dehumidifier, and finally exhausted to the ambient air. These processes are shown in the psychrometric chart of Fig. 9.18, where the numbers correspond to the bulk airstream conditions at the numbered system locations shown in Fig. 9.17.

In the recirculation system configuration the supply and regenerating airflow streams are separate. The supply stream air is removed from the conditioned space, adiabatically dehumidified, sensibly cooled, evaporatively cooled, and then reintroduced into the conditioned space. In the regenerating stream, ambient air is evaporatively cooled, then heated as it cools the supply

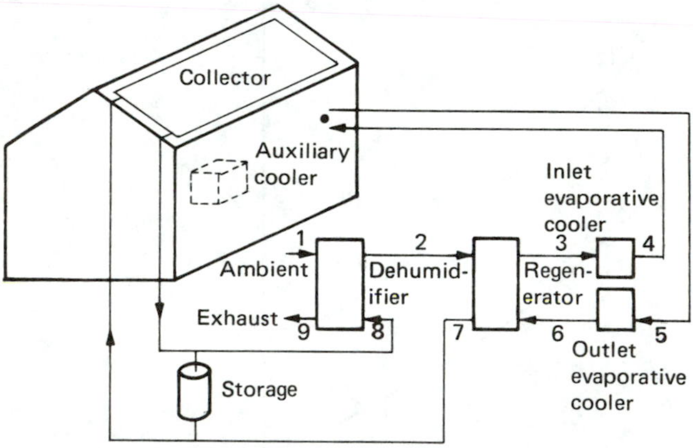

Figure 9.17 Solar desiccant ventilation system configuration. Stage 1–2, dehumidification of supply air; 2–3, regenerative cooling; 3–4, evaporative cooling; 4–5, heating by mixing with air in building; 5–6, evaporative cooling of building airstream; 6–7, regenerative heating of building airstream; 7–8, solar heating of building airstream in collector; and 8–9, cooling and humidification of solar-heated air in dehumidifier for supply stream (this step regenerates the desiccant bed).

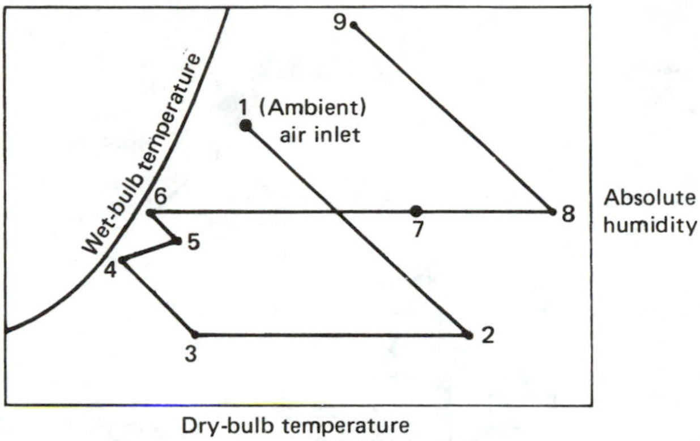

Figure 9.18 Ventilation mode psychrometric diagram.

stream, heated again by solar energy, humidified and cooled as it regenerates the dehumidifier, and exhausted to the ambient air.

Closed-Cycle Adsorption Systems

Another desiccant cooling approach utilizes an adsorption material in the same role as the ammonia or lithium bromide in an absorption system, i.e., for promoting the phase changes required in the cycle through the surface energy involved in the sorption processes. Closed-cycle adsorption cooling systems have not been developed as fully as their counterparts based on the absorption process, probably because the heat exchange is more difficult and more material is required as the sorbent. However, closed-cycle adsorption refrigeration systems using silica gel as the sorbent and sulfur dioxide as the refrigerant were in commercial use in the late 1920s and early 1930s.

The schematic diagram in Fig. 9.19 demonstrates the basic principles of a closed cycle using a combination of natural zeolite and water to create the required pressure ratio for cycling the refrigerant between the liquid and vapor phases. This is a once-a-day cycle. The adsorbent is dried by solar heat, with the effluent vapor condensed to liquid water by rejecting the heat of condensation to the ambient air. Saturation pressure (and therefore the percentage of moisture retention) is established by the heat sink temperature. Zeolite has a relatively flat characteristic of retained moisture versus vapor pressure. Thus the penalty for an air-cooled air conditioner does not seem significant. The cooling occurs in the night cycle. The adsorbent bed, now cooler because it is not insolated and the ambient temperature is lower, is in equilibrium with a much lower water vapor pressure in the enclosed space. The water condensed in the day cycle can be evaporated by heat absorbed from the building space air, which is passed through an evaporative heat exchanger. The heat of adsorption of the vapor in the bed must be rejected to the ambient air in order to keep the bed at its minimum temperature and thus maximum moisture retention capacity. The details of this system are described in [8].

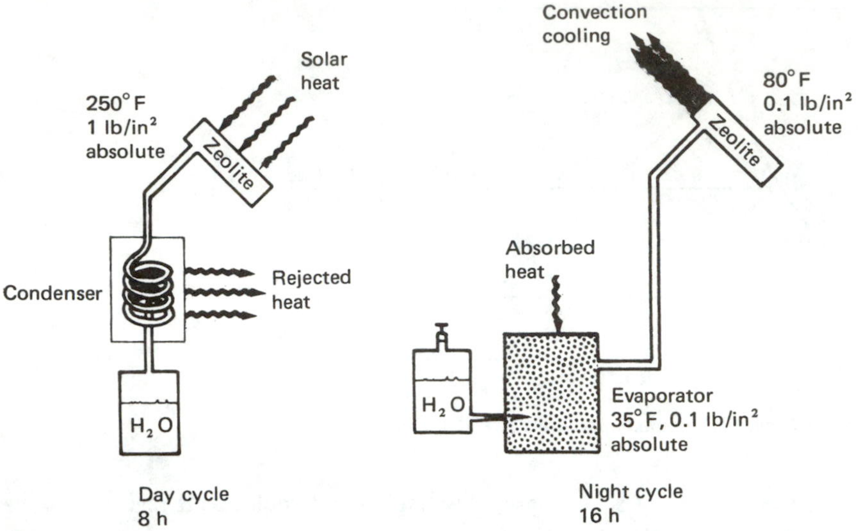

Figure 9.19 Schematic diagram of a closed-cycle adsorption cooling system.

NONMECHANICAL SYSTEMS

Australian Rock System

A nocturnal cooling-storage system first tested in 1955 in a desert in the southwest United States has been more recently tested and developed in Australia. The system consists of a large bed of rocks cooled by drawing cool night air across them and exposing them to night sky radiation. During the day, warm inside air may be cooled by circulating it through the rock bed. Augmented cooling can be achieved by drawing the night air through a porous surface having a high emittance at low temperature and facing the night sky. Such a system operates best in a desert climate, where the night skies are clear and humidity is low. In a desert climate, the diurnal temperature variation may be 45°F. Although this system is not an active solar cooling system, it uses the same pebble bed storage that an air-cooled solar heating system uses for heat storage.

Use of a pebble bed to provide energy storage for both heating and cooling cycles has been described by Close [9] and by Dunkle [10]. A solar air heater can be combined with pebble bed storage to provide a heat source for a building. The cycles in Fig. 9.20 illustrate the operation of the same pebble bed when it is used for cooling. During the night, cool air from the outside is brought through an evaporative cooler into the pebble bed and is cooled at approximately wet-bulb temperature to a condition approaching saturation. The entire bed is eventually brought to this temperature by passing the cooled air through it for several hours, as shown in Fig. 9.20*a*. On the next day, when cooling is required in the building, outside air is drawn vertically downward through the bed and cooled to

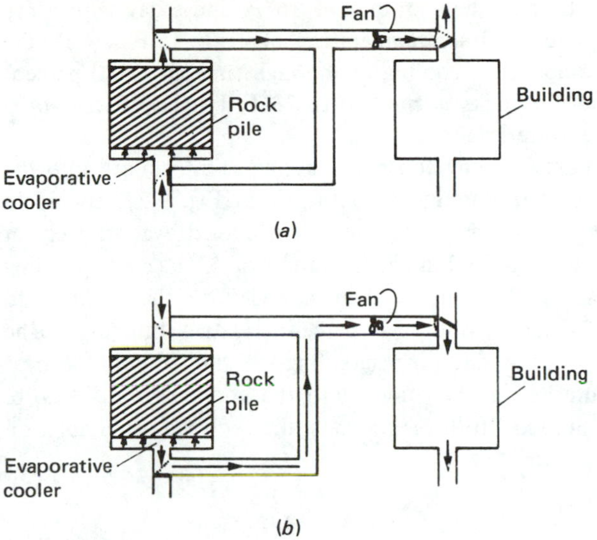

Figure 9.20 Operation of a pebble bed thermal storage as a source of air conditioning.

the bed temperature. As the air leaves the bed it may be cooled further by evaporation and then passed into the building as shown in Fig. 9.20*b*.

The mode of operation of the cycle just described can be illustrated quantitatively by means of an example. The psychrometric chart in Fig. 9.21 indicates the state points of the air passing through the cycle during a 24-h period with environmental conditions corresponding to a night with a wet-bulb

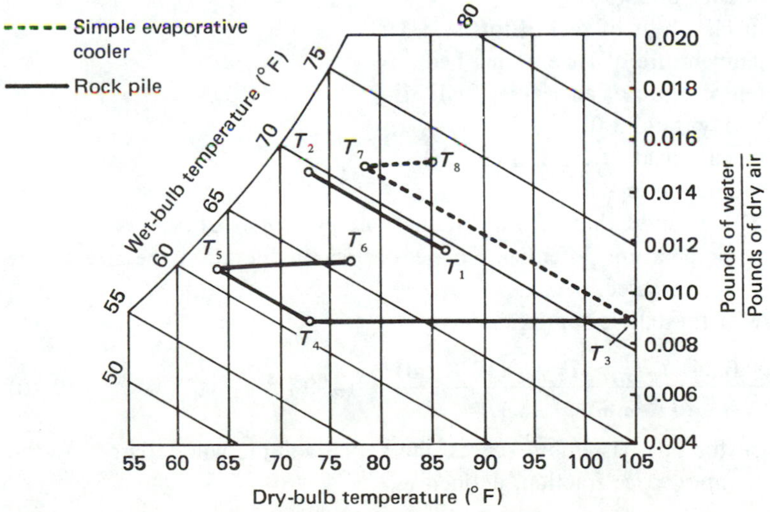

Figure 9.21 Cooling cycles for pebble bed with and without evaporative cooler.

temperature of 70°F and a dry-bulb temperature of 86°F, and a day with a dry bulb temperature of 105°F and wet-bulb temperature of 71°F. In Fig. 9.21, the line T_1-T_2 corresponds to evaporative cooling of the nighttime air to 80 percent of saturation, which will cool the rocks in the bed to 73°F if steady state can be achieved at the lowest temperature level.

When the building requires cooling the next day, air is introduced into the pebble bed under conditions corresponding to point T_3 in Fig. 9.21, cooled in the bed to 73°F, corresponding to point T_4, and then cooled evaporatively at constant wet-bulb temperature to a dry-bulb temperature of 63°F, corresponding again to 80 percent saturation. This air is then passed into the building to maintain conditions of 77°F and about 57 percent relative humidity. The increase in internal energy of the air, corresponding to the amount of heat transferred from the hot interior of the building to the coolant, is 3.5 Btu/lb. Thus for a cooling load of 30,000 Btu/h or 2.5 tons, the required air circulation rate is about

$$\frac{(30{,}000 \text{ Btu/h}) \times (14 \text{ ft}^3/\text{lb air})}{(3.5 \text{ Btu/lb air}) \times (60 \text{ min/h})} = 2000 \text{ ft}^3/\text{min}$$

For comparison, the dotted line in Fig. 9.21 corresponds to a simple evaporative cooler. To maintain a temperature of 85°F in the building, the required airflow rate is about 30 percent larger than for a rock bed that maintains the building at 77°F.

To determine the size of the pebble bed necessary for a cooling period of 12 h, a heat balance must be made on the rocks, or

$$V_r \rho_r c_r (T_{a,\text{in}} - T_{a,\text{out}}) = \dot{m}_a c_a (T_{a,\text{in}} - T_{a,\text{out}})\theta \tag{9.19}$$

where V_r = volume of rock in bed, ft^3
ρ_r = density of rock, lb/ft^3
c_r = specific heat of rock, Btu/(lb·°F)
$T_{a,\text{in}}$ = temperature of air entering bed, °F
$T_{a,\text{out}}$ = temperature of air leaving bed, °F
$\dot{m}_a$ = airflow rate, lb/h
c_a = specific heat of air, Btu/(lb·°F)
θ = cooling period, h

Equation (9.19) assumes that all the thermal energy stored can be extracted at the maximum temperature potential. For less efficient operation the size of the rock pile must be increased.

The size of the pile is approximately

$$V_r = \frac{(2000 \times 0.24)(T_{a,\text{in}} - T_{a,\text{out}})(12 \times 60)}{(85 \times 0.21)(T_{a,\text{in}} - T_{a,\text{out}})14} = 1400 \text{ ft}^3 \tag{9.20}$$

However, since the pile has empty spaces, the actual volume will be larger by the inverse of the empty-space fraction, defined as

$$\frac{\text{Volume of rock pile} - \text{volume of voids}}{\text{Volume of rock pile}}$$

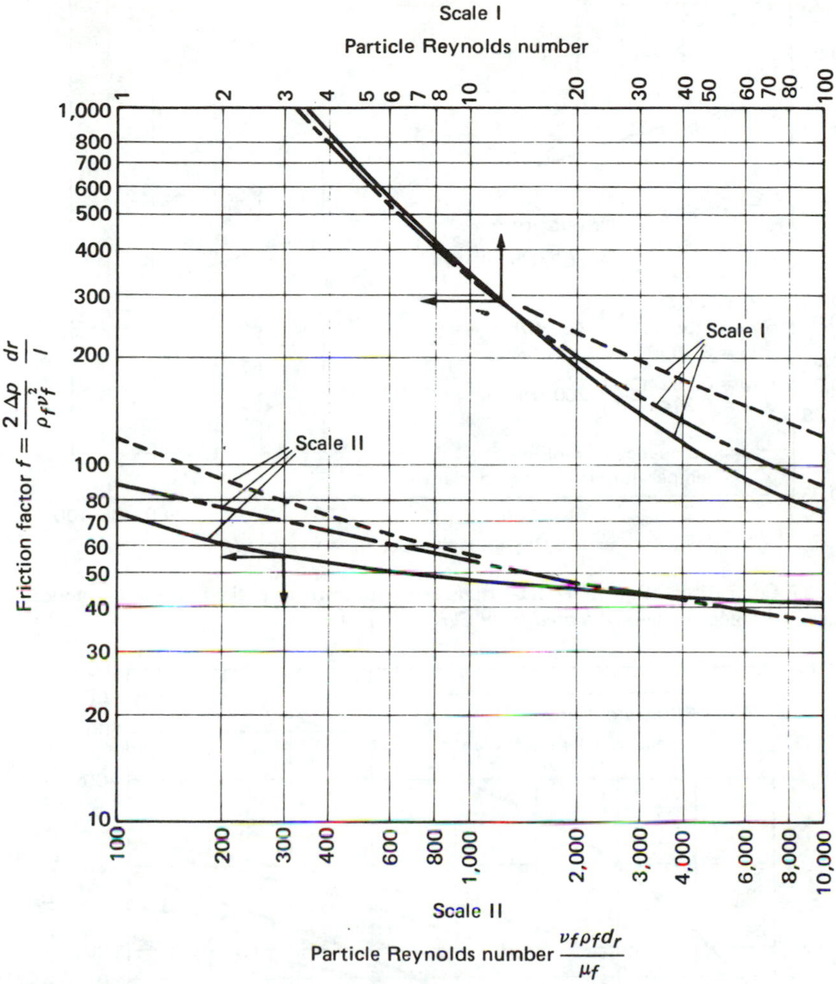

Figure 9.22 Friction factors for packed beds as a function of particle Reynolds number.

To design a rock pile storage system completely, the friction factor for packed beds and the heat transfer coefficient for air passing through the pile must be known. In Fig. 9.22 the friction fractor for packed beds f (defined as twice the pressure drop Δp divided by $\rho_f v_f^2$ times the ratio d_r/l, where l is the packed bed length) is plotted as a function of the particle Reynolds number Re (defined as $v_f \rho_f d_r/\mu_f$). In Fig. 9.22 v_f is the superficial air velocity (airflow rate per pile cross-sectional area), d_r is the equivalent spherical diameter of the rocks $(6\ V_p/\pi)^{1/3}$, V_p is the rock particle volume, μ_f is the viscosity of the air in the pile, ρ_f is the density of the air, and Δp is the pressure drop across the pile. These parameters must be in a consistent set of units to make the abscissa and the ordinate dimensionless. This correlation, proposed by Dunkle [10], agrees well with other correlations, but is simpler to use. Experimentally measured heat

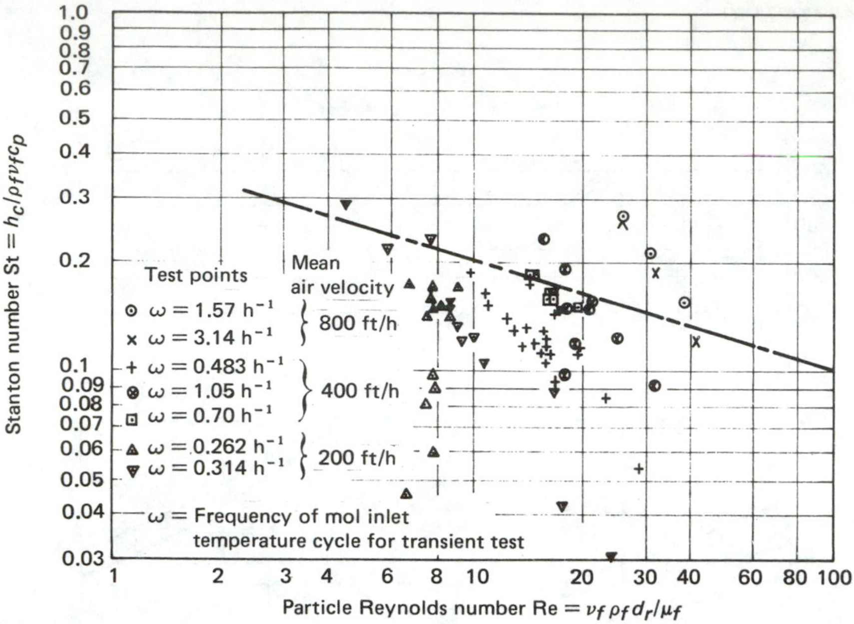

Figure 9.23 Comparison of measured heat transfer coefficients for packed beds as a function of particle Reynolds number. (*Adapted from Close [9].*)

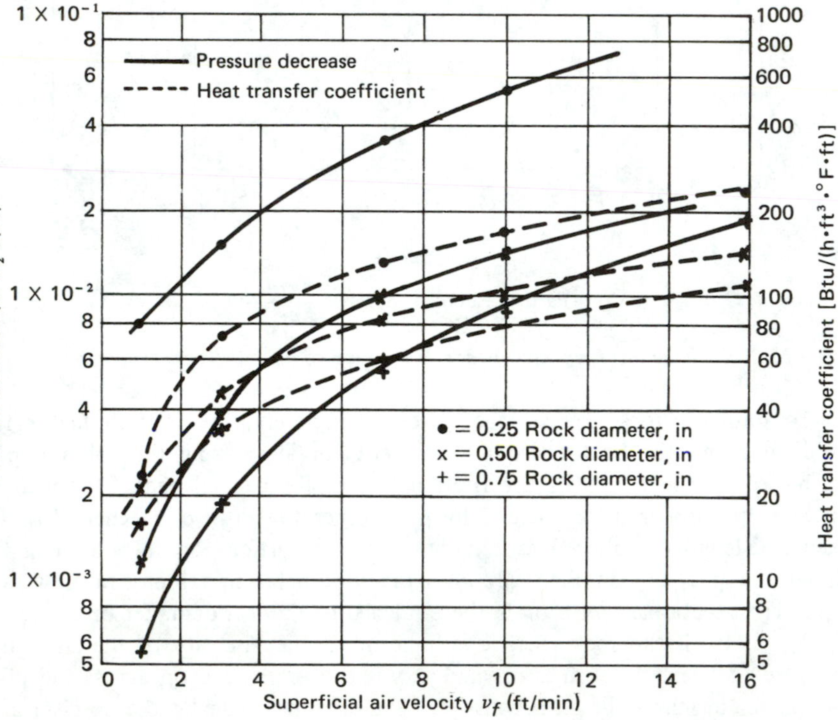

Figure 9.24 Heat transfer coefficients and pressure drops for various air flows and rock sizes in a pebble bed.

transfer coefficients from Close [9] are shown in Fig. 9.23 as a function of particle Reynolds number. In Fig. 9.23 the Stanton number St, defined as a heat transfer coefficient divided by the product of air density, velocity, and specific heat, is plotted again as a function of the particle Reynolds number. Figure 9.24 shows the pressure decrease (inches of water per foot of depth of the pile) and heat transfer coeffiient as a function of superficial air velocity (feet per minute) for various rock sizes. The following example illustrates the calculations for a rock pile storage system. For a more detailed analysis the variation in temperature during a 24-h period must be taken into account.

EXAMPLE

A rock pile is required to store 1 million Btu. Charging and extraction are at constant rates, and each lasts approximately 10 h. The temperature difference is to be 60°F (70°F minimum and 130°F maximum). The maximum allowable pressure decrease is 0.1 in H_2O. Approximately 80 percent of the energy stored in the pile can be extracted. Determine the fluid flow rate and ratio of energy storage to energy required to move the air through the bed.

SOLUTION

Assume the rock has a density of 85 lb/ft^3 and a specific heat of 0.21 Btu/(lb·°F); the total volume of rock required for the storage would then be 934 ft^3. If this storage is to be charged in 10 h, the mass flow rate of air $\dot{m}_a$ may be obtained from heat balance. If it can be assumed that all the heat is extracted at the stored temperature

$$\dot{m}_a = \frac{10^6 \text{ Btu}}{10 \text{ h} \times [(0.24 \text{ Btu})/(\text{lb}\cdot{}^\circ\text{F})] \times 60^\circ\text{F}} = 7000 \text{ lb/h}$$

The volumetric flow rate for air at 70°F is thus about 1600 ft^3/min. If the maximum allowable pressure decrease is 0.1 in H_2O, the ratio of heat stored to energy required to operate the pile is approximately 400:1.

Table 9.3 summarizes the dimensions and air velocities for various pebble beds as compiled by Close [9] for engineering design. It should be noted that best performance is obtained when the rocks are as small as possible and the bed sized for the maximum allowable pressure decrease. The smaller the rocks, however, the more the pile is subject to blockage by dust, and the installation of a dust removal screen may be necessary.

Sky-Therm* System

H. R. Hay has built a home in Atascadero, California, that uses his patented, passive, solar heating and cooling arrangement. It consists of a

*Registered trademark, Skytherm Processes and Engineering, Los Angeles, California.

Table 9.3 Dimensions and Air Velocities for Various Rock Beds

Bed dimensions and air velocity	Rock diameter (in) 0.25	0.5	0.75	Design basis
Bed height, ft	3.4	5.9	7.4	Maximum allowable
Bed area, ft^2	275.5	159	126	pressure drop
Air velocity, ft/min	6	10.5	13.1	(maximum height)
Bed height, ft	2.75	4.0	5.0	Intermediate height
Bed area, ft^2	340	233.5	186.8	
Air velocity, ft/min	4.85	7.07	8.8	
Bed height, ft	2	2.0	2.0	Minimum height
Bed area, ft^2	467	467	467	
Air velocity, ft/min	3.53	3.53	3.53	

combined collector-radiator-storage system mounted on the horizontal roof of a one-story building. The combined energy medium consists of 8-in-deep pools of water enclosed in plastic bags located between beams of the house and on top of a black plastic liner used to further waterproof the roof. Solar energy heats the water to 85°F, an intentionally maintained low temperature. Insulated shutters are then placed over the panels during winter nights to reduce heat loss [11, 12].

During the summer the shutters cover the panels to prevent heating during the day. However, the collection area used for service water heating remains uncovered during daytime. At night the primary ponds are uncovered to permit radiation of the heat collected from the building during the day. A smaller building was maintained at a temperature of 70 ± 2°F for 1 year in Phoenix, Arizona, by this method.

The summer function of such a system relies on nocturnal radiation loss. Such losses are usually in the range 10 to 35 Btu/(ft^2·h). To provide 1 ton·h of heat removal, 500 ft^2 of radiator surface is required. If the entire roof of a 2000-ft^2 house were used as a radiator, only 4 tons of heat would be removed, assuming the water ponds do not reach an equilibrium temperature and night heat loss averages 25 Btu/(ft^2·h). For an 8-h night this cooling effect amounts to 32 ton·h. In the Southwest, where the system is to be applied, additional daytime cooling might be necessary to maintain comfort. In the 2000-ft^2 house, the water ponds would create a total physical load of 42 tons.

SUMMARY

This chapter describes methods by which solar energy can be used to provide building cooling. The first section presents the procedure for calculating the cooling load on a building during the cooling season. A design study of a flat-plate collector–lithium bromide air conditioner showed that the three most important design parameters for the solar collection system are the tilt angle, number of covers, and collector area.

Several thermodynamic cooling cycles are considered in detail in the remainder of the chapter. Performance criteria for the Rankine cycle-powered vapor compression cycle are described and the use of vapor compression heat pumps with solar assist is explained. The absorption cycle is also discussed, and the method of analyzing the cycle thermodynamically is shown in detail for a lithium bromide system. Open and closed cycle desiccant cooling systems as well as a nonmechanical cooling system that uses pebble bed storage as a means of summer cooling are also described. Design correlations for pressure drop and heat transfer coefficient for this nonmechanical system are given.

REFERENCES

1. American Society of Heating, Refrigerating, and Air-Conditioning Engineers, *Handbook of Fundamentals,* ASHRAE, New York, 1980.
2. G. O. G. Löf and R. A. Tybout, "Cost of House Heating with Solar Energy," *Solar Energy,* vol. 14, 1973, p. 23.
3. G. O. G. Löf, *Design and Construction of a Residential Solar Heating and Cooling System,* National Technical Information Service, Springfield, Va., 1974.
4. E. A. Kush, "Performance of Heat Pumps at Elevated Evaporating Temperatures with Application to Solar Input," *J. Solar Energy Eng.,* vol. 102, 1980, pp. 203–210.
5. J. H. Keenan and F. G. Keyes, *Thermodynamic Properties of Steam,* Wiley, New York, 1936.
6. G. Grossman, J. R. Bourne, J. Ben-Dror, Y. Kimchi, and I. Vardi, "Design Improvements in LiBr Absorption Chillers for Solar Applications," *J. Solar Energy Eng.,* vol. 103, 1981, pp. 56–61.
7. B. Shelpuk, *Proceedings of the Desiccant Cooling Conference of November 16, 1977,* SERI Rep. 22, Solar Energy Research Institute, Golden, Colo., 1978.
8. D. I. Tchernev, "Exploration of Molecular Sieve Zeolites for the Cooling of Buildings with Solar Energy," Lincoln Laboratory, Massachusetts Institute of Technology, Lexington, Mass., 1977.
9. D. J. Close, "Rock Pile Thermal Storage for Comfort Air Conditioning," *Mech. Chem. Eng. Trans. Inst. Eng. (Australia),* vol. MC-1, 1965, p. 11.
10. R. V. Dunkle, "A Method of Solar Air Conditioning," *Mech. Chem. Eng. Trans. Inst. Eng. (Australia),* vol. MC-1, 1965, p. 73.
11. H. R. Hay and J. I. Yellott, "A Naturally Air Conditioned Building," *Mech. Eng.,* vol. 92, 1970, p. 19.
12. H. R. Hay, "Energy Technology and Solarchitecture," *Mech. Eng.,* vol. 94, 1973, p. 18.

10 Special Topics

This chapter introduces some special topics related to solar design: photovoltaic energy systems, solar ponds, and wood stoves. The first of these technologies, photovoltaics, is technically ready but is still too expensive for widespread use. The second, solar ponds, has a great potential for district heating and process heat, but operating experience with solar ponds to date is quite limited, especially in the United States. Wood stoves are, of course, an old technology that has recently become popular again in many parts of the country.

PHOTOVOLTAIC ENERGY SYSTEMS

Although photovoltaic energy is not yet cost effective, the price of photovoltaic arrays has decreased markedly over the past 5 years and is expected to continue this trend, as shown in Fig. 10.1. Consequently, it is expected that photovoltaically generated electricity will soon become available on the market at a price where it will be competitive for some applications, such as use in remote locations where natural gas or electricity is unavailable. In order to assist the designer of buildings to assess the potential of a photovoltaic power system, this chapter will present a brief outline of how photovoltaic cells work, describe the components of a complete photovoltaic power system, present the performance characteristics of such power systems, and then give a generalized approach for estimating the cost of photovoltaic power in the United States.

How Solar Cells Work

A solar cell is a photovoltaic semiconductor device that converts solar energy directly into electrical energy. Photons of solar radiation impinging on the

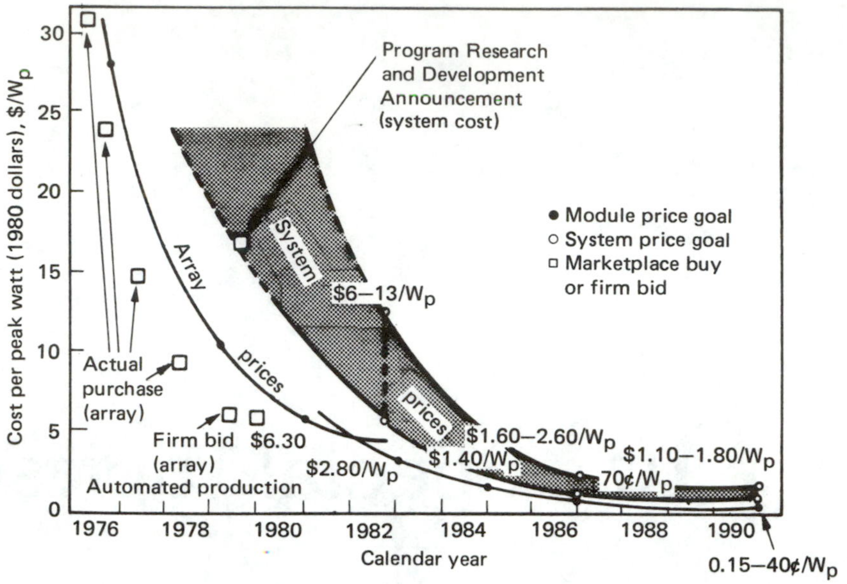

Figure 10.1 Past and projected prices of photovoltaic arrays. (*From U.S. Department of Energy, Photovoltaics Energy Systems Division, Washington, D.C., 1980.*)

cell impart energy to the semiconductor material, cause electrons to flow across a semiconductor junction, usually called a *p-n* junction, and thus convert the solar energy into electrical energy, as shown in Fig. 10.2. At present, most solar cells are manufactured from silicon. A single crystal of pure silicon is first grown and then sliced into wafers approximately 0.05 cm thick. The *n* and *p* junctions of the solar cell are produced by introducing impurities into the raw silicon. The silicon wafer can thus be made to conduct either negative (*n*) or positive (*p*) charges. Arsenic, phosphorus, or antimony impurities create an *n*-type wafer, whereas boron makes the material a *p*-type carrier. These wafers are arranged in a module to yield the desired current and voltage. When a photon strikes a *p-n* junction, a negative and a positive charge are created and an electron moves toward the *n*-type region, initiating an electric current. By connecting contacts and leads to the surfaces of the wafer, a circuit is formed that allows electrons to return to

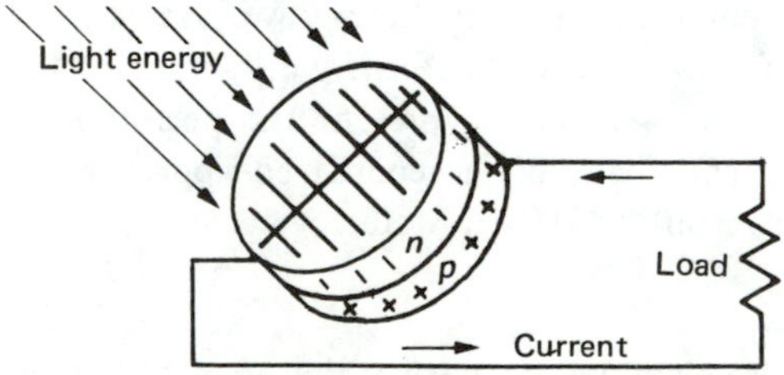

Figure 10.2 Schematic diagram of *p-n* junction.

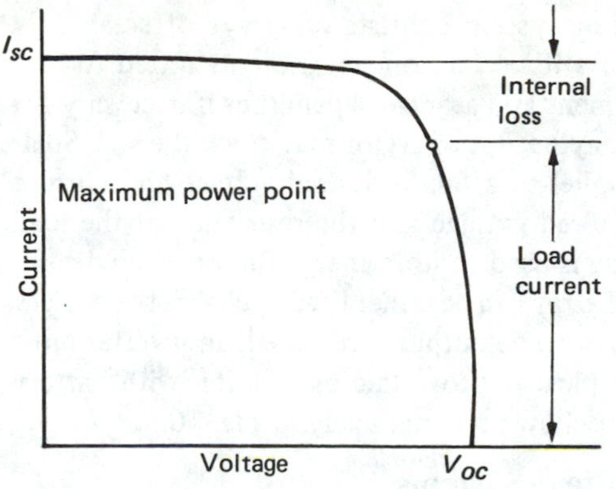

Figure 10.3 Current-voltage characteristics of a typical solar cell.

the *p*-type region through an external load. The current-voltage characteristics of a typical cell are shown in Fig. 10.3 for an equivalent circuit shown in Fig. 10.4. If the load resistance is small, the cell acts like a constant-current generator with the current output equal to the short-circuit current I_{sc}. In this region the current output is approximately proportional to the intensity of radiation incident on the cell. But as the load resistance is increased, the cell current decreases and more current flows through the internal diode. For very large values of the load resistance, the voltage across the cell terminals approaches the open-circuit voltage V_{OC}.

Under standard test conditions (STC) of 1000 W/m^2 insolation and a cell temperature of 27°C, the short-circuit current is approximately 1 to 1.2 A per square centimeter of cell and the open-circuit voltage is between 0.55 and 0.6 V. The open-circuit voltage is determined primarily by the cell temperature. It decreases at the rate of 0.022 V/°C. The standard test conditions used in terrestrial solar applications were chosen because they are easy to obtain in production testing. Since, on a clear day, the solar intensity is about 1000 W/m^2 and the average temperature is 27°C (80°F), they also yield useful results for actual cell operation.

The basic unit of a photovoltaic system is the module. To build a module, individual cells are assembled in series and/or parallel combinations. The number of cells in series determines the module voltage and the number in parallel determines the module current capability. Modules are assembled together in a frame or support to form an array.

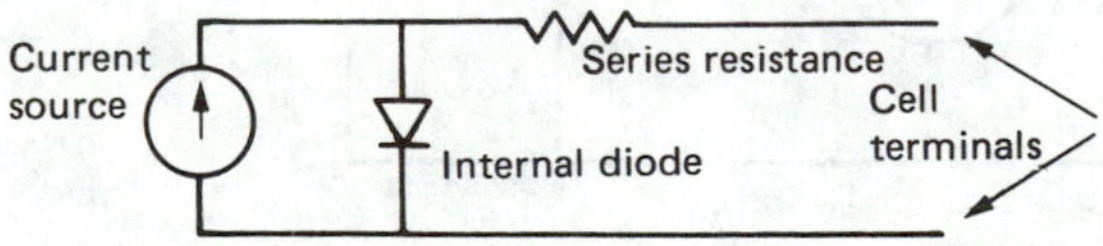

Figure 10.4 Equivalent circuit of solar cell.

A typical photovoltaic conversion system consists of arrays of solar cells, which may be either flat or combined with solar concentrators mounted to face the sun, plus power conditioning equipment and storage capabilities if necessary for the particular application. The mount may be stationary or may track the sun. Solar cell arrays are arranged in series or parallel to attain a desired voltage and current output. A controller maintains the required voltage and the resistance of the load determines the current. Usually a battery is used to store energy for use when the sun is down. The dc output from a solar cell array can be either used to charge the battery or transmitted directly to the load. When an ac output is required, an inverter must be incorporated into the system. A typical photovoltaic installation with battery interconnections, load, and solar panel is shown schematically in Fig. 10.5.

Storage Battery for Photovoltaic Systems

Batteries for use in photovoltaic systems should be in perfect condition to operate reliably under diverse environmental conditions of temperature, humidity, and dust with little or no maintenance. Use of a calcium-lead alloy, instead of the conventional antimony-lead alloy employed in lead acid batteries, offers significant advantages. First, it reduces the open-circuit losses appreciably and thus the battery has more available capacity when operated for long periods in a partially charged condition. Second, it permits longer intervals between additions of water, which reduces maintenance costs. Third, by eliminating the poisoning effect of antimony transfer, it permits lower charge currents to be used, and since the charge current is a major factor in determining a battery's life, the life expectancy increases. Finally, when calcium-lead batteries are operated at the recommended float potential, they never require equalizing charges. All of these factors serve to provide the kind of storage capacity needed in photovoltaic applications, especially in remote locations.

In addition to having calcium-lead alloy grids, photovoltaic batteries differ from conventional lead acid batteries in the plate insulation and the specific gravity of the cell. The insulation of a photovoltaic battery is similar to that of a motive power or traction battery. It usually incorporates a retentive glass matte, which minimizes shedding of positive active materials and permits the battery to operate cyclically.

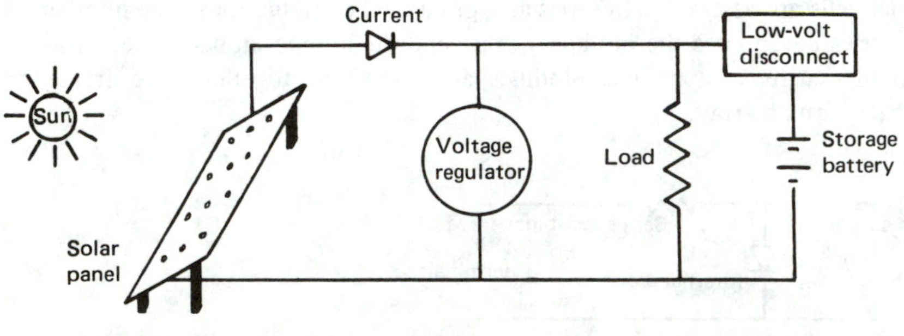

Figure 10.5 Diagram of photovoltaic installation showing interconnection of battery, load, voltage regulator, and solar panel.

In practice, battery rating and service life are specified by the following parameters:

1 *Amperes-hour capacity*. This is an indication of the amount of electricity that can be drawn from a battery over a period of time. It depends on the number and design of the plates in a cell. Lead acid batteries for photovoltaic systems are rated at 16 A/h. Thus a 6-cell, 19-plate lead acid battery will deliver a total of 480 A·h if it is discharged at a continuous rate of 30 A·h for 16 h.

2 *Kilowatt-hour capacity*. This is the total amount of power (volts × amperes × hours/1000) that can be drawn from a battery.

3 *Battery life*. This is a measure of the usefulness of a battery, determined by the number of times it can be charged and discharged before it is no longer able to perform. A cycle consists of one complete discharge to approximately 20 percent of rated capacity and one complete charge. The number of complete cycles a battery will provide depends on the construction of the battery, type of charging, maintenance, and use. The normal service life for a lead acid battery is up to 5 years with a daily discharge not exceeding 80 percent of rated capacity, or up to 10 years with a daily discharge not exceeding 25 percent of rated capacity. The need for constant topping of the cells can be reduced by expanding the electrolyte capacity.

The state of charge (SOC) of a photovoltaic battery varies seasonally, as shown in Fig. 10.6. During the summer, when days are long and sunlight is most intense, the energy output of a properly designed solar array should exceed the demand of the load, and the excess energy is used to maintain the battery in a fully charged state. However, during the winter, when the days grow shorter and less energy is produced by the array, the reserve capacity of the battery is called on to make up any energy deficiencies that the array cannot supply. Thus the battery's SOC gradually decreases through the winter and remains low until the lengthening of days reverses the trend and provides more energy than is removed by the load.

Cost of Photovoltaic Power

The homeowner's annual amortized payments for a photovoltaic solar system are given by the relation

$$Q = \frac{CR + M}{\epsilon UH} \times 10^4 \tag{10.1}$$

where Q = homeowner's annual amortized payments for the photovoltaic solar energy system (¢/kWh).

C = installed system cost per unit panel area ($\$/m^2$). This includes materials, processing, labor, and balance-of-system costs.

R = fixed charge rate for homeowners. A rate of 12% less a 20% tax write-off gives $R = 10\%$ for effective principal, interest, taxes, and insurance. An adjustment for the current 40% federal tax credit would put the effective annual mortgage rate at 8%.

M = maintenance and operation yearly cost [\$/(m^2·year)].
ϵ = system efficiency normalized to 1 kW/m^2 insolation (%).
U = energy utilization (%) = $[1 - f(1 - S)] \times 100$.
f = fraction of energy out of phase, typically $f = \frac{1}{3}$.
S = sellback rate ($0 < S < 1$), most probably $S \approx 0$. For a system without storage or sellback, $U = 67\%$.
H = average hours per year of 1 kW/m^2 insolation, typically 2080 h/year for the sunbelt zones.

For a full interpretation of U, one needs to consider systems with and without utility hookup. Off-peak power purchase can enhance the usefulness of storage by increasing U. Suggested ranges of U without sellback ($S = 0$) are shown in Table 10.1.

A simple procedure for estimating the energy cost and storage of photovoltaic systems by means of a nomograph has been worked out by Bawa [1]. Bawa's methodology is illustrated by an example below. It is recommended that the reader follow the example, step by step in order to become acquainted with this quick and easy-to-use methodology.

Energy Cost of the System

The nomograph shown in Fig. 10.7 is for a typical sunbelt location that receives 2080 h annual insolation. Equation (10.1) was used in constructing the nomograph. The following example illustrates the costing procedures if the installed cost of the system is assumed to be \$120/m^2, or approximately \$12/ft^2:

1 Find the intercept on the vertical X line reading up from \$120/m^2 and across from the 10 percent line. This provides an effective net mortgage rate of 10 percent per year after income tax rebates, etc.

2 Join this point on the X line with the maintenance cost line at the far right on the diagram (M = \$1/m^2 per year in this example). Then mark the point of intersection on the Y scale.

3 For an assumed overall efficiency of 8 percent for conversion of insolation to electrical energy to storage to electrical energy, and for 100 percent utilization of the electrical energy, join 8 on the ϵ percentage scale with 100 on the U percentage scale and extend it to intersect the vertical Z line. Join the two

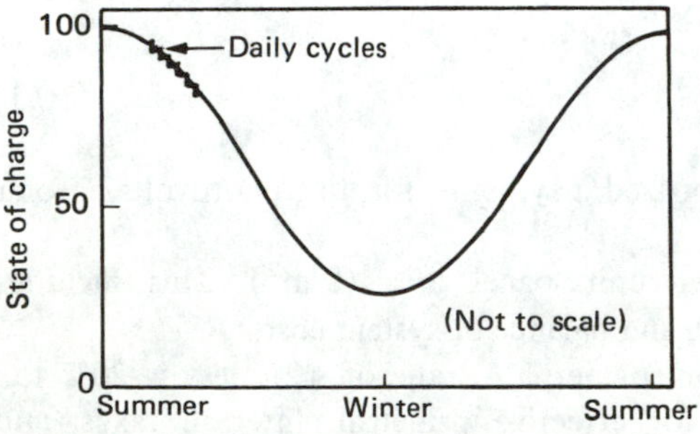

Figure 10.6 Seasonal state of charge of battery.

Table 10.1 Suggested Ranges of Energy Utilization

Installation size	Storage size	U (%)
	With utility hookup	
Daytime demand	0	100
>Daytime demand	0	$100 > U \geqslant 67$
Total demand (day and night)	0	$67 > U \geqslant 33$
Total demand (day and night)	1 day	$100 > U \geqslant 67$
	Without utility hookup	
Total demand	4 days	100

Source: Bawa [1].

points on the Y and Z lines and the line will intersect the diagonal Q line (¢/kWh). The assumption of 100 percent utilization implies that the system has adequate storage capability or is small enough to meet the requirement in full or part during periods of sunshine.

The point of intersection, 7.2 c/kWh, is the energy cost for the example.

Economic Worth of Storage

A nomograph can also be used to determine the economic worth of a storage subsystem in a solar energy system. Typically, set f at $\frac{1}{3}$ for a system without storage or sellback and U at 67 percent. For a system without storage, the efficiency on the basis of photovoltaic collection alone is about 10 percent.

In the second example, Fig. 10.8, the contemplated system is large enough to supply most of the electrical needs. How much storage should be installed? It is assumed in this example that the efficiency of conversion from insolation to

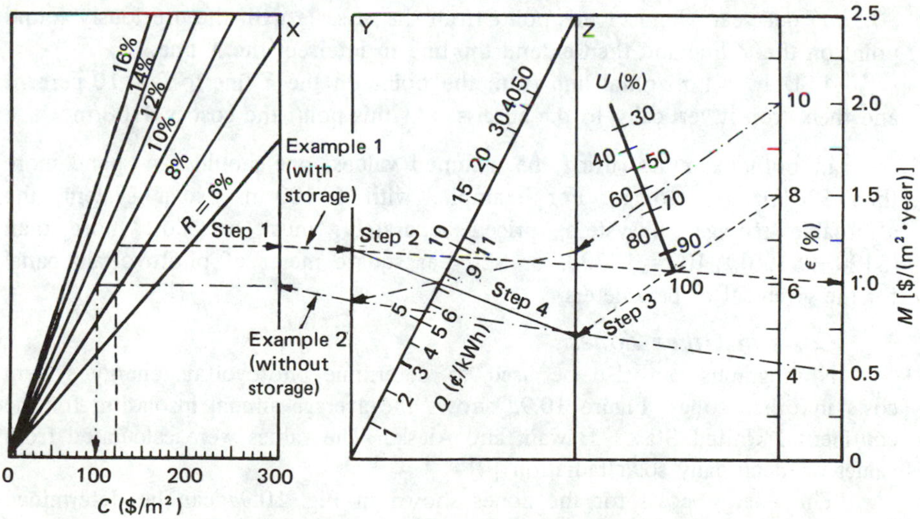

Figure 10.7 Nomograph for photovoltaic energy cost. (*From [1]*.)

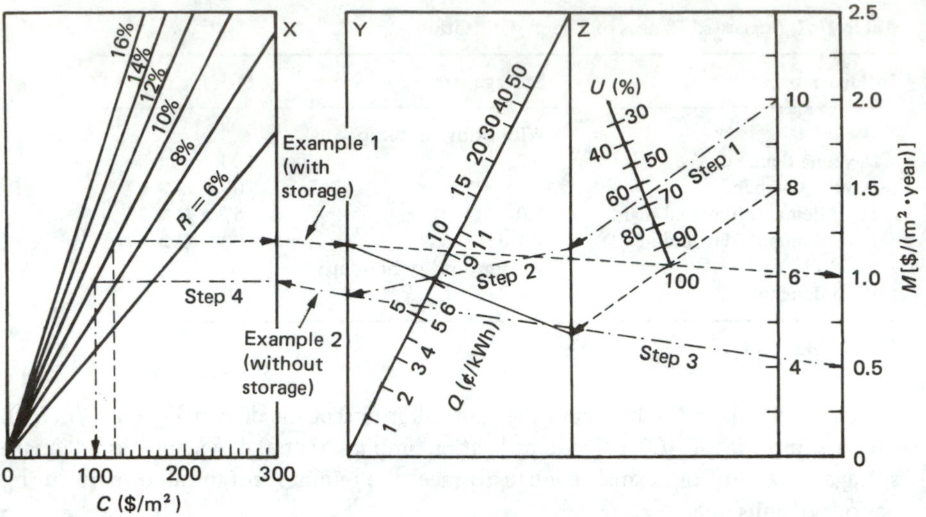

Figure 10.8 Nomograph for photovoltaic storage worth. (*From [1]*.)

electrical energy without the inefficiency of a storage system will be about 10 percent. The utilization will be lower; i.e., demand will be two-thirds in phase and one-third out of phase.

1 Join 10 percent on the ϵ percentage scale with U at 67 percent and extend the line to intersect the Z line.

2 Join this point with 7.2 ¢/kWh and extend the line to intersect the Y line.

3 For a system without storage, the maintenance cost is assumed to be $0.5/m^2 per year. Connect this point from the M scale with the previously found point on the Y line and then extend the line to intersect the X line.

4 Draw a horizontal line from the point on the X line to $R = 10$ percent and then drop it vertically to the abscissa. At this point the cost is $100/m^2.

In both examples, using the assumed values, one should not spend more than $20/m^2 on storage. For example, with a 100 m^2 solar system, the affordable storage subsystem, priced separately, must not cost more than ($120 − $100) × 100 = $2000, or $20 per square meter of photovoltaic panel for the given set of parameters.

Costs in Other Zones

Nomographs can also be used to determine photovoltaic energy system costs in other zones. Figure 10.9*a* shows the average annual insolation for the continental United States, Hawaii, and Alaska. The values were calculated from values of mean daily solar radiation [2].

The energy costs for the zones shown in Fig. 10.9*a* can be determined from Fig. 10.9*b* by reading up and across from energy cost Q, which is determined

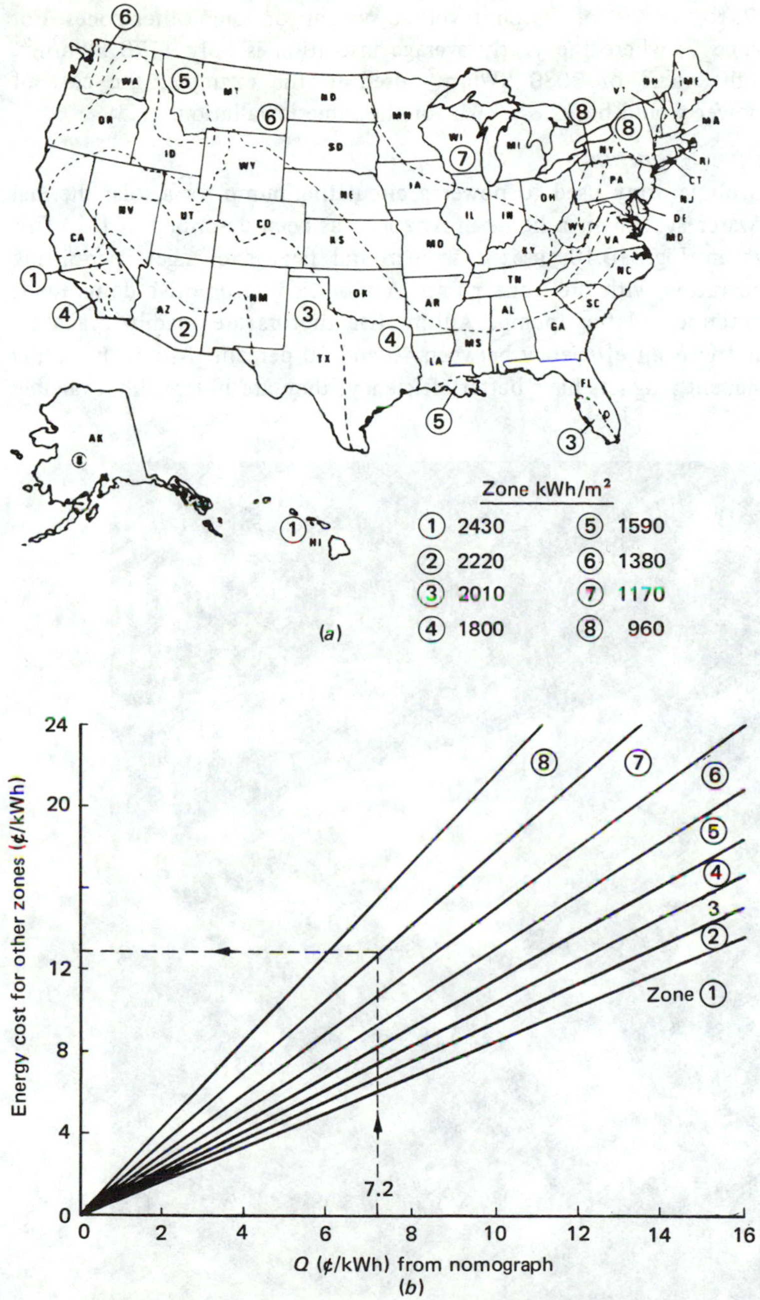

Figure 10.9 (*a*) Average insolation zones and (*b*) energy cost correction. (*From [1].*)

from Fig. 10.7, to the cost of the photovoltaic system for some other zones. For example, in zone 7, where the yearly average insolation is only 1170 kWh/m^2, compared to the value of 2080 kWh/m^2 used in the example, the cost of photovoltaic power would be 12.8 ¢/kWh for the same installation.

Illustrative Example

A photovoltaic array used to power a circulating pump for a solar thermal domestic hot-water system in a home in Arizona has been described in [3]. This system, shown in Fig. 10.10, uses a wraparound heat exchanger, permitting double-wall separation with only one pump. A modified automotive dc motor is used to drive the circulating pump. Automotive motors are readily available, inexpensive, and give an efficiency between 40 and 50 percent. Although motors based on permanent magnets have better efficiency, they are not readily available

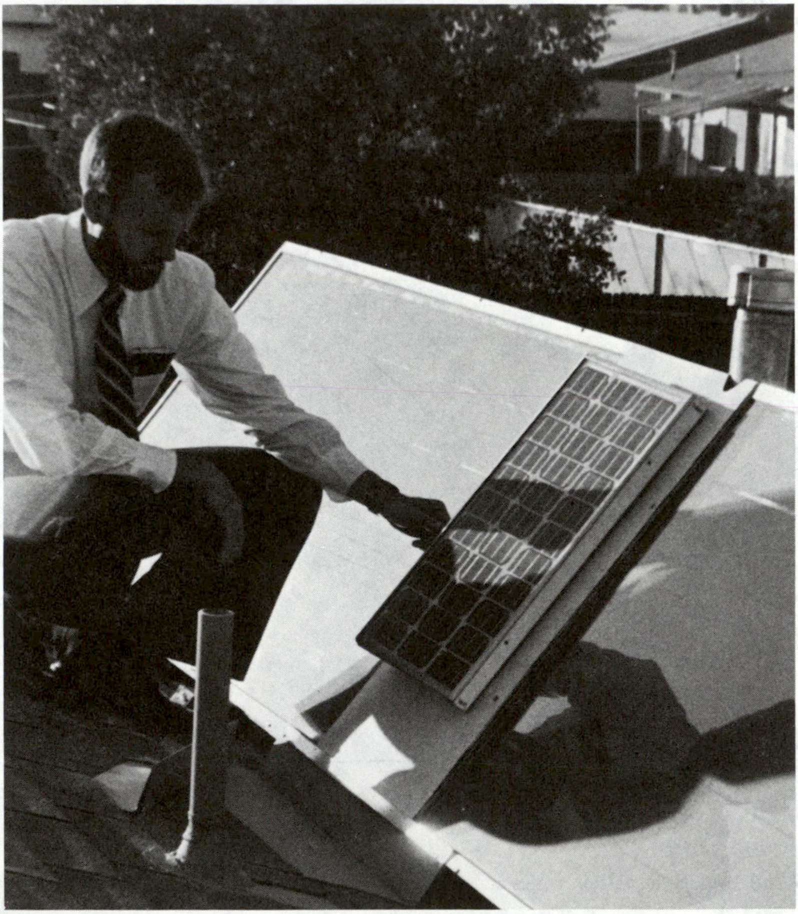

Figure 10.10 Photograph of photovoltaic-powered circulating pump system for domestic hot water. (*Courtesy of B. Hammond and Motorola Inc.*)

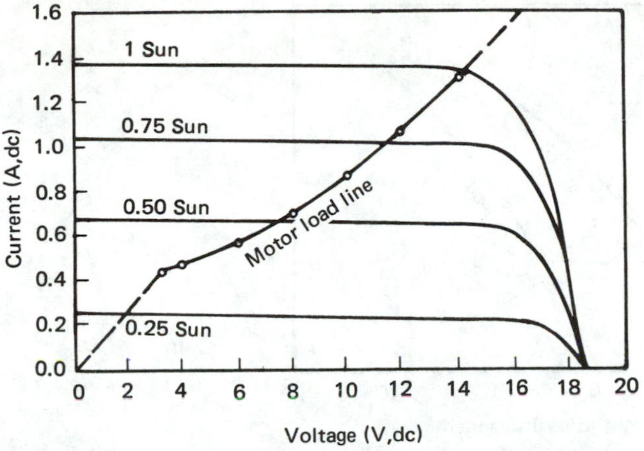

Figure 10.11 Composite photovoltaic module and motor characteristics at $T_a = 32°F$. (*From [3].*)

and cost four to five times more than series-wound automotive motors. Special brushless motors with permanent magnets are being developed for photovoltaic applications.

Figure 10.11 shows the motor load line and the current-voltage characteristics of the photovoltaic array. The motor operating points are at the intersections of the module curves and the motor line. The motor pump performance is shown in Fig. 10.12. Operational characteristics of the unit are plotted in Fig. 10.13, where the flow rate is shown as a function of insolation for heads of 0 and 3 ft, respectively. As the insolation decreases, the flow rate also decreases—a relationship that provides inherent proportional control. Consequently, no differential controller is required for the system, and this results in a substantial cost saving.

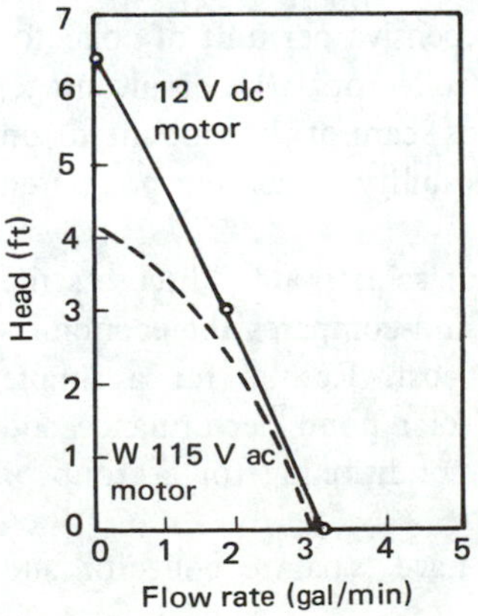

Figure 10.12 Pump characteristics: head versus flow rate. (*From [3].*)

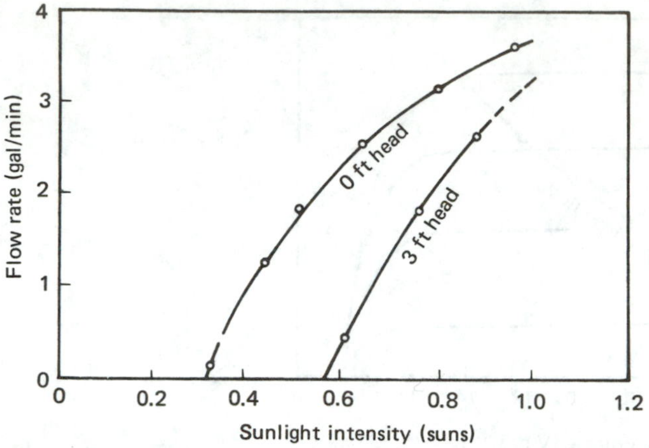

Figure 10.13 System characteristics: flow versus insolation for two pressure heads. (*From [3]*.)

SOLAR PONDS*

The solar pond is one of the simplest devices for direct thermal conversion of solar energy. Moreover, it is simultaneously a collector of solar radiation and a thermal storage device. Any pond converts insolation to heat, but most natural ponds quickly lose that heat through vertical convection within the pond and evaporation and convection at the surface. The solar pond artificially prevents either vertical convection or surface evaporation and convection, or both. Because of its massive thermal storage and measures taken to retard heat loss, a typical pond takes several weeks for a 10°C temperature drop, even in the absence of insolation. Thus the solar pond converts intermittent solar radiation into a steady source of thermal energy.

Solar ponds can be operated at most habitable latitudes. In some locations their surfaces may freeze in winter, but storage temperatures generally remain high enough for low-temperature applications such as heat pumps, and some insolation penetrates the ice. Solar ponds are less expensive per unit of collector area and per unit of thermal output than flat-plate collectors. One disadvantage, compared with flat-plate collectors, is that ponds cannot be mounted on rooftops. For this reason they have little or no applicability in densely populated areas.

This section describes several different types of solar ponds, discusses the availability and cost of salts for salt gradient ponds, and compares the economics of salty and saltless ponds as a function of salt cost. Results for a simple computational model are presented to approximate solar pond performance and to size ponds for district heating to provide space conditioning for a group of homes in different regions of the United States.

Most low-temperature solar thermal systems have separate collector and

*This section was abstracted from *Solar Ponds* by Jayadev and Edesess [4].

storage units, and storage is usually only sufficient to last a night. With solar ponds the same body of water serves as solar collector and storage medium. This body of water is usually large enough to provide long-term storage.

Solar pond concepts may be classified in two categories: (1) those that reduce heat loss by preventing convection within the water storage medium and (2) those that reduce heat loss by covering the pond surface. Combinations are possible, of course, but one method is usually the primary determinant of the pond design.

Nonconvecting Ponds

Nonconvecting ponds prevent heat loss by inhibiting thermal buoyancy convection. In natural ponds insolation is converted to heat within the pond, but the warmed water from the bottom rises to the surface and most of the heat is lost to the atmosphere. Nonconvecting ponds employ a salt gradient to prevent the warmed water from rising to the surface.

Salt gradient ponds have been studied extensively [5-10]. The salt concentration in such a pond is highest near the bottom and lowest near the surface. The salts most commonly used are NaCl and $MgCl_2$, although there are other possibilities.

As solar radiation enters the pond, whatever is not absorbed in the water is absorbed on the dark bottom. As a result, heat is collected at the bottom and the deeper waters warm up. Pure water, when warmed, becomes less dense. If there were no salt concentration gradient in the pond there would be continuous convection of the warmed water from the bottom of the pond to the cooler layers near the top. However, the increased density created by the salt prevents this thermal buoyancy convection. Heat transfer to the surface of the pond therefore occurs primarily by conduction, which is slow enough to enable the lower regions of the pond to maintain a high temperature (100°C has been measured in actual ponds).

The salt gradient pond has three layers, as shown in Fig. 10.14. Vertical convection takes place in the top layer due to the effects of wind and evaporation. This layer serves no useful purpose and is kept as thin as possible. The next layer, which may be approximately 3 ft (1 m) thick, has a salt concentration that increases with depth; this layer is nonconvecting. The bottom layer, which is convecting, provides most of the thermal storage and facilitates heat extraction.

Salt gradient ponds have been built and operated in such diverse locations

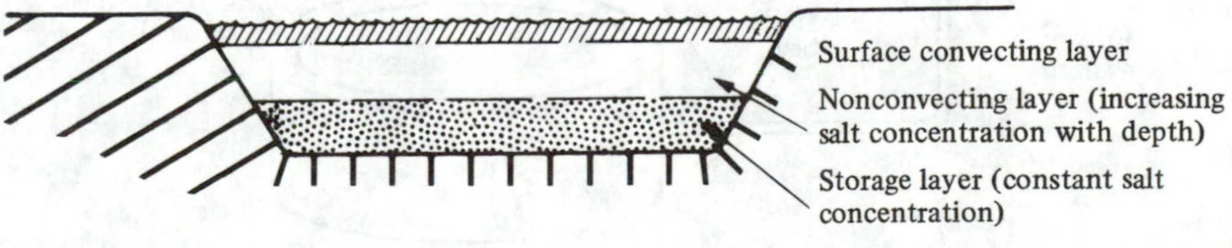

Figure 10.14 Salt gradient solar pond. (*From [4]*.)

as Israel [11], Canada [12], and Ohio [8] and New Mexico [10] in the United States.

Convecting Ponds

This section is devoted to two types of convecting ponds: shallow and deep saltless ponds. Shallow convecting solar ponds have been designed and built at Lawrence Livermore Laboratory [13, 14]. A typical shallow solar pond (see Fig. 10.15) has about a 10-cm depth of pure water enclosed in a large water bag, typically 3.5 m by 60 m, with a blackened bottom, insulated below with foam insulation and on top with glazings. The water from many such ponds is pumped into a large storage tank for night storage and back into the water bags each morning, in an operating method called the batch mode. The shallow solar pond may also be operated in the flow-through mode, in which the water flows continuously through the water bags in such a way as to maintain control over the outlet temperature. When operated in the flow-through mode, the shallow solar pond is similar to a flat-plate collector with water storage. The main difference is that the solar pond collector is fixed in a horizontal position and is less costly than the usual flat-plate collector.

Several shallow solar ponds 3.5 m wide and 15 m long were built and tested at Lawrence Livermore Laboratory, and different designs and materials were investigated to develop the design illustrated in Fig. 10.15. The thermal performance of all the ponds tested, with a single top glazing, was found to be essentially the same. Measurements of η_i, the instantaneous collection efficiency, were made over the noon hour during several days near the summer solstice. A spread of $\Delta T/I$ values was achieved by varying the water depth each day over the range 2 to 12 cm. The results of these measurements are shown in Fig. 10.16*a*.

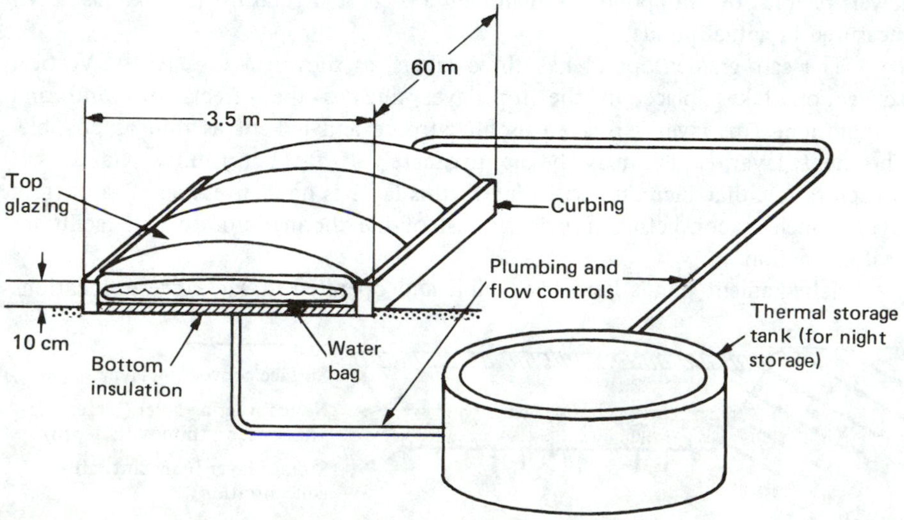

Figure 10.15 Shallow solar pond design. (*From [14]*.)

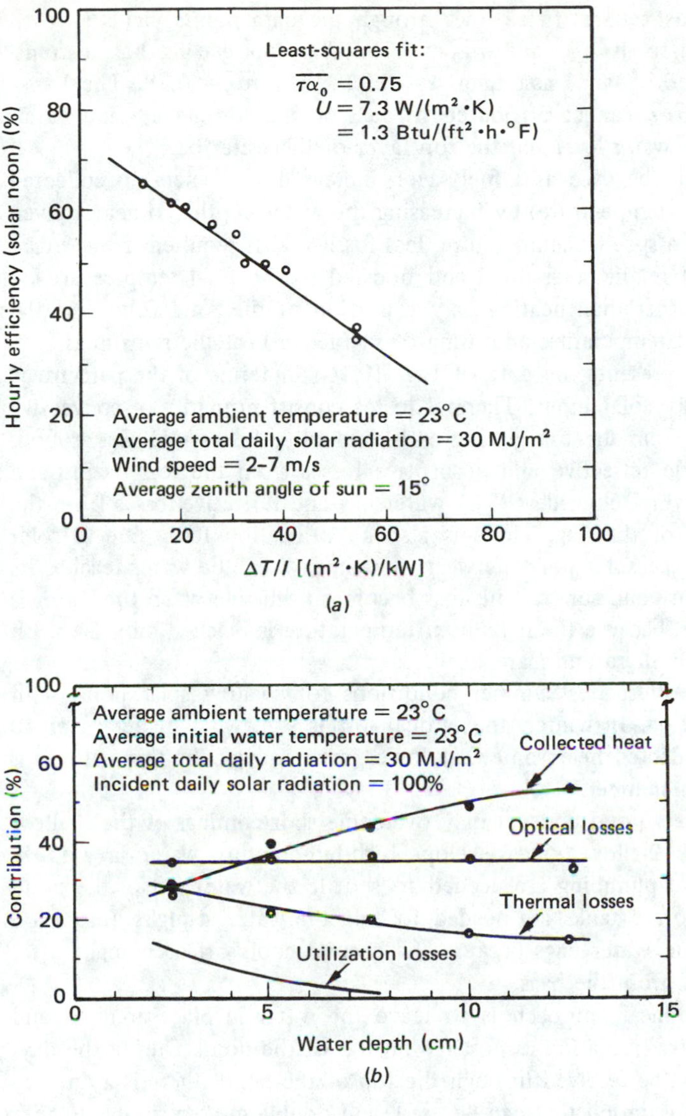

Figure 10.16 (*a*) Hourly shallow solar pond efficiency at solar noon as a function of $\Delta T/I$, where ΔT is the average difference between water and ambient temperatures and I is the total incident radiation between 11:30 a.m. and 12:30 p.m. Data were taken over a 10-day period in July 1975 in Livermore, California. (*b*) Distribution of the total incident daily solar radiation among collected heat, optical losses, thermal losses, and utilization losses as a function of pond water depth, from data obtained in July 1975 in Livermore. (*From [15]*.)

The slope of the least-squares fitted line through the data points yields $U = 7.3$ W/m$^2\cdot$°K (1.3 Btu/ft$^2\cdot$h$\cdot$°F), and $\overline{\tau\alpha_0} = 0.75$. The top glazing has a transmissivity of about 0.85 and, assuming $\alpha = 0.92$, then $\overline{\tau\alpha_0} = 0.78$. The lower measured value of $\overline{\tau\alpha_0}$ can be attributed to dust on the top glazing and to air bubbles between the water level and the top layer of the water bag.

If the SSP is to be used as a fuel saver, increased solar energy is collected (although at a lower temperature) by increasing the water depth. To heat a given volume of water to a specific temperature, less fossil fuel is required if the entire volume is preheated in the solar pond and boosted to the final temperature by use of fossil fuel, rather than heating only a portion of the total volume in the pond to the required temperature and using fossil fuel to heat the remainder.

Figure 10.16*b* presents the data of Fig. 10.16*a* in terms of the percentage contributions of daily solar input. Thermal losses consist primarily of convective and radiative losses from the top glazing and conductive losses into the ground. Optical losses include reflective and absorptive losses from the top glazing and the top of the polyvinyl chloride (PVC) water bag and reflective losses from the bottom black layer of the bag. There is also a "utilization loss" due to solar radiation falling on the solar pond between the time when the water reaches its maximum temperature and sunset. This loss becomes negligible when the water is 10 cm or more deep, because the maximum temperature is reached only 2 to 3 h before sunset when both $\overline{\tau\alpha}$ and I are small.

Figure 10.16*b* illustrates summer conditions for shallow solar pond with good bottom insulation. In winter, the optical loss is somewhat higher (near 40 percent) and the collected heat somewhat lower, even though the thermal loss is about the same as in summer.

The deep saltless pond concept may overcome shortcomings of the shallow pond. Although the shallow pond develops high-temperature water in a fairly short time, pipes and plumbing are needed to shuttle the water out of the pond each evening and storage tanks are needed to hold the water at night. Insulation is required under the water bags because the ground cools off each night after the water is removed from the bags.

A more economical approach is to leave the water in place at night and provide as much extra insulation as possible on top of the pond. During the day, when insolation must be received through the top of the pond, there is a limit to the amount of top insulation that can be used, and double glazing similar to that used in the shallow solar pond would be employed. But at night or during periods of low insolation, additional top insulation could be provided. An obvious and simple method would be to lay extra insulation over the top of the pond, either automatically or manually, whenever insolation falls below a prescribed level. Another possibility is to spray foam insulation between the glazings and between glazing and pond when insolation drops below a prescribed level (Fig. 10.17). In the morning, the spray foam insulation would be allowed to settle and run off, leaving a negligible residue. The capital cost of using spray foam to provide supplemental night insulation is estimated as less than \$1 per square meter of pond. A spray foam has been used successfully to provide night

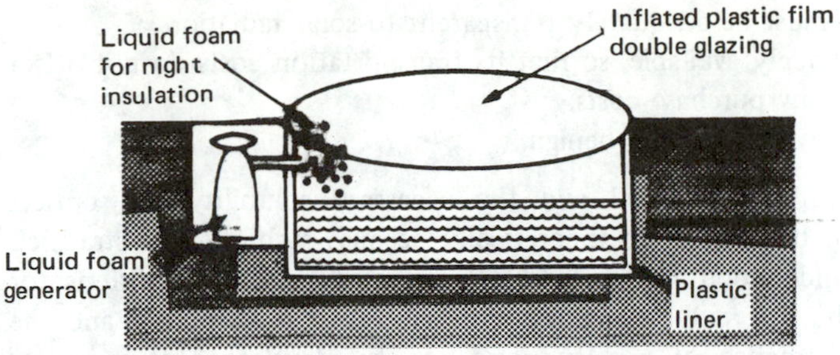

Figure 10.17 Example of deep saltless pond design. (*From [4].*)

insulation for greenhouses [16]. It was found in practice to reduce the heat loss by at least 50 percent, although in theory an 85 percent reduction should be attainable. The spray foam used for the greenhouses was a material normally used for firefighting; it seems likely that improvements could be made in the material for the purpose of pond insulation.

Besides eliminating the need for pipes, pumps, and plumbing to transport the water to nighttime storage, this "stationary" pond would not require bottom insulation. After a warmup period, the temperature of the ground would approach that of the pond water, providing good insulation. The only additional insulation that might be desired would be along the sides of the pond to prevent edge losses.

To provide sufficient storage to even out daily and seasonal temperature fluctuations, the stationary convecting pond would have to be a deep pond, not a shallow one. The deep saltless pond concept, proposed by Taylor [16], has been much less researched than has the salt gradient pond. In the following section the projected costs and performance of these two pond types are compared.

Cost and Performance Comparison

For the salt gradient pond and the deep saltless pond the chief costs are for earth moving, bottom liner, and salt in the case of the salt gradient pond or surface glazings and additional insulation in the case of the deep saltless pond. The cost of salt varies widely, and the relative attractiveness of a salt gradient or a saltless pond is primarily a function of this highly site-dependent cost.

The cost of salt for a solar pond represents a sizable fraction of the total initial investment. Depending on the design details and the proximity to a source of salt, a typical NaCl salt pond may require 30 to 60 percent of the initial investment for the initial charge of NaCl [17, 18]. Therefore identification of suitable, low-cost alternative salts could strongly affect the overall economic attractiveness of a salt pond.

A suitable salt must meet several criteria:

1 It must be adequately soluble (with a solubility that increases with temperature).

2 Its solution must be adequately transparent to solar radiation.

3 It must be widely available, so that its transportation costs do not offset the advantages of its low purchase costs.

4 It must be environmentally benign.

The amount of salt required and the necessary solubility and optical characteristics cannot be established theoretically because stability in a stratified pond is not well understood [19]. However, certain sufficient conditions for pond stability can be inferred by analogy with successful NaCl ponds, and the overall thermal performance of a salty pond can be simulated by computer modeling when the solubility and optical properties of the alternative salt are known.

A typical NaCl pond has a solution concentration ranging from nearly zero at the surface to a maximum of 17 wt% in the storage layer. This corresponds to a density gradient of only about 0.05 g/cm^3 per meter of depth. An alternative salt with a similar or lower diffusivity that can provide a similar density gradient at the operating temperatures should also produce stable stratification. Figure 10.18 shows the solubilities of some candidate salts. In all cases, the diffusivities of the alternative salts are lower than that of NaCl and the temperature dependence of solubility is greater. Therefore a concentration sufficient to produce a density gradient of 0.05 g/(cm^3·m) at a typical operating temperature gradient (20°C/m) should provide at least as much pond stability as NaCl.

Table 10.2 summarizes some properties of candidate salts. The costs are approximate since they depend on location; however, it is clear that only salts that can be obtained as waste products offer a substantial economic advantage. Magnesium chloride "bitterns" are available from plants that refine NaCl, and the

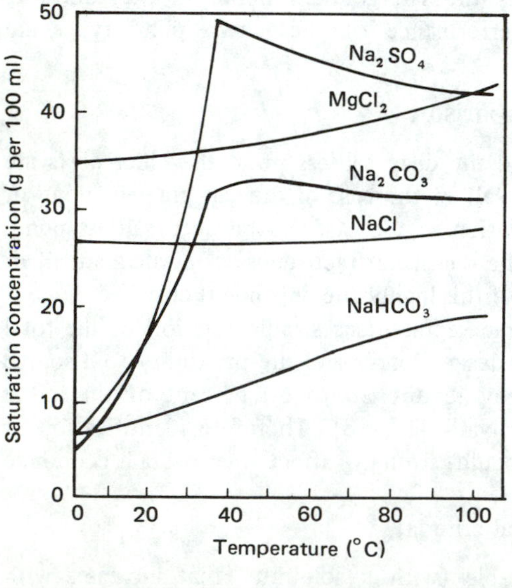

Figure 10.18 Solubilities of candidate salts.for solar ponds. (*From [20].*)

Table 10.2 Properties of Candidate Salts

Salt	Formula	Source	Cost ($ per 10^3 kg)	Comments
Sodium chloride	NaCl	See Fig. 10.19	20	
Sodium carbonate	$Na_2CO_3 \cdot H_2O$	Synthetic (Solvay process)	96	East Coast price
		As trona (Green River, Wyoming)	61	Wyoming price
		As trona (Green River, Wyoming)	70	California price
Sodium bicarbonate	$NaHCO_3$	Nahcolite (Piceance Creek Basin, Wyoming)	(~35)*	By-product of oil shale mining (not yet in production)
Sodium sulfate	Na_2SO_4	"Salt cake"	47	East Coast price
		"Salt cake"	45	West Coast price
		As flue gas desulfurization waste	(~0)*	Price depends on proximity of other markets
Magnesium chloride	$MgCl_2$	Salt plants (see Fig. 10.19)	140	99% pure, hydrated salt
		As bitterns (see Fig. 10.19)	(~2)*	Waste product also containing other salts (not normally sold)

*Estimated prices.
Source: Jayadev and Edesess [4].

sites where this is done are numerous (Fig. 10.19). Sodium sulfate, however, has the potential for much more widespread availability in the next few years, as it is a waste product of flue gas desulfurization at coal-fired power plants.

A performance comparison of salt gradient and saltless ponds was made at the Solar Energy Research Institute (SERI) by means of computer analyses. First, a computer simulation was performed for a hypothetical salty solar pond at Barstow, California. A finite-element model of the pond was employed [21], and the simulation took into account edge losses and ground storage as well as losses through the surface, losses to the ground, and pond storage. The pond was assumed to be 30 m in diameter (roughly the size that could be used to heat a small group of houses) and to have a storage layer 1 m deep, a nonconvecting layer 1.5 m thick, and a surface convecting layer 0.3 m thick. (The surface convecting layer is due to wind turbulence and evaporation and cannot be avoided.) No insulation around the pond was considered except that provided by the ground. It was further assumed that a constant load of 35,343 W (50 W per square meter of pond surface area) was extracted from the pond. The simulation showed that the average annual temperature of the pond's storage layer would be

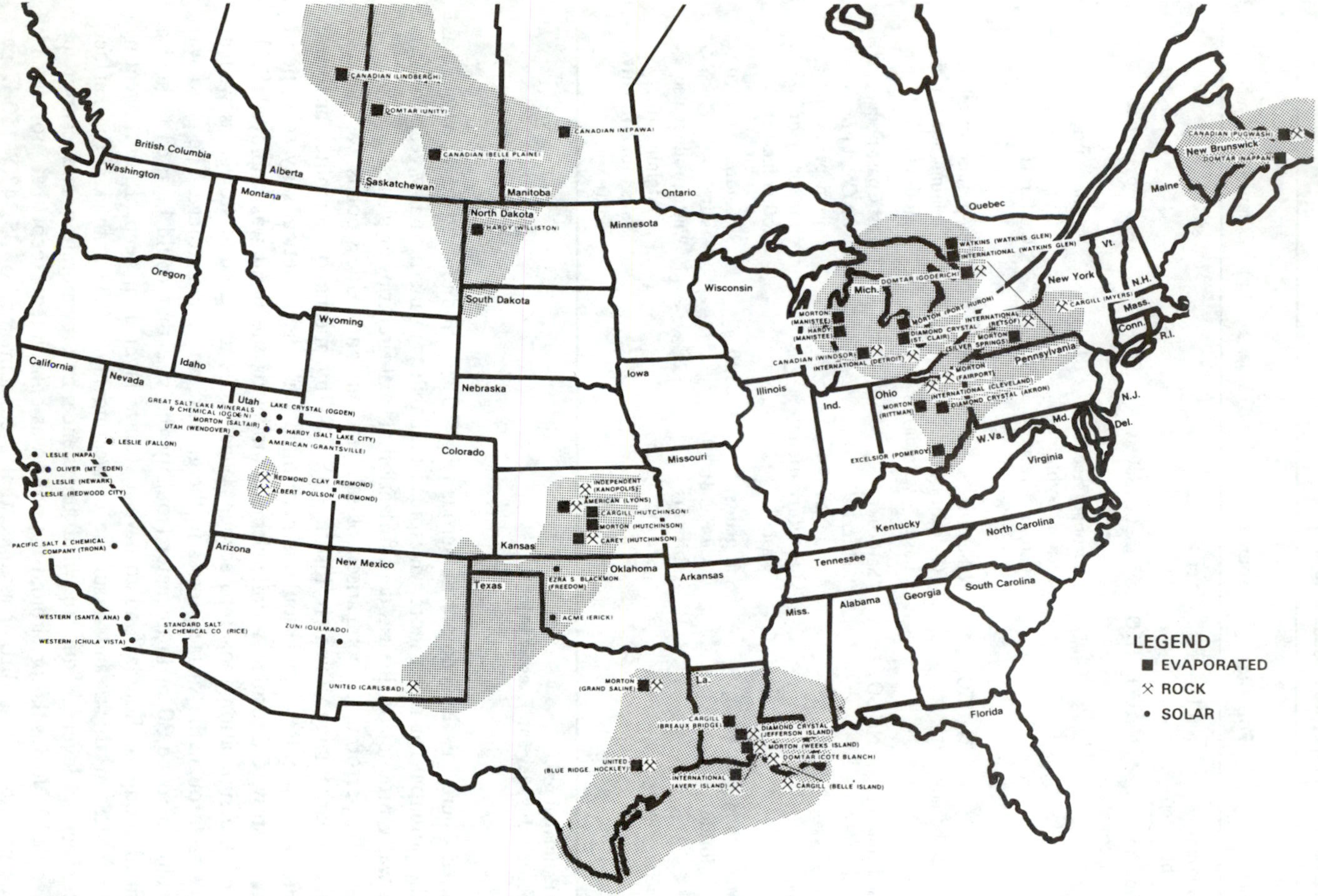

Figure 10.19 Major salt deposits in the United States and Canada. (*From [4]*.)

61°C, reaching a maximum of 81°C around mid-August and a minimum of 41°C in mid-February.

Next, a saltless solar pond was simulated at the same location. The saltless pond was assumed to be convecting with the same temperature maintained throughout. It was also assumed to have glazings over the top with a heat loss coefficient of 3 W/(m^2·°C) and additional night insulation resulting in a nighttime heat loss coefficient of 1 W/(m^2·°C). Therefore the surface heat loss coefficient averaged about 2 W/(m^2·°C). Transmissivity of the surface glazing to solar radiation was assumed to be 0.65.

By an iterative modeling process, it was found that the saltless solar pond would have nearly the same temperature profile, under the same 50 W/m^2 constant load, as a salty pond. The saltless pond is assumed to be 30 m in diameter, to have only ground insulation, like the salty pond, but to be 10 m deep, much deeper than the salty pond. The additional depth (additional thermal mass) is required to even out the temperature fluctuations in the saltless pond.

Computer simulation of the saltless pond showed that its average temperature would be 60°C. Its maximum temperature, reached in August, would be 80°C, and its minimum temperature, in mid-February, would be 40°C. Thus its temperature profile throughout the year would be much like that of the salty pond. Figure 10.20 shows the average annual temperatures of the two ponds.

At the present stage of development of solar ponds, costs can be only roughly estimated. The estimates below can serve, however, as preliminary economic comparisons of salty and saltless ponds.

Capital expenses for the salty solar pond include the costs of excavation, of a blackened liner for the bottom of the pond, and of the salt. For a typical salty pond 30 m in diameter and 2.8 m deep, with an excavation cost of \$2/m^3, the total excavation cost would be \$4000, or about \$5.60 per square meter of pond surface area. The liner for the bottom of the pond must be a durable material

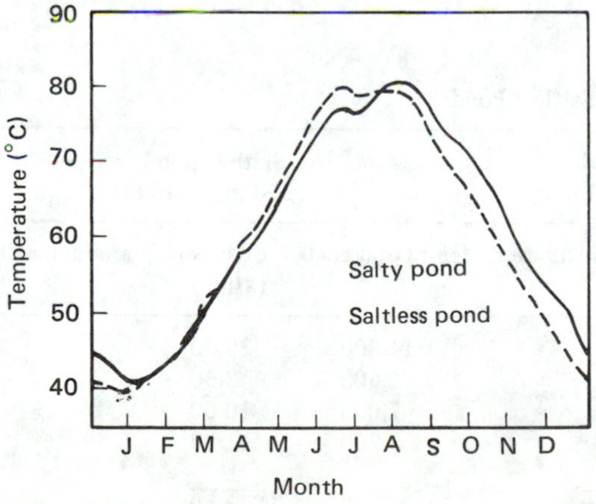

Figure 10.20 Calculated average annual temperature for salty and saltless ponds in Barstow, California. (*From [21]*.)

like Hypalon, which would cost $\$10/m^2$ or about $8000 for the entire pond, including sides.

The salty pond used in the simulations would require about 0.5 ton of salt per square meter of pond surface area. Because of its variation with location, the cost of salt may be treated as a variable in economic comparisons with the saltless pond. Capital expenses for the saltless pond include excavation expense, the cost of the liner, the cost of the surface structure and glazings, and the cost of night insulation.

The saltless pond that yielded approximately the same output as the salty pond was 10 m deep. At a cost of $\$2/m^3$ the excavation expense is about $14,000, or $20 per square meter of pond surface area–$\$14.40/m^2$ more than the salty pond. However, the cost of the liner could be reduced to about $\$2/m^2$ because of the much reduced requirement for retardation of leakage. For the entire pond, the liner cost would be about $1600.

The cost of the surface structure and glazings depends on the means of implementation. In one possible scheme a lattice structure would be placed over the top of the pond, and sections of double-layered plastic film glazing would be fastened to it and inflated by air at low pressure. For this design a conservative cost of $\$10/m^2$ is assumed.

If liquid foam insulation were used at night, it could be sprayed into the space between the layers of inflated plastic film. The cost of the liquid foam generating equipment averages less than $\$1/m^2$.

Table 10.3 summarizes the rough costs for the salty and saltless ponds. At a salt cost of $16.40 per square meter of pond surface area, the cost of the salty pond equals that of the saltless pond. Since 0.5 ton of salt is required for each square meter of pond surface area, the break-even price for salt is $32.80 a ton. At a lower cost of salt, the salty pond would be more economical, and at a higher cost, the saltless pond would be favored. For the $\$33.30/m^2$ cost of the saltless pond, the capital cost of energy at the 50 W/m^2 extraction rate is $666/kW/thermal.

Table 10.3 Estimated Costs of Salty and Saltless Ponds

	Salty pond ($1250\ m^2 \times 2.8$ m)		Saltless pond ($1250\ m^2 \times 10$ m)	
Pond component	Total cost ($)	Cost per square meter ($\$/m^2$)	Total cost ($)	Cost per square meter ($\$/m^2$)
Excavation	4,000	5.60	14,000	20.00
Liner	8,000	11.30	1,600	2.30
Glazings			7,000	10.00
Night insulation			700	1.00
Salt		x		
Total		$16.90 + x		$33.30

Source: Jayadev and Edesess [4].

Working experience with solar ponds has been insufficient to provide a good estimate of operation and maintenance costs. The salty pond requires frequent maintenance to preserve the salt concentration gradient and maintain water clarity. There is no reason to expect higher operation and maintenance costs with the saltless pond; in fact, there is reason to expect these costs to be lower, since the saltless pond is covered and has no salt gradient to maintain.

Simplified Solar Pond Performance Model

A simple method to calculate solar pond sizes and outputs has been developed by Edesess et al. [22]. He assumed that storage in the pond is so large that daily fluctuations in ambient temperature and insolation have a negligible effect on the pond temperature and that only seasonal variations in the environment need be considered. He also assumed that the heat loss from the pond is related linearly to the temperature differences between the pond and the ambient air and between the pond and the ground. This means that there must be effective heat loss coefficients U_a and U_g such that the rate of heat loss per unit area is $U_a(T - T_a) + U_g(T - T_g)$, where T_a is the ambient temperature, T_g is the ground temperature (presumably equal to $\bar{T}_a$, the average annual ambient temperature), and T is the temperature of the storage layer of the pond. In the saltless pond T is assumed to be the temperature at any point.

Suppose that characteristic heat loss coefficients U_s, U_e, and U_b can be identified for a pond of surface area A, perimeter P, and depth D, where U_s is the coefficient of heat loss from the surface of the pond, U_e from the edges of the pond, and U_b from the bottom of the pond. The heat loss coefficients have units of watts per square meter per degrees Celsius, A is measured in square meters, and P and D are measured in meters. Then the coefficients of heat loss to the ambient air U_a and to the ground U_g can be expressed in terms of U_s, U_e, U_b, A, and P as follows:

$$U_a = AU_s + PU_e \qquad U_g = AU_b$$

It is a reasonable approximation to model the insolation and the ambient temperature as sine waves. If it is also assumed that the load can be represented as a sine wave,

$$T_a(t) = \bar{T}_a + \tilde{T}_a \sin 2\pi(t - \phi_T)$$

$$I(t) = \bar{I} + \tilde{I} \sin 2\pi(t - \phi_I)$$

$$L(t) = \bar{L} + \tilde{L} \sin 2\pi(t - \phi_L)$$

where time t and phase angles ϕ_T, ϕ_I, ϕ_L are measured in years. If insolation peaks in June, then ϕ_I is approximately 0.22; if ambient temperature peaks about a month later, then ϕ_T is approximately 0.30.

If A is the solar collection area, $\overline{\tau\alpha}$ the fraction of insolation transmitted to the storage area of the pond, and $\rho V c_p$ the total heat capacity of storage (where ρ is the density of water, V the volume of storage, and c_p the heat capacity per unit mass), an energy balance yields

$$\overline{\tau\alpha}AI(t) = L(t) + U_a[T(t) - T_a(t)] + U_g[T(t) - \overline{T}_a] + \rho Vc_p T(t) \quad (10.1)$$

or
$$\dot{T}(t) + \frac{U_a + U_g}{\rho Vc_p}T(t) = \frac{1}{\rho Vc_p}[\overline{\tau\alpha}A\overline{I} + (U_a + U_g)\overline{T}_a - \overline{L} + \overline{\tau\alpha}A\tilde{I}\sin 2\pi(t - \phi_I) + U_a\tilde{T}_a \sin 2\pi(t - \phi_T) - \tilde{L}\sin 2\pi(t - \phi_L)] \quad (10.2)$$

The solution to this differential equation is

$$T(t) = T + \psi(t) - C(t_0)e^{-\sigma t} \quad (10.3)$$

where

$$\overline{T} = \overline{T}_a + \frac{\overline{\tau\alpha}A\overline{I} - \overline{L}}{U_a + U_g}$$

$$\psi(t) = \frac{S}{\rho Vc_p}[\overline{\tau\alpha}A\tilde{I}\,h(t - \phi_I) + U_a\tilde{T}_a\,h(t - \phi_T) - \tilde{L}\,h(t - \phi_L)]$$

$$h(t - \phi) = \frac{\sigma \sin 2\pi(t - \phi) - 2\pi \cos 2\pi(t - \phi)}{(2\pi)^2 + \sigma^2}$$

$$\sigma = \frac{S(U_a + U_g)}{\rho Vc_p}$$

$$C(t_0) = \overline{T} - \overline{T}_a + \psi(t_0)e^{\sigma t_0}$$

t_0 = the start-up time for the pond (in years from January 1)

S = the number of seconds in a year if I and L are expressed in watts

It is assumed that $T = \overline{T}_a$ at time t_0.

Note that Eq. (10.1) expresses the pond storage temperature as the sum of the long-term average pond temperature $\overline{T}$, a periodic temperature deviation $\psi(t)$, and a transient term $C(t_0)e^{\sigma t}$. Setting the derivative of Eq. (10.1) equal to zero, one finds that in the steady state, extreme temperatures occur at the times

$$t_{\text{extreme } T} = \frac{1}{2}\pi \tan^{-1}\frac{\psi(0.25)}{\psi(0)} \quad (10.4)$$

By substituting these times into Eq. (10.1), one can find the maximum and minimum temperatures.

Example

For a circular salty pond with a 12-m radius and 2-m depth, simulated by Nielsen [8], wall losses were 3573 W and floor losses were 2920 W when the pond temperature was 50°C and the ambient temperature 10°C. (Note that only earth insulation was used in this simulation.) Assuming that the coefficient of heat loss to ambient is $U_a = 3573/(50\text{–}10) = 89.3$ W/°C and the coefficient of heat loss to ground is $U_g = 2920/(50\text{–}10) = 73$ W/°C, the projected pond temperatures shown in Table 10.4 are obtained with the formulas just developed. The pond is assumed to have been started on April 1. Transmission through the

Table 10.4 Projected Pond Temperatures Obtained with the Developed Formulas

		Projected temperature (°C)			
Year	Month	No load	5 kW constant load	5 ± 3 kW summer peaking	5 ± 3 kW winter peaking
	July 1	51.0	44.1	40.8	47.4
	October 1	66.3	56.7	53.7	59.7
1					
	January 1	53.7	43.0	45.1	40.9
	April 1	49.8	38.7	41.2	36.1
	July 1	67.1	55.7	53.4	58.0
	October 1	72.8	61.4	58.7	64.0
2					
	January 1	56.3	44.9	47.1	42.6
	April 1	50.9	39.4	42.0	36.8
	July 1	67.5	56.0	53.7	58.3
	October 1	73.0	61.5	58.9	64.1
3					
	January 1	56.4	44.9	47.2	42.7
	April 1	50.9	39.4	42.0	36.8
Steady state	Average	62.0	50.5	50.5	50.5
	Minimum	49.6	38.1	41.4	34.8
	Maximum	74.4	62.9	59.6	66.2

Source: Jayadev and Edesess [4].

nonconvective layer is assumed to be 25 percent, ambient temperature 10 ± 15°C, and insolation 200 ± 50 W/m^2.

Estimating the Area Required for a Salt Gradient Pond

The formulas developed in the preceding section can be applied to estimate the required size of a solar pond. For the simplest version of the solar pond sizing method, a "base-case" salt gradient pond with a surface convecting layer 0.3 m thick and a nonconvecting layer 1.2 m thick is assumed. These parameters are not necessarily optimal for every location and application, but they provide a conservative estimate of required pond size.

For the base-case salt gradient pond, an average optical transmission of 0.31 through the surface convecting and nonconvecting layers is assumed. Surface heat losses are assumed to be 0.4 $W/(m^2 \cdot °C)$; bottom losses, 0.1 $W/(m^2 \cdot °C)$ (differential between pond and ground temperatures); and edge losses, 2.2 W/°C per meter of pond perimeter (this would be reduced substantially if the edges were insulated). These assumptions are summarized in Table 10.5. Note that heat loss coefficients and optical transmissions vary with local conditions and pond construction. If better estimates of these parameters than those assumed for the

Table 10.5 Assumptions for the Base-Case Salt Gradient Pond

Parameter	Value	Comments
Surface convecting layer thickness	0.3 m	Varies with surface conditions
Nonconvecting layer thickness	1.2 m	May not be optimal
Average optical transmission through top two layers	0.31	Should be lower at high latitudes
Heat loss from pond surface through nonconvecting layer	0.4 W/(m²·°C)	
Edge losses	2.2 W/°C per meter of perimeter	Varies with soil content, distance of pond surface above or below grade, and presence of edge insulation
Losses from pond bottom to ground	0.1 W/(m²·°C)	Varies with soil content and existence or depth of ground water

Source: Jayadev and Edesess [4].

base case can be obtained, the expanded method described in Edesess et al. [22] should be used.

The required surface area for the solar pond is a function of desired annual average pond temperature, annual average ambient temperature, annual insolation, annual load, and latitude. The surface area increases as either the desired average pond temperature or the annual load increases, and it decreases as the annual average ambient temperature or insolation increases. Latitude indicates only the average elevation angle of the sun and therefore the surface reflection losses, which are greater at higher latitudes. Because of larger reflection losses and the likelihood of decreased ambient temperature and insolation at higher latitudes, the required pond surface area tends to increase with latitude.

Input data required to estimate pond size are:

$\bar{T}$ = annual average pond temperature desired (°C; if in °F, subtract 32 and multiply by 5/9)
$\bar{T}_a$ = annual average ambient temperature (°C)
$\bar{I}$ = annual average insolation (W/m²; if in langleys per day, multiply by 0.4845)
$\bar{L}$ = annual average load (W; if in Btu's per year, multiply by 3.34×10^{-5})
ϕ = latitude of location (deg)

The procedure for calculating pond area is:

1 Multiply insolation $\bar{I}$ by adjustment factor f to obtain $\bar{I}_r$, the insolation received after adjustment for surface reflection losses. The factor f is a function of latitude ϕ, as shown in Table 10.6.

2 Multiply $\bar{I}_r$ by 0.31 to obtain $\bar{I}_p$, the insolation received in the pond after adjustment for reflection and transmission losses.

3 Let $T_d = \bar{T} - \bar{T}_a$. Then the equation for the radius r (m) of a circular pond to meet the requirements is

$$r = \frac{2.2T_d + [4.84T_d^2 + \bar{L}(0.3183\bar{I}_p - 0.1592T_d)]^{1/2}}{\bar{I}_p - 0.5T_d}$$

4 Once the radius is determined, use $A = \pi r^2$ to find the required surface area (m^2). To obtain the required area in acres, multiply by 0.000247.

Some specimen pond areas calculated by the method of the preceding section are given in Table 10.7. Pond depths and outputs may also be estimated

Table 10.6 Reflection Loss Adjustment Factors

Latitude ϕ (deg)	Reflection loss adjustment factor f
0 to 29	0.98
30 to 43	0.97
44 to 49	0.96
50 to 53	0.95
54 to 56	0.94
57 to 58	0.93
59 to 60	0.92
61 to 62	0.91
63	0.90
64	0.89
65	0.88
66	0.87
67	0.86
68	0.85
69	0.84
70	0.83
71	0.81
72	0.80
73	0.78
74	0.76
75	0.74
76	0.71
77	0.69
78	0.66
79	0.63
80	0.59
81	0.56
82	0.52
83	0.47
84	0.42
85	0.37

Source: Jayadev and Edesess [4].

Table 10.7 Required Solar Pond Surface Areas and Depths at Various Locations in the United States

						Pond sizes for 50 kWt avg/70 kWt max load*			
						Winter peaking		Summer peaking	
Region	Location	Latitude (°N)	Insolation (W/m^2) avg/min	Ambient temperature (°C) avg/min	Pond temperature (°C) avg/min	Area (acres)	Depth (m)	Area (acres)	Depth (m)
Pacific	Los Angeles	34	209/112	16.5/12.5	75/50	0.52	3.5	0.52	2.6
		34	209/112	16.5/12.5	60/40	0.38	4.2	0.38	2.7
Mountain	Denver	39	206/96	10.1/−1.2	75/50	0.63	3.7	0.63	3.0
		39	206/96	10.1/−1.2	60/40	0.44	4.5	0.44	3.3
West North Central	Omaha	41	174/67	9.7/−6.6	75/50	1.04	3.6	1.04	3.2
		41	174/67	9.7/−6.6	60/40	0.64	4.3	0.64	3.4
West South Central	Dallas	33	193/103	19.0/7.4	75/50	0.59	3.4	0.59	2.6
		33	193/103	19.0/7.4	60/40	0.42	4.2	0.42	2.8
East North Central	Chicago	41	160/53	10.3/−4.3	75/50	1.37	3.5	1.37	3.1
		41	160/53	10.3/−4.3	60/40	0.76	4.2	0.76	3.4
East South Central	Jackson,	32	185/93	18.3/8.4	75/50	0.66	3.4	0.66	2.7
	Mississippi	32	185/93	18.3/8.4	60/40	0.45	4.1	0.45	3.3
New England	Boston	42	145/53	10.7/−1.6	75/50	2.07	3.2	2.07	2.9
		42	145/53	10.7/−1.6	60/40	0.96	3.8	0.96	3.2
Middle Atlantic	Philadelphia	40	154/62	12.6/0.2	75/50	1.42	3.2	1.42	2.9
		40	154/62	12.6/0.2	60/40	0.77	3.9	0.77	3.1
South Atlantic	Miami	25	194/134	24.2/19.6	75/50	0.50	2.9	0.50	1.9
		25	194/134	24.2/19.6	60/40	0.37	3.6	0.37	1.9

*Approximately the demand of 25 to 50 households.

Source: Jayadev and Edesess [4].

by the formulas developed there. Detailed methods for doing so are given by Edesess et al. [22].

Solar Pond Applications

Solar ponds are readily applicable for such low-temperature uses as residential or commercial heat and hot water, low-temperature industrial or agricultural process heat, or preheat for higher temperature industrial process heat (IPH) application. Combined with organic Rankine cycle engines or thermoelectric devices, solar ponds may be used for electric power generation. By using the heat to run an absorption chiller, solar ponds may be used for cooling.

Solar ponds can be used for district heating in many parts of the world. To minimize heat losses at the pond edges, it is best to maximize the ratio of pond area to pond perimeter. Therefore a small pond will not be as efficient as a larger one, and it is better for residential heating applications to build one large pond for a group of houses than to build a small pond for each house.

Table 10.7 shows the results of sizing the base-case salt gradient solar pond, by the simple technique described in this chapter at various locations in the United States. The load is assumed to be 50 kWt on the average, attaining a maximum of 70 kWt during the peak demand period. Sizing calculations were performed for winter peaking and summer peaking loads. Summer peaking loads are more likely at lower latitudes, where solar ponds may be used for cooling. The surface area requirement is unaffected by the timing of the peak demand. The depth requirement is affected, however, and greater depth is required for a winter peaking load. Sizing was performed for both a "hot pond" (75°C average/50°C minimum) and a "warm pond" (60°C average/40°C minimum) at each location.

The surface area requirement for the hot pond to serve the specified load ranges from about $\frac{1}{2}$ acre in Miami, Florida, and Los Angeles, California, to a little over 2 acres in Boston, Massachusetts. Surface area requirements for the warm pond range from a little over $\frac{1}{3}$ acre in Miami and Los Angeles to almost 1 acre in Boston. Depth requirements range from 1.9 m for a summer peaking load in Miami for both hot and warm ponds to 4.5 m for a winter peaking load and a warm pond in Denver. The depth requirement may be relaxed by increasing the surface area and thereby raising the entire temperature profile of the pond. The pond sized in each case, with allowance for different climates and user loads, would be sufficient to serve about 25 to 50 households.

WOOD STOVES

Wood stoves for space heating and cooking are based on a technology that dates from the days of Benjamin Franklin in the 18th century [23]. Design of these devices has been, and still is, more of an art than a science. As a result, quantitative information is sparse compared to that available for other solar techologies presented in this book. In this section we discuss wood fuel characteristics; stove sizing, design, and testing; and practical matters involved

Figure 10.21 Typical modern wood stove. (*Courtesy of Novan Energy Inc.*)

in wood burning. A photograph of a modern wood stove is shown in Fig. 10.21.

Heat Content and Cost of Wood

Although the heat content of wood depends on the mass or weight involved, wood is sold in the United States almost exclusively on a volume basis. The most common unit is the *standard cord*, which is 128 ft^3. As shown in Fig. 10.22, it is measured as a 4 by 8 ft stack of 4-ft-long logs. Of the 128-ft^3 volume only 80 to 90 ft^3 is wood, the balance being void. Another unit of wood measure is the *face cord*, which is a 4 by 8 ft stack of logs of any length, sometimes as short as 12 in. Since the log length is not defined uniquely in the face cord, it is not a useful unit of measure.

Table 10.8 shows the approximate heating value of many types of wood expressed in units of millions of Btu's per cord. Heat content varies from forest to forest, hence some variation even within species is to be expected. As a rule, evergreens and softwoods have lower heating values and burn quickly; deciduous woods have higher heating values and burn more slowly.

Moisture content also affects heating value. Part of the heat released by the combustion of green wood is used to evaporate moisture (up to 80 percent of the total weight) and is lost to stove sensible heat production. The net heating

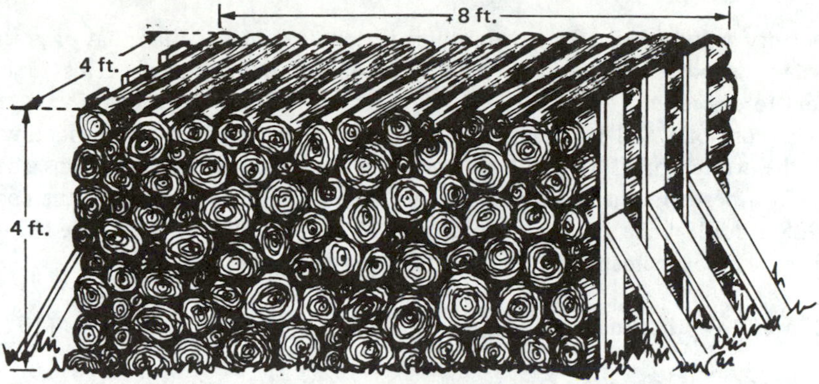

Figure 10.22 Sketch of the standard cord volume unit.

Table 10.8 Approximate Heating Values per Cord of Wood*

High (24–31 MMBtu)	Medium (20–24 MMBtu)	Low (16–20 MMBtu)
Live oak	Holly	Black spruce
Shagbark hickory	Pond pine	Hemlock
Black locust	Nut pine	Catalpa
Dogwood	Loblolly pine	Red sider
Slash pine	Tamarack	Tulip poplar
Hop hornbean	Shortleaf pine	Red fir
Persimmon	Western larch	Sitka spruce
Shadbush	Juniper	Black willow
Apple	Paper birch	Large-tooth aspen
White oak	Red maple	Butternut
Honey locust	Cherry	Ponderosa pine
Black birch	American elm	Noble fir
Yew	Black gum	Redwood
Blue beech	Sycamore	Quaking aspen
Red oak	Gray birch	Sugar pine
Rock elm	Douglas fir	White pine
Sugar maple	Pitch pine	Balsam fir
American beech	Sassafras	Cottonwood
Yellow birch	Magnolia	Basswood
Longleaf pine	Red cedar	Western red cedar
White ash	Norway pine	Balsam poplar
Oregon ash	Bald cypress	White spruce
Black walnut	Chestnut	

*Assuming 80 ft^3 of solid wood per cord and 8600 Btu per pound of oven-dry wood.
Source: U.S. Department of Energy [24].

value of dry wood (~25 percent moisture) is therefore higher than that of green wood. Dry wood is preferred. Seasoning wood for one summer seems to be sufficient to dry it.

The cost of wood varies widely depending on location, species, and time. During the winter of 1981–1982 hardwood in the Appalachian Mountains cost about $75 a cord. In the Rocky Mountains, where hardwood is scarce, the cost was $225 a cord, while indigenous pine cost $95 a cord. If wood is burned with an efficiency of 50 percent, the net cost of wood *heat* is

$$\text{Net cost (\$/MMBtu)} = \frac{2 \times \text{cost per cord}}{\text{MMBtu/cord}} \tag{10.5}$$

The cost of wood *heating* is the cost above plus the amortized stove cost, the cost of chimney and stove maintenance, and any taxes.

Stove Sizing

Currently there is no single accepted method for stove sizing. The presentation here gives a rough sizing guide, assuming that the required heat load L (Btu's per hour) is known. Heat output of wood stoves varies (see next section) with air supply rate ("draft"), wood firing rate, wood surface-to-volume ratio, moisture content, and wood species. For average conditions, the following values are assumed [25]:

Density: 26 lb/ft^3 (including voids)
Heating value: 6200 Btu/lb (dry wood)
Stove efficiency: 50 percent
Desired firing interval: 8 h

For these average conditions the required size, i.e., *net firebox volume* V_{fb}, is

$$V_{fb}\ (\text{in}^3) = 0.172L \tag{10.6}$$

If a 2000-ft^2 home has a heat load of 50,000 Btu/h, a 8600-in^3 or 5-ft^3 firebox would be selected.

If a stove is oversized, the wood is oxygen-starved and slow burning results. This produces creosote buildup, which is costly to remove and can lead to chimney fires. Oversized stoves cost more than properly sized ones, but undersized stoves cannot heat the space properly to maintain comfort.

Stove Testing

Equations (10.5) and (10.6) are based on a steady-state combustion efficiency of 50 percent, which is typical of many stoves. Sizing and economic benefit calculations depend critically on stove efficiency—hence the need for a standard performance test and definition of efficiency. At present, there is no such standard.

Several difficulties have impeded the adoption of a standard test code:

1 Efficiency depends nonlinearly on wood supply (firing) rate.

2 Efficiency varies with wood surface-to-volume ratio, species, and water content.

3 Air supply and stack geometry are critical determinants of airflow.

4 Efficiency—the heat output divided by the input heating value—cannot be measured directly, since conduction and radiation modes of heat transfer coexist. (A large calorimeter could be used, but it is expensive.) It is usually measured indirectly by measuring wood supply rate and stack losses, i.e., velocity, composition, and temperature of flue gases.

5 The chemical composition of the wood varies over the burn cycle; therefore the length of the test is important.

These and other difficulties currently preclude adoption of a uniform test code. Manufacturers use many definitions of efficiency; perhaps a nominal value of 45 to 55 percent for design is the only recourse at present. Efficiencies measured in the laboratory have fallen in this range. As is the case for other solar systems, efficiency is not the final selection criterion; economics and system considerations will dictate the final approach.

The American Society of Heating, Refrigerating and Air-Conditioning Engineers (ASHRAE) is currently developing a stove testing standard. If it follows the course of many other ASHRAE standards, it will eventually become a consensus industry standard. However, this standard will deal only with thermal matters. Air pollutant emissions testing will also be required in the future. Mountain valleys in the West presently have air pollution problems owing to the wide use of wood fuel.

Checklist for Wood Stoves

Proper and safe function of wood stoves depends on many practical details. The checklist below summarizes the major points.

Things to do:

1 Install a proper shield with metal backing (with asbestos use mill board, not cement board).

2 Use an airspace (1 in for walls and 1.5 in for floors) between the shield and a combustible wall or floor.

3 Paint all asbestos shields to control fiber release or, better, avoid asbestos completely.

4 Use UL-listed shields.

5 Install a spark screen.

6 Use several (4 to 5) feet of stovepipe to release heat to the space more efficiently.

7 Control firing rate so that stove outlet temperature is between 190 and 450°F. Use of heat-sensitive colored tape on the stovepipe provides an easy way to do this.

8 Use heavy-gauge stovepipes, which last longer.

9 Design stovepipe wall penetrations carefully.

10 Consider the economics of a massive, central chimney for heat storage.

11 Preferably, use tile-lined chimneys.

12 Clean the chimney regularly (do not use a chimney fire for this purpose). Slow-burning hardwoods can produce more creosote than softwoods.

13 Extend the chimney at least 3 ft above a flat roof and 2 ft above a pitched roof.

14 Install smoke detectors.

15 Keep a fire extinguisher available.

Things not to do:

1 Do not use stovepipe for a chimney.

2 Do not use excessive bends and turns in the stovepipe.

3 Do not oversize the chimney relative to the stovepipe; this can lead to back drafts.

4 Do not use the stove as an incinerator.

For further study of wood stove design and use, the reader should consult [23, 26–29].

REFERENCES

1. M. S. Bawa, "Using Nomograph Methods for Costing PV Systems," *Solar Engineering,* vol. 5, Sept. 1980, p. 26.
2. U.S. Department of Commerce, "Climates of the United States" (Map 55), Washington, D.C., 1974.
3. "PV Power Pump in Solar Thermal System," *Solar Engineering,* Sept. 1980, pp. 28–29.
4. T. S. Jayadev and M. Edesess, "Solar Ponds," SERI/TR-731-587, Solar Energy Research Institute, Golden, Colo., 1980.
5. H. Tabor, "Solar Ponds: Large Area Collectors for Power Production," *Solar Energy,* vol. 7, 1963, pp. 189–194.
6. H. Tabor and R. Matz, "A Status Report on Solar Pond Projects," *Solar Energy,* vol. 9, 1965, pp. 177–182.
7. H. Tabor and Z. Weinberg, "Nonconvecting Solar Ponds," in J. F. Kreider & F. Kreith (eds.), *Solar Energy Handbook,* McGraw-Hill, New York, 1980, Chap. 10.
8. C. E. Nielsen, "Nonconvective Salt Gradient Solar Ponds," in W. C. Dickinson and P. N. Cheremisinoff (eds.), *Solar Energy Handbook,* Dekker, New York, 1979.
9. A. Rabl and C. E. Nielsen, "Solar Ponds for Space Heating," *Solar Energy,* vol. 17, 1975, pp. 1–12.
10. F. Zangrando and H. C. Bryant, "A Salt Gradient Solar Pond," *Solar Age,* vol. 3, April 1978, pp. 21–36.
11. S. L. Sargent, "An Overview of Solar Pond Technology," presented at the Solar Industrial Process Heat Conference, Oakland, Calif., Oct. 31 to Nov. 2, 1979.
12. B. Saulnier, S. Savage, and N. Chepurniy, "Experimental Testing of a Solar Pond," presented at the International Solar Energy Conference, Los Angeles, Brace Institute Rep. R-119, 1975.
13. W. C. Dickinson, A. F. Clark, J. A. Day, and L. F. Wouters, "The Shallow Solar Pond Energy Conversion System," *Solar Energy,* vol. 18, 1976, pp. 3–10.
14. A. B. Casamajor and R. E. Parsons, "Design Guide for Shallow Solar Ponds," UCRL-52385 Rev. 1, Lawrence Livermore Laboratory, Livermore, Calif., Jan. 1979.
15. W. C. Dickinson, A. F. Clark, and A. Iantuono, "Shallow Solar Ponds for Industrial Process Heat," UCRL-78288, Lawrence Livermore Laboratory, Livermore, Calif., 1976.
16. J. E. Groh, "Liquid Foam–Greenhouse Insulation and Shading Techniques," presented at the International Symposium on Controlled-Environment Agriculture, Tucson, Ariz., April 1977.
17. A. Apte, "Solar Pond Power Plant," PRC Energy Analysis Co., Washington, D.C. briefing, Sept. 1978.

18. Battelle Pacific Northwest Laboratories, "The Nonconvecting Solar Pond: An Overview of Technological Status and Possible Application," BNWL-1891, Richland, Wash., Jan. 1975.
19. J. P. Leshuk, R. J. Zaworski, D. L. Styris, and O. K. Harling, "Solar Pond Stability Experiments," *Solar Energy,* vol. 21, 1978, pp. 237–244.
20. W. F. Linke (ed.), *Solubility of Inorganic and Metal-Organic Compounds,* American Chemical Society, Washington, D.C., vol. 2, 1965.
21. T. S. Jayadev and J. Henderson, "Salt Concentration Gradient Solar Ponds: Modeling and Optimization," presented at the 1979 International Solar Energy Society Conference, Atlanta, Ga., May 28, 1979.
22. M. Edesess, J. Henderson, and T. S. Jayadev, "A Simple Design Tool for Sizing Solar Ponds," SERI/RR-351-347, Solar Energy Research Institute, Golden, Colo., Dec. 1979.
23. O. Wik, *Wood Stoves,* Alaska Northwest, Anchorage, Alaska, 1977.
24. Department of Energy, *Heating with Wood,* DOE/CS-0158, Government Printing Office, Washington, D.C., 1980.
25. *Wood and Energy Solid Fuel Journal,* vol. 1, no. 3, 1981.
26. J. Vivian, *Wood Heat*, Rodale, Emmaus, Pa., 1976.
27. J. W. Shelton and A. B. Shapiro, *Woodburner's Encyclopedia,* Vermont Crossroads, Waitsfield, Vt., 1976.
28. J. W. Shelton, *Wood Heat Safety,* Garden Way, Charlotte, Vt., 1979.
29. G. Harrington, *The Wood Burning Stove Book,* Macmillan, New York, 1979.

11 State Approaches to Solar Legislation: A Survey

Stephen B. Johnson*

INTRODUCTION

The constitutional framework of the United States is based on state government, the source and final repository of all political power not otherwise delegated to the federal government or reserved to the people [1]. This system of federalism is responsible for the popular conception of the states as "laboratories of democracy." Except for federal efforts in solar research, demonstration [2-4], and standards [5, 6], virtually all of the pre-1979 incentives for the commercialization and exploitation of solar energy originated with state and local governments. This phenomenon is appropriate considering the popular appeal of solar energy,† the distinctly local and decentralizing impact of most available solar technologies, climatic variations, and even the diffuse nature of sunlight. While this trend may decline, especially in the case of financial incentives, due to the recent passage of laws collectively referred to as the National Energy Act [9], the states and their citizens remain the leaders in encouraging solar energy development.

The following description and comparison of the legislative efforts of the

*Attorney-at-law in Boulder; law clerk for Hon. Robert A. Behrman, Chief Judge of 19th Judicial District of Colorado. The author is indebted to George Morgan and to Prof. Jan Laitos of the SERI Law and Government Program for providing the opportunity and inspiration to write this paper.

†The Harris Survey reports that Americans favor expansion of solar energy use by 94.2 percent, the highest priority granted to any energy source. Quoted in [8].

Abstracted from *Solar Law Reporter*, vol. 1, no. 1, 1979, pp. 55–98, with permission of the author. It is recommended that the reader refer to later issues of the *Reporter* for the most current information.

states in the areas of property, income, excise, and franchise taxation; loans; standards and warranties; building codes; solar access; utilities; and promotional activities is conceptually oriented, systematically assessing the laws that define some of the parameters of the solar "experiments" in the states. Administration of solar legislation is, with minor exceptions, not discussed.*

FINANCIAL INCENTIVES

Financial incentives offer the fastest and the most effective means by which to encourage solar commercialization. Solar technologies are disadvantaged when they must compete economically with artificially cheap "conventional" energy sources such as coal, oil, or natural gas. These energy sources have received federal subsidies since 1918 of over $217.42 billion [11].

The existence of such subsidies alone can be used to justify solar subsidies at any governmental level. But even if one believes that the subsidies to conventional energy sources should instead be reduced in order to enhance competition in the market, powerful policy arguments remain for assisting an infant solar industry. To the extent that solar energy can reduce reliance on dirty, expensive, capital-intensive, inefficient, or imported conventional sources of energy, beneficial effects may be expected with regard to national security, balance of payments, value of the dollar, capital availability, employment, and the environment. The ecological principle that species diversity enhances systemic stability is applicable to societal energy needs as well: multiple independent sources of solar thermal energy are virtually immune, given ample storage capacity, to breakdowns.

Since the oil embargo of 1973 and 1974, large numbers of state legislatures have asserted a public interest in the use of their revenue and police powers to encourage the harnessing of solar energy. This encouragement has most often taken the form of financial incentives, or subsidies. The mechanisms include property tax exemptions or reduced assessments; income tax credits and deductions; rapid depreciation or amortization; franchise, sales, and use (excise) tax deductions; and low-interest or guaranteed loans. States vary considerably in providing incentives. Some states provide only one incentive, others offer combinations. Of course, there is also variation among states in handling the individual incentives, which are often interdependent.

Property-related Incentives

Passage of Laws

Solar property tax exemptions are the most popular financial incentive device. Twenty-eight states are currently offering real property tax exemptions

*J. Ashworth, B. Green et al. have prepared a case study on the administration of state legislation in selected states [10]. This author agrees that "The implementation process is an important determinant of the final form and of the effectiveness of a state incentive for solar energy. Implementation is particularly important for determining which technologies and components are eligible for an incentive."

for solar energy systems. In the November 1978 elections Florida and Texas voters approved constitutional amendment referenda allowing property tax exemptions [12].

The necessity for such amendments in some states results from constitutional provisions that all occupation or real property taxes shall be levied or assessed in an "equal and uniform" manner, with certain charitable exceptions [13]. Georgia ratified such an amendment in 1976 [14] allowing local adoption of the exemption. California voters rejected a proposed exemption in June 1978 [15], paralleling the Proposition 13 property tax revolt. Nebraska voters rejected a similar proposal in November 1978 [16].

Most states, however, are able to legislate exemptions to property taxes. This can be done under constitutional authorization, or by legislative definition of solar systems as personal property or as a separate classification of property, as Louisiana has done [17]. Perhaps the failure of more state legislatures to pass property tax exemptions reflects a hesitancy to disturb local revenues. Montana replaced its property incentives with income tax credits [18], but not all states have income taxes.

All property tax exemptions can be portrayed as to some extent destructive of local governmental revenues, as well as discriminatory, rather than equal and uniform in effect. The New York legislature dealt with this concern by finding that the exemption would not reduce "tax income to the community [19]. Obviously, as solar use increases, the increase in growth of local revenues may be slowed, but this is not necessarily for the worse. It has, for example, been pointed out that solar installation could reduce a community's overall expenditures by lessening net pollution, reducing fossil fuel expenditures, and encouraging more energy-efficient zoning [20].

Eligible Technologies

The threshold issue in the case of property tax exemptions is the proper choice and definition of eligible solar technology. This issue has two components. First, it must be determined whether the solar system is "real property" subject to taxation. Second, the legislature must choose which technologies should be eligible.

Most common solar heating and cooling devices become permanent fixtures, and thus real property subject to property taxes when they are attached to a building, if subsequent removal would damage the building's structure. Devices incorporated in the design or construction are also taxable realty by definition. Side yard collectors and mobile solar devices are classified as tangible personal property, and their taxable status varies among the states. If permanently attached, they may in some cases be considered a real property improvement.

All such state statutes include solar heating and cooling of buildings (SHACOB) equipment in their definition of solar energy "system" or "device," "alternative energy device," or "renewable energy resource," at least where heating and cooling is accomplished by means of active flat-plate collector systems. Normally, the entire active system (collectors, pipes, pumps, storage

Table 11.1 Real Property Tax Incentives in January 1979[a]

	Ariz.	Colo.	Conn.	Fla.	Ga.	Hawaii
Year	74	75	76, 77	78	76	76
Law-chapter	165	344	409, 490	354	SR. 284	189
Constitutional amendment				•	•	
Local adoption			•			
Legal codification (primary)	ARS 42-123.01	CRS 39-1-104 39-5-105	CSA 12-81 (56)	FSA 193.622	GCA 2-4604	HRS 246-34.7
System						
SHACOB (collectors only)	•	•	•	•	•	•
Passive	•	•		•	•	
Hot water	•	•	•	•	•	•
Wind	•		•			•
Bio						•
PV	•					•
Hydro/TE			•			•
Building						
Residential	•		•		•	
Commercial	•		•		•	
Other					•[b]	
Procedure						
Apply		•	•			•
Automatic	•				•	
Install						
By			10/1/91			12/31/81
After			10/1/77			6/30/76
Duration or termination	12/31/84		15 yr		1986	Open
Tax exemption (deduction) or credit (refund)	•	•	•		•	•
Valuation						
Separate		•				
Nonassessed	•					
Measurement formula (key)	2	5[c]	3	3	2	2
Standards or certification			•			
Actual use minimum energy requirement						•
Federal preemption						

(*See footnotes on p. 314.*)

Ill.	Ind.	Iowa	Kans.	La.	Me.	Md.	Md.	Mass.
75, 77	74–77	78	77	78	77	76	75, 78	75
943, 430	15–68	1056	345	591	542	740	509, 509	734
						•		
ICA 120-§501 96½-§7301	ISA 6.1. 1-12-26	IC 93.2	KSA 79-45a01	L.R.S.A. 47-1706	MRS 36-656	ACM 81-12-F(5)	ACM 81-14(b) (4)	ALM 59-5
•	•	•	•	•	•	•	•	•
•				•	•	•	•	
•				•	•			•
•			•					•
•			•	•				•
•								
	•	•	•	•	•	•	•	•
	•		•	•	•	•	•	•
				Pool				
	•				•	•		•
				•			•	
					1/1/83			7/1/76
						Open		
Open	Open	1/1/79 12/31/85	4 yr 12/31/85	Open	5 yr 1/1/83	3 yr	Open	20 yr
	•			•	•		•	•
			•			•		
•	•						•	•
				•				
2	2		5[d]	1	2	8	2 or 6	2
			•					
	•				•			
			•[e]					

Table 11.1 Real Property Tax Incentives in January 1979 (*continued*)

	Mich.	Minn.	Nev.	N.H.	N.J.	N.Y.	N.C.
Year	76	78	77	75, 75	77	77	77
Law-chapter	135	786	345	256, 465	256	322, 618	965
Constitutional amendment							
Local adoption				•			
Legal codification (primary)	MSA 7.7(4e)	MSA 273.11	NRS 361	RSA 72:61-64	NJSA 54:4-3.120	R.P.T.L. 487	NCGS 105-277(g)
System							
SHACOB (collectors only)	•	•	•	•		•	•
Passive	•		•		•	•	
Hot water	•			•	•		
Wind	•		•	•	•		
Bio			•		•	•	
PV	•				•		
Hydro/TE	•[f]		•		•		
Building							
Residential	•	•	•	•	•		•
Commercial	•[g]	•		•	•		•
Other							
Procedure							
Apply	•		•	•	•	•	
Automatic		•					•
Install							
By	6/30/85	1/1/84				7/1/88	12/1/85
After							
Duration or termination	Open 1/1/85			Permanent	12/31/82	15 yr	12/1/85
Tax exemption	•	•		•			•
(deduction) or						•	
credit (refund)			•				
Valuation							
Separate			•		•	•	
Nonassessed	•	•					•
Measurement formula (key)	2	2	4	5	2	2	3
Standards or			•		•		
certification						•	
Actual use minimum energy requirement			•		•	•	
Federal preemption							

(*See footnotes on p. 314.*)

N.D.	Ore.	R.I.	S.D.	Tenn.	Tex.	Vt.	Va.	Wash.
75	75, 77	77	78	78	77	76	77	77
508	460, 196	202	74	837	SJR 53	226	561	364
					•			
						•	•	
N.D.C. 57-02-08	O.RS 307.175	R.I. 44-3-19	S.D.C.L. 10-6-35.8-.18	T.C.A. 67-511	Tex.Con. art.VIII 52(a)	V.S.A. 53-15	R.C. 58-16.4	R.C.W.A. 84.36-410
•	•	•	•	•	•	•	•	•
								•
	•		•			•	•	•
	•		•			•	•	•
	•		•	•	•	•		
			•			•		
	•		•	•		•	•	
•	•	•	•			•		•
•	•	•	•			•		•
			•				•	•
•	•	•				•		
						Open	Open	12/31/81
5 yr Open	1/1/98	4/1/97	5,3[j] yr	1/1/88		Open	5 yr	7 yr
•		•		•		•	•	•
•	•	•	•	•		•	•	•
2	2	6	1,5[k]	2		5	5[l]	2
	•	•[i]		•			•	•
				•			•	•
			•					

Table 11.1 Real Property Tax Incentives in January 1979 (*continued*)

[a]Definitions:

SHACOB: solar heating and cooling of buildings. Includes controls, wiring, pumps, storage tanks, exchangers, etc. present in active solar systems.

Passive: structural building elements absorbing and radiating solar thermal energy, usually via nonmechanical systems. Examples include eaves, high thermal mass walls, roof ponds, roof vents, greenhouse walls, thermosiphon walls.

Hot water: heating of water, as opposed to space heating.

Wind: wind energy conversion producing mechanical or electrical power.

Bio: bioconversion; commonly in the form of methane generation from agricultural products and wastes, or direct combustion of wastes.

PV: photovoltaics, the production of electricity directly from sunlight.

Hydro/ocean: technologies ranging from water wheels, low-level water osmosis, temperature-gradient exploitation (especially in the ocean.)

TE: thermal electric; electrical power produced from steam heating, usually by reflection of sunlight from heliostats into a central receiver.

Measurement formula—exemption equals:

1. Total cost/value/price (C/V/P) (as assessed).
2. Difference between C/V/P of solar and conventional systems.
3. Lesser of 2 and set dollar amount.
4. Set percentage of C/V/P.
5. Other—local variation or annually declining rate.
6. Maximum assessment equals value of conventional system necessary to serve the building.

(Note: The author's interpretation may not be consistent with administrative interpretation and/or regulations, especially in the area of eligible technologies.)

[b]Manufacturing equipment.

[c]5 percent of actual value.

[d]35 percent of tax refunded.

[e]70 percent heating load capability for buildings or additions.

[f]Excludes water wheels.

[g]Excludes corps in solar business.

[h]HUD standards.

[i]Residential, commercial; then a 3-year declining rate (75, 50, 25 percent) applies.

[j]Residential, minimum actual installed cost; commercial, 50 percent of actual installed cost.

[k]Excludes either whole or partial amount of assessed value, which includes installation costs.

[l]For supplemental (49 percent of maximum) solar energy systems.

tanks, wiring, etc.) is included in the SHACOB exemption, while elements of a supplementary or conventional backup system are excluded. Only New Hamsphire has limited its exemption for solar energy systems to collectors [21]. On the other hand, some state statutes exempt the more exotic solar technologies such as wind, biomass, photovoltaics, thermal electric, low-head hydroelectric, and water temperature gradient (see Table 11.1).

In addition to active SHACOB systems, many states have provided exemptions for the next most common solar technologies—passive heating and cooling and water heaters.* The passive issue is particularly sensitive, for some of the techniques employed in passive design (high-mass walls, ventilation eaves, movable walls, earth ceilings, berms, film windows) incorporate structures or

*See tax and loan tables (Tables 11.1 and 11.4) in this chapter.

modifications of structures that are present in most houses. The problem is to differentiate for tax incentive purposes between the uniquely "solar" characteristics and those that are ordinarily present in an unmodified structure or in one not designed to capture and conserve solar energy.

Many states have restricted their definitions of eligible solar energy systems to those using incident sunlight "directly and exclusively" for solar heating and cooling. North Carolina explicitly excludes passive solar designs [22]. On the other hand, several states explicitly recognize and allow passive systems [23]. New Jersey includes passive solar design [24] in its definition as "evaporative cooling," or "nocturnal heat radiation," typically deriving from large thermal masses like those found in Harold Hay's roof pond or the Trombe Wall* (Fig. 11.1). Some states can exclude passive design through a requirement that the solar system "collect, transfer, and store" incident solar radiation. Only active solar heating systems transfer energy between collection and storage. Other indications of exclusion of passive systems are requirements of separate storage, or specific reference to equipment or devices that exclude purely passive structural building components.

To avoid confusion, express inclusion of water-heating systems in incentive statutes is recommended where there is a desire to include such devices. In states that define solar equipment in accordance with federal performance standards, hot-water systems are eligible because they are included in the federal definition of solar heating or solar heating and cooling [27]. Some conventional heating systems heat both space and water at a single point, which further confuses the issue. A possible reason for the paucity of explicit incentives for water heaters is that such systems are already economically viable in most parts of the country, based on life-cycle costs. However, it is more likely that the omission of water-heating systems from incentive statutes is due to ambiguous definitions of eligible technology.

Eligible Buildings

With the exception of Nevada [28] and Louisiana [29], all state property tax incentives apply to commercial as well as residential buildings or structures. The term commercial includes industrial and, occasionally, agricultural structures. Georgia is alone in exempting equipment used directly in the manufacture of solar devices [20], a real boost to the numerous small businesses that compose much of the solar industry. By contrast, Michigan disallows an exemption for the property of corporations involved in the design or building of solar devices for resale [30]. Indiana has amended its property tax statute to exempt solar improvements on mobile homes [31], although statutory definitions of real

*The Harold Hay Sky Pond is in essence a large bag of water perched on the roof of a structure. During the day it blocks sunlight and absorbs heat, cooling the house. At night it is covered (a "semipassive" feature) and radiates its heat into the house below. The Trombe Wall is a French invention having windows that allow exposure of a darkened heavy cinder block or concrete wall to the sunlight. A vent system allows air, heated by the wall, to circulate throughout the structure when desired [25, 26]. (See Chaps. 7 and 9 for details.)

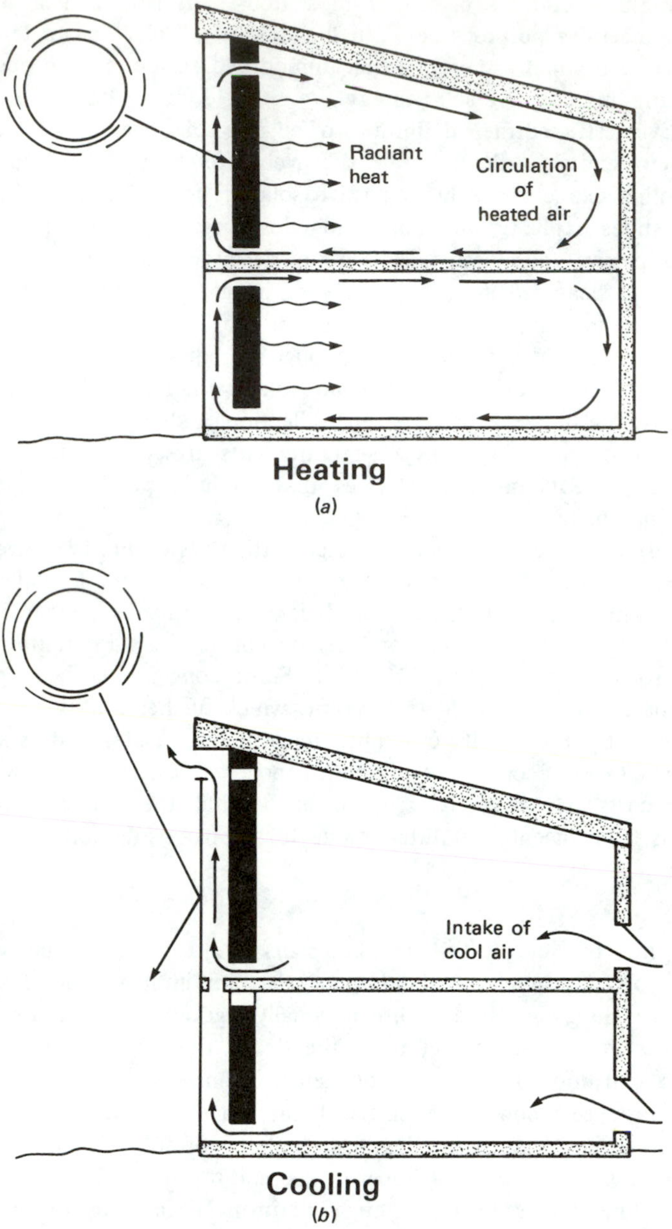

Figure 11.1 (*a*) A Trombe wall is a passive solar collection system named after its inventor, Felix Trombé, which incorporates a transparent glass or plastic heat-retaining surface on the outside of a darkened massive wall. (*b*) Warm air rising between the surfaces can be ducted inside or outside depending on the thermal requirements of the structure.

property or administrative regulations could achieve the same result in some states.*

Application Procedures and Eligible Recipients

Some states automatically exempt solarized structures from increased property taxes. In such states, lessees, tenants, or other occupants, as well as owners, could conceivably benefit, depending on the identity of the ultimate property taxpayer. Other states require an application for exemption by the owner, on either an annual or permanent basis. South Dakota eliminates the exemption if title is transferred, unless the transferee is the first occupant in a residential building [33], though originally the only eligible recipient was the owner who also occupied the building [34]. In that state the administrative procedure is typically complex: the owner must send for permanent application forms from the department of revenue; include information about the type of system, cost, etc.; attach receipts; and file two copies with the county auditor and one with the department of revenue. The county assessor then verifies the statement, whereupon the auditor makes the deduction from the tax rolls.

Many states have strict application dates or deadlines. The value of such an application process is that it can generate information about use of the incentive and allow monitoring of system conformance to state regulations and standards. In states that have complex procedures for assessing solar energy systems, red tape may be hard to eliminate. However, some states provide an incentive by simply not assessing the solar energy system. In these states government red tape and expenses should be lowered. Usually, differences in assessment procedures are caused by variations in constitutional treatment of real property.†

Installation Timing and Incentive Duration

Almost two-thirds of the 28 states with property tax exemptions have one of several kinds of time limitations. Such limitations on eligibility for exemption may precipitate solar investments sooner than would otherwise be the case, as well as protect the state revenue from suffering into the indefinite future. Connecticut, Georgia, Maine, Michigan, Minnesota, New York, North Carolina, and Washington have enacted installation date limitations. While no taxpayer has a guarantee of future availability of an incentive, since states or localities can always repeal exemptions, built-in installation date limitations, a fixed duration of the incentive, or a certain termination date all serve notice to potential solar purchasers that early acquisitions may be more cost effective.

Another rationale for such time limitations, especially for the termination provisions, is that most eligible solar systems will eventually become more economically competitive with conventional energy sources as fuel prices rise due to deregulation, cartel action, and supply scarcity, or as mass production and technical advances lower solar system costs. When solar technologies like water

*For example, Maine exempts solar energy equipment without reference to where it is located.

†See text accompanying [12–20].

heaters demonstrate independent viability in the marketplace, the purely economic justification for the subsidy ceases. However, other barriers such as lack of public knowledge or acceptance may persist several years after economic viability. Therefore termination dates should reflect the date on which a solar energy technology attains independent economic viability. Most termination dates are around the mid-1980s. Oregon's provision does not expire until 1998 [35]. Connecticut would allow the exemption for 15 years after 1991 [36], while New Hampshire offers a permanent exemption [37].

Limitations on duration range from 3 to 20 years and average 7 years. Massachusetts amended its duration limit from 10 to 20 years in 1978 [38]. The most restrictive duration period–3 years–is found in Maryland, but this applies only to its credit program [39]. South Dakota allows only 3 years of participation by corporations [40]. Both of these duration limits accompany credits instead of the more usual exemptions.

Several states have targeted the tax exemptions for installations made after a certain date, probably because most early purchasers who could afford solar devices were considered to have little need for an incentive applied retroactively to systems already in place. However, equitable considerations have prevailed in most states against such exclusions of prior installations.

Exemptions Versus Credits

As noted, the usual method of structuring incentives is by an exemption from increased property tax assessment. Besides Maryland and South Dakota, only Kansas and Nevada offer credits or refunds to property taxpayers. The 1977 Nevada statute appropriated $32,000 for reimbursement to applicants [41]. While such a system generates increased administrative costs and delays the actual benefit, credits under any form of taxation are generally more efficacious and popular than are deductions.

Valuation

States vary in their property tax assessment procedures, relying on market data, income production, replacement-depreciation, or other indicia of property value. The procedures are not often apparent from the bill language; further code research for individual states is necessary to ascertain which technique is used. Most states that rely on market data could simply choose to ignore the increase in value due to solar installations. In other words, a system would simply not be assessed, rather than being separately assessed and exempted. This technique is administratively simple and is recommended in the model property tax exemption statute proposed by the authors of the American Bar Foundation study [20, p. 79].

Measurement Formulas

Technically, exemptions from property value improvements are classified as part of the body of *ad valorem* taxation laws. *Ad valorem* taxes charge a fixed proportion of the value of the property [42]. The exemption is a privilege granted to a taxpayer/debtor on grounds of public policy, allowing retention of property or earnings [43].

The most common exemption (found in 16 states) is for the entire cost, assessed value, or price of the solar system, depending on the valuation procedure in the particular state. This measurement (formula 2 in Table 11.1) is often computed as the assessed value of the property with a solar energy system minus the assessed value without the system. This is the most generous formula, assuming that the system works well enough that it does not *decrease* the market value of the property. Theoretically, the exemption could be largest for new homes with eligible passive systems, as opposed to retrofits of active systems onto homes with usable conventional heating and cooling systems. This is so because without the passive system, which also reduces the heating requirements of a home, large heating costs at low levels of efficiency would become necessary, thereby reducing present marketability.

Most amendments to property and income exemptions broaden, rather than restrict, the size of the incentive. New York recently amended its formula (formula 1) to exempt the entire assessed value of the solar system, instead of subtracting the assessed value of a typical conventional system from the increased valuation due to the solar system (formula 2).*

Formula 2, which subtracts the value of a conventional system from total valuation, is sometimes phrased to require the assessor to value the property "as if it had a conventional system." In contrast to the exemption-from-increased-valuation formula (formula 1, Table 11.1) considered above, this seems to discriminate relatively against new homes with passive systems that have only supplementary conventional heating and cooling systems, in favor of retrofit of older properties with full-size conventional backup systems. Only three states now use this formula. A variation of this formula used by Rhode Island [45] provides for a maximum assessment equal to the value of a conventional system necessary to serve the building (formula 6, Table 11.1). In contrast to formula 2, this favors new buildings with passive design.

A formula (formula 4, Table 11.1) occasionally used exempts a set percentage of the value of the system. Colorado was the first state to use this formula, requiring valuation at no more than 5 percent of actual value [46]. Kansas allows a 35 percent credit for 5 years [47].

A major variation of interest is the annual declining rate (formula 5, Table 11.1). South Dakota allows a complete exemption for residential buildings for 5 years, then applies a 3-year schedule decreasing the exemption to 75, 50, and 25 percent of the base credit [48]. For commercial property the declining schedule applies after only 3 years. Such declining rates are more frequently used in income tax incentives. They are premised on anticipated future increases in the prices of other competing fuel sources, which act to lessen the economic need for solar subsidies.

Standards or Certification

Certification of eligible systems is a common provision designed to ensure that exemptions are nonfraudulent, or to ensure that the system will perform well enough to justify the subsidy. In the latter case, certification is dependent

*This is listed in Table 11.1 as formula 2 [44].

on material or performance standards. The standards are usually promulgated by state revenue agencies with assistance from state energy offices. Only two states have adopted the HUD reference standards, discussed below.

The Federal Role

South Dakota [48] has decided that federal energy income tax incentives [9] might reduce the need for state-level property tax incentives. Some other states are expected to reduce this type of subsidy as federal assistance becomes publicized and is perceived to displace state or local property tax relief by making total governmental support seem to be too generous.

Income Tax Incentives

Introduction

While most property tax exemptions are designed to preclude additional financial burdens on those installing solar systems by reducing the increase in taxable property value, income tax incentives serve as a positive reward for solar investments. Instead of merely removing potential barriers, income tax measures are positive inducements to "go solar" and may actually redistribute income in the process. This factor, combined with the limited number of states with income taxes, may account for the fact that there are fewer income tax than property tax incentives. Only 22 separate solar income tax laws have been passed in a total of 16 states (see Table 11.2).

Eligible Structures and Taxpayers

With the exception of express solar equipment depreciation and amortization deductions [49] that apply exclusively to corporations (in Massachusetts, Arizona, and Kansas), virtually all the income tax incentive laws seem to be aimed primarily at occupants of residential structures or individual homeowners. Many of the states (Alaska, Arizona, Idaho, Montana, Oklahoma, and Oregon) offering income tax incentives restrict the availability of straight credits or deductions to residential dwellings, excluding industrial or commercial buildings. While this can be justified for populist or budgetary reasons, or out of sympathy for the residential energy consumer faced with expensive fossil fuels and limited income, some doubt exists about the efficacy of exclusively noncommercial incentives in causing substantial change in building design. A 1974 study found that successful innovations in building design have always been introduced first in public and commercial buildings and then in custom houses, before the mass residential housing market is affected [50]. Only a small fraction of housing is custom-built. States considering income legislation to encourage investment in solar might attempt to take advantage of this trickle-down demonstration effect and allow commercial as well as residential deductions.

In addition to residential and, occasionally, commercial buildings, several states have provided deductions or credits for other types of structures and uses of solar energy. An Arkansas law broadly refers to "any structure using solar heating" [51]. California explicitly recognizes condominium owners as eligible on a prorated basis [52]. New Mexico includes solar-powered irrigation systems, while California includes solar-heated pools [53]. North Carolina has recognized individually metered family units in multidwelling buildings as eligible, if the installer owns or controls the unit and has paid the majority of the proportional cost [54]. Most states limit residential eligibility to principal dwellings, occasionally specifying that the dwelling situs be in the state.

To be eligible, a taxpayer must usually reside in the solar-equipped dwelling. Often the taxpayer must stand in a certain legal or financial relation to the property, such as being an owner or contract purchaser. Residency must generally coincide with completed installation of a solar system for most deductions, especially in the case of deductions for primary dwellings.

Oregon is unique in explicitly providing incentives for both nonresidents and secondary dwellings [55], although the laws of several other states could be interpreted to include secondary dwellings, since no limitation to principal dwellings is specified and taxpayers are not expressly limited to one exemption a year. Alaska has joined Oregon in allowing for credits by nonresident taxpayers [56]. About half such states have made provisions for claiming deductions when the property is held jointly or communally and when taxes are filed separately. Usually the deduction or credit can be claimed in whole or proportionately by any co-owner. New Mexico is an exception to this rule [57].

Few states have explicitly considered continuing incentives for subsequent purchasers or transferees, probably on the theory that once the initial installation is made, the main purpose of the incentive has been achieved. Most states still require installation by the taxpayer as a precondition, thereby excluding transferee eligibility. However, California has rejected its earlier "original use" requirement [58].

One side effect of a provision favoring transferees might be to raise the market value of the affected property, which could be marginally counterproductive as an incentive unless a solar property tax exemption also exists. Only eight states (Arizona, Colorado, Hawaii, Kansas, Massachusetts, North Carolina, North Dakota, and Oregon) currently offer both income and property tax measures. However, a policy allowing transferees to take advantage of income tax incentives offered in conjunction with property incentives could prove beneficial, since an owner who was considering moving might be more likely to invest in a solar system if the tax benefits were marketable.

Statutory provisions accompanying the incentive can deter the initial installation in cases where the recipient must be the installer, or where the installer is a developer but is unable to claim multiple credits. California allows a builder of unsold housing either to take the credit or to pass it through to the purchaser [59]. Assistance for housing built by developers promises to be of enormous benefit to solar commercialization, and deserves greater consideration. Eligibility provisions such as Oregon's, limiting a taxpayer to a single credit per year, could deter developers from installing solar systems in large-scale projects.

Eligible Systems

Of the various solar technologies eligible for income tax credits or deductions, solar heating and cooling of buildings (SHACOB) and water heating predominate.

Table 11.2 Income Tax Incentives

	Alas.	Ariz.	Ariz.
Year	77	75, 76, 78	77, 78
Law-chapter	94	93, 129, 112	81, 112
Legal codification (primary)	AS 43.20 039(d)(4)	ARS 43-123.37	ARS 43-128.03-.04
System			
SHACOB	•	•	•
Passive		•	•
Hot water	•	•	•
Wind	•	•	•
Bio	•	•	•
PV		•	•
Hydro/TE	•	•	•
Criteria/regulations			•
Building			
Residential	•	•	•
Commercial		•	
Other			
Duration or termination	12/31/76 12/31/82	12/31/84	12/31/77 12/31/84
Deduction or		•	
credit	•		•
Measurement formula[a]	*b*	36-month amortization	*c*
Actual energy use			
Minimum energy requirement			
Federal displacement			
Eligible taxpayer	•		
Residential individual		•	•
Installer/owner			
Depreciation/		• or	
amortization		•	
In lieu of			•
Carryover			5 yr
Costs			
Installation	•	•	•
Supplemental equipment			
Remodeling	•	•	•
Situs in state	•	•	•
Expenditures/installation/after	12/31/76		
Apply			

(*See footnotes on p. 326.*)

Ark.	Cal.	Colo.	Hawaii	Idaho	Kansas
77	76, 77, 78	77	76	76	76
535	168, 1082, 1159	512	189	212	434
ASA 84-2016.8-.10	Rev. & Tax § 17052.5	CRS 39-22-113(4)(c)	HRS 235-12	IC 63-3022 C	KSA 79-32, 166-167
•	•	•	•	•	•
	•	•	•		
•	•	•	•	•	•
	•	•	•	•	•
				•	
	•			•	•
•	•				
•	•	•	•	•	•
•	•	•	•		
•	•				
1/1/77 Open	1/1/81	1/1/77 Open	6/30/76 12/31/81		12/31/75 7/1/83
•		•		•	•
	•		•		
d	*e*	*d*	No. 5[f]	No. 6[g]	No. 7[h]
	•	•	•		•
	•				
•		•	•		•
	•	•	•	•	•
•	•	•			•
	•	•	•		
No	Yes				4-yr-individual
•	•	•		•	•
	•	•		•	
		•			•
•	•				
		1/1/77	12/31/74		

Table 11.2 Income Tax Incentives (*continued*)

	Kansas	Mass.	Mont.	Mont.
Year	77, 78	76	77	77
Law-chapter	346, 409	487	574	576
Legal codification (primary)	KSA 79-32, 168; 79-1118	ALM c63 §38H	RMC 87-7414	RMC 84-7403
System				
SHACOB	•	•	•	•
Passive			•	•
Hot water	•	•	•	•
Wind	•		•	•
Bio			•	•
PV	•			
Hydro/TE			•	•
Criteria/regulations	•	• Certified	•	•
Building				
Residential			•	•
Commercial	•	•		•
Other				
Duration or termination	12/31/75 7/1/83		12/31/76 12/31/82	12/31/76 Open
Deduction/or credit	• •	•	•	• • Additional
Measurement formula[a]	*h*	*d*	No. 8[i]	No. 9[j]
Actual energy use				
Minimum energy requirement	•			
Federal displacement			•	
Eligible taxpayer			•	•
Residential individual		•	•	•
Installer/owner	•	•	•	•
Depreciation/ amortization	• 60-month • amortization	• •		
In lieu of	No	•		
Carryover	No	4 yr		
Costs				
Installation	•	•	•	•
Supplemental equipment				
Remodeling	•	•		
Situs in state		•		
Expenditures/installation/after				
Apply			•	•

(*See footnotes on p. 326.*)

N.M.	N.C.	N.D.	Okla.	Ore.	Wisc.
75, 77	77	77	77	77	77
12, 170, 114	792	537	209	196	313
NMSA 72-15A-11.3-11.4	NCGS §§105-151.2, -130.23	NDCC 57-38-01.8	68 OS §2357.1-.3	ORC §§469.010 -.140,316-116	W.S.A. §71.04 (16a), 71.09(12)
•	•	•	•	•	•
•		•	•	•	•
•	•	•	•	•	•
		•		•	
		•		•	•
				•	•
				•	
• HUD certified	• HUD		•	• Certified	
•	•	•	•	•	• Credit
• Irrig.	•	•			• Dept.
• Pool					
Open Pool 1/1/78	1/1/77 to Open	Open	12/31/77	1/1/78 1/1/85	12/31/84
•	•	•	•	•	• (Res.) • (Comm.)
Pool max. $25,000[h]	*h*	*k*	No. 11[l]	No. 7[h]	*m*
			•	• 10%	
•					
•	•	•	•		
•			•	•	
•	•	•	•	•	
					• 5 yr or •
Refund	3 yr		5 yr	5 yr	Refund
	•	•	•	•	•
		No			•
•	•				•
					4/20/77
				•	•

Table 11.2 Income Tax Incentives (*continued*)

[a]Measurement formulas are discussed in the text and outlined below. Note that the author's statutory interpretation may not always be consistent with administrative interpretations and/or regulations, especially in the areas of eligible technologies.
[b]10% of expenses of installation, maximum $200. Also 5% of residential fuel costs (wood) (Alaska).
[c]35% of cost, declining 5% per year, until 1984, maximum $1000. Also 25% of cost of insulation, wind-driven turbine, ventilation, or passive roof vent, maximum $100 (Arizona).
[d]Deduct entire cost; exclude interest, finance charges, acquisition, installation (Arkansas).
[e]55% of cost, maximum $3000. In multidwelling houses or commercial buildings where cost exceeds $12,000, 25% of the cost. Federal credits claimed shall reduce California credits to equal total credit of 55%. Grants are excluded from cost basis.
[f]10% of total cost of device.
[g]40% of cost for year of purchase, 20% for 3 years thereafter, maximum $5000 for any one year (entire cost over 4 years).
[h]25% of cost, maximum $1000 for individuals. Maximum $1000 per building or metered residence in North Carolina. Maximum of $2000 for Oregon, $3000 for businesses in Kansas.
[i]10% of first $1000, 5% of next $3000 of cost, less grants received. Similar federal credits result in halving Montana's contributions, whether or not the credits are claimed.
[j]100% of first $1000 (residential) or $2000 (commercial) expended. 50% of next $1000 (res.) or $2000 (comm.) expended. 20% of next $1000 (res.) or $2000 (comm.) expended. 10% of next $1000 (res.) or $2000 (comm.) expended. Expenditures shall not include offsetting grants. Maximum $1800 (res.) or $3600 (comm.).
[k]5% of cost per year for 2 years.
[l]25% of cost, maximum $2000 per individual.
[m]The following percentage of total costs (if they exceed $500) for a minimum credit of $10,000 for structures appearing on the tax roll:

	Prior to 4/20/77	After 4/20/77
77 and 78	30%	20%
79 and 80	24%	16%
81 and 82	18%	12%
83 and 84	12%	8%

The two types of systems are either explicitly listed or implied in the various definitions of "solar energy systems," "alternative energy devices," and "solar equipment." SHACOB and solar hot water have achieved greater market penetration to date than the other, less well known solar technologies. The subsidies for solar hot water should be particularly effective, because the cost of an average system for a one- or two-family building ($1000 to $2000) is low enough that even a small price offset will stimulate consumer purchase.*

Wind energy conversion systems are potentially eligible for incentives in 11 states. Bioconversion is generally included by the northern and western states (including Alaska, Arizona, Hawaii, Idaho, Montana, North Dakota, Oregon, and Wisconsin), where appropriate agricultural applications and organic materials or wastes are readily available. Idaho provides deductions for the costs of heat

*A figure of $1700 for 84 ft^2 of double-glazed collector area with liquid storage and gas or electric backup (1977 dollars) is estimated in Flaim et al. [60].

pumps and burning of wood or wood products, as well as fireplaces with built-in metal heat exchangers. Hydropower, as an indirect form of solar energy, is allowed for by Alaska (tidal), Montana (low-head hydroelectricity), and Oregon (sea-thermal gradient) (see income tax table).

Duration of Incentives

About half of the states with income tax incentives limit the duration of availability of the incentive by including termination dates. These dates generally apply to the existence of the incentive and not to installation or application timing. On the average, these states have chosen a termination date of 1983. California's program has the shortest statutory existence and expires on January 1, 1981 [61]. Oklahoma has the longest program, lasting until December 31, 1987 [62]. Although termination dates can be criticized because they require further legislation to continue the incentive, they do serve notice to potential solar purchasers to act quickly or risk losing the incentive. They may thus stimulate consumer demand faster than an open-ended program. Another function of built-in expiration dates, analogous to sunset laws, is to force a review of the economic rationale for the existence of the subsidy. Proponents of continuation must then justify the program in order to continue the status quo.

States offering open schedules might consider statutory provisions for monitoring and reporting on the utilization and efficacy of the incentives, including numbers and amounts of credits used, their distribution by income class, dwelling type, location of solar equipment, and administrative costs. Enforcement agencies in such states may informally assess the value of the incentive, but formal reporting requirements could form a rational basis for amendments or termination. Only California [63] and Arizona [64] statutorily provide for such monitoring and reporting, and these are states with definite termination dates.

Deductions versus Credits

Income tax incentives occur in two primary forms: deductions and credits. The primary distinction between the two is that the credit is applied against the total tax bill or net tax, while the deduction is generally applied against the gross taxable income.

Deductions are also used for depreciation, or the expensing of value lost through use of capital, and for amortization, the extinguishing of tax liability through periodic payments. When deductions other than accelerated depreciation or amortization are allowed to corporations, they are known as investment tax incentives. Five states have expressly stated that use of either investment credits or deductions is in lieu of use of normal depreciation by businesses. Kansas has expressly allowed rapid amortization in addition to the deduction it offers [65].

Credits are favored by 12 states; general (nondepreciation) deductions of various kinds by 5. There are several possible reasons for this preference for credits. First, the credits do not require itemization. Second, the amount of incentive available under a deduction will vary with factors such as tax rates, tax brackets, and other available income that are unrelated to the solar device expense. An applicant in a high income bracket could benefit more by a deduction than one who needs the incentive more if tax rates are progressive. In contrast, credits are considered more egalitarian. Third, credits are more commonly carried over if they exceed the tax liability of the applicant, which is a real possibility for low-income taxpayers.

Whenever the credit exceeds tax liability, potential for a rebate is created. A rebate is a powerful and immediate reinforcement for the solar consumer. Rebates could eliminate the need for a carry-over and are most effective in quickly reducing front-end costs, the greatest obstacle to solar purchasers. Only New Mexico [66] and Wisconsin [67] currently offer refunds when the tax credit exceeds the tax liability. Rebates accompanying tax credits are also favored by the Solar Energy Industries Association [68].

Measurement Formulas

This section compares the 12 different approaches used by the states having income tax credits and deductions, as well as several depreciation/ amortization schemes. These basic formulas are set out and discussed below. The formula numbers referred to are described in the key to Table 11.2.

Formula 1

Alaska offers a residential fuel conservation credit for 10 percent of system expenses, with a maximum of $200. Expenses include "installation of alternate sources of power generation not dependent upon fossil fuel." Presumably this includes costs such as labor, remodeling, and financing. The bill became effective after December 31, 1976. This latter type of provision is found elsewhere in only three other states–Colorado, Hawaii, and Wisconsin. Nonretroactive provisions are criticized by some authors [20, p. 80]. Alaska also offers a 5 percent residential fuel credit for the cost of wood, among other conventional fuels [69].

Formula 2

Arizona offers nondependent taxpayers one credit per residence equal to 35 percent of solar system cost, declining at 5 percent per year until December 31, 1984, with a total maximum of $1000. A 5-year carry-over is provided [70]. In addition, a credit of 25 percent ($100 maximum) is provided for residential insulation, wind-driven turbine ventilators, and passive roof vents [71]. The distinctive features of this formula are twofold. First, it is the only formula to distinguish among solar technologies by providing different benefits based on different costs for active and passive heating and cooling devices. Second, the incentive is spread over time, with two results. Solar consumers purchasing relatively inexpensive solar devices (e.g., water heaters) receive a potentially greater benefit by purchasing early enough (at least 3 years before termination) for the formula to yield the maximum benefit. Also, a formula that only reduces tax liability a small amount each year but extends the benefit over time will favor lower-income groups, since without a carryover provision the credit might exceed their tax liability. But it will achieve that result at the price of failing to help as much with front-end costs.

Arizona also allows a 36-month period to amortize the adjusted basis of the costs of acquisition. This period was originally set at 60 months [72]. The election of amortization is in lieu of both depreciation and the credit offered [72].

Formula 3

Arkansas, Colorado, and Massachusetts allow deductions from gross income of the entire cost (including installation, but excluding interest and finance charges) of the eligible solar device. Arkansas restricts the deduction for "energy saving equipment" to the year of purchase, and allows extremely broad administrative discretion in agency rule-making [73]. Lack of definitions (or examples of qualifying solar heating and cooling equipment) is typical of many of the statutes considered here.

Colorado offers a deduction for the cost of "alternative energy devices" (including installation, construction, remodeling, and acquisition) that are in service in an occupied building. Heat pumps and fluid reservoirs are included [74]. The income tax deduction is uniquely defined as "standard" for resident individuals [75], for most such deductions are "itemized." Corporations are not entitled to depreciation if the deduction is taken [76].

The Massachusetts deductions for solar- or wind-powered "climatic control" or "water heating" units apply to corporations only and are in lieu of depreciation [77]. Provision is made for taxation of capital gain or loss on sale of the solar-equipped building, whereby the deduction is disregarded and straight-line depreciation is assumed. Labor costs are included, and performance certification is required.

Formula 4

California offers a generous but short-lived income tax package [81]. The state allows a credit of 55 percent (maximum $3000) of the cost, including labor, installation, and energy conservation measures such as insulation, but excluding interest charges. Conservation measures are eligible if they reduce backup or total costs, and can be installed up to 2 years after the solar system. In a multiple-family dwelling where the cost exceeds $12,000 the credit is 25 percent. An open carryover is provided until the credit expires on January 1, 1981. In 1976 California began offering a credit of 10 percent, with a maximum of $1000, but amended the law in 1977 and 1978 to allow the greater credits [79]. Credits for the cost of purchasing solar easements are included since the passage of the Solar Rights Act of 1978 [80].

Formula 5

Hawaii simply allows a credit of 10 percent of the total cost of a "solar energy device," with an open carryover period. The device must have been installed after December 31, 1974 in order to qualify [81].

Formula 6

Idaho allows a deduction of the entire cost of the solar device over a 4-year period [82]. Forty percent of the cost may be deducted in the year of purchase, and 20 percent per year in each of the following 3 years. There is a maximum limit of $5000 for any one year's deduction. The high first-year deduction is well suited to lowering front-end costs, and the extended time period assists low-income purchasers. The deduction is limited to residential uses because state law already allows favorable depreciation treatment for income-producing property.

Formula 7

Kansas, New Mexico, North Carolina, and Oregon offer variations of a formula based on a 25 percent credit with a maximum of $1000. Kansas increases the maximum to $3000 for commercial applications [83]. The 4-year carryover for principal dwellings is not available for the commercial credits. Owners of income-producing property are expressly encouraged to amortize over 60 months in addition to use of the credits.

New Mexico [84] limits the applicability of its 25 percent ($1000 maximum) credit to costs of equipment for solar heating or cooling of a residence, or solar heating of a swimming pool. Alternatively, the state offers a credit of $25,000 maximum, available over a 3-year period, for a solar-powered irrigation pump system of approved design, constructed on the taxpayer's or a partnership's property. The excess of credits over tax liability may be rebated in all cases.

North Carolina [85] does not distinguish between commercial and residential applications in its formula. It allows installation costs and a 3-year carryover for any single building or separately metered residence where the owner or controller pays the fuel bills.

In contrast to North Carolina, Oregon [86] limits its credit to one per taxpayer, rather than one per building. To certify a system under this statute, the system must provide 10 percent of the total energy requirements of the house. In addition, Oregon, as well as New Mexico and North Carolina, requires certification under the federal performance criteria [87]. Oregon's 10 percent contribution requirement will require excellent and detailed monitoring of solar energy installations. Oregon is thus likely to generate comprehensive data on solar installations. With a larger volume of installations, however, there are likely to be administrative problems in reviewing such a large mass of material.

Formulas 8 and 9

In 1977 Montana simultaneously passed two bills [88] that created both income tax credits and deductions while terminating that state's solar property tax incentive. The legislative purpose was to encourage use of "alternative energy sources" through incentives requiring individual "initiative" [89].

Formula 8, Montana's credit system, allows a residential tax credit of 10 percent of the first $1000 and 5 percent of the next $3000 of the cost (including installation) of any "recognized non-fossil form of energy generation" which is "non-nuclear" [90]. The maximum available credit is $250, with a 4-year carryover. However, federal credits "similar in kind" have been passed, reducing the allowable credit to 5 percent on $1000 and $2\frac{1}{2}$ percent on the next $3000, whether or not the federal credits are claimed.

Formula 9, the Montana deduction statute, provides deductions on application for residential or commercial "energy-related investment." The schedule for residential capital expenses is 100 percent of the first $1000, 50 percent of the next $1000, 20 percent of the next $1000, and 10 percent of the next $1000 expended, with a maximum of $1800. For nonresidential buildings the business investment tax schedule becomes 100 percent of the first $2000, 50 percent of the next $2000, 20 percent of the next $2000, and 10 percent of the next $2000, with a maximum of $3600. Deductible expenditures do not include offsetting grants or federal credits [91]. Presumably subsidies may not be claimed under both formulas 8 and 9 by a residential owner.

Formula 10

North Dakota [92] allows a credit of 5 percent per year for 2 years of the cost of a "solar or wind energy device" that converts the "natural energy of the sun or wind." Eligible costs of acquisition and installation do not include alterations to the building structure. There is neither a maximum credit nor a carryover provision. The relatively simple statute is the shortest such measure on record, creating no additional administrative machinery. The North Dakota credit, however, is one of the smallest available. By comparison, Hawaii offers 10 percent in 1 year, Alaska's credit is a maximum of $200, and Montana's credit is a maximum of $250, in addition to deductions. A study for the Federal Energy Administration by the National Conference of State Legislatures concluded that "regardless of system costs, incentives are not effective unless they create perceived discounts in excess of 25%" [93].

Formula 11

Oklahoma [94] offers a direct credit of 25 percent of the cost (including installation) of a solar energy system. This is a variation of formula 7; it is distinguished by a $2000 maximum instead of $1000. The bill as introduced allowed only an 8 percent credit [95]. A 5-year carryover for the credit is accompanied by a lenient definition of costs, including remodeling. Certification is required by the Department of Revenue, and field inspections by the Corporations Commission are authorized.

Federal Action

The pertinent feature of the National Energy Act (NEA) is the residential energy credit [96]. Its most important feature is the credit it allows for a "renewable energy source," which by definition includes solar, wind, and geothermal energy equipment used in a taxpayer's principal residence. This credit refunds 30 percent of the first $2000 and 20 percent of the next $8000 of system cost. A credit is also allowed for energy conservation expenditures such as insulation that were incurred from April 20, 1977 to January 1, 1986. This credit allows 15 percent of the first $2000 expended.

California, New Mexico, and Montana are the only states that explicitly anticipated the availability of federal income tax credits under the 1978 National Energy Act. Nine other states passed income tax incentive legislation in 1977, just prior to the NEA. They either ignored the possibility of federal action, decided state incentives were independently needed, or preferred to respond retroactively, if at all.

California provides that if federal credits are forthcoming for eligible systems, the state's share will be reduced so that the maximum credit will not exceed 55 percent (including carryovers) [97]. New Mexico denies the credit for residential buildings, pools, or irrigation pumps if any form of federal benefit is claimed on a federal tax return [98]. Montana simply halves its credit to 5 percent on the first $1000, and $2\frac{1}{2}$ percent on the next $3000 if a federal credit similar in kind (though not necessarily in amount) is made available, whether or not the federal credit is claimed [99]. Montana's deduction is unaffected, since only grants (federal, state, or private) are excluded from the costs eligible for the deduction.

Excise Tax Incentives: Sales, Use, Transaction, and Franchise Taxes

Excise taxes are generally synonymous with privilege taxes*, and include sales, use, franchise, and transaction taxes. The sales tax is levied by state and local governments on a purchase or on gross receipts from the sale of goods sold in the state. The use and consumption tax applies to goods being transported or stored in the state that are not amenable to taxation through a sale in the chain of distribution. The franchise tax is defined as a tax on the privilege of corporate existence in a state. The transaction tax is closely related; it applies to the privilege of conducting business in the state [101].

Sales taxes are of particular importance because they are not often subject to the uniformity clauses present in most state constitutions [102], thus avoiding the constitutional obstacles confronting property tax exemptions. Although large-scale solar technologies may be included in existing sales tax exemptions for building materials, utility energy supplies, industrial plants, energy use in manufacturing, or retail sales of energy (commonly natural gas and electricity), it is usually held that exemptions must be conferred in plain terms [103]. The extent to which such exemptions apply is therefore not clear. Further, sales taxes do not generally apply to services [104], so labor costs for transportation, installation, and remodeling are automatically untaxed. In contrast, property and income taxes generally include such costs, so explicit exemptions, deductions, or credits are required.

The amount of any excise tax exemption varies considerably with the tax structures of the individual states. Such incentives are generally less substantial than the property, income, and loan incentives. Their economic impact becomes more significant when they are viewed in combination with the other incentives, especially when it is realized that excise tax exemptions, like income tax credits, generally reduce front-end costs, while most property tax incentives generally reduce constant costs.

*Except insofar as the license (franchise) taxes are enacted pursuant to the police power in addition to the taxing power [100].

Only eight states (Arizona, Connecticut, Georgia, Maine, Massachusetts, Michigan, New Jersey, and Texas) have passed forms of excise tax exemptions. Except for Arizona and Michigan, the other six states have also provided for some type of sales tax exemption. Only Arizona and Michigan offer transaction or business activities tax exemptions. Texas is the only state to exempt corporations from franchise privilege taxes. Use, storage, and consumption tax exemptions are provided for in five of the eight states. The salient characteristics of the excise tax exemptions are listed in Table 11.3. Unique aspects and pertinent details of the laws of each state are discussed below.

Arizona's transaction privilege tax exemption [105] allows corporations to avoid this tax. It applies to a wide variety of solar technologies and applications, including heat storage and irrigation. The use tax exemption [106] on tangible personal property repeats this same broad definition of "solar energy devices."

Connecticut offers a combination of sales and use tax exemptions [107], which apply only to solar "collectors" [108]. The state sales tax is currently 7 percent and the exemption has open duration.

Georgia passed one of the earliest sales tax exemptions in 1976. Its measure originally favored solar "purchasers," but was amended to apply to the "owner" of the property to which the solar equipment is attached. This provision ensures that Georgia will not subsidize systems that could be installed out of state, or that dealers in solar energy devices would be exempt from sales taxes on every device purchased for resale. To qualify, the owner must have the system certified and apply for a rebate [109].

Maine [110] passed a sales and use tax refund with its property tax exemption. Until 1983, the Office of Energy Resources will certify "solar energy equipment" on application. The applicant must include information on type of equipment, manufacturer, cost, seller, and use as a prerequisite to this exemption.

Massachusetts restricts its sales exemption to retail sales of solar equipment for use in an individual's principal residence [111]. The system may be either primary or auxiliary.

Michigan offers a sales/use tax combination [112] similar to Maine's and Connecticut's. The package has the longest statutorily determined duration (until 1985) and was one of the earliest such measures, passed in 1976. Eligible passive designs must not be similar to those found on unmodified (nonsolar) homes. (Massachusetts has a similar qualification.) The measure excludes equipment used on buildings that are owned by corporations engaged in the resale of solar devices.

New Jersey provides a sales and use tax exemption [113] for solar devices or systems that meet standards set by the New Jersey's Department of Energy.

Texas is the only state offering only a franchise tax exemption [114]. It applies to corporations engaged "exclusively" in the manufacture, sale, or installation of "solar energy devices," which are broadly defined. The "exclusive" condition would seem to deny the benefit to many large corporations.

Table 11.3 Excise Taxes

	Ariz.	Conn.	Ga.
Law-chapter	42, 112	457	1030, 1309
Year	77, 78	77	76, 78
Codification (primary)	ARS § §42-1312.01(A)(9), 42-1409(B)(9)	CGSA 12-412(dd)	GACA §92-3403a(c)(2)(2.1)
Eligible system			
SHACOB	•	•	•
Passive	•		
Hot water	•	•	•
Wind	•		
PV/TE	•		
Hydro	•		
Pump/heat			•
Storage	•		•
Type excise			
Sales		•	•
Transaction	•		
Franchise			
Use	•	•	
Eligible			
Owner			•
Purchaser	•	•	
Seller	•		
Corporation			
Refund			•
Duration			
Effective	1977	1977	1978
Expires	12/31/84	Open	Open
Eligible use			
Residential	•	•	•
(Principal)	•		
Agricultural	(Irrigation)	•	•
Industrial	•	•	•
Commercial	•	•	•
Apply			•
Certification			•

Me.	Mass.	Mich.	N.J.	Texas
542	989	132, 133	465	584
77	77	76	77	77
MRS § 36-1760-37	ALMc64H § b(dd)	MSA § § 7.525 (8), 7.555(4e)	N.J.S.A. C.54:32-B(ff)	Tax-Gen. titl. 122A, § 12.03.1(r)
•	•	•	•	•
•	•	•	•	•
•	•	•	•	•
	•	•	•	•
		•	•	•
		•		•
•	•			
	•			•
•	•	•	•	•
•	(Retail)			•
		•	•	
•	•	•	•	
		•		•
•				
1977 1/1/83	1/30/78 Open	5/27/76 1/1/85	7/1/78 Open	8/29/77 Open
•		•	•	•
	•			
•			•	•
•			•	•
•		•[a]	•	•
•				
•			•	

[a]Excludes buildings owned by businesses in the solar resale business.

LOAN AND GRANT PROGRAMS

Introduction

Loan incentives in the form of subsidized low-interest loans and guaranteed or insured loans have a great potential for reducing both front-end and life-cycle costs of solar systems.* Low-interest loans administered by private or government agencies and financed by tax funds or tax-exempt revenue bonds can substantially cut long-term financing costs associated with the more expensive solar technologies. Guaranteed or insured loans can increase capital availability, allowing solar consumers to finance a larger share of the purchase and thereby reducing the critical initial cost. Low-interest loans can also more equitably benefit low-income groups, which are often charged higher interest rates. This benefit may be partially offset because when similar interest payments are deducted from taxes, the benefit is relatively greater for persons in the higher tax brackets.

State Programs

Eight states currently offer 10 different loan or grant programs. Except in Montana, all the programs were enacted in 1977 or 1978. Important features of each program are listed in Table 11.4; additional details and unique aspects are discussed below.

Alaska has established an alternative power resource revolving loan fund in its Department of Commerce and Economic Development [115]. The loans are targeted for development of means of energy production, including "windmills, water, and solar power devices." Thirty percent of purchase, construction, and installation costs may be loaned. The department is required to develop eligibility guidelines.

California's loan program [116] is a demonstration program benefiting victims of disasters that result in a state of emergency being declared on or after July 1, 1977. The 1978 legislation [117] was a response to the Santa Barbara fire in the summer of 1977. A $200,000 loan fund is administered by the Department of Housing and Community Development for solar improvements to damaged property on the original site. The program stresses water heaters, and seeks to generate data for the State Energy Resources Conservation and Planning Department on the energy-saving potential of solar equipment. Any tax credits available are assigned to reducing the loan balance. California had earlier attempted to implement a more comprehensive program financing residential installations of insulation and solar systems, to be administered by the State Energy Resources Conservation Development Commission and the Housing Finance Authority [118]. However, a $25 million bond proposal to finance the program was defeated in the November 1976 election.

California also has a purchase program for military veterans [119]. Under the 1978 act, the market value limit of $43,000 on buildings purchased by the

*Income tax provisions for depreciation or amortization are functionally similar to loans in that the taxpayer defers the tax and keeps the money longer than otherwise possible.

Department of Veterans' Affairs for resale to veterans is raised to $48,000 when "solar energy heating devices or equipment" are present as improvements in the building.

Iowa has authorized its housing finance authority to make "property improvement loans" for the improvement or rehabilitation of (1) housing deemed substandard in its "structural, plumbing, heating, cooling, or electrical systems, or (2) any housing to increase its "energy efficiency" or to "finance solar or other renewable energy systems for use in that housing" [120]. The state may condition the loan on bringing the building into compliance with the thermal efficiency standards in the state building code.

Massachusetts' loan provision [121] increases the maximum amounts and repayment period for mortgage loans financed by private institutions other than banks on the condition that at least $2000 of the loan is for the "purposes of financing the purchasing and installation of a solar or wind-powered system or heat pump system." The maximum repayment period for loans is extended to 10 years. Maximum amounts for qualifying solar loans are increased to $7000 for cooperative banks and trusts, and to $9500 for credit unions. In 1978 Massachusetts authorized bank loans of up to $15,000 for purchase and installation of "solar or wind-powered system or heat pump systems" [122].

Minnesota amended the specific powers of its Housing Finance Agency to allow grants for repayment of loans, or to provide loans for rehabilitation of residential buildings occupied by persons of low or moderate income, if one purpose of the loan is to "accomplish energy conservation related improvements" [123]. This phrase is not statutorily defined, but a good argument can be made that it is meant to include solar systems. A related bond measure raises $175,000 for rehabilitation loans in general [124].

Since 1975, Montana has allowed electric or natural gas utilities to extend credit to customers to pay for "installation of energy conservation materials in a dwelling" [125]. The customer can repay the debt through monthly bills, with a maximum interest charge of 7 percent on the declining balance. The utility is reimbursed for the difference between 7 percent and the prevailing interest rates on home improvement loans through a credit against its license tax liability. An "energy conservation purpose" is defined as "reducing waste or dissipation of energy, or reducing the amount of energy required to accomplish a given quantity of work" [126]. While this certainly applies to weatherization measures such as insulation and storm windows, conceivably energy-efficient solar technologies or passive methods could fit the latter part of the definition.

Oregon has enabled a recipient of a 30-year, $42,500 war veterans' loan to receive a subsequent loan up to a maximum of $3000 for the "purpose of installing an alternative energy device for a home" [127]. The device must be certified for compliance with standards authorized under the act that require the system to produce at least 10 percent of the total energy consumption of the home. The Oregon legislature also passed a constitutional amendment [128] that would have put into effect a law [129] changing the limits on offering general obligation bonds for state power development. However, the bond issue did not

Table 11.4 Loan and Grant Programs

	Alaska	Calif.	Calif.	Iowa
Year	78	78	78	78
Law-chapter	29	1	1, 1243	1086
Codification (primary)	§§45.88. 010-.040	Health & Safety §§41260-65, 50680-85	Mil. & Vet. §§987.64-92	I.C. §§220.1-.12
Loan		(DEMO) o		
Maximum	$10,000	$2,000		
Interest	8% maximum	0		
Fee		2%		
Repayment	20 yr	30 mo		
Grant			•	
Source and type	Dept. Comm. Mtg.			
Tax fund	•		•	•
Bonds				•
Private		•		•
Insure				
Eligible Recipient	Open	Disaster victims (7/1/77 on)	Veterans	Low/medium income, elderly, disabled
Eligible systems				
SHACOB	•	•	•	•
Hot water	•	•	•	•
Wind	•			
PV	•	•	•	•
Bio/hydro	•			
Heat pump				•
Passive/conservation		•	•	•
Building[a]		Res.	Res., Ag.	Res. Rehab.
Duration of program		1980	Open	Open
Minimum energy requirement				
Standards	•			• State

[a]Res., residential; Rehab., rehabilitation; Ag., agricultural; Ind., industrial; Fed., federal; Cert., certification; Mtg., mortgage.

Mass.	Minn.	Mont.	Ore.	Ore.	Tenn.
77	77	75	77	77	78
28, 73	401	548	315	732	884
ALM c168 §35(10)	MSA 462A .05(14-15)	MRCA 84 7405	ORS §407(2)	Const. Amend. Uncodified	TCS §§13-2303, 2316
•		•		•	•
$15,000 (banks)			$3000		
$2000 add. (other)	Market		(add.)		
Prevailing	value varies		Varies		
Maximum 10 yr			30 yr		Varies
	•				
		•	•		
	•			•	
•		Utility			
					•
Anyone	Low/ medium income; owners		Loanee, war veteran fund	Utility and/or designee	Low or moderate income
•			•	•	•
•			•	•	•
•			•	•	
•			•		
				•	
•		•	•		•
	•	•		•	•
	Res., Rehab.	Res.	Res.	Res., Ag.	Ind. Res./ Res. Rehab.
Open	Open		Open	30 yr	Open
			10%		•
			Cert.		• Fed.

receive voter approval. The bonds would have been available to individuals or electric utilities, alone or in combination, for undertaking "alternative energy projects." Eligible projects would have included "conversion or development of an energy resource into a usable non-electric form of energy" [130].

Tennessee has empowered its Housing Development Agency to facilitate or directly make loans for energy-saving improvements and solar water heaters to low- and moderate-income persons and families in residential housing [131]. For insured residential rehabilitation construction loans, the costs are to be included in the loan only "if feasible" [132]. Repayment schedules for loans to low- or moderate-income recipients are intended to conform to savings reflected in their utility bills. Energy-conserving designs are to be given "maximum consideration." Standards identifying reliable and efficient solar water heaters are to be provided by the appropriate federal agency.

National Energy Act

The National Energy Act has several provisions designed to stimulate private lending and lessen the need for additional state loan programs [133]. In addition to $800 grants to low-income families and farmers for weatherization, loan insurance for "energy conserving improvements" (including active or passive solar methods) will be available under Title I of the National Housing Act. Limits on federal mortgage insurance are increased 20 percent where a building has solar equipment installed.

Another section of the NEA authorizes the Governmental National Mortgage Association (GNMA), under the direction of the Secretary of Energy, to purchase loans that are made to low- and moderate-income families for the installation of energy-conserving improvements in one- to four-family dwellings owned by such families. The loan cannot exceed $2500 and must have a repayment term of 5 to 15 years. The total amount of outstanding loans the secretary is authorized to purchase will be determined by appropriation acts, but shall not exceed $3 billion. The GNMA is further authorized to purchase similar loans when the secretary finds that insufficient credit for such is available on a national basis. For these unsubsidized loans the amount appropriated cannot exceed $2 billion [134].

REFERENCES

1. U.S. Const. amend. X.
2. *Solar Heating and Cooling Demonstration Act*, Pub. L. 93-409, Sept. 3, 1974, 42 U.S.C. §§5501–5517.
3. *Solar Energy Research Development and Demonstration Act*, Pub. L. 93-473, Oct. 26, 1974, 42 U.S.C. §§5551–5566.
4. *Federal Non-Nuclear Energy Research and Development Act*, Pub. L. 93-577, Dec. 1974, 42 U.S.C. §§5901–5917.
5. Department of Commerce, *Intermediate Minimum Property Standards for Solar Heating and Domestic Hot Water Systems,* Washington, D.C., March 1977; NBSIR 77-1226 (unpublished). Authorized under 42 U.S.C. §3535(d), 12 U.S.C. §17516, 5 U.S.C. §552(a); 24 C.F.R. §200.929(6)(4), (incorporated by reference); amended in 42 Fed. Reg. 33898 (July 1, 1977).

6. National Bureau of Standards, *Intermediate Minimum Property Standards Supplement: Solar Heating and Domestic Hot Water Systems,* vol. 5 (supersedes NBSIR 77-1226), Government Printing Office, SD Catalog No. 4930.2, Washington, D.C., 1977.
7. National Bureau of Standards, *Intermediate Standards for Solar Domestic Hot Water Systems/HUD Initiative,* NBSIR 77-1272, National Technical Information Service, PB 271 758, Springfield, Va., 1977.
8. *Solar Energy Intelligence Report,* vol. 4, no. 24, June 12, 1978, p. 175.
9. *Energy Tax Act of 1978,* Pub. L. 95-618, 92 Stat. 3173, in *Congressional Index,* no. 97, Nov. 16, 1978.
10. J. Ashworth, B. Green, et al., *The Implementation of State Solar Incentives: A Preliminary Assessment,* Solar Energy Research Institute, Golden, Colo., 1979.
11. B. W. Cone, D. L. Brenchley, V. L. Brix, et al., *An Analysis of Federal Incentives Used to Stimulate Energy Production,* Battelle Pacific Northwest Laboratories, Richland, Wash., March 1978, PNL-2410, p. 276; National Technical Information Service, Springfield, Va.
12. S.J.R. 53, 1977 Texas Gen. Laws, amending Tex. Const. art. VIII, §2(a); Ch. 354, 1977, Fla. Gen. Laws, amending Fla. Const. art. 7, §4(b)(3), codified at Fla. Stat. §193.622 (Supp. 1978).
13. Tex. Const. art. VIII, §2(a).
14. S.R. 284, Res. 167, Ch. 1465, 1976 Ga. Laws. Passage amended Ga. Const. art. VII, §1, Par. IV.
15. California Proposition 3, June 6, 1978; Senate Const. Amend. 15, 1977; Ch. 103, 1977 Cal. Stats.
16. Ch. 323, 1978 Neb. Laws.
17. Ch. 591, 1978 La. Acts; to be codified at La. Rev. Stat. Ann. §47-1706.
18. Ch. 548, 1975 Mont. Laws, repealed by Ch. 576, 1977 Mont. laws; to be codified at Mont. Rev. Codes Ann. §84-7403.
19. Ch. 322, §1, 1977 N.Y. Laws; N.Y. Real Prop. Tax Law §487 (Supp. 1977).
20. W. A. Thomas, A. Miller, and R. A. Robbins, *Overcoming Legal Uncertainties about Use of Solar Energy Systems,* American Bar Foundation, Chicago, 1978, pp. 75–76.
21. N.H. Rev. Stat. Ann. §§72:61-68, I (Supp. 1977).
22. Ch. 965, 1977 N.C. Sess. Laws; N.C. Gen. Stat. §§105–277(g) (Cum. Supp. 1972–1977).
23. Cal. Rev. & Tax Code §17052.5 (West Cum. Supp. 1977–1978).
24. Ch. 256, §1, 1977 Laws of New Jersey; N.J. Rev. Stat. §§54:4–3:113 (West Supp. 1977).
25. Solar Energy Research Institute, Program Evaluation, *Annual Review of Solar Energy,* SERI/TR-54-066, Solar Energy Research Institute, Golden, Colo., 1978, pp. 81–82.
26. D. L. Buchanan, *A Review of the Economics of Selected Passive and Hybrid Systems,* SERI/TR-61-144, Solar Energy Research Institute, Golden, Colo., 1979.
27. 42 U.S.C. §§5502.1–.2, 5506 (1977).
28. Ch. 345, §1, 1977 Nev. Stats.; Nev. Rev. Stat. §361.795(1) (1977).
29. Ch. 591, 1978 La. Sess. Laws; to be codified at §47-1706.
30. Ch. 135, §4, 1976 Mich. Pub. Acts; Mich. Stats. Ann. §7.7(4e)(11) (Callaghan 1978).
31. Ch. 68, 1977 Ind. Acts; amending Ch. 15, 1974 Ind. Acts; Ind. Code §6-1.1-12-27 (1977).
32. Ch. 542 1977, Me. Acts, Me. Rev. Stat. Ann. tit. 36 §656(1)/H(West Cum. Supp. 1964–1978).
33. Ch. 74, §8, 1978 S.D. Sess. Laws; S.D. Comp. Laws Ann. §10-6-35.8.
34. Ch. 111, 1975 S.D. Sess. Laws.
35. Ch. 196, 1977 Or. Laws; Or. Rev. Stat. §307.175 (1977).
36. Ch. 409, §2, 1976 Conn. Pub. Acts, as amended by Ch. 490, 1977 Conn. Pub. Acts; Conn. Gen. Stat. Ann. §§12-81(56-57) (West Cum. Supp. 1972–1977).
37. N.H. Rev. Stat. Ann. §72:61-68, I (Supp. 1977).
38. Ch. 388, 1978 Sess. Laws, Mass. Ann. Laws, Ch. 9, §5.
39. Ch. 740, 1976 Md. Laws; Md. Ann. Code (1975) art. 81, §12-F(5) (Michie Cum. Supp., 1975–1977).

40. Ch. 74, §6, 1978 S.D. Sess. Laws; S.D. Comp. Laws Ann. §10-6-35.8 to 35.18 (Supp. 1978).
41. Ch. 345, 1977 Nev. Stats., to be codified at §199.120(3).
42. 51 Am. Jur. 1st Tax §26.
43. 31 Am. Jur. 2d Exemptions § § 1, 2 at 329.
44. 1977 N.Y. Laws; N.Y. Real Prop. Tax Law §487 (Supp. 1978).
45. Ch. 202, 1977 R.I. Pub. Laws; R.I. Gen. Laws §44-3-19 (Cum. Supp. 1970–1977).
46. Ch. 344, 1975 Colo. Sess. Laws; Colo. Rev. Stat. §39-1-104 (Cum. Supp. 1973–1976).
47. Ch. 345, §1, 1977 Kansas Sess. Laws; Kan. Stat. Ann. §79(a)(01-02) (1977).
48. Ch. 74, §6, 1978 S.D. Sess. Laws; S.D. Comp. Laws Ann. §10-6-35.8 to 35.18 (Supp. 1978).
49. Ch. 487, 1976 Mass. Laws; Mass. Ann. Laws Ch. 63 §38H (Michie/Law Co-op 1978).
50. Booz, Allen, and Hamilton, "Solar Energy Utilization in Florida," cited in *Solar Energy Commercialization at the State Level: The Florida Solar Energy Water Heater Program,* FEA/G-77/270, Florida Solar Energy Center, March 1977, p. 24; PB 270158, National Technical Information Center, Springfield, Va.
51. Ark. Stat. Ann. §84-2016.8 (Cum. Supp. 1960–1977).
52. Cal. Rev. & Tax Code §17052.5(4) (West Cum. Supp. 1977–1978).
53. N.M. Stat. Ann. §72-15A-11.3–11.4 (Cum. Supp. 1953–1977); Cal. Rev. & Tax Code §234 (West. Supp. 1970–1977).
54. N.C. Gen. Stat. § §105–151.2 (Cum. Supp. 1962–1977).
55. Or. Rev. Stat. §316 (1977).
56. Alaska Stat. §43.20.039(c).
57. N.M. Stat. Ann. §72-15A-11.3 (Supp. 1975).
58. Ch. 1082, 1977 Cal. Stats., amending Ch. 168, 1976 Cal. Stats.
59. Ch. 1159, 1978 Cal. Stats.; Cal. Rev. & Tax Code §17052.5 (West Cum. Supp. 1977–1978).
60. S. J. Flaim et al., "Economic Feasibility and Market Readiness of Eight Solar Technologies," Interim Draft Rept., June 1978, SERI-34, p. 28.
61. Ch. 1082, §4, 1977 Cal. Stats.
62. Ch. 209, §3, 1977 Okla. Sess. Laws; Okla. Stat. Ann. tit. 68 §2357.3.
63. Ch. 1082, §5, 1977 Cal. Stats.
64. Ch. 81, §6, 1977 Ariz. Sess. Laws; Ariz. Rev. Stat. §42-123.01 (Supp. 1957–1978).
65. Kan. Stat. Ann. §79-32-168 (1977).
66. N.M. Stat. Ann. §72-15A-11.3(F) to 11.4(G) (Supp. 1975).
67. Wis. Stat. Ann. §71.09(12)(b) (West Cum. Supp. 1969–1977).
68. Testimony of Sheldon Butt, President, Solar Energy Industries Assoc.; *Joint Hearing before the Subcommittee on Energy Production and Supply, Subcommittee on Energy Research and Development of the Committee on Energy and National Resources, and the Select Committee on Small Business,* U.S. Senate, 95th Cong., 1st Sess., June 1, 1977.
69. Alaska Stat. Ann. §43.20.038(a)(b) (1977).
70. Ariz. Rev. Stat. §43-128.03(a)–(c) (Supp. 1957–1978).
71. Ariz. Rev. Stat. §43-128.04.
72. Ch. 93, 1975 Ariz. Sess. Laws, amended by Ch. 129, 1976 Ariz. Sess. Laws; Ariz. Rev. Stat. §43-123.37 (Supp. 1957–1978).
73. Ark. Stat. Ann. §84-2016.8-.10 (Supp. 1977).
74. Colo. Rev. Stat. §39-22-113(4)(c)(I)–(III) (Supp. 1977).
75. Colo. Rev. Stat. §39-22-112.
76. Colo. Rev. Stat. §39-22-304(3).
77. Mass. Ann. Laws ch. 63, §38H(b)(1) (Michie/Law Co-op 1978).
78. Cal. Rev. & Tax Code §17052.5 (West Cum. Supp. 1977–1978).
79. Ch. 168, 1976 Cal. Stats., amended by Ch. 1082, 1977 Cal. Stats., amended by Ch. 1159, 1978 Cal. Stats.; Cal. Rev. & Tax Code §17052.5.
80. Ch. 1154, 1978 Cal. Stats.; Cal. Civ. Code § §801, 801.5 (West Cum. Supp. 1977–1978).

81. Haw. Rev. Stat. §235-12 (1976).
82. Idaho Code §63-3022(c) (Supp. 1978).
83. Kan. Stat. Ann. § §79-32-166, -167 (1977).
84. N.M. Stat. Ann. § §72-15A-11.2, -11.4 (Cum. Supp. 1953–1977).
85. N.C. Gen. Stat. § §105-130.23, 151.2 (Cum. Supp. 1962–1977).
86. Or. Rev. Stat. § §469.160–.180, 316.116 (1977).
87. Promulgated pursuant to Pub. L. 93-409, *Solar Heating and Cooling Demonstration Act of 1974*, 42 U.S.C. §5506.
88. Ch. 576, 1977 Mont. Laws (deductions); Ch. 574, 1977 Mont. Laws (credits); Mont. Rev. Codes Ann. § §84-7403 (deductions), 84-7414 (credits) (Cum. Supp. 1947–1977).
89. Mont. Rev. Codes Ann. §84-7401 (Cum. Supp. 1947–1977).
90. Mont. Rev. Codes Ann. §84-7414.
91. Mont. Rev. Codes Ann. §84-7402.
92. N.D. Cent. Code §57-38-01.8 (Supp. 1977).
93. R. G. Jones, H. M. Sramek, and J. M. Pelster, *Analysis of State Solar Energy Policy Options,* National Conference of State Legislatures, Washington, D.C., June 1976, p. I-1; PB-254 730, National Technical Information Service, Springfield, Va.
94. Okla. Stat. Ann. tit. 68 §2357.1-.2 (West Cum. Supp. 1971–1977).
95. Ch. 209, 1977 Okla. Sess. Laws.
96. *Energy Tax Act of 1978*, Pub. L. 95-618, §101, 92 Stat. 3174 in *Cong. Index* No. 97, Nov. 16, 1978. Amending the Internal Revenue Code of 1954 by adding Code §44c.
97. Cal. Rev. & Tax Code § §17052.5(j), 23601(h) (West Cum. Supp. 1975–1977).
98. N.M. Stat. Ann. §72-15A-11.3(C), -11.4(D) (West Cum. Supp. 1953–1977).
99. Mont. Rev. Codes Ann. §84-7414(1) (Cum. Supp. 1947–1977).
100. 71 Am. Jur. 2d, State and Local Taxation § §28–30 at 360.
101. 71 Am. Jur. 2d, State and Local Taxation §163 at 484.
102. 68 Am. Jur. 2d, Sales and Use Taxes §129 at 177.
103. 68 Am. Jur. 2d, Sales and Use Taxes at 190, §135.
104. 71 Am. Jur. 2d, State and Local Taxation, §266 at 583.
105. Ariz. Rev. Stat. §42-1312.01(A)(9) (Cum. Supp. 1957–1978).
106. Ariz. Rev. Stat. §42-1409(B)(9) (Cum. Supp. 1957–1977).
107. Conn. Gen. Stat. Ann. §12-412(dd) (West Cum. Supp. 1974–1978).
108. Conn. Gen. Stat. Ann. §12-81(56) (West Cum. Supp. 1974–1978).
109. Ch. 1030, Ga. Laws of 1976, amended by Ch. 1309, §1, Ga. Laws of 1978; to be codified at §92-3403a(C)(2)(aa).
110. Me. Rev. Stat. tit. 36, §1760(37) (1977).
111. Mass. Ann. Laws ch. 64H, §6(dd) (Michie/Law Co-op 1978).
112. Mich. Stat. Ann. §7.525(8) (sales), §7.555(4e) (use).
113. N.J. Stat. Ann. §54:32B-8(ff) (West Supp. 1978–1979).
114. Tex. Tax Gen. Code Ann. tit. 122A, §12.03(1)(r) (Vernon Supp. 1978).
115. Ch. 1154, 1978 Alaska Sess. Laws; Alaska Stat. § §45.88.010-.040 (1978).
116. Ch. 1, 1978 Cal. Stats.; to be codified at Cal. Health & Safety Code § §41260–65, 50680–85.
117. Ch. 1, §8, 1978 Cal. Stats.
118. Ch. 264, 1976 Cal. Stats.; Cal. Pub. Res. Code §25410 (West Cum. Supp. 1963–1978).
119. Ch. 1243, 1978 Cal. Stats.; Cal. Mil. & Vet. Code § §987.64, 987.92 (West Cum. Supp. 1955–1977).
120. Iowa Code § §221.1–.12.
121. Mass. Ann. Laws Ch. 168, §168, §35(10) (banks); Ch. 170, §26(6) (cooperative banks); Ch. 171, §24(D) (credit unions); Ch. 172, §55 (trust companies) (Michie/Law Co-op Supp. 1977).
122. Ch. 73, 1978 Mass. Laws; amending Mass. Ann. Laws Ch. 168 §35(10) (Michie/Law Co-op Supp. 1978).

123. Minn. Stat. Ann. §462A.05(14-15) (West Cum. Supp. 1963–1978).
124. Minn. Stat. Ann. §462A.22 (West Cum. Supp. 1963–1978).
125. Mont. Rev. Codes Ann. §84-7405(1)–(2) (Cum. Supp. 1947–1977).
126. Mont. Rev. Codes Ann. §84-7402(3).
127. Or. Rev. Stat. §407.048 (1977).
128. S.J.R. 32, 1977 Or. Laws, amending Or. Const. art. XI-D.
129. Ch. 732, §6, 1977 Or. Laws (uncodified).
130. Ch. 732, §2, 1977 Or. Laws (uncodified).
131. Tenn. Code Ann. §§13-2303(19), -2315(29), new section (Cum. Supp. 1973–1977).
132. Tenn. Code Ann. §13-2316(a)–(d).
133. Pub. L. 95-315, 92 Stat. 377 (1978); Pub. L. 95-476, 92 Stat. 1497 (1978); Pub. L. 95-113, 91 Stat. 996 (1977); Pub. L. 95-619, 92 Stat. 3235 (1978).
134. *National Energy Conservation Policy Act of 1978,* Pub. L. 95-619, 92 Stat. 3206 (1978).

Appendix 1 Conversion Factors

Conversion Factors

Physical quantity	Symbol	Conversion factor
Area	A	1 ft^2 = 0.0929 m^2
		1 in^2 = 6.452 × 10^{-4} m^2
Density	ρ	1 lb_m/ft^3 = 16.018 kg/m^3
		1 $slug/ft^3$ = 515.379 kg/m^3
Heat, energy, or work	Q or W	1 Btu = 1055.1 J
		1 cal = 4.186 J
		1 ft-lb_f = 1.3558 J
		1 hp/h = 2.685 × 10^6 J
Force	F	1 lb_f = 4.448 N
Heat flow rate	q	1 Btu/h = 0.2931 W
		1 Btu/s = 1055.1 W
Heat flux	q/A	1 Btu/(h·ft^2) = 3.1525 W/m^2
Heat transfer coefficient	h	1 Btu/(h·ft^2·°F) = 5.678 W/(m^2·K)
Length	L	1 ft = 0.3048 m
		1 in = 2.54 cm
		1 mile = 1.6093 km
Mass	m	1 lb_m = 0.4536 kg
		1 slug = 14.594 kg
Mass flow rate	$\dot{m}$	1 lb_m/h = 0.000126 kg/s
		1 lb_m/s = 0.4536 kg/s
Power	$\dot{W}$	1 hp = 745.7 W
		1 ft/(lb_f·s) = 1.3558 W
		1 Btu/s = 1055.1 W
		1 Btu/h = 0.293 W

Conversion Factors (*continued*)

Physical quantity	Symbol	Conversion factor
Pressure	p	1 lb_f/in^2 = 6894.8 Pa (N/m^2)
		1 lb_f/ft^2 = 47.88 Pa (N/m^2)
		1 atm = 101.325 Pa (N/m^2)
Radiation	I	1 langley (Ly) = 41.860 J/m^2
Specific heat capacity	c	1 Btu/(lb_m·°F) = 4187 J/kg/K
Internal energy or enthalpy	e or h	1 Btu/lb_m = 2326.0 J/kg
		1 cal/g = 4184 J/kg
Temperature	T	$T(°R) = (\frac{9}{5})T(K)$
		$T(°F) = \frac{9}{5}\,[T(°C)] + 32$
		$T(°F) = \frac{9}{5}\,[T(K) - 273.15] + 32$
Thermal conductivity	k	1 Btu/(h·ft·°F) = 1.731 W/(m·K)
Thermal resistance	R_{th}	$[Btu/(h \cdot ft^2 \cdot °F)]^{-1}$ = 1.8958 W/(m^2·K)
Velocity	V	1 ft/s = 0.3048 m/s
		1 mile/h = 0.44703 m/s
Viscosity, dynamic	μ	1 lb_m/(ft·s) = 1.488 N/(s·m^2)
		1 cP = 0.00100 N/(s·m^2)
Viscosity, kinematic	ν	1 ft^2/s = 0.09029 m^2/s
		1 ft^2/h = 2.581×10^{-5} m^2/s
Volume	V	1 ft^3 = 0.02832 m^3
		1 in^3 = 1.6387×10^{-5} m^3
		1 gal (U.S. liq.) = 0.003785 m^3
Volumetric flow rate	$\dot{Q}$	1 ft^3/min = 0.000472 m^3/s

Appendix 2 Glossary

Absorber Component of a solar collector (generally metallic), whose function is to collect and retain as much of the radiation from the sun as possible. A heat transfer fluid flows through the absorber or the conduits attached to the absorber.

Absorptance Ratio of absorbed to incident solar radiation. Absorptivity is the property of absorbing radiation possessed by all materials to varying extents.

Absorption air conditioning Cooling by an absorption-desorption process without a large shaft work input; desorption is caused by heat input.

Ambient conditions Conditions of the surroundings.

Aperture Opening through which radiation passes before absorption in a solar collector.

ASHRAE Acronym for American Society of Heating, Refrigerating, and Air Conditioning Engineers. ASHRAE handbooks are sources of basic data on heating and air conditioning.

Auxiliary system System that acts as a backup to the solar system during extended periods of extremely cold and/or cloudy weather.

Azimuth Angle between the south-north line at a given location and the projection of the earth-sun line in the horizontal plane; see Chap. 2.

Btu British thermal unit; amount of heat required to raise the temperature of 1 pound of water (at 4°C) 1°F.

Assembled with assistance from Karen George and Albert Nunez, University of Colorado, Denver, Colorado.

C Celsius temperature scale; water freezes at 0°C and boils at 100°C at 1 atmosphere pressure.

Capital cost Cost of equipment, construction, land, and other items required to construct a facility; different from recurrent operating and maintenance costs.

Collector Device for gathering the sun's radiation and converting it to useful energy.

Collector efficiency Ratio of energy collected by a solar collector to radiant energy incident on the collector.

Collector tilt angle Angle at which the collector aperture is slanted up from the horizontal plane.

Comfort zone Range of temperatures, humidities, and airflow rates at which most persons are comfortable; varies with activity.

Conduction Heat transfer through matter by exchange of kinetic energy from particle to particle.

Constant dollars Dollars (or other currency) expressed, net of inflation, in terms of today's dollars.

Convection Heat transfer resulting from fluid motion in the presence of a temperature difference.

Cooling load Amount of heat and humidity that must be removed from a building to maintain occupant comfort.

COP Coefficient of performance; see Chap. 9 for uses of this term.

Cover plate Transparent material used to cover collector-absorber plate so that solar energy is "trapped" by the greenhouse effect (primarily a convection suppression effect).

CPC Compound parabolic concentrator.

Current dollars Dollars (or other currency) expressed in future terms that include inflation effects.

Degree-day A convenient measure of monthly or annual heating demands; see Chap. 7.

Depreciation Allocation of initial cost of a facility or system to the time period during which it is used.

Design conditions Indoor and outdoor temperatures, humidities, and wind speeds used to predict maximum expected heating and cooling loads for a building.

Diffuse radiation Scattered radiation from the sun that falls on a plane of stated orientation over a stated period; in the case of an inclined surface, includes radiation reflected from the ground.

Direct radiation Radiation from the sun that falls on a plane of stated orientation over a stated period, received from a narrow solid angle measured from the earth's surface.

Discounting Adjusting cash flows for the time value of money; see Chap. 5.

Discount rate Opportunity cost of making an investment; see Chap. 5.

Efficiency Ratio of the measure of a desired effect to the measure of the input causing the effect, both expressed in the same units of measure.

Emittance Ratio of radiation emitted by a real surface to radiation emitted by a perfect radiator at the same temperature. Normal emittance is the value measured at 90° to the plane of the sample; hemispheric emittance is the total amount emitted in all directions.

Evacuated-tube collector Collector manufactured from specially coated concentric glass tubes with an evacuated space between the outer two tubes.

Evaporator Heat exchanger in which a fluid undergoes a liquid-to-vapor phase change.

Fan coil Heat exchanger used to transfer heat from a liquid to air; used in space heating or cooling.

Flat-plate collector Thermal collector of diffuse and beam solar radiation, with five basic component parts; see Chap. 3.

Fossil fuels Fuels derived from the remains of carbonaceous flora and fauna, e.g., petroleum, natural gas, coal, oil shale, and tar sands. Fossil fuels had their origin millions of years ago when ancient plants trapped solar energy and were buried under layers of sediment. About 28 million tons of carbon form new fossil sediments every year, but current consumption of fossil fuels uses about 6 billion tons of carbon per year.

Glazing Glass, plastic, or other transparent covering of a collector-absorber surface.

Greenhouse effect Heat transfer effect in which heat loss from surfaces is controlled by suppressing the convection loss; frequently incorrectly attributed to suppression of radiation from an enclosure.

Gross area Total frontal area of a collector, including framing and structural supports.

Heat In thermodynamics, energy in transition due to a temperature difference. In common language, also denotes energy stored in a body by virtue of its temperature.

Heat exchanger Device used to transfer heat between two fluid streams without mixing them.

Heating load Amount of heat and humidity that must be added to a building to maintain occupant comfort.

Heat pump Device that transfers heat from a relatively low-temperature reservoir to one at a relatively higher temperature by input of shaft work.

Hydronic Trade term for liquid-based space-heating system.

Incidence angle Angle between the sun's rays and a line normal to the irradiated surface. In the case of tubular collectors, the angle between the normal to the tube centerline and the sun's rays.

Infiltration Uncontrolled air leakage into a building through cracks and pores.

Infrared radiation Wavelengths longer than those at the red end of the visible spectrum; thermal radiation at these wavelengths.

Insolation Radiation from the sun received by a surface.

Interest rate Cost of borrowing money; see discount rate.

ISES Acronym for International Solar Energy Society, a worldwide society of professionals, tinkerers, and enthusiasts interested in the utilization of solar energy.

Joule Energy unit equal to 1 newton-meter.

Langley Unit of solar radiation intensity equivalent to 1.0 gram-calorie per square centimeter.

Latitude Angular distance north (+) or south (–) of the equator, measured in degrees.

Laws of thermodynamics:

First law In nonrelativistic processes, energy is neither created nor destroyed; it is only transformed from one form to another.

Second law When free exchange of heat takes place between two bodies, heat is always transformed from the warmer to the cooler body.

Local solar time, LST System of astronomical time in which the sun always crosses the true north-south meridian at 12 noon. Differs from local time according to longitude, time zone, and equation of time.

Net area Area of a solar collector aperture through which radiation may pass; gross area net of all opaque area.

Normal Perpendicular, in the geometric sense.

Pascal Unit of pressure equal to 1 newton per square meter.

Passive systems Systems using the sun's energy without mechanical means; achieved by properly placed storage masses, overhangs, and fenestration, use of vegetation, proper building materials, and energy-conserving design.

Payback period Length of time required to recover an investment in a project from benefits accruing from the investment; payback period is an incomplete economic index; see Chap. 5.

Present value Value of a future cash flow discounted to the present; based on the premise that a dollar today is worth more than a dollar received in the future by virtue of the amount of interest (or return) it earns.

Pyranometer A solar radiometer that measures total diffuse and direct radiation.

Quad, Q Unit of energy equivalent to 10^{15} Btu's (occasionally defined as 10^{18} Btu's).

Radiometer Instrument for measuring the intensity of any kind of radiation.

Reflectance Ratio of radiation reflected from a surface to that incident on the surface. Relectivity is the property of reflecting radiation possessed by all materials to varying extents; called albedo in atmospheric references.

Retrofit Installation of solar energy systems in existing structures.

Selective surface A surface that absorbs more or less energy than it emits; from the original term *wavelength selective.*

Solar altitude angle Angle between the line joining the center of the solar disk to the point of observation at any given instant and the horizontal plane through that point of observation.

Solar constant Intensity of solar radiation beyond the earth's atmosphere, at the average earth-sun distance, on a surface perpendicular to the sun's rays. The value of the solar constant is 1353 W/m^2, 1.940 $cal/(cm^2 \cdot min)$, or 429.2 $Btu/(ft^2 \cdot h)(\pm 1.6$ percent).

Solar energy That energy, in the form of radiation, emitted from the sun and generated by a fusion reaction within the sun.

Solar radiation Radiant energy received from the sun both directly as a beam component and diffusely by scattering from the sky and reflection from the ground.

Solar right The right of a person who uses a solar energy device not to have sunlight blocked by another structure.

Specific heat Amount of heat required to raise the temperature of a unit mass a unit amount.

Spectral energy distribution Curve showing the variation of spectral irradiance with wavelength.

Spectral irradiance Monochromatic irradiance of a surface per unit bandwidth at a particular wavelength; often expressed in watts per square meter per nanometer bandwidth.

Spectrum Range of wavelengths that encompasses all types of electromagnetic radiation, including thermal radiation.

Specular reflection Mirrorlike reflection in which incident and reflected angles are equal.

Sun time Time measured on a basis of the sun's virtual motion.

Thermal conductivity Amount of heat that can be transferred by conduction through a material of unit area and thickness per unit temperature difference.

Thermostat Temperature sensor used to monitor the temperature in a space; provides a signal to a building's thermal control system.

Thermosyphon The convective circulation of fluid in a closed system in which less dense, warm fluid rises, displaced by denser, cooler fluid in the same fluid loop.

Ton of refrigeration Removal of heat at the rate of 12,000 Btu's per hour; derived from melting 1 ton (2000 lb) of ice, requiring 2000×144 Btu's over a 24-h period.

Transmittance Ratio of the radiant energy transmitted by a material to the radiant energy incident on a surface of that material; depends on the angle of incidence.

U Overall heat-loss coefficient; frequently called the *U* value.

Watt Energy rate of 1 joule per second.

Appendix 3 Sunpath Diagrams

A description of the method for calculating true solar time is given together with meteorological charts for computing solar altitude and azimuth angles. (*a*) Description of method; (*b*) chart, 25°N; (*c*) chart, 30°N; (*d*) chart, 35°N; (*e*) chart, 40°N; (*f*) chart, 45°N; (*g*) chart, 50°N.

(*a*) Description of the Method

The altitude and azimuth of the sun are given by

$$\sin a = \sin \phi \sin \delta + \cos \phi \cos \delta \cos h \tag{1}$$

and

$$\sin \alpha = -\frac{\cos \delta \sin h}{\cos a} \tag{2}$$

where a = altitude of the sun (angular elevation above the horizon)
ϕ = latitude of the observer
δ = declination of the sun
h = hour angle of sun (angular distance from the meridian of the observer)
α = azimuth of the sun (measured eastward from north)

From Eqs. (1) and (2) it can be seen that the altitude and azimuth of the sun are functions of the latitude of the observer, the time of day (hour angle), and the date (declination).

Description and charts are based on the *Smithsonian Meteorological Tables* with permission from the Smithsonian Institute, Washington, D.C.

A series of charts are used, one for each 5° of latitude (except 5°, 15°, 75°, and 85°) giving the altitude and azimuth of the sun as a function of the true solar time and the declination of the sun in a form originally suggested by Hand.* Linear interpolation for intermediate latitudes will give results within the accuracy to which the charts can be read.

On these charts, a point corresponding to the projected position of the sun is determined from the heavy lines corresponding to declination and solar time.

To find the solar altitude and azimuth:

1 Select the chart or charts appropriate to the latitude.

2 Find the solar declination δ corresponding to the date.

3 Determine the *true solar time* as follows:

(a) To the *local standard time* (zone time) add 4 minutes for each degree of longitude by which the station is east of the standard meridian or subtract 4 minutes for each degree west of the standard meridian to get the *local mean solar time*.

(b) To the local mean solar time add algebraically the equation of time; the sum is the *true solar time*.

4 Read the required altitude and azimuth at the point determined by the declination and the true solar time. Interpolate linearly between two charts for intermediate latitudes.

It should be emphasized that the solar altitude determined from these charts is the true geometric position of the center of the sun. At low solar elevations, terrestrial refraction may considerably alter the apparent position of sun. Under average atmospheric refraction the sun will appear on the horizon when it actually is about 34′ below the horizon; the effect of refraction decreases rapidly with increasing solar elevation. Since sunset or sunrise is defined as the time when the upper limb of the sun appears on the horizon, and the semidiameter of the sun is 16′, sunset or sunrise occurs under average atmospheric refraction when the sun is 50′ below the horizon. In polar regions especially, unusual atmospheric refraction can produce considerable variation in the time of sunset or sunrise.

The 90°N chart is included for interpolation; the azimuths lose their directional significance at the pole.

To compute solar altitude and azimuth for southern latitudes, change the sign of the solar declination and proceed as above. The resulting azimuths will indicate angular distance from *south* (measured eastward) rather than from north.†

*I. F. Hand, *Heating and Ventilating*, vol. 45, p. 86, 1948.

†Note that the angular convention used by the Smithsonian differs from that used in Chap. 2 and elsewhere in this book.

(*b*) Chart for 25° N

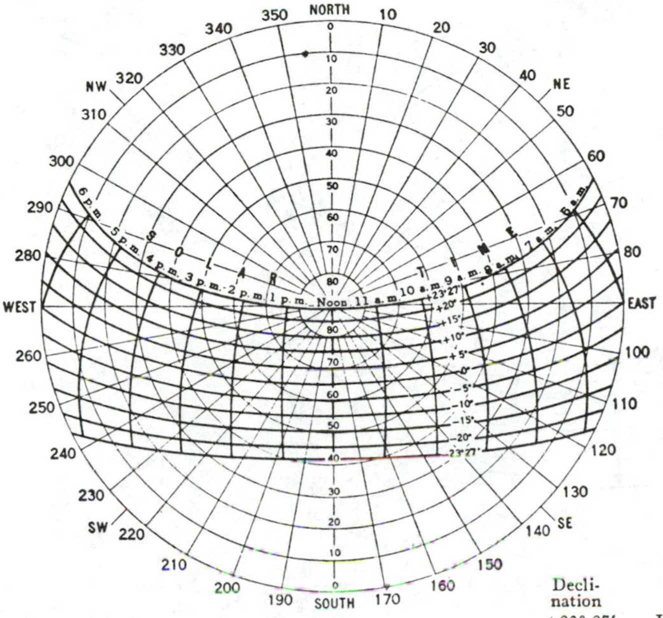

(b)

Declination	Approx. dates
+23° 27′	June 22
+20°	May 21, July 24
+15°	May 1, Aug. 12
+10°	Apr. 16, Aug. 28
+ 5°	Apr. 3, Sept. 10
0°	Mar. 21, Sept. 23
− 5°	Mar. 8, Oct. 6
−10°	Feb. 23, Oct. 20
−15°	Feb. 9, Nov. 3
−20°	Jan. 21, Nov. 22
−23° 27′	Dec. 22

(*c*) Chart for 30° N

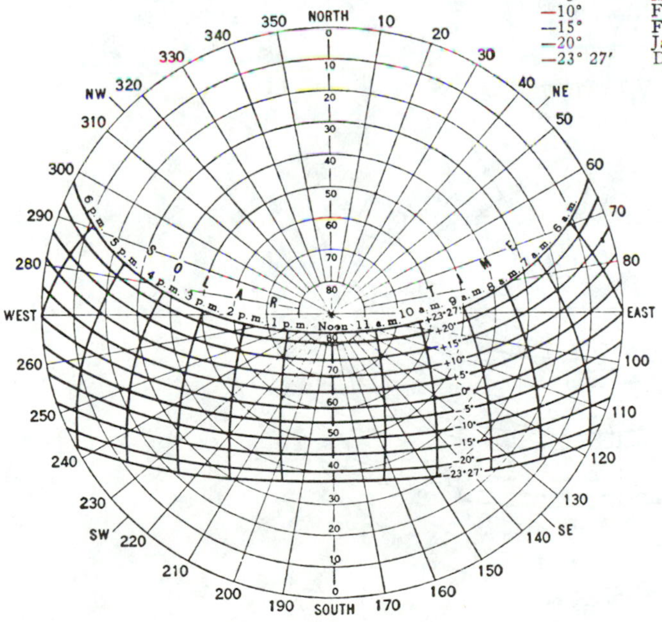

(c)

(*d*) Chart for 35° N

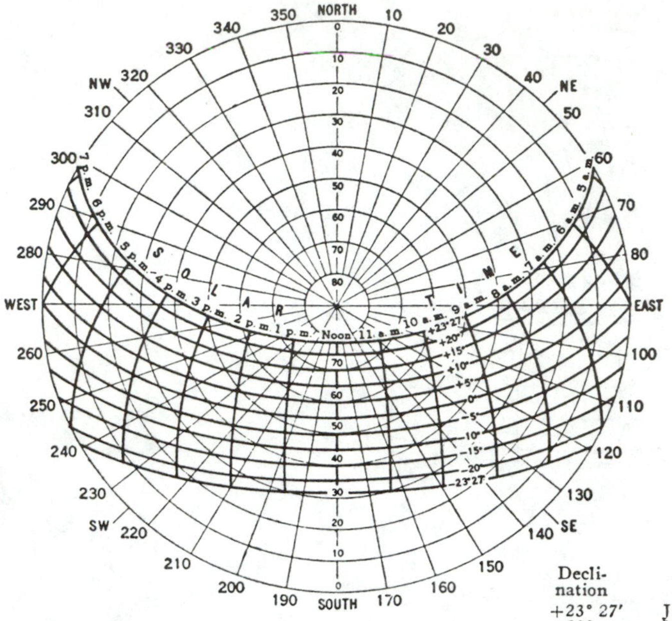

(d)

Declination	Approx. dates
+23° 27′	June 22
+20°	May 21, July 24
+15°	May 1, Aug. 12
+10°	Apr. 16, Aug. 28
+ 5°	Apr. 3, Sept. 10
0°	Mar. 21, Sept. 23
− 5°	Mar. 8, Oct. 6
−10°	Feb. 23, Oct. 20
−15°	Feb. 9, Nov. 3
−20°	Jan. 21, Nov. 22
−23° 27′	Dec. 22

(*e*) Chart for 40° N

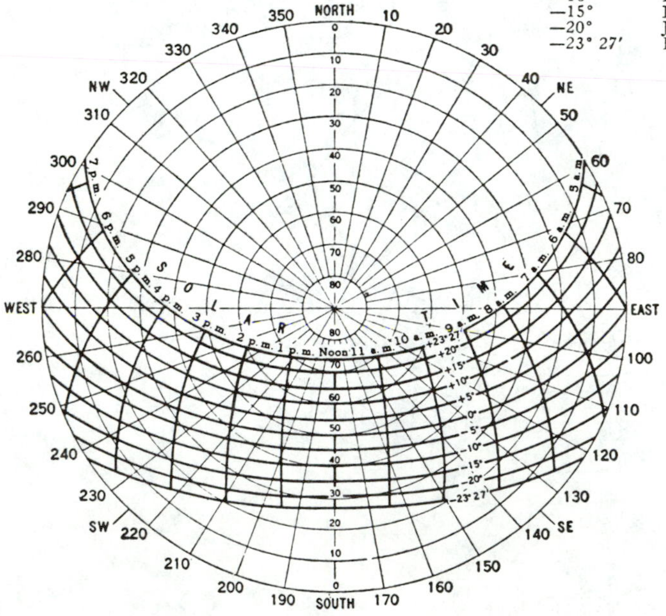

(e)

(*f*) Chart for 45° N

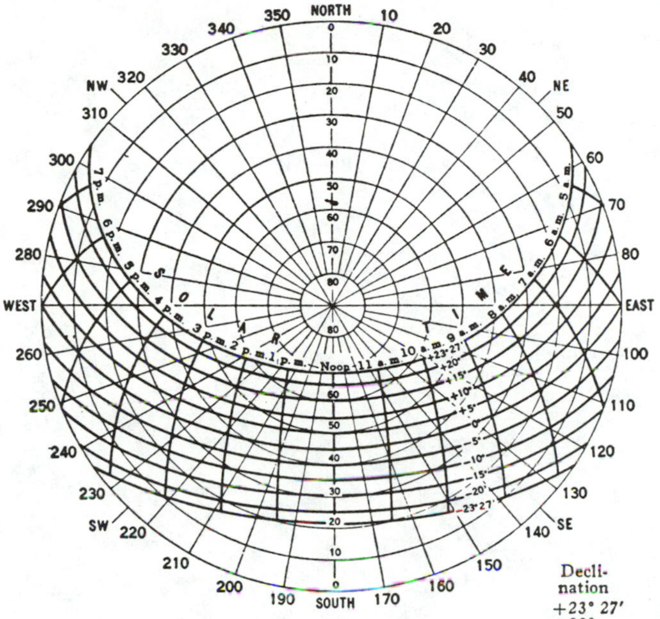

(f)

Declination	Approx. dates
+23° 27′	June 22
+20°	May 21, July 24
+15°	May 1, Aug. 12
+10°	Apr. 16, Aug. 28
+ 5°	Apr. 3, Sept. 10
0°	Mar. 21, Sept. 23
− 5°	Mar. 8, Oct. 6
−10°	Feb. 23, Oct. 20
−15°	Feb. 9, Nov. 3
−20°	Jan. 21, Nov. 22
−23° 27′	Dec. 22

(*g*) Chart for 50° N

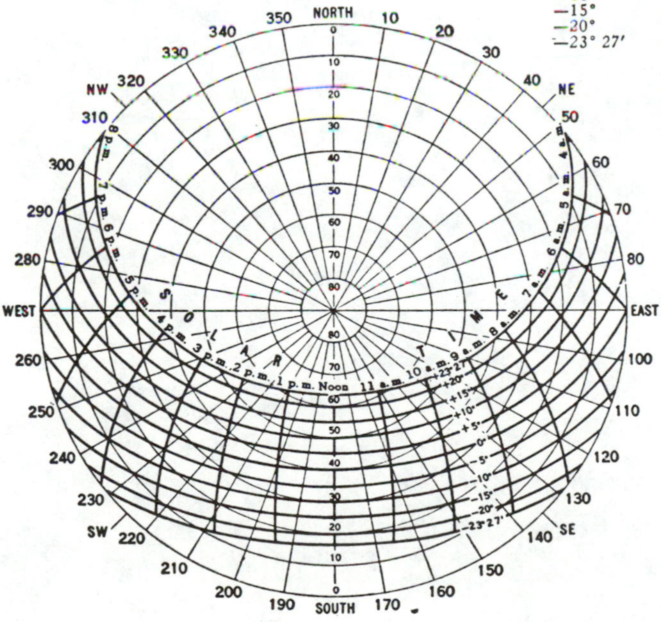

(g)

Appendix 4 Monthly Insolation Maps

The annual and monthly maps represent the mean daily solar radiation on a horizontal surface in megajoules per square meter* for the continental United States. (Courtesy of W. H. Hoecker, Air Resources Laboratory, National Oceanic and Atmospheric Administration.)

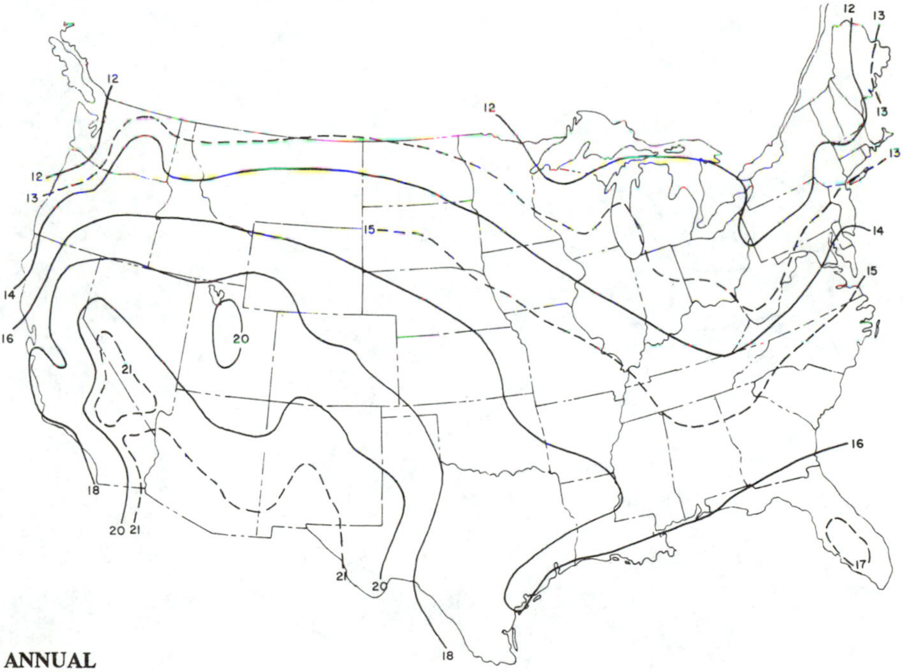

ANNUAL

*1 $mJ/m^2 = 88.1\ Btu/ft^2$.

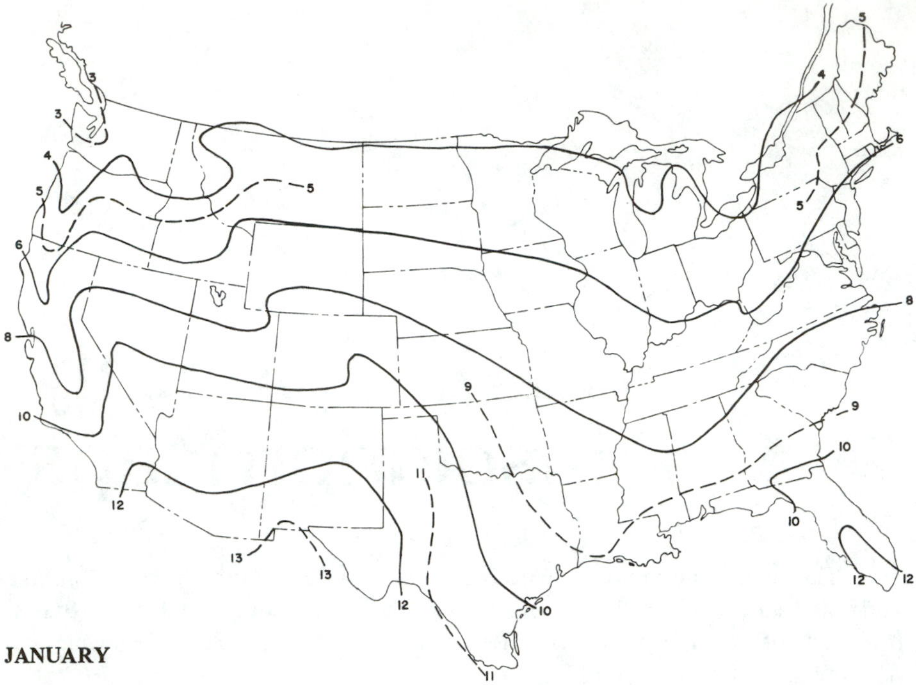

JANUARY

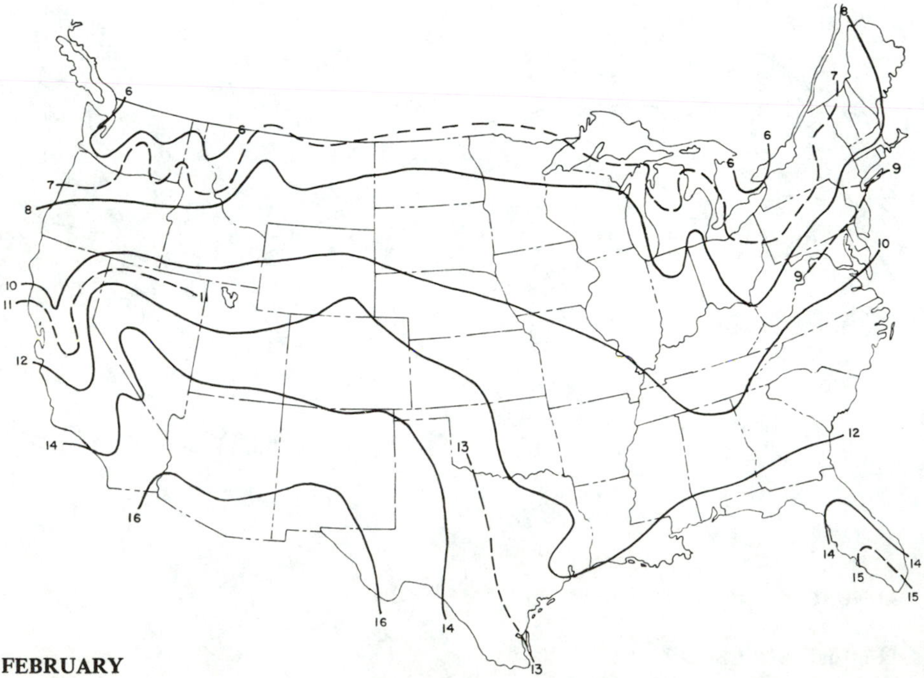

FEBRUARY

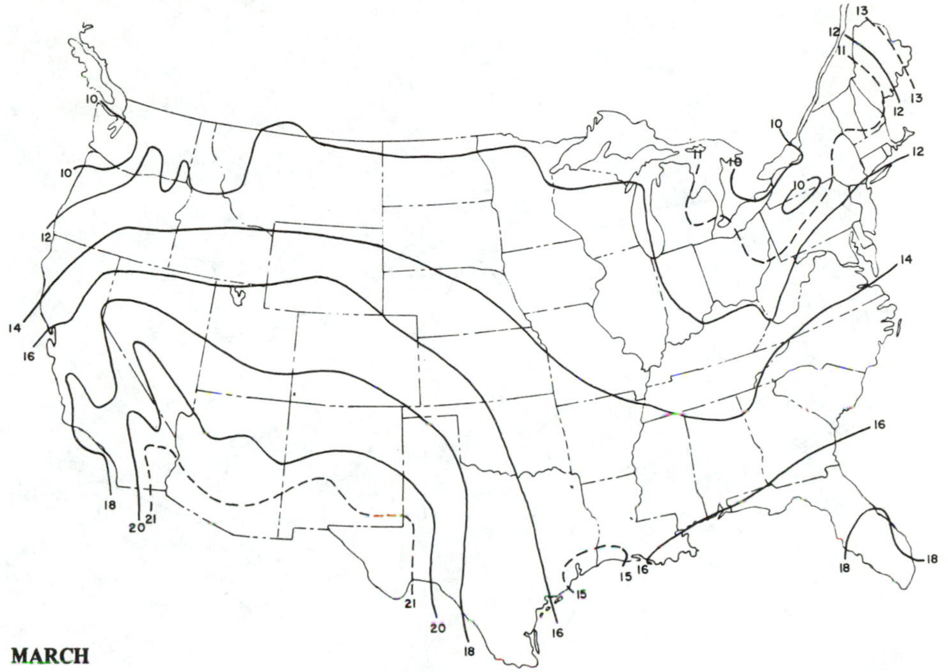
13
12
11
13
12
10
12
10
11
10
10
10
12
14
14
16
16
18
18
20
21
21
20
18
16
15
15
16
18

MARCH

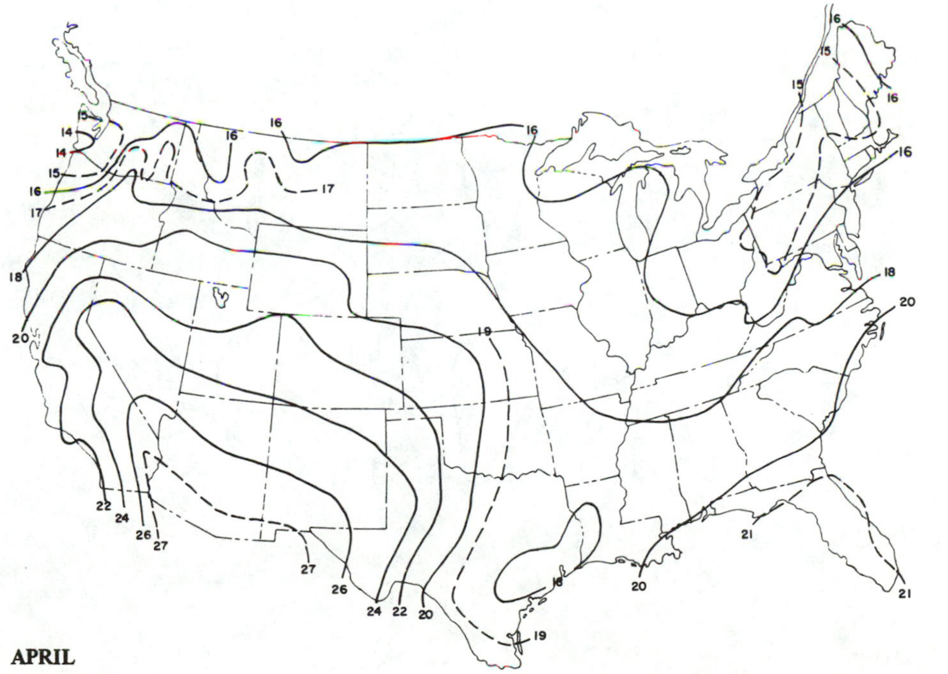
16
15
16
15
15
14
14
16
16
16
16
15
16
17
17
18
20
19
18
20
22
24
26
27
27
26
24
22
20
19
18
20
21
21

APRIL

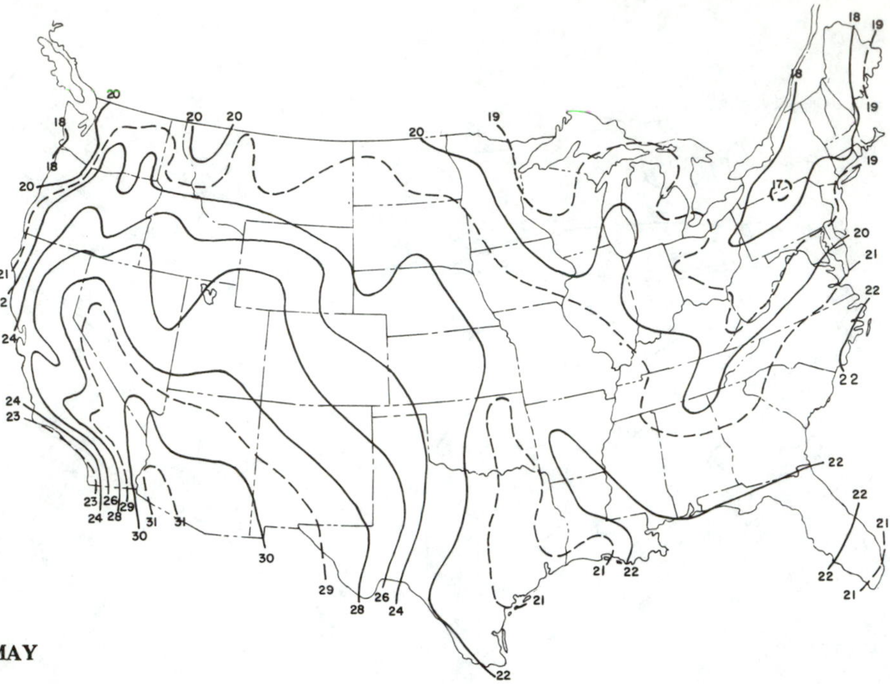
18
19
18
19
19
17
20
20
20
18
18
20
20
19
20
21
22
21
22
24
22
24
23
23
26
29
24
28
31
31
30
30
29
28
26
24
22
21
22
22
22
22
21
21
21
22

MAY

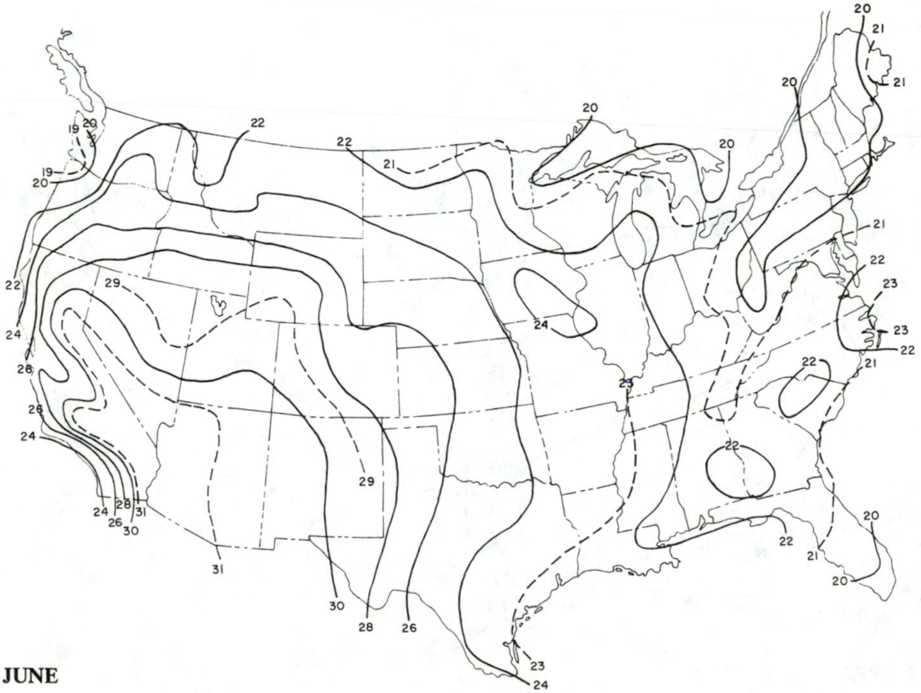
20
21
21
20
20
22
19
20
19
20
22
21
20
21
22
22
23
23
22
21
24
24
22
29
26
26
24
24
28
31
26
30
23
22
22
31
29
30
28
26
22
21
20
20
23
24

JUNE

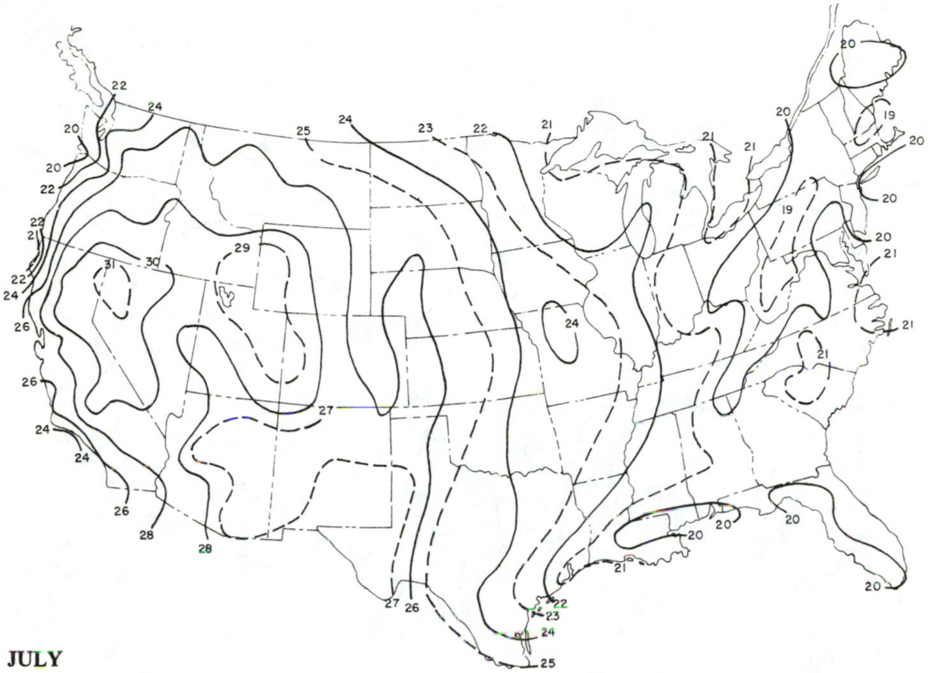

22
24
20
20
22
22
21
22
24
26
26
24
24
26
28
28
31
30
29
27
25
24
23
22
21
24
21
21
20
20
19
20
20
19
20
20
21
21
21
20
20
20
21
20
22
23
24
25
27
26

JULY

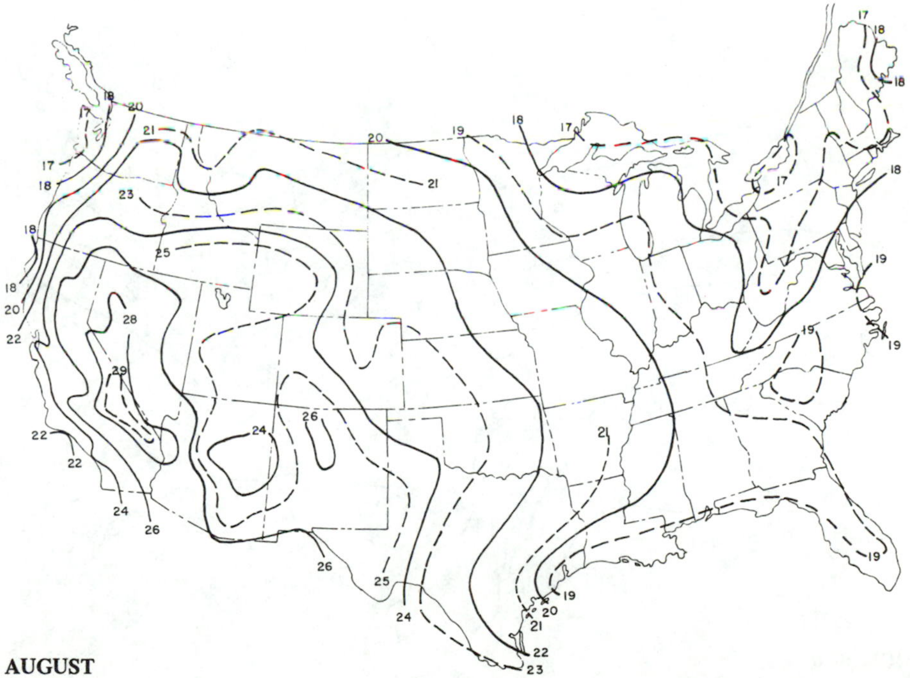

18
20
17
21
17
18
23
18
18
20
22
25
28
29
22
22
24
26
24
26
26
25
24
20
21
19
18
17
17
18
18
18
17
19
19
19
21
19
20
21
22
23

AUGUST

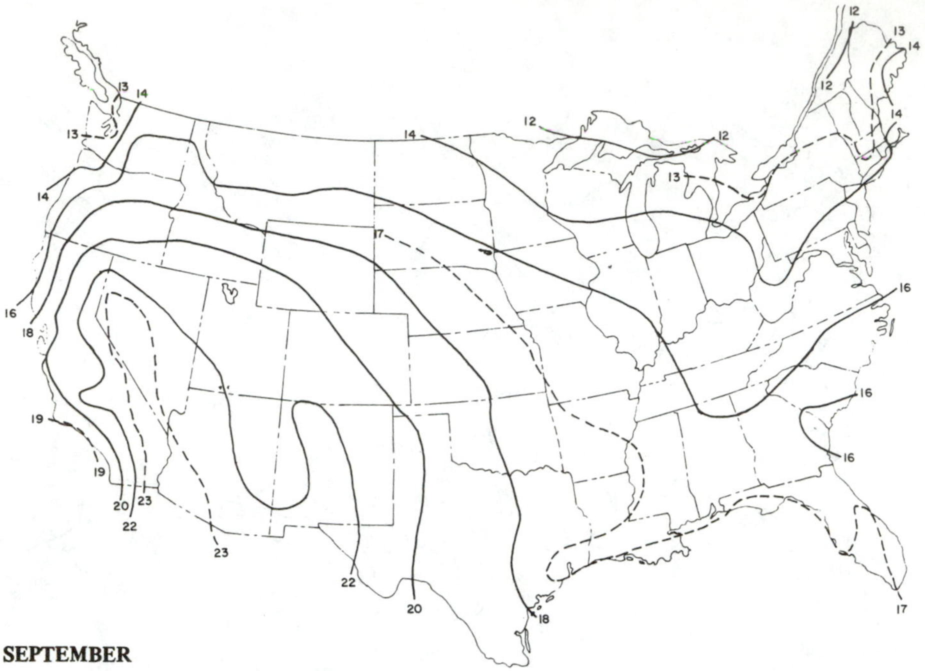

SEPTEMBER

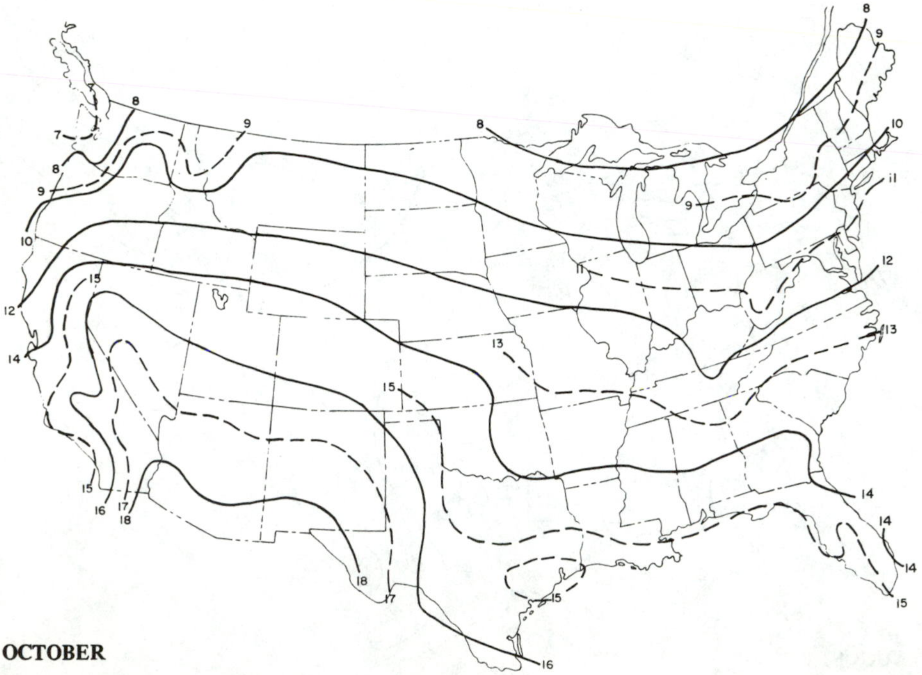

OCTOBER

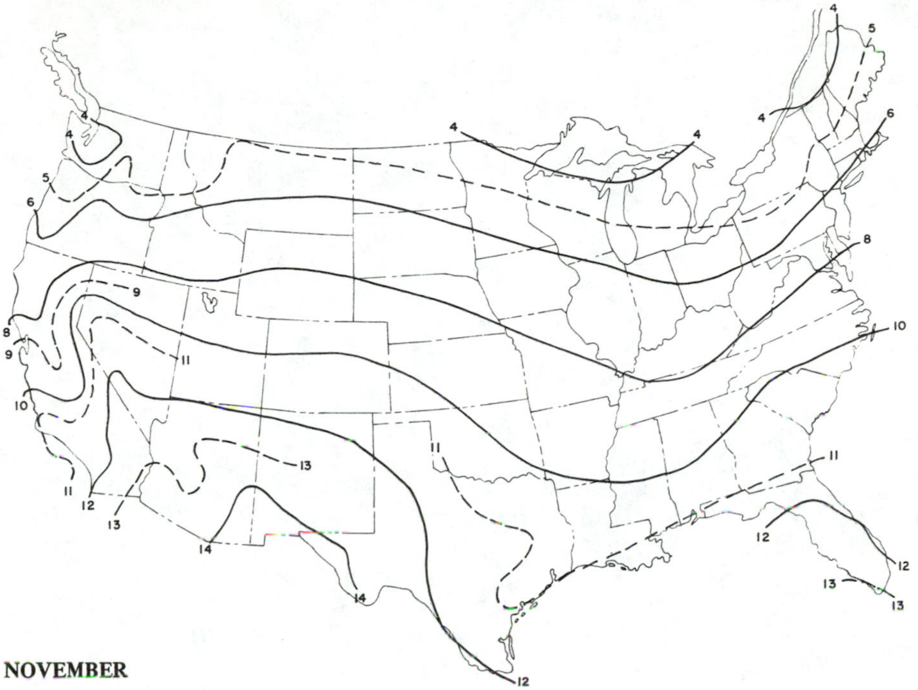

NOVEMBER

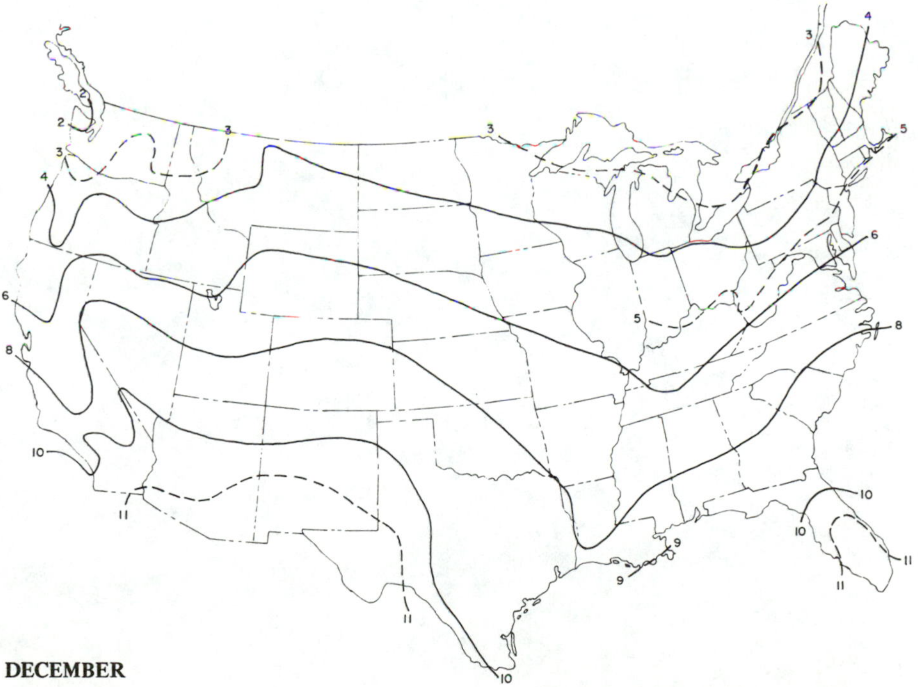

DECEMBER

Appendix 5 Hourly Clear-Sky Solar Radiation

(a) Solar Position and Insolation Values for 24°N

Date	Solar time		Solar position		Total insolation on surfaces* [Btu/(h·ft²)]						
							South-facing surface angle with horizontal				
	a.m.	p.m.	Altitude	Azimuth	Normal†	Horizontal	14	24	34	44	90
Jan. 21	7	5	4.8	65.6	71	10	17	21	25	28	31
	8	4	16.9	58.3	239	83	110	126	137	145	127
	9	3	27.9	48.8	288	151	188	207	221	228	176
	10	2	37.2	36.1	308	204	246	268	282	287	207
	11	1	43.6	19.6	317	237	283	306	319	324	226
	12		46.0	0.0	320	249	296	319	332	336	232
Surface daily totals					2766	1622	1984	2174	2300	2360	1766
Feb. 21	7	5	9.3	74.6	158	35	44	49	53	56	46
	8	4	22.3	67.2	263	116	135	145	150	151	102
	9	3	34.4	57.6	298	187	213	225	230	228	141
	10	2	45.1	44.2	314	241	273	286	291	287	168
	11	1	53.0	25.0	321	276	310	324	328	323	185
	12		56.0	0.0	324	288	323	337	341	335	191
Surface daily totals					3036	1998	2276	2396	2436	2424	1476
Mar. 21	7	5	13.7	83.3	194	60	63	64	62	59	27
	8	4	27.2	76.8	267	141	150	152	149	142	64
	9	3	40.2	67.9	295	212	226	229	225	214	95
	10	2	52.3	54.8	309	266	285	288	283	270	120
	11	1	61.9	33.4	315	300	322	326	320	305	135
	12		66.0	0.0	317	312	334	339	333	317	140
Surface daily totals					3078	2270	2428	2456	2412	2298	1022
Apr. 21	6	6	4.7	100.6	40	7	5	4	4	3	2
	7	5	18.3	94.9	203	83	77	70	62	51	10
	8	4	32.0	89.0	256	160	157	149	137	122	16
	9	3	45.6	81.9	280	227	227	220	206	186	46
	10	2	59.0	71.8	292	278	282	275	259	237	61
	11	1	71.1	51.6	298	310	316	309	293	269	74
	12		77.6	0.0	299	321	328	321	305	280	79
Surface daily totals					3036	2454	2458	2374	2228	2016	488
May 21	6	6	8.0	108.4	86	22	15	10	9	9	5
	7	5	21.2	103.2	203	98	85	73	59	44	12
	8	4	34.6	98.5	248	171	159	145	127	106	15
	9	3	48.3	93.6	269	233	224	210	190	165	16
	10	2	62.0	87.7	280	281	275	261	239	211	22
	11	1	75.5	76.9	286	311	307	293	270	240	34
	12		86.0	0.0	288	322	317	304	281	250	37
Surface daily totals					3032	2556	2447	2286	2072	1800	246
June 21	6	6	9.3	111.6	97	29	20	12	12	11	7
	7	5	22.3	106.8	201	103	87	73	58	41	13
	8	4	35.5	102.6	242	173	158	142	122	99	16
	9	3	49.0	98.7	263	234	221	204	182	155	18
	10	2	62.6	95.0	274	280	269	253	229	199	18
	11	1	76.3	90.8	279	309	300	283	259	227	19
	12		89.4	0.0	281	319	310	294	269	236	22
Surface daily totals					2994	2574	2422	2230	1992	1700	204

(See footnotes on page 379.)

(a) Solar Position and Insolation Values for 24° N (*Continued*)

Date	Solar time a.m.	Solar time p.m.	Solar position Alti-tude	Solar position Azi-muth	Total insolation on surfaces [Btu/(h·ft²)] Normal	Hori-zontal	South-facing surface angle with horizontal 14	24	34	44	90
July 21	6	6	8.2	109.0	81	23	16	11	10	9	6
	7	5	21.4	103.8	195	98	85	73	59	44	13
	8	4	34.8	99.2	239	169	157	143	125	104	16
	9	3	48.4	94.5	261	231	221	207	187	161	18
	10	2	62.1	89.0	272	278	270	256	235	206	21
	11	1	75.7	79.2	278	307	302	287	265	235	32
	12		86.6	0.0	280	317	312	298	275	245	36
Surface daily totals					2932	2526	2412	2250	2036	1766	246
Aug. 21	6	6	5.0	101.3	35	7	5	4	4	4	2
	7	5	18.5	95.6	186	82	76	69	60	50	11
	8	4	32.2	89.7	241	158	154	146	134	118	16
	9	3	45.9	82.9	265	223	222	214	200	181	39
	10	2	59.3	73.0	278	273	275	268	252	230	58
	11	1	71.6	53.2	284	304	309	301	285	261	71
	12		78.3	0.0	286	315	320	313	296	272	75
Surface daily totals					2864	2408	2402	2316	2168	1958	470
Sept. 21	7	5	13.7	83.8	173	57	60	60	59	56	26
	8	4	27.2	76.8	248	136	144	146	143	136	62
	9	3	40.2	67.9	278	205	218	221	217	206	93
	10	2	52.3	54.8	292	258	275	278	273	261	116
	11	1	61.9	33.4	299	291	311	315	309	295	131
	12		66.0	0.0	301	302	323	327	321	306	136
Surface daily totals					2878	2194	2342	2366	2322	2212	992
Oct. 21	7	5	9.1	74.1	138	32	40	45	48	50	42
	8	4	22.0	66.7	247	111	129	139	144	145	99
	9	3	34.1	57.1	284	180	206	217	223	221	138
	10	2	44.7	43.8	301	234	265	277	282	279	165
	11	1	52.5	24.7	309	268	301	315	319	314	182
	12		55.5	0.0	311	279	314	328	332	327	188
Surface daily totals					2868	1928	2198	2314	2364	2346	1442
Nov. 21	7	5	4.9	65.8	67	10	16	20	24	27	29
	8	4	17.0	58.4	232	82	108	123	135	142	124
	9	3	28.0	48.9	282	150	186	205	217	224	172
	10	2	37.3	36.3	303	203	244	265	278	283	204
	11	1	43.8	19.7	312	236	280	302	316	320	222
	12		46.2	0.0	315	247	293	315	328	332	228
Surface daily totals					2706	1610	1962	2146	2268	2324	1730
Dec. 21	7	5	3.2	62.6	30	3	7	9	11	12	14
	8	4	14.9	55.3	225	71	99	116	129	139	130
	9	3	25.5	46.0	281	137	176	198	214	223	184
	10	2	34.3	33.7	304	189	234	258	275	283	217
	11	1	40.4	18.2	314	221	270	295	312	320	236
	12		42.6	0.0	317	232	282	308	325	332	243
Surface daily totals					2624	1474	1852	2058	2204	2286	1808

(b) Solar Position and Insolation Values for 32° N

Date	Solar time a.m.	Solar time p.m.	Solar position Altitude	Solar position Azimuth	Total insolation on surfaces [Btu/(h·ft²)] Normal	Horizontal	South-facing surface angle with horizontal 22	32	42	52	90
Jan. 21	7	5	1.4	65.2	1	0	0	0	0	1	1
	8	4	12.5	56.5	203	56	93	106	116	123	115
	9	3	22.5	46.0	269	118	175	193	206	212	181
	10	2	30.6	33.1	295	167	235	256	269	274	221
	11	1	36.1	17.5	306	198	273	295	308	312	245
	12		38.0	0.0	310	209	285	308	321	324	253
Surface daily totals					2458	1288	1839	2008	2118	2166	1779
Feb. 21	7	5	7.1	73.5	121	22	34	37	40	42	38
	8	4	19.0	64.4	247	95	127	136	140	141	108
	9	3	29.9	53.4	288	161	206	217	222	220	158
	10	2	39.1	39.4	306	212	266	278	283	279	193
	11	1	45.6	21.4	315	244	304	317	321	315	214
	12		48.0	0.0	317	255	316	330	334	328	222
Surface daily totals					2872	1724	2188	2300	2345	2322	1644
Mar. 21	7	5	12.7	81.9	185	54	60	60	59	56	32
	8	4	25.1	73.0	260	129	146	147	144	137	78
	9	3	36.8	62.1	290	194	222	224	220	209	119
	10	2	47.3	47.5	304	245	280	283	278	265	150
	11	1	55.0	26.8	311	277	317	321	315	300	170
	12		58.0	0.0	313	287	329	333	327	312	177
Surface daily totals					3012	2084	2378	2403	2358	2246	1276
Apr. 21	6	6	6.1	99.9	66	14	9	6	6	5	3
	7	5	18.8	92.2	206	86	78	71	62	51	10
	8	4	31.5	84.0	255	158	156	148	136	120	35
	9	3	43.9	74.2	278	220	225	217	203	183	68
	10	2	55.7	60.3	290	267	279	272	256	234	95
	11	1	65.4	37.5	295	297	313	306	290	265	112
	12		69.6	0.0	297	307	325	318	301	276	118
Surface daily totals					3076	2390	2444	2356	2206	1994	764
May 21	6	6	10.4	107.2	119	36	21	13	13	12	7
	7	5	22.8	100.1	211	107	88	75	60	44	13
	8	4	35.4	92.9	250	175	159	145	127	105	15
	9	3	48.1	84.7	269	233	223	209	188	163	33
	10	2	60.6	73.3	280	277	273	259	237	208	56
	11	1	72.0	51.9	285	305	305	290	268	237	72
	12		78.0	0.0	286	315	315	301	278	247	77
Surface daily totals					3112	2582	2454	2284	2064	1788	469
June 21	6	6	12.2	110.2	131	45	26	16	15	14	9
	7	5	24.3	103.4	210	115	91	76	59	41	14
	8	4	36.9	96.8	245	180	159	143	122	99	16
	9	3	49.6	89.4	264	236	221	204	181	153	19
	10	2	62.2	79.7	274	279	268	251	227	197	41
	11	1	74.2	60.9	279	306	299	282	257	224	56
	12		81.5	0.0	280	315	309	292	267	234	60
Surface daily totals					3084	2634	2436	2234	1990	1690	370

(b) Solar Position and Insolation Values for 32° N (*Continued*)

Date	Solar time a.m.	Solar time p.m.	Solar position Altitude	Solar position Azimuth	Total insolation on surfaces [Btu/(h·ft²)] Normal	Horizontal	South-facing surface angle with horizontal 22	32	42	52	90
July 21	6	6	10.7	107.7	113	37	22	14	13	12	8
	7	5	23.1	100.6	203	107	87	75	60	44	14
	8	4	35.7	93.6	241	174	158	143	125	104	16
	9	3	48.4	85.5	261	231	220	205	185	159	31
	10	2	60.9	74.3	271	274	269	254	232	204	54
	11	1	72.4	53.3	277	302	300	285	262	232	69
	12		78.6	0.0	279	311	310	296	273	242	74
Surface daily totals					3012	2558	2422	2250	2030	1754	458
Aug. 21	6	6	6.5	100.5	59	14	9	7	6	6	4
	7	5	19.1	92.8	190	85	77	69	60	50	12
	8	4	31.8	84.7	240	156	152	144	132	116	33
	9	3	44.3	75.0	263	216	220	212	197	178	65
	10	2	56.1	61.3	276	262	272	264	249	226	91
	11	1	66.0	38.4	282	292	305	298	281	257	107
	12		70.3	0.0	284	302	317	309	292	268	113
Surface daily totals					2902	2352	2388	2296	2144	1934	736
Sept. 21	7	5	12.7	81.9	163	51	56	56	55	52	30
	8	4	25.1	73.0	240	124	140	141	138	131	75
	9	3	36.8	62.1	272	188	213	215	211	201	114
	10	2	47.3	47.5	287	237	270	273	268	255	145
	11	1	55.0	26.8	294	268	306	309	303	289	164
	12		58.0	0.0	296	278	318	321	315	300	171
Surface daily totals					2808	2014	2288	2308	2264	2154	1226
Oct. 21	7	5	6.8	73.1	99	19	29	32	34	36	32
	8	4	18.7	64.0	229	90	120	128	133	134	104
	9	3	29.5	53.0	273	155	198	208	213	212	153
	10	2	38.7	39.1	293	204	257	269	273	270	188
	11	1	45.1	21.1	302	236	294	307	311	306	209
	12		47.5	0.0	304	247	306	320	324	318	217
Surface daily totals					2696	1654	2100	2208	2252	2232	1588
Nov. 21	7	5	1.5	65.4	2	0	0	0	1	1	1
	8	4	12.7	56.6	196	55	91	104	113	119	111
	9	3	22.6	46.1	263	118	173	190	202	208	176
	10	2	30.8	33.2	289	166	233	252	265	270	217
	11	1	36.2	17.6	301	197	270	291	303	307	241
	12		38.2	0.0	304	207	282	304	316	320	249
Surface daily totals					2406	1280	1816	1980	2084	2130	1742
Dec. 21	8	4	10.3	53.8	176	41	77	90	101	108	107
	9	3	19.8	43.6	257	102	161	180	195	204	183
	10	2	27.6	31.2	288	150	221	244	259	267	226
	11	1	32.7	16.4	301	180	258	282	298	305	251
	12		34.6	0.0	304	190	271	295	311	318	259
Surface daily totals					2348	1136	1704	1888	2016	2086	1794

(c) Solar Position and Insolation Values for 40° N

Date	Solar time a.m.	Solar time p.m.	Solar position Altitude	Solar position Azimuth	Total insolation on surfaces [Btu/(h·ft²)] Normal	Horizontal	South-facing surface angle with horizontal 30	40	50	60	90
Jan. 21	8	4	8.1	55.3	142	28	65	74	81	85	84
	9	3	16.8	44.0	239	83	155	171	182	187	171
	10	2	23.8	30.9	274	127	218	237	249	254	223
	11	1	28.4	16.0	289	154	257	277	290	293	253
	12		30.0	0.0	294	164	270	291	303	306	263
Surface daily totals					2182	948	1660	1810	1906	1944	1726
Feb. 21	7	5	4.8	72.7	69	10	19	21	23	24	22
	8	4	15.4	62.2	224	73	114	122	126	127	107
	9	3	25.0	50.2	274	132	195	205	209	208	167
	10	2	32.8	35.9	295	178	256	267	271	267	210
	11	1	38.1	18.9	305	206	293	306	310	304	236
	12		40.0	0.0	308	216	306	319	323	317	245
Surface daily totals					2640	1414	2060	2162	2202	2176	1730
Mar. 21	7	5	11.4	80.2	171	46	55	55	54	51	35
	8	4	22.5	69.6	250	114	140	141	138	131	89
	9	3	32.8	57.3	282	173	215	217	213	202	138
	10	2	41.6	41.9	297	218	273	276	271	258	176
	11	1	47.7	22.6	305	247	310	313	307	293	200
	12		50.0	0.0	307	257	322	326	320	305	208
Surface daily totals					2916	1852	2308	2330	2284	2174	1484
Apr. 21	6	6	7.4	98.9	89	20	11	8	7	7	4
	7	5	18.9	89.5	206	87	77	70	61	50	12
	8	4	30.3	79.3	252	152	153	145	133	117	53
	9	3	41.3	67.2	274	207	221	213	199	179	93
	10	2	51.2	51.4	286	250	275	267	252	229	126
	11	1	58.7	29.2	292	277	308	301	285	260	147
	12		61.6	0.0	293	287	320	313	296	271	154
Surface daily totals					3092	2274	2412	2320	2168	1956	1022
May 21	5	7	1.9	114.7	1	0	0	0	0	0	0
	6	6	12.7	105.6	144	49	25	15	14	13	9
	7	5	24.0	96.6	216	214	89	76	60	44	13
	8	4	35.4	87.2	250	175	158	144	125	104	25
	9	3	46.8	76.0	267	227	221	206	186	160	60
	10	2	57.5	60.9	277	267	270	255	233	205	89
	11	1	66.2	37.1	283	293	301	287	264	234	108
	12		70.0	0.0	284	301	312	297	274	243	114
Surface daily totals					3160	2552	2442	2264	2040	1760	724
June 21	5	7	4.2	117.3	22	4	3	3	2	2	1
	6	6	14.8	108.4	155	60	30	18	17	16	10
	7	5	26.0	99.7	216	123	92	77	59	41	14
	8	4	37.4	90.7	246	182	159	142	121	97	16
	9	3	48.8	80.2	263	233	219	202	179	151	47
	10	2	59.8	65.8	272	272	266	248	224	194	74
	11	1	69.2	41.9	277	296	296	278	253	221	92
	12		73.5	0.0	279	304	306	289	263	230	98
Surface daily totals					3180	2648	2434	2224	1974	1670	610

(c) Solar Position and Insolation Values for 40° N (*Continued*)

Date	Solar time		Solar position		Total insolation on surfaces [Btu/(h·ft²)]						
							South-facing surface angle with horizontal				
	a.m.	p.m.	Altitude	Azimuth	Normal	Horizontal	30	40	50	60	90
July 21	5	7	2.3	115.2	2	0	0	0	0	0	0
	6	6	13.1	106.1	138	50	26	17	15	14	9
	7	5	24.3	97.2	208	114	89	75	60	44	14
	8	4	35.8	87.8	241	174	157	142	124	102	24
	9	3	47.2	76.7	259	225	218	203	182	157	58
	10	2	57.9	61.7	269	265	266	251	229	200	86
	11	1	66.7	37.9	275	290	296	281	258	228	104
	12		70.6	0.0	276	298	307	292	269	238	111
Surface daily totals					3062	2534	2409	2230	2006	1728	702
Aug. 21	6	6	7.9	99.5	81	21	12	9	8	7	5
	7	5	19.3	90.9	191	87	76	69	60	49	12
	8	4	30.7	79.9	237	150	150	141	129	113	50
	9	3	41.8	67.9	260	205	216	207	193	173	89
	10	2	51.7	52.1	272	246	267	259	244	221	120
	11	1	59.3	29.7	278	273	300	292	276	252	140
	12		62.3	0.0	280	282	311	303	287	262	147
Surface daily totals					2916	2244	2354	2258	2104	1894	978
Sept. 21	7	5	11.4	80.2	149	43	51	51	49	47	32
	8	4	22.5	69.6	230	109	133	134	131	124	84
	9	3	32.8	57.3	263	167	206	208	203	193	132
	10	2	41.6	41.9	280	211	262	265	260	247	168
	11	1	47.7	22.6	287	239	298	301	295	281	192
	12		50.0	0.0	290	249	310	313	307	292	200
Surface daily totals					2708	1788	2210	2228	2182	2074	1416
Oct. 21	7	5	4.5	72.3	48	7	14	15	17	17	16
	8	4	15.0	61.9	204	68	106	113	117	118	100
	9	3	24.5	49.8	257	126	185	195	200	198	160
	10	2	32.4	35.6	280	170	245	257	261	257	203
	11	1	37.6	18.7	291	199	283	295	299	294	229
	12		39.5	0.0	294	208	295	308	312	306	238
Surface daily totals					2454	1348	1962	2060	2098	2074	1654
Nov. 21	8	4	8.2	55.4	136	28	63	72	78	82	81
	9	3	17.0	44.1	232	82	152	167	178	183	167
	10	2	24.0	31.0	268	126	215	233	245	249	219
	11	1	28.6	16.1	283	153	254	273	285	288	248
	12		30.2	0.0	288	163	267	287	298	301	258
Surface daily totals					2128	942	1636	1778	1870	1908	1686
Dec. 21	8	4	5.5	53.0	89	14	39	45	50	54	56
	9	3	14.0	41.9	217	65	135	152	164	171	163
	10	2	20.0	29.4	261	107	200	221	235	242	221
	11	1	25.0	15.2	280	134	239	262	276	283	252
	12		26.6	0.0	285	143	253	275	290	296	263
Surface daily totals					1978	782	1480	1634	1740	1796	1646

(d) Solar Position and Insolation Values for 48°N

Date	Solar time a.m.	Solar time p.m.	Solar position Altitude	Solar position Azimuth	Total insolation on surfaces [Btu/(h·ft²)] Normal	Horizontal	South-facing surface angle with horizontal 38	48	58	68	90
Jan. 21	8	4	3.5	54.6	37	4	17	19	21	22	22
	9	3	11.0	42.6	185	46	120	132	140	145	139
	10	2	16.9	29.4	239	83	190	206	216	220	206
	11	1	20.7	15.1	261	107	231	249	260	263	243
	12		22.0	0.0	267	115	245	264	275	278	255
Surface daily totals					1710	596	1360	1478	1550	1578	1478
Feb. 21	7	5	2.4	72.2	12	1	3	4	4	4	4
	8	4	11.6	60.5	188	49	95	102	105	106	96
	9	3	19.7	47.7	251	100	178	187	191	190	167
	10	2	26.2	33.3	278	139	240	251	255	251	217
	11	1	30.5	17.2	290	165	278	290	294	288	247
	12		32.0	0.0	293	173	291	304	307	301	258
Surface daily totals					2330	1080	1880	1972	2024	1978	1720
Mar. 21	7	5	10.0	78.7	153	37	49	49	47	45	35
	8	4	19.5	66.8	236	96	131	132	129	122	96
	9	3	28.2	53.4	270	147	205	207	203	193	152
	10	2	35.4	37.8	287	187	263	266	261	248	195
	11	1	40.3	19.8	295	212	300	303	297	283	223
	12		42.0	0.0	298	220	312	315	309	294	232
Surface daily totals					2780	1578	2208	2228	2182	2074	1632
Apr. 21	6	6	8.6	97.8	108	27	13	9	8	7	5
	7	5	18.6	86.7	205	85	76	69	59	48	21
	8	4	28.5	74.9	247	142	149	141	129	113	69
	9	3	37.8	61.2	268	191	216	208	194	174	115
	10	2	45.8	44.6	280	228	268	260	245	223	152
	11	1	51.5	24.0	286	252	301	294	278	254	177
	12		53.6	0.0	288	260	313	305	289	264	185
Surface daily totals					3076	2106	2358	2266	2114	1902	1262
May 21	5	7	5.2	114.3	41	9	4	4	4	3	2
	6	6	14.7	103.7	162	61	27	16	15	13	10
	7	5	24.6	93.0	219	118	89	75	60	43	13
	8	4	34.7	81.6	248	171	156	142	123	101	45
	9	3	44.3	68.3	264	217	217	202	182	156	86
	10	2	53.0	51.3	274	252	265	251	229	200	120
	11	1	59.5	28.6	279	274	296	281	258	228	141
	12		62.0	0.0	280	281	306	292	269	238	149
Surface daily totals					3254	2482	2418	2234	2010	1728	982
June 21	5	7	7.9	116.5	77	21	9	9	8	7	5
	6	6	17.2	106.2	172	74	33	19	18	16	12
	7	5	27.0	95.8	220	129	93	77	59	39	15
	8	4	37.1	84.6	246	181	157	140	119	95	35
	9	3	46.9	71.6	261	225	216	198	175	147	74
	10	2	55.8	54.8	269	259	262	244	220	189	105
	11	1	62.7	31.2	274	280	291	273	248	216	126
	12		65.5	0.0	275	287	301	283	258	225	133
Surface daily totals					3312	2626	2420	2204	1950	1644	874

(d) Solar Position and Insolation Values for 48° N (*Continued*)

Date	Solar time a.m.	Solar time p.m.	Solar position Altitude	Solar position Azimuth	Total insolation on surfaces [Btu/(h·ft²)] Normal	Horizontal	South-facing surface angle with horizontal 38	48	58	68	90
July 21	5	7	5.7	114.7	43	10	5	5	4	4	3
	6	6	15.2	104.1	156	62	28	18	16	15	11
	7	5	25.1	93.5	211	118	89	75	59	42	14
	8	4	35.1	82.1	240	171	154	140	121	99	43
	9	3	44.8	68.8	256	215	214	199	178	153	83
	10	2	53.5	51.9	266	250	261	246	224	195	116
	11	1	60.1	29.0	271	272	291	276	253	223	137
	12		62.6	0.0	272	279	301	286	263	232	144
Surface daily totals					3158	2474	2386	2200	1974	1694	956
Aug. 21	6	6	9.1	98.3	99	28	14	10	9	8	6
	7	5	19.1	87.2	190	85	75	67	58	47	20
	8	4	29.0	75.4	232	141	145	137	125	109	65
	9	3	38.4	61.8	254	189	210	201	187	168	110
	10	2	46.4	45.1	266	225	260	252	237	214	146
	11	1	52.2	24.3	272	248	293	285	268	244	169
	12		54.3	0.0	274	256	304	296	279	255	177
Surface daily totals					2898	2086	2300	2200	2046	1836	1208
Sept. 21	7	5	10.0	78.7	131	35	44	44	43	40	31
	8	4	19.5	66.8	215	92	124	124	121	115	90
	9	3	28.2	53.4	251	142	196	197	193	183	143
	10	2	35.4	37.8	269	181	251	254	248	236	185
	11	1	40.3	19.8	278	205	287	289	284	269	212
	12		42.0	0.0	280	213	299	302	296	281	221
Surface daily totals					2568	1522	2102	2118	2070	1966	1546
Oct. 21	7	5	2.0	71.9	4	0	1	1	1	1	1
	8	4	11.2	60.2	165	44	86	91	95	95	87
	9	3	19.3	47.4	233	94	167	176	180	178	157
	10	2	25.7	33.1	262	133	228	239	242	239	207
	11	1	30.0	17.1	274	157	266	277	281	276	237
	12		31.5	0.0	278	166	279	291	294	288	247
Surface daily totals					2154	1022	1774	1860	1890	1866	1626
Nov. 21	8	4	3.6	54.7	36	5	17	19	21	22	22
	9	3	11.2	42.7	179	46	117	129	137	141	135
	10	2	17.1	29.5	233	83	186	202	212	215	201
	11	1	20.9	15.1	255	107	227	245	255	258	238
	12		22.2	0.0	261	115	241	259	270	272	250
Surface daily totals					1668	596	1336	1448	1518	1544	1442
Dec. 21	9	3	8.0	40.9	140	27	87	98	105	110	109
	10	2	13.6	28.2	214	63	164	180	192	197	190
	11	1	17.3	14.4	242	86	207	226	239	244	231
	12		18.6	0.0	250	94	222	241	254	260	244
Surface daily totals					1444	446	1136	1250	1326	1364	1304

(e) Solar Position and Insolation Values for 56° N

Date	Solar time a.m.	Solar time p.m.	Solar position Altitude	Solar position Azimuth	Total insolation on surfaces [Btu/(h·ft²)] Normal	Horizontal	South-facing surface angle with horizontal 46	56	66	76	90
Jan. 21	9	3	5.0	41.8	78	11	50	55	59	60	60
	10	2	9.9	28.5	170	39	135	146	154	156	153
	11	1	12.9	14.5	207	58	183	197	206	208	201
	12		14.0	0.0	217	65	198	214	222	225	217
Surface daily totals					1126	282	934	1010	1058	1074	1044
Feb. 21	8	4	7.6	59.4	129	25	65	69	72	72	69
	9	3	14.2	45.9	214	65	151	159	162	161	151
	10	2	19.4	31.5	250	98	215	225	228	224	208
	11	1	22.8	16.1	266	119	254	265	268	263	243
	12		24.0	0.0	270	126	268	279	282	276	255
Surface daily totals					1986	740	1640	1716	1742	1716	1598
Mar. 21	7	5	8.3	77.5	128	28	40	40	39	37	32
	8	4	16.2	64.4	215	75	119	120	117	111	97
	9	3	23.3	50.3	253	118	192	193	189	180	154
	10	2	29.0	34.9	272	151	249	251	246	234	205
	11	1	32.7	17.9	282	172	285	288	282	268	236
	12		34.0	0.0	284	179	297	300	294	280	246
Surface daily totals					2586	1268	2066	2084	2040	1938	1700
Apr. 21	5	7	1.4	108.8	0	0	0	0	0	0	0
	6	6	9.6	96.5	122	32	14	9	8	7	6
	7	5	18.0	84.1	201	81	74	66	57	46	29
	8	4	26.1	70.9	239	129	143	135	123	108	82
	9	3	33.6	56.3	260	169	208	200	186	167	133
	10	2	39.9	39.7	272	201	259	251	236	214	174
	11	1	44.1	20.7	278	220	292	284	268	245	200
	12		45.6	0.0	280	227	303	295	279	255	209
Surface daily totals					3024	1892	2282	2186	2038	1830	1458
May 21	4	8	1.2	125.5	0	0	0	0	0	0	0
	5	7	8.5	113.4	93	25	10	9	8	7	6
	6	6	16.5	101.5	175	71	28	17	15	13	11
	7	5	24.8	89.3	219	119	88	74	58	41	16
	8	4	33.1	76.3	244	163	153	138	119	98	63
	9	3	40.9	61.6	259	201	212	197	176	151	109
	10	2	47.6	44.2	268	231	259	244	222	194	146
	11	1	52.3	23.4	273	249	288	274	251	222	170
	12		54.0	0.0	275	255	299	284	261	231	178
Surface daily totals					3340	2374	2374	2188	1962	1682	1218
June 21	4	8	4.2	127.2	21	4	2	2	2	2	1
	5	7	11.4	115.3	122	40	14	13	11	10	8
	6	6	19.3	103.6	185	86	34	19	17	15	12
	7	5	27.6	91.7	222	132	92	76	57	38	15
	8	4	35.9	78.8	243	175	154	137	116	92	55
	9	3	43.8	64.1	257	212	211	193	170	143	98
	10	2	50.7	46.4	265	240	255	238	214	184	133
	11	1	55.6	24.9	269	258	284	267	242	210	156
	12		57.5	0.0	271	264	294	276	251	219	164
Surface daily totals					3438	2526	2388	2166	1910	1606	1120

(e) Solar Position and Insolation Values for 56° N (*Continued*)

Date	Solar time a.m.	Solar time p.m.	Solar position Altitude	Solar position Azimuth	Total insolation on surfaces [Btu/(h·ft²)] Normal	Horizontal	South-facing surface angle with horizontal 46	56	66	76	90
July 21	4	8	1.7	125.8	0	0	0	0	0	0	0
	5	7	9.0	113.7	91	27	11	10	9	8	6
	6	6	17.0	101.9	169	72	30	18	16	14	12
	7	5	25.3	89.7	212	119	88	74	58	41	15
	8	4	33.6	76.7	237	163	151	136	117	96	61
	9	3	41.4	62.0	252	201	208	193	173	147	106
	10	2	48.2	44.6	261	230	254	239	217	189	142
	11	1	52.9	23.7	265	248	283	268	245	216	165
	12		54.6	0.0	267	254	293	278	255	225	173
Surface daily totals					3240	2372	2342	2152	1926	1646	1186
Aug. 21	5	7	2.0	109.2	1	0	0	0	0	0	0
	6	6	10.2	97.0	112	34	16	11	10	9	7
	7	5	18.5	84.5	187	82	73	65	56	45	28
	8	4	26.7	71.3	225	128	140	131	119	104	78
	9	3	34.3	56.7	246	168	202	193	179	160	126
	10	2	40.5	40.0	258	199	251	242	227	206	166
	11	1	44.8	20.9	264	218	282	274	258	235	191
	12		46.3	0.0	266	225	293	285	269	245	200
Surface daily totals					2850	1884	2218	2118	1966	1760	1392
Sept. 21	7	5	8.3	77.5	107	25	36	36	34	32	28
	8	4	16.2	64.4	194	72	111	111	108	102	89
	9	3	23.3	50.3	233	114	181	182	178	168	147
	10	2	29.0	34.9	253	146	236	237	232	221	193
	11	1	32.7	17.9	263	166	271	273	267	254	223
	12		34.0	0.0	266	173	283	285	279	265	233
Surface daily totals					2368	1220	1950	1962	1918	1820	1594
Oct. 21	8	4	7.1	59.1	104	20	53	57	59	59	57
	9	3	13.8	45.7	193	60	138	145	148	147	138
	10	2	19.0	31.3	231	92	201	210	213	210	195
	11	1	22.3	16.0	248	112	240	250	253	248	230
	12		23.5	0.0	253	119	253	263	266	261	241
Surface daily totals					1804	688	1516	1586	1612	1588	1480
Nov. 21	9	3	5.2	41.9	76	12	49	54	57	59	58
	10	2	10.0	28.5	165	39	132	143	149	152	148
	11	1	13.1	14.5	201	58	179	193	201	203	196
	12		14.2	0.0	211	65	194	209	217	219	211
Surface daily totals					1094	284	914	986	1032	1046	1016
Dec. 21	9	3	1.9	40.5	5	0	3	4	4	4	4
	10	2	6.6	27.5	113	19	86	95	101	104	103
	11	1	9.5	13.9	166	37	141	154	163	167	164
	12		10.6	0.0	180	43	159	173	182	186	182
Surface daily totals					748	156	620	678	716	734	722

(f) Solar Position and Insolation Values for 64° N

Date	Solar time a.m.	Solar time p.m.	Solar position Altitude	Solar position Azimuth	Total insolation on surfaces [Btu/(h·ft²)] Normal	Horizontal	South-facing surface angle with horizontal 54	64	74	84	90
Jan. 21	10	2	2.8	28.1	22	2	17	19	20	20	20
	11	1	5.2	14.1	81	12	72	77	80	81	81
	12		6.0	0.0	100	16	91	98	102	103	103
Surface daily totals					306	45	268	290	302	306	304
Feb. 21	8	4	3.4	58.7	35	4	17	19	19	19	19
	9	3	8.6	44.8	147	31	103	108	111	110	107
	10	2	12.6	30.3	199	55	170	178	181	178	173
	11	1	15.1	15.3	222	71	212	220	223	219	213
	12		16.0	0.0	228	77	225	235	237	232	226
Surface daily totals					1432	400	1230	1286	1302	1282	1252
Mar. 21	7	5	6.5	76.5	95	18	30	29	29	27	25
	8	4	20.7	62.6	185	54	101	102	99	94	89
	9	3	18.1	48.1	227	87	171	172	169	160	153
	10	2	22.3	32.7	249	112	227	229	224	213	203
	11	1	25.1	16.6	260	129	262	265	259	246	235
	12		26.0	0.0	263	134	274	277	271	258	246
Surface daily totals					2296	932	1856	1870	1830	1736	1656
Apr. 21	5	7	4.0	108.5	27	5	2	2	2	1	1
	6	6	10.4	95.1	133	37	15	9	8	7	6
	7	5	17.0	81.6	194	76	70	63	54	43	37
	8	4	23.3	67.5	228	112	136	128	116	102	91
	9	3	29.0	52.3	248	144	197	189	176	158	145
	10	2	33.5	36.0	260	169	246	239	224	203	188
	11	1	36.5	18.4	266	184	278	270	255	233	216
	12		97.6	0.0	268	190	289	281	266	243	225
Surface daily totals					2982	1644	2176	2082	1936	1736	1594
May 21	4	8	5.8	125.1	51	11	5	4	4	3	3
	5	7	11.6	112.1	132	42	13	11	10	9	8
	6	6	17.9	99.1	185	79	29	16	14	12	11
	7	5	24.5	85.7	218	117	86	72	56	39	28
	8	4	30.9	71.5	239	152	148	133	115	94	80
	9	3	36.8	56.1	252	182	204	190	170	145	128
	10	2	41.6	38.9	261	205	249	235	213	186	167
	11	1	44.9	20.1	265	219	278	264	242	213	193
	12		46.0	0.0	267	224	288	274	251	222	201
Surface daily totals					3470	2236	2312	2124	1898	1624	1436
June 21	3	9	4.2	139.4	21	4	2	2	2	2	1
	4	8	9.0	126.4	93	27	10	9	8	7	6
	5	7	14.7	113.6	154	60	16	15	13	11	10
	6	6	21.0	100.8	194	96	34	19	17	14	13
	7	5	27.5	87.5	221	132	91	74	55	36	23
	8	4	34.0	73.3	239	166	150	133	112	88	73
	9	3	39.9	57.8	251	195	204	187	164	137	119
	10	2	44.9	40.4	258	217	247	230	206	177	157
	11	1	48.3	20.9	262	231	275	258	233	202	181
	12		49.5	0.0	263	235	284	267	242	211	189
Surface daily totals					3650	2488	2342	2118	1862	1558	1356

(f) Solar Position and Insolation Values for 64°N (*Continued*)

Date	Solar time a.m.	Solar time p.m.	Solar position Altitude	Solar position Azimuth	Total insolation on surfaces [Btu/(h·ft²)] Normal	Horizontal	South-facing surface angle with horizontal 54	64	74	84	90
July 21	4	8	6.4	125.3	53	13	6	5	5	4	4
	5	7	12.1	112.4	128	44	14	13	11	10	9
	6	6	18.4	99.4	179	81	30	17	16	13	12
	7	5	25.0	86.0	211	118	86	72	56	38	28
	8	4	31.4	71.8	231	152	146	131	113	91	77
	9	3	37.3	56.3	245	182	201	186	166	141	124
	10	2	42.2	39.2	253	204	245	230	208	181	162
	11	1	45.4	20.2	257	218	273	258	236	207	187
	12		46.6	0.0	259	223	282	267	245	216	195
Surface daily totals					3372	2248	2280	2090	1864	1588	1400
Aug. 21	5	7	4.6	108.8	29	6	3	3	2	2	2
	6	6	11.0	95.5	123	39	16	11	10	8	7
	7	5	17.6	81.9	181	77	69	61	52	42	35
	8	4	23.9	67.8	214	113	132	123	112	97	87
	9	3	29.6	52.6	234	144	190	182	169	150	138
	10	2	34.2	36.2	246	168	237	229	215	194	179
	11	1	37.2	18.5	252	183	268	260	244	222	205
	12		38.3	0.0	254	188	278	270	255	232	215
Surface daily totals					2808	1646	2108	1008	1860	1662	1522
Sept. 21	7	5	6.5	76.5	77	16	25	25	24	23	21
	8	4	12.7	72.6	163	51	92	92	90	85	81
	9	3	18.1	48.1	206	83	159	159	156	147	141
	10	2	22.3	32.7	229	108	212	213	209	198	189
	11	1	25.1	16.6	240	124	246	248	243	230	220
	12		26.0	0.0	244	129	258	260	254	241	230
Surface daily totals					2074	892	1726	1736	1696	1608	1532
Oct. 21	8	4	3.0	58.5	17	2	9	9	10	10	10
	9	3	8.1	44.6	122	26	86	91	93	92	90
	10	2	12.1	30.2	176	50	152	159	161	159	155
	11	1	14.6	15.2	201	65	193	201	203	200	195
	12		15.5	0.0	208	71	207	215	217	213	208
Surface daily totals					1238	358	1088	1136	1152	1134	1106
Nov. 21	10	2	3.0	28.1	23	3	18	20	21	21	21
	11	1	5.4	14.2	79	12	70	76	79	80	79
	12		6.2	0.0	97	17	89	96	100	101	100
Surface daily totals					302	46	266	286	298	302	300
Dec. 21	11	1	1.8	13.7	4	0	3	4	4	4	4
	12		2.6	0.0	16	2	14	15	16	17	17
Surface daily totals					24	2	20	22	24	24	24

*1 Btu/(h·ft²) = 3.152 W/m².

†"Normal" table entries do not include either diffuse sky or reflected radiation.

Note: (1) Based on data in Table 1, p. 26.2 in *ASHRAE Handbook of Fundamentals*, ASHRAE, Inc., 345 E 47th St., New York, NY, 1977; 0% ground reflectance; 1.0 clearness factor. (2) See Fig. 4, p. 26.9 in *ASHRAE Handbook of Fundamentals* for typical regional clearness factors. (3) Ground reflection not included on normal or horizontal surfaces.

Appendix 6

Monthly Solar and Climatic Data for Locations in the Conterminous United States

STA NO: 3133	HORIZ INSOL:	953.	1273.	1764.	2246.	2577.	2734.	2653.	2382.	2022.	1516.	1041.	853.	1835.
YUCCA FLATS	TILT = LAT:	1635.	1851.	2147.	2265.	2259.	2247.	2247.	2266.	2306.	2126.	1722.	1534.	0
NV	TILT = LAT+15:	1748.	1897.	2078.	2036.	1908.	1841.	1867.	1986.	2174.	2151.	1824.	1657.	0
LAT = 36.95	TILT = 90:	1541.	1499.	1363.	983.	680.	552.	607.	850.	1281.	1629.	1572.	1499.	0
	KT:	.62	.64	.68	.71	.72	.73	.73	.72	.72	.70	.64	.61	0
	AMB TEMP:	0	0	0	0	0	0	0	0	0	0	0	0	0
	HTG DEG DAYS:	0	0	0	0	0	0	0	0	0	0	0	0	0
STA NO: 3812	HORIZ INSOL:	722.	971.	1306.	1668.	1804.	1854.	1776.	1627.	1361.	1147.	849.	658.	1312.
ASHEVILLE	TILT = LAT:	1086.	1285.	1492.	1642.	1594.	1564.	1532.	1527.	1460.	1471.	1266.	1032.	0
NC	TILT = LAT+15:	1139.	1297.	1429.	1479.	1368.	1314.	1301.	1348.	1365.	1468.	1323.	1094.	0
LAT = 35.43	TILT = 90:	961.	983.	905.	718.	541.	479.	498.	608.	787.	1065.	1099.	950.	0
	KT:	.45	.47	.49	.52	.50	.50	.49	.49	.48	.51	.49	.44	0
	AMB TEMP:	37.9	39.4	45.9	55.9	63.7	70.6	73.5	72.8	66.7	56.8	46.3	38.7	55.7
	HTG DEG DAYS:	840	717	592	279	100	14	0	0	50	269	561	815	4237
STA NO: 3813	HORIZ INSOL:	769.	1020.	1363.	1736.	1885.	1919.	1785.	1718.	1439.	1247.	940.	729.	1379.
MACON	TILT = LAT:	1087.	1295.	1526.	1702.	1671.	1630.	1550.	1613.	1526.	1552.	1335.	1077.	0
GA	TILT = LAT+15:	1139.	1307.	1464.	1535.	1436.	1371.	1321.	1427.	1430.	1551.	1396.	1143.	0
LAT = 32.70	TILT = 90:	932.	954.	881.	686.	509.	444.	459.	586.	777.	1084.	1129.	967.	0
	KT:	.44	.46	.50	.53	.52	.52	.49	.51	.49	.53	.50	.45	0
	AMB TEMP:	47.8	50.4	56.5	65.8	73.5	79.6	81.4	80.9	75.8	65.7	55.2	48.3	65.1
	HTG DEG DAYS:	543	423	298	66	6	0	0	0	0	82	304	518	2240
STA NO: 3820	HORIZ INSOL:	751.	1015.	1338.	1728.	1865.	1904.	1803.	1667.	1410.	1220.	917.	721.	1362.
AUGUSTA	TILT = LAT:	1075.	1303.	1504.	1696.	1652.	1614.	1563.	1565.	1498.	1528.	1317.	1084.	0
GA	TILT = LAT+15:	1125.	1316.	1442.	1529.	1419.	1357.	1329.	1384.	1403.	1526.	1377.	1151.	0
LAT = 33.37	TILT = 90:	927.	970.	879.	698.	517.	453.	471.	585.	774.	1076.	1121.	981.	0
	KT:	.43	.47	.49	.53	.52	.51	.49	.49	.49	.52	.50	.45	0
	AMB TEMP:	45.8	48.3	54.6	63.8	71.7	78.2	80.4	79.6	74.2	64.1	53.7	46.4	63.4
	HTG DEG DAYS:	601	475	346	90	10	0	0	0	0	104	344	577	2547
STA NO: 3822	HORIZ INSOL:	795.	1044.	1399.	1761.	1852.	1844.	1783.	1621.	1364.	1217.	941.	754.	1365.
SAVANNAH	TILT = LAT:	1117.	1318.	1562.	1725.	1645.	1572.	1551.	1521.	1435.	1496.	1319.	1105.	0
GA	TILT = LAT+15:	1170.	1331.	1500.	1557.	1415.	1327.	1322.	1347.	1344.	1493.	1378.	1173.	0
LAT = 32.13	TILT = 90:	954.	965.	893.	684.	495.	431.	451.	554.	723.	1031.	1106.	988.	0
	KT:	.44	.47	.51	.54	.51	.50	.49	.48	.46	.51	.50	.45	0
	AMB TEMP:	49.9	52.1	58.0	66.1	73.3	79.1	81.1	80.6	76.2	67.1	57.1	50.4	65.9
	HTG DEG DAYS:	483	379	256	63	0	0	0	0	0	60	253	458	1952
STA NO: 3860	HORIZ INSOL:	526.	757.	1067.	1448.	1710.	1844.	1769.	1560.	1306.	1004.	638.	467.	1176.
HUNTINGTON	TILT = LAT:	787.	1002.	1219.	1425.	1505.	1542.	1515.	1466.	1424.	1322.	958.	726.	0
WV	TILT = LAT+15:	815.	1002.	1160.	1279.	1289.	1292.	1282.	1307.	1329.	1315.	990.	760.	0
LAT = 38.37	TILT = 90:	691.	775.	770.	680.	569.	521.	540.	641.	813.	989.	832.	660.	0
	KT:	.36	.39	.42	.46	.48	.49	.48	.48	.48	.48	.41	.35	0
	AMB TEMP:	34.3	36.1	44.3	55.7	64.5	72.4	75.3	73.9	67.7	57.1	45.5	36.0	55.2
	HTG DEG DAYS:	952	809	649	293	115	11	0	0	46	265	585	899	4624
STA NO: 3870	HORIZ INSOL:	730.	982.	1329.	1697.	1833.	1913.	1830.	1699.	1406.	1180.	880.	670.	1347.
GREENVILLE-SPARTANBU	TILT = LAT:	1083.	1288.	1513.	1671.	1625.	1618.	1579.	1597.	1508.	1507.	1305.	1038.	0
SC	TILT = LAT+15:	1136.	1301.	1450.	1504.	1394.	1358.	1340.	1409.	1411.	1505.	1366.	1101.	0
LAT = 34.90	TILT = 90:	953.	979.	910.	718.	538.	477.	497.	619.	804.	1085.	1130.	951.	0
	KT:	.44	.47	.50	.53	.51	.51	.50	.51	.49	.52	.50	.44	0
	AMB TEMP:	42.3	44.4	50.9	61.0	69.1	75.9	78.3	77.5	71.7	61.7	51.0	42.9	60.6

	HTG DEG DAYS:	704	577	450	144	29	0	0	0	9	145	420	685	3163
STA NO: 3927	HORIZ INSOL:	805.	1069.	1409.	1616.	1890.	2153.	2155.	1983.	1621.	1293.	938.	766.	1475.
FORT WORTH	TILT = LAT:	1157.	1373.	1585.	1580.	1676.	1816.	1856.	1865.	1739.	1622.	1336.	1151.	0
TX	TILT = LAT+15:	1215.	1390.	1522.	1426.	1440.	1518.	1567.	1646.	1633.	1624.	1398.	1225.	0
LAT = 32.83	TILT = 90:	1001.	1020.	919.	650.	512.	456.	491.	649.	884.	1139.	1133.	1044.	0
	KT:	.46	.49	.51	.50	.53	.58	.59	.59	.56	.55	.51	.47	0
	AMB TEMP:	44.8	48.7	55.0	65.2	72.5	80.6	84.8	84.9	77.7	67.6	55.8	47.9	65.5
	HTG DEG DAYS:	626	456	335	88	0	0	0	0	0	60	287	530	2382
STA NO: 3928	HORIZ INSOL:	784.	1058.	1405.	1783.	2036.	2264.	2239.	2032.	1616.	1250.	871.	690.	1502.
WICHITA	TILT = LAT:	1302.	1497.	1663.	1774.	1730.	1880.	1907.	1924.	1798.	1699.	1401.	1196.	0
KS	TILT = LAT+15:	1379.	1522.	1598.	1593.	1526.	1559.	1599.	1690.	1686.	1705.	1472.	1279.	0
LAT = 37.65	TILT = 90:	1207.	1198.	1055.	812.	622.	556.	597.	773.	1014.	1287.	1261.	1147.	0
	KT:	.52	.54	.55	.56	.57	.60	.61	.61	.58	.59	.54	.51	0
	AMB TEMP:	31.3	36.3	43.6	56.6	66.1	75.8	80.7	79.7	70.6	59.6	44.8	34.5	56.6
	HTG DEG DAYS:	1045	804	671	275	90	7	0	0	32	211	606	946	4687
STA NO: 3937	HORIZ INSOL:	728.	1010.	1313.	1570.	1849.	1970.	1788.	1657.	1485.	1381.	917.	706.	1365.
LAKE CHARLES	TILT = LAT:	954.	1228.	1434.	1528.	1649.	1684.	1563.	1556.	1558.	1682.	1220.	962.	0
LA	TILT = LAT+15:	990.	1236.	1375.	1382.	1422.	1419.	1335.	1381.	1462.	1686.	1269.	1011.	0
LAT = 30.12	TILT = 90:	775.	866.	788.	589.	463.	408.	424.	530.	748.	1135.	987.	822.	0
	KT:	.38	.43	.46	.47	.51	.53	.49	.49	.50	.56	.46	.40	0
	AMB TEMP:	52.3	55.1	60.3	68.9	75.2	80.7	82.4	82.2	78.4	70.0	60.2	54.3	68.3
	HTG DEG DAYS:	415	306	200	26	0	0	0	0	0	36	177	338	1498
STA NO: 3940	HORIZ INSOL:	754.	1026.	1369.	1708.	1941.	2024.	1909.	1780.	1509.	1271.	902.	709.	1409.
JACKSON	TILT = LAT:	1049.	1296.	1528.	1672.	1721.	1716.	1655.	1672.	1604.	1578.	1257.	1028.	0
MS	TILT = LAT+15:	1096.	1308.	1466.	1509.	1478.	1440.	1406.	1479.	1504.	1578.	1311.	1087.	0
LAT = 32.32	TILT = 90:	890.	950.	876.	671.	509.	443.	465.	594.	808.	1097.	1050.	912.	0
	KT:	.42	.46	.50	.52	.54	.54	.52	.53	.51	.54	.48	.43	0
	AMB TEMP:	47.1	49.8	56.1	65.7	72.7	79.4	81.7	81.2	76.0	65.8	55.3	48.9	65.0
	HTG DEG DAYS:	569	442	313	74	6	0	0	0	0	91	301	504	2300
STA NO: 3945	HORIZ INSOL:	612.	875.	1179.	1526.	1880.	2090.	2116.	1878.	1450.	1101.	703.	523.	1328.
COLUMBIA	TILT = LAT:	981.	1215.	1376.	1510.	1652.	1736.	1802.	1778.	1609.	1492.	1105.	865.	0
MO	TILT = LAT+15:	1028.	1226.	1314.	1354.	1410.	1444.	1513.	1561.	1505.	1490.	1150.	913.	0
LAT = 38.82	TILT = 90:	892.	967.	882.	723.	613.	561.	604.	751.	928.	1135.	982.	808.	0
	KT:	.43	.46	.47	.4[illegible]	.53	.56	.58	.57	.53	.53	.46	.40	0
	AMB TEMP:	29.3	33.6	41.7	55.0	64.4	73.0	77.3	76.0	68.3	58.0	43.9	32.8	54.4
	HTG DEG DAYS:	1107	879	730	314	117	11	0	5	42	247	633	998	5083
STA NO: 3947	HORIZ INSOL:	648.	895.	1203.	1575.	1873.	2080.	2102.	1862.	1452.	1092.	737.	562.	1340.
KANSAS CITY	TILT = LAT:	1080.	1265.	1417.	1564.	1645.	1726.	1789.	1764.	1619.	1493.	1196.	975.	0
MO	TILT = LAT+15:	1136.	1278.	1355.	1403.	1403.	1436.	1502.	1548.	1514.	1492.	1250.	1035.	0
LAT = 39.30	TILT = 90:	998.	1017.	917.	755.	620.	568.	612.	756.	942.	1144.	1078.	929.	0
	KT:	.46	.47	.48	.50	.52	.55	.58	.57	.54	.53	.49	.44	0
	AMB TEMP:	27.1	32.3	40.7	54.2	64.1	73.0	77.5	76.5	68.0	57.6	42.3	31.3	53.7
	HTG DEG DAYS:	1175	916	753	336	127	15	0	0	50	259	681	1045	5357
STA NO: 4725	HORIZ INSOL:	386.	576.	861.	1242.	1496.	1681.	1659.	1425.	1131.	779.	414.	297.	996.
BINGHAMTON	TILT = LAT:	591.	768.	986.	1221.	1309.	1396.	1411.	1339.	1249.	1044.	604.	445.	0
NY	TILT = LAT+15:	606.	760.	932.	1091.	1120.	1168.	1191.	1174.	1160.	1031.	613.	456.	0
LAT = 42.22	TILT = 90:	524.	605.	653.	639.	568.	548.	573.	642.	763.	805.	517.	394.	0
	KT:	.31	.33	.36	.41	.42	.45	.46	.44	.43	.41	.31	.27	0

	AMB TEMP:	22.0	22.8	31.3	44.7	55.1	64.8	69.1	67.3	60.2	50.3	38.2	25.4	46.0
	HTG DEG DAYS:	1333	1182	1045	609	320	75	21	40	172	456	804	1228	7285
STA NO:12832	HORIZ INSOL:	853.	1126.	1474.	1879.	2091.	1998.	1814.	1683.	1535.	1371.	1040.	818.	1474.
APALACHICOLA	TILT = LAT:	1150.	1387.	1624.	1836.	1856.	1709.	1586.	1585.	1611.	1660.	1409.	1145.	0
FL	TILT = LAT+15:	1205.	1403.	1562.	1659.	1595.	1439.	1355.	1407.	1514.	1663.	1476.	1215.	0
LAT = 29.73	TILT = 90:	957.	986.	886.	668.	475.	403.	420.	530.	765.	1112.	1156.	1000.	0
	KT:	.44	.48	.52	.57	.58	.54	.50	.50	.51	.55	.52	.45	0
	AMB TEMP:	53.7	55.8	60.7	68.3	74.9	80.0	81.4	81.5	78.6	70.8	61.1	55.2	68.5
	HTG DEG DAYS:	368	290	175	30	0	0	0	0	0	22	158	318	1361
STA NO:12834	HORIZ INSOL:	958.	1213.	1548.	1884.	1968.	1826.	1784.	1682.	1478.	1251.	1036.	870.	1458.
DAYTONA BEACH	TILT = LAT:	1308.	1500.	1706.	1839.	1755.	1573.	1564.	1580.	1542.	1485.	1385.	1219.	0
FL	TILT = LAT+15:	1380.	1522.	1643.	1663.	1511.	1332.	1338.	1403.	1449.	1482.	1449.	1298.	0
LAT = 29.18	TILT = 90:	1101.	1067.	921.	657.	457.	392.	411.	520.	728.	976.	1126.	1068.	0
	KT:	.49	.51	.54	.57	.55	.50	.49	.49	.49	.50	.51	.48	0
	AMB TEMP:	58.4	59.6	63.9	69.7	75.0	79.4	81.0	81.1	79.5	73.3	65.1	59.6	70.5
	HTG DEG DAYS:	241	210	120	17	0	0	0	0	0	5	97	212	902
STA NO:12839	HORIZ INSOL:	1057.	1314.	1603.	1859.	1844.	1708.	1763.	1630.	1456.	1303.	1119.	1019.	1473.
MIAMI	TILT = LAT:	1369.	1568.	1730.	1807.	1660.	1495.	1563.	1534.	1498.	1495.	1421.	1361.	0
FL	TILT = LAT+15:	1445.	1594.	1670.	1639.	1438.	1275.	1342.	1369.	1410.	1493.	1487.	1455.	0
LAT = 25.80	TILT = 90:	1111.	1063.	869.	581.	399.	354.	369.	461.	653.	930.	1109.	1163.	0
	KT:	.50	.52	.54	.55	.51	.47	.49	.48	.47	.49	.51	.51	0
	AMB TEMP:	67.2	67.8	71.3	75.0	78.0	81.0	82.3	82.9	81.7	77.8	72.2	68.3	75.5
	HTG DEG DAYS:	53	67	17	0	0	0	0	0	0	0	13	56	206
STA NO:12841	HORIZ INSOL:	999.	1244.	1582.	1898.	1989.	1831.	1801.	1673.	1497.	1304.	1096.	926.	1487.
ORLANDO	TILT = LAT:	1356.	1529.	1739.	1851.	1775.	1581.	1581.	1572.	1559.	1544.	1463.	1294.	0
FL	TILT = LAT+15:	1432.	1553.	1676.	1675.	1529.	1339.	1353.	1397.	1465.	1543.	1534.	1382.	0
LAT = 28.55	TILT = 90:	1138.	1079.	927.	647.	448.	384.	404.	509.	722.	1008.	1188.	1135.	0
	KT:	.50	.52	.55	.57	.55	.50	.50	.49	.49	.52	.53	.50	0
	AMB TEMP:	60.3	61.5	65.9	71.3	76.4	80.2	81.4	81.8	80.1	74.3	66.6	61.5	71.8
	HTG DEG DAYS:	197	184	94	13	0	0	0	0	0	0	75	170	733
STA NO:12842	HORIZ INSOL:	1011.	1259.	1594.	1908.	1998.	1847.	1753.	1653.	1492.	1346.	1108.	935.	1492.
TAMPA	TILT = LAT:	1357.	1538.	1745.	186[illegible].	1785.	1597.	1543.	1553.	1550.	1590.	1464.	1291.	0
FL	TILT = LAT+15:	1433.	1563.	1682.	1684.	1538.	1353.	1323.	1382.	1457.	1590.	1535.	1378.	0
LAT = 27.97	TILT = 90:	1131.	1077.	918.	637.	438.	377.	394.	496.	709.	1030.	1180.	1124.	0
	KT:	.50	.52	.55	.57	.56	.50	.48	.48	.49	.53	.53	.49	0
	AMB TEMP:	60.4	61.8	66.0	72.0	77.2	81.0	81.9	82.2	80.8	74.7	66.8	61.6	72.2
	HTG DEG DAYS:	203	176	90	9	0	0	0	0	0	0	71	169	718
STA NO:12844	HORIZ INSOL:	1000.	1233.	1556.	1814.	1845.	1706.	1779.	1663.	1419.	1224.	1060.	958.	1438.
WEST PALM BEACH	TILT = LAT:	1304.	1475.	1685.	1764.	1657.	1488.	1571.	1564.	1462.	1407.	1356.	1290.	0
FL	TILT = LAT+15:	1374.	1496.	1625.	1599.	1435.	1269.	1348.	1394.	1375.	1402.	1416.	1376.	0
LAT = 26.68	TILT = 90:	1063.	1008.	863.	590.	411.	362.	379.	479.	653.	884.	1064.	1105.	0
	KT:	.48	.50	.53	.54	.51	.47	.49	.49	.46	.47	.49	.49	0
	AMB TEMP:	65.5	66.1	69.8	73.9	77.5	80.5	81.9	82.3	81.5	77.2	71.0	66.8	74.5
	HTG DEG DAYS:	83	91	25	0	0	0	0	0	0	0	22	78	299
STA NO:12907	HORIZ INSOL:	959.	1196.	1516.	1727.	1952.	2073.	2131.	2009.	1705.	1408.	1041.	890.	1551.
LAREDO	TILT = LAT:	1263.	1440.	1647.	1679.	1747.	1782.	1862.	1888.	1782.	1663.	1348.	1203.	0
TX	TILT = LAT+15:	1328.	1459.	1586.	1522.	1507.	1500.	1580.	1673.	1679.	1667.	1408.	1278.	0
LAT = 27.53	TILT = 90:	1036.	995.	859.	586.	429.	371.	394.	541.	796.	1074.	1069.	1030.	0

	KT:	.47	.49	.52	.52	.54	.57	.59	.59	.56	.55	.49	.46	0
	AMB TEMP:	56.5	60.9	67.6	76.3	81.3	86.0	87.9	87.7	82.9	75.5	65.2	58.6	73.9
	HTG DEG DAYS:	299	177	87	0	0	0	0	0	0	8	74	231	876
STA NO:12916	HORIZ INSOL:	835.	1112.	1415.	1780.	1968.	2004.	1814.	1717.	1514.	1335.	973.	779.	1437.
NEW ORLEANS	TILT = LAT:	1126.	1372.	1555.	1738.	1751.	1712.	1585.	1612.	1588.	1615.	1307.	1086.	0
LA	TILT = LAT+15:	1179.	1388.	1494.	1571.	1507.	1441.	1354.	1430.	1492.	1617.	1364.	1150.	0
LAT = 29.98	TILT = 90:	937.	978.	853.	647.	471.	406.	424.	540.	760.	1084.	1065.	944.	0
	KT:	.44	.48	.50	.54	.55	.54	.50	.50	.51	.54	.49	.44	0
	AMB TEMP:	52.9	55.6	60.7	68.6	75.1	80.4	81.9	81.9	78.2	69.8	60.1	54.8	68.3
	HTG DEG DAYS:	403	299	188	29	0	0	0	0	0	40	179	327	1465
STA NO:12917	HORIZ INSOL:	800.	1071.	1353.	1610.	1871.	2011.	1846.	1736.	1527.	1321.	953.	754.	1404.
PORT ARTHUR	TILT = LAT:	1067.	1312.	1480.	1566.	1668.	1718.	1613.	1630.	1603.	1596.	1274.	1042.	0
TX	TILT = LAT+15:	1114.	1324.	1421.	1417.	1438.	1446.	1376.	1445.	1506.	1596.	1328.	1100.	0
LAT = 29.95	TILT = 90:	881.	930.	811.	598.	463.	406.	425.	543.	766.	1069.	1035.	899.	0
	KT:	.42	.46	.48	.49	.52	.54	.51	.51	.51	.53	.48	.42	0
	AMB TEMP:	52.0	55.1	60.1	68.9	75.0	80.8	83.0	83.1	78.9	69.9	60.2	54.2	68.5
	HTG DEG DAYS:	420	302	202	33	0	0	0	0	0	35	184	342	1518
STA NO:12919	HORIZ INSOL:	913.	1135.	1458.	1737.	1927.	2115.	2213.	2027.	1694.	1439.	1055.	862.	1548.
BROWNSVILLE	TILT = LAT:	1153.	1330.	1563.	1687.	1732.	1827.	1938.	1905.	1757.	1672.	1329.	1118.	0
TX	TILT = LAT+15:	1206.	1343.	1505.	1531.	1497.	1537.	1643.	1691.	1657.	1676.	1386.	1182.	0
LAT = 25.90	TILT = 90:	913.	888.	788.	559.	403.	348.	364.	509.	753.	1050.	1029.	927.	0
	KT:	.43	.45	.49	.52	.54	.58	.61	.59	.55	.55	.48	.43	0
	AMB TEMP:	60.3	63.4	67.7	74.9	79.3	82.8	84.4	84.4	81.6	75.7	68.1	62.8	73.8
	HTG DEG DAYS:	225	151	89	0	0	0	0	0	0	5	35	145	650
STA NO:12921	HORIZ INSOL:	895.	1154.	1450.	1612.	1895.	2069.	2121.	1947.	1638.	1350.	1009.	847.	1499.
SAN ANTONIO	TILT = LAT:	1214.	1423.	1592.	1568.	1690.	1767.	1843.	1829.	1725.	1626.	1352.	1190.	0
TX	TILT = LAT+15:	1276.	1441.	1531.	1420.	1457.	1485.	1562.	1619.	1622.	1628.	1414.	1265.	0
LAT = 29.53	TILT = 90:	1016.	1012.	865.	592.	458.	400.	428.	573.	812.	1084.	1101.	1043.	0
	KT:	.46	.49	.51	.49	.53	.56	.58	.57	.54	.54	.50	.47	0
	AMB TEMP:	50.7	54.5	60.8	69.	76.0	82.2	84.7	84.7	79.3	70.5	59.7	53.2	68.8
	HTG DEG DAYS:	451	310	194	31	0	0	0	0	0	32	179	373	1570
STA NO:12924	HORIZ INSOL:	898.	1147.	1430.	1642.	1866.	2094.	2186.	1991.	1687.	1416.	1043.	845.	1521.
CORPUS CHRISTI	TILT = LAT:	1173.	1379.	1549.	1595.	1672.	1798.	1906.	1870.	1764.	1679.	1357.	1136.	0
TX	TILT = LAT+15:	1229.	1394.	1489.	1446.	1445.	1511.	1615.	1657.	1661.	1683.	1418.	1203.	0
LAT = 27.77	TILT = 90:	954.	951.	812.	570.	429.	374.	397.	543.	793.	1089.	1080.	967.	0
	KT:	.44	.47	.49	.49	.52	.57	.60	.58	.55	.55	.49	.44	0
	AMB TEMP:	56.3	59.6	64.9	72.8	77.9	82.4	84.8	85.1	81.0	73.9	64.9	59.1	71.9
	HTG DEG DAYS:	304	199	120	0	0	0	0	0	0	7	81	219	930
STA NO:12928	HORIZ INSOL:	912.	1161.	1435.	1663.	1864.	2036.	2111.	1922.	1625.	1390.	1034.	849.	1500.
KINGSVILLE	TILT = LAT:	1189.	1393.	1552.	1615.	1671.	1752.	1845.	1805.	1693.	1639.	1338.	1137.	0
TX	TILT = LAT+15:	1247.	1409.	1493.	1465.	1444.	1476.	1567.	1602.	1594.	1641.	1397.	1204.	0
LAT = 27.52	TILT = 90:	966.	958.	810.	571.	425.	371.	394.	529.	760.	1057.	1059.	965.	0
	KT:	.45	.47	.49	.50	.52	.56	.58	.56	.53	.54	.49	.44	0
	AMB TEMP:	0	0	0	0	0	0	0	0	0	0	0	0	0
	HTG DEG DAYS:	0	0	0	0	0	0	0	0	0	0	0	0	0
STA NO:12960	HORIZ INSOL:	772.	1034.	1297.	1522.	1775.	1898.	1828.	1666.	1471.	1276.	924.	730.	1351.
HOUSTON	TILT = LAT:	1023.	1261.	1414.	1479.	1585.	1627.	1597.	1563.	1541.	1533.	1228.	1000.	0
TX	TILT = LAT+15:	1066.	1270.	1355.	1339.	1369.	1373.	1363.	1405.	1446.	1532.	1278.	1054.	0

LAT = 29.98	TILT = 90:	839.	890.	774.	573.	454.	404.	424.	534.	738.	1024.	993.	858.	0
	KT:	.40	.44	.46	.46	.49	.51	.50	.50	.49	.52	.46	.41	0
	AMB TEMP:	52.1	55.3	60.8	69.4	75.8	81.1	83.3	83.4	79.2	70.9	61.1	54.6	68.9
	HTG DEG DAYS:	416	294	189	23	0	0	0	0	0	24	155	333	1434
STA NO:13721	HORIZ INSOL:	608.	862.	1181.	1538.	1763.	1893.	1817.	1627.	1357.	1021.	707.	537.	1243.
PATUXENT RIVER	TILT = LAT:	955.	1178.	1371.	1520.	1551.	1582.	1555.	1531.	1487.	1347.	1093.	878.	0
MD	TILT = LAT+15:	998.	1187.	1309.	1364.	1327.	1323.	1315.	1347.	1388.	1340.	1138.	926.	0
LAT = 38.28	TILT = 90:	860.	928.	871.	719.	579.	526.	547.	656.	848.	1008.	965.	817.	0
	KT:	.42	.45	.47	.49	.49	.51	.50	.49	.49	.49	.45	.40	0
	AMB TEMP:	0	0	0	0	0	0	0	0	0	0	0	0	0
	HTG DEG DAYS:	0	0	0	0	0	0	0	0	0	0	0	0	0
STA NO:13722	HORIZ INSOL:	694.	943.	1276.	1644.	1808.	1864.	1776.	1611.	1377.	1105.	812.	636.	1295.
RALEIGH-DURHAM	TILT = LAT:	1046.	1250.	1459.	1620.	1546.	1570.	1530.	1513.	1484.	1418.	1212.	1002.	0
NC	TILT = LAT+15:	1096.	1261.	1397.	1457.	1369.	1318.	1298.	1335.	1387.	1413.	1265.	1062.	0
LAT = 35.87	TILT = 90:	926.	959.	891.	717.	549.	486.	504.	610.	807.	1029.	1053.	923.	0
	KT:	.43	.46	.48	.51	.50	.50	.49	.48	.49	.50	.48	.43	0
	AMB TEMP:	40.5	42.2	49.2	59.5	67.4	74.4	77.5	76.5	70.6	60.2	50.0	41.2	59.1
	HTG DEG DAYS:	760	638	502	180	48	0	0	0	12	186	450	738	3514
STA NO:13723	HORIZ INSOL:	715.	970.	1313.	1683.	1868.	1953.	1864.	1697.	1418.	1141.	839.	659.	1343.
GREENSBORO	TILT = LAT:	1096.	1300.	1511.	1661.	1648.	1640.	1602.	1595.	1535.	1478.	1272.	1058.	0
NC	TILT = LAT+15:	1151.	1313.	1448.	1494.	1411.	1373.	1357.	1406.	1435.	1476.	1330.	1125.	0
LAT = 36.08	TILT = 90:	979.	1004.	928.	737.	563.	499.	519.	639.	838.	1081.	1114.	985.	0
	KT:	.45	.47	.50	.53	.52	.52	.51	.51	.50	.52	.50	.45	0
	AMB TEMP:	38.7	40.6	47.8	58.6	67.1	74.4	77.2	76.0	69.7	59.2	48.3	39.6	58.1
	HTG DEG DAYS:	815	683	544	203	59	0	0	0	24	209	501	787	3825
STA NO:13737	HORIZ INSOL:	678.	932.	1281.	1677.	1888.	2000.	1853.	1680.	1396.	1083.	811.	624.	1325.
NORFOLK	TILT = LAT:	1051.	1259.	1482.	1659.	1663.	1674.	1590.	1581.	1517.	1409.	1249.	1016.	0
VA	TILT = LAT+15:	1102.	1271.	1419.	1490.	1422.	1399.	1346.	1392.	1418.	1404.	1306.	1079.	0
LAT = 36.90	TILT = 90:	942.	980.	923.	751.	581.	517.	531.	649.	842.	1038.	1102.	949.	0
	KT:	.44	.46	.49	.53	.53	.53	.51	.51	.50	.50	.49	.44	0
	AMB TEMP:	40.5	41.4	48.1	57.8	66.7	74.5	78.3	76.9	71.8	61.7	51.6	42.3	59.3
	HTG DEG DAYS:	760	661	532	226	53	0	0	0	9	141	402	704	3488
STA NO:13739	HORIZ INSOL:	555.	795.	1108.	1434.	1660.	1811.	1758.	1575.	1281.	959.	619.	470.	1169.
PHILADELPHIA	TILT = LAT:	898.	1104.	1297.	1417.	1456.	1510.	1501.	1483.	1412.	1288.	970.	783.	0
PA	TILT = LAT+15:	938.	1110.	1236.	1270.	1247.	1263.	1268.	1302.	1316.	1280.	1005.	824.	0
LAT = 39.88	TILT = 90:	817.	882.	843.	699.	580.	539.	560.	664.	829.	981.	860.	732.	0
	KT:	.40	.43	.45	.46	.46	.48	.48	.48	.48	.47	.42	.38	0
	AMB TEMP:	32.3	33.9	41.9	52.9	63.2	72.3	76.8	74.8	68.1	57.4	46.2	35.2	54.6
	HTG DEG DAYS:	1014	871	716	367	122	0	0	0	38	249	564	924	4865
STA NO:13740	HORIZ INSOL:	632.	877.	1210.	1566.	1762.	1872.	1774.	1601.	1348.	1033.	733.	567.	1248.
RICHMOND	TILT = LAT:	977.	1183.	1398.	1546.	1552.	1569.	1523.	1504.	1466.	1345.	1118.	915.	0
VA	TILT = LAT+15:	1022.	1192.	1336.	1388.	1329.	1315.	1290.	1325.	1369.	1338.	1163.	967.	0
LAT = 37.50	TILT = 90:	874.	922.	877.	716.	566.	512.	528.	634.	823.	995.	980.	849.	0
	KT:	.42	.44	.47	.49	.49	.50	.49	.48	.49	.48	.46	.41	0
	AMB TEMP:	37.5	39.4	46.9	57.8	66.5	74.2	77.9	76.3	70.0	59.3	49.0	39.0	57.8
	HTG DEG DAYS:	853	717	569	226	64	0	0	0	21	203	480	806	3939
STA NO:13741	HORIZ INSOL:	661.	899.	1236.	1581.	1764.	1882.	1796.	1620.	1358.	1080.	765.	591.	1270.
ROANOKE	TILT = LAT:	1029.	1216.	1429.	1561.	1554.	1578.	1542.	1523.	1477.	1416.	1173.	960.	0

VA	TILT = LAT+15:	1079.	1226.	1367.	1402.	1331.	1322.	1305.	1341.	1379.	1411.	1224.	1017.	0
LAT = 37.32	TILT = 90:	925.	948.	895.	719.	564.	510.	528.	637.	826.	1050.	1032.	895.	0
	KT:	.44	.45	.48	.50	.49	.50	.49	.49	.49	.50	.47	.43	0
	AMB TEMP:	36.4	38.1	45.3	55.9	64.4	71.7	75.2	74.1	68.0	57.8	46.7	37.4	55.9
	HTG DEG DAYS:	887	753	611	283	101	0	0	0	32	235	549	856	4307
STA NO:13754	HORIZ INSOL:	757.	1025.	1387.	1794.	1925.	1939.	1830.	1634.	1427.	1170.	907.	718.	1376.
CHERRY POINT	TILT = LAT:	1135.	1357.	1588.	1770.	1700.	1634.	1578.	1534.	1534.	1491.	1354.	1135.	0
NC	TILT = LAT+15:	1193.	1373.	1524.	1593.	1456.	1371.	1340.	1355.	1435.	1489.	1419.	1209.	0
LAT = 34.90	TILT = 90:	1004.	1037.	957.	754.	552.	478.	497.	601.	817.	1072.	1178.	1051.	0
	KT:	.46	.49	.52	.56	.54	.52	.50	.49	.50	.52	.52	.47	0
	AMB TEMP:	0	0	0	0	0	0	0	0	0	0	0	0	0
	HTG DEG DAYS:	0	0	0	0	0	0	0	0	0	0	0	0	0
STA NO:13781	HORIZ INSOL:	571.	827.	1149.	1480.	1710.	1883.	1823.	1615.	1318.	984.	645.	489.	1208.
WILMINGTON	TILT = LAT:	925.	1156.	1349.	1465.	1502.	1568.	1556.	1521.	1455.	1324.	1016.	818.	0
DE	TILT = LAT+15:	968.	1164.	1288.	1313.	1284.	1309.	1313.	1337.	1357.	1317.	1054.	862.	0
LAT = 39.67	TILT = 90:	843.	925.	876.	717.	589.	546.	569.	675.	851.	1008.	903.	767.	0
	KT:	.41	.44	.46	.47	.48	.50	.50	.49	.49	.48	.43	.39	0
	AMB TEMP:	32.0	33.6	41.6	52.3	62.4	71.4	75.8	74.1	67.9	57.2	45.7	34.7	54.0
	HTG DEG DAYS:	1023	879	725	381	128	0	0	0	32	254	579	939	4940
STA NO:13865	HORIZ INSOL:	744.	1012.	1328.	1662.	1860.	1963.	1823.	1739.	1454.	1258.	897.	699.	1370.
MERIDIAN	TILT = LAT:	1033.	1276.	1477.	1625.	1650.	1667.	1583.	1633.	1540.	1559.	1249.	1011.	0
MS	TILT = LAT+15:	1079.	1286.	1416.	1466.	1420.	1401.	1348.	1445.	1444.	1558.	1302.	1068.	0
LAT = 32.33	TILT = 90:	875.	933.	846.	656.	500.	440.	458.	584.	777.	1083.	1043.	895.	0
	KT:	.42	.45	.48	.51	.52	.53	.50	.51	.50	.53	.48	.42	0
	AMB TEMP:	46.9	49.8	56.1	65.4	72.4	79.2	81.2	80.7	75.3	64.8	54.2	47.9	64.5
	HTG DEG DAYS:	575	443	312	79	7	0	0	0	0	111	331	530	2388
STA NO:13866	HORIZ INSOL:	498.	707.	1010.	1356.	1639.	1776.	1682.	1514.	1272.	972.	613.	440.	1123.
CHARLESTON	TILT = LAT:	730.	917.	1142.	1329.	1443.	1488.	1443.	1422.	1383.	1271.	909.	668.	0
WV	TILT = LAT+15:	753.	913.	1085.	1192.	1238.	1248.	1224.	1252.	1290.	1262.	938.	695.	0
LAT = 38.37	TILT = 90:	634.	702.	719.	638.	553.	511.	524.	619.	789.	947.	785.	598.	0
	KT:	.34	.37	.40	.43	.46	.47	.46	.46	.46	.46	.39	.33	0
	AMB TEMP:	34.5	36.5	44.5	55.9	64.5	72.0	75.0	73.6	67.5	57.0	45.4	36.2	55.2
	HTG DEG DAYS:	946	798	642	287	113	10	0	0	46	267	588	893	4590
STA NO:13874	HORIZ INSOL:	718.	969.	1304.	1686.	1854.	1914.	1812.	1709.	1422.	1200.	883.	674.	1345.
ATLANTA	TILT = LAT:	1022.	1238.	1464.	1654.	1642.	1621.	1569.	1605.	1515.	1506.	1266.	1004.	0
GA	TILT = LAT+15:	1068.	1248.	1402.	1491.	1410.	1362.	1334.	1418.	1419.	1503.	1322.	1061.	0
LAT = 33.65	TILT = 90:	879.	920.	859.	689.	520.	457.	476.	600.	787.	1064.	1077.	902.	0
	KT:	.42	.45	.48	.52	.52	.51	.50	.51	.49	.52	.49	.43	0
	AMB TEMP:	42.4	45.0	51.1	61.1	69.1	75.6	78.0	77.5	72.3	62.4	51.4	43.5	60.8
	HTG DEG DAYS:	701	560	443	144	27	0	0	0	8	137	408	667	3095
STA NO:13876	HORIZ INSOL:	707.	967.	1296.	1674.	1857.	1919.	1810.	1724.	1455.	1211.	858.	661.	1345.
BIRMINGHAM	TILT = LAT:	1000.	1234.	1453.	1641.	1645.	1625.	1567.	1619.	1552.	1520.	1219.	976.	0
AL	TILT = LAT+15:	1043.	1243.	1392.	1479.	1413.	1366.	1333.	1431.	1454.	1518.	1271.	1031.	0
LAT = 33.57	TILT = 90:	856.	915.	851.	683.	519.	456.	474.	602.	804.	1073.	1031.	873.	0
	KT:	.41	.45	.48	.51	.52	.52	.50	.51	.50	.52	.47	.42	0
	AMB TEMP:	44.2	46.9	53.3	63.2	70.5	77.4	79.9	79.2	73.9	63.3	52.1	45.2	62.4
	HTG DEG DAYS:	654	517	389	116	20	0	0	0	6	137	391	614	2844
STA NO:13880	HORIZ INSOL:	744.	995.	1339.	1732.	1860.	1844.	1799.	1585.	1394.	1193.	934.	721.	1345.

CHARLESTON	TILT = LAT:	1049.	1262.	1498.	1698.	1649.	1568.	1561.	1487.	1476.	1478.	1332.	1068.	0
SC	TILT = LAT+15:	1097.	1272.	1436.	1532.	1418.	1322.	1329.	1317.	1382.	1475.	1393.	1133.	0
LAT = 32.90	TILT = 90:	897.	930.	868.	690.	509.	442.	464.	557.	755.	1030.	1129.	960.	0
	KT:	.42	.45	.49	.53	.52	.50	.49	.47	.48	.51	.50	.44	0
	AMB TEMP:	48.6	50 5	56.5	64.6	72.1	77.9	80.2	79.6	75.2	66.1	56.3	49.3	64.7
	HTG DEG DAYS:	521	419	200	69	5	0	0	0	0	74	271	487	2146
STA NO:13881	HORIZ INSOL:	719.	971.	1318.	1695.	1856.	1921.	1831.	1695.	1416.	1173.	866.	672.	1344.
CHARLOTTE	TILT = LAT:	1073.	1279.	1503.	1670.	1639.	1619.	1578.	1593.	1523.	1505.	1289.	1054.	0
NC	TILT = LAT+15:	1125.	1291.	1441.	1503.	1405.	1358.	1339.	1405.	1425.	1503.	1348.	1119.	0
LAT = 35.22	TILT = 90:	946.	975.	909.	724.	546.	482.	502.	624.	817.	1088.	1119.	971.	0
	KT:	.44	.46	.50	.53	.52	.51	.50	.51	.50	.52	.50	.45	0
	AMB TEMP:	42.1	44.0	50.6	60.6	68.8	75.9	78.5	77.7	72.0	61.7	51.0	42.5	60.5
	HTG DEG DAYS:	710	588	461	145	34	0	0	0	10	152	420	698	3218
STA NO:13882	HORIZ INSOL:	631.	859.	1176.	1550.	1732.	1831.	1735.	1630.	1336.	1108.	773.	580.	1245.
CHATTANOOGA	TILT = LAT:	900.	1096.	1319.	151[illegible].	1532.	1548.	1499.	1530.	1426.	1402.	1112.	863.	0
TN	TILT = LAT+15:	935.	1099.	1260.	1368.	1317.	1302.	1275.	1351.	1333.	1396.	1155.	906.	0
LAT = 35.03	TILT = 90:	772.	819.	791.	665.	523.	471.	487.	602.	763.	1005.	946.	771.	0
	KT:	.38	.41	.44	.48	.48	.49	.48	.49	.47	.49	.44	.38	0
	AMB TEMP:	40.2	42.9	49.8	60.5	68.5	76.0	78.8	78.0	71.9	60.8	48.9	41.2	59.8
	HTG DEG DAYS:	769	625	483	165	51	0	0	0	9	182	483	738	3505
STA NO:13883	HORIZ INSOL:	762.	1021.	1355.	1747.	1895.	1947.	1842.	1703.	1439.	1211.	921.	722.	1380.
COLUMBIA	TILT = LAT:	1113.	1325.	1533.	1717.	1676.	1646.	1592.	1599.	1538.	1530.	1346.	1107.	0
SC	TILT = LAT+15:	1167.	1339.	1470.	1547.	1438.	1381.	1352.	1413.	1440.	1529.	1409.	1177.	0
LAT = 33.95	TILT = 90:	971.	996.	906.	717.	531.	464.	484.	604.	804.	1087.	1157.	1012.	0
	KT:	.45	.47	.50	.54	.53	.52	.50	.51	.50	.53	.51	.46	0
	AMB TEMP:	45.4	47.6	54.2	64.1	72.1	78.8	81.2	80.2	74.5	64.2	53.8	46.0	63.5
	HTG DEG DAYS:	608	493	360	63	12	0	0	0	0	112	341	589	2598
STA NO:13889	HORIZ INSOL:	900.	1164.	1522.	1856.	1956.	1885.	1802.	1694.	1442.	1223.	996.	818.	1438.
JACKSONVILLE	TILT = LAT:	1250.	1460.	1692.	1815.	1740.	1614.	1573.	1591.	1512.	1471.	1360.	1168.	0
FL	TILT = LAT+15:	1316.	1480.	1629.	1640.	1496.	1362.	1343.	1411.	1419.	1467.	1423.	1242.	0
LAT = 30.50	TILT = 90:	1063.	1055.	938.	678.	479.	411.	430.	544.	734.	987.	1122.	1033.	0
	KT:	.48	.50	.54	.55	.54	.51	.50	.50	.48	.50	.50	.47	0
	AMB TEMP:	54.6	56.3	61.2	68.1	74.3	79.2	81.0	81.0	78.2	70.5	61.2	55.4	68.4
	HTG DEG DAYS:	348	282	176	24	0	0	0	0	0	19	161	317	1327
STA NO:13891	HORIZ INSOL:	621.	863.	1191.	1599.	1803.	1902.	1804.	1666.	1383.	1121.	759.	569.	1273.
KNOXVILLE	TILT = LAT:	903.	1120.	1348.	1573.	1592.	1600.	1554.	1566.	1491.	1440.	1110.	864.	0
TN	TILT = LAT+15:	939.	1125.	1288.	1415.	1365.	1342.	1318.	1381.	1394.	1436.	1154.	909.	0
LAT = 35.82	TILT = 90:	783.	848.	820.	698.	547.	489.	507.	626.	810.	1047.	953.	780.	0
	KT:	.39	.42	.45	.50	.50	.51	.49	.50	.49	.51	.45	.39	0
	AMB TEMP:	40.6	42.8	49.9	60.3	68.4	75.5	78.2	77.3	71.6	60.9	49.2	41.5	59.7
	HTG DEG DAYS:	756	630	484	173	47	0	0	0	10	175	474	729	3478
STA NO:13893	HORIZ INSOL:	683.	945.	1278.	1639.	1885.	2045.	1972.	1824.	1471.	1205.	817.	629.	1366.
MEMPHIS	TILT = LAT:	999.	1233.	1450.	1611.	1665.	1718.	1696.	1717.	1567.	1548.	1193.	959.	0
TN	TILT = LAT+15:	1043.	1242.	1388.	1450.	1427.	1437.	1434.	1513.	1486.	1547.	1243.	1014.	0
LAT = 35.05	TILT = 90:	870.	934.	873.	699.	548.	483.	516.	656.	848.	1120.	1024.	871.	0
	KT:	.42	.45	.48	.51	.52	.55	.54	.54	.52	.54	.47	.42	0
	AMB TEMP:	40.5	43.8	51.0	62.5	70.9	78.6	81.6	80.4	73.6	63.0	50.9	42.7	61.6
	HTG DEG DAYS:	760	594	457	131	22	0	0	0	7	142	423	691	3227

STA NO:13894	HORIZ INSOL:	828.	1100.	1408.	1722.	1872.	1869.	1715.	1642.	1449.	1299.	955.	759.	1385.
MOBILE	TILT = LAT:	1134.	1369.	1555.	1681.	1666.	1599.	1500.	1541.	1522.	1580.	1298.	1070.	0
AL	TILT = LAT+15:	1188.	1385.	1493.	1519.	1435.	1350.	1283.	1367.	1427.	1580.	1355.	1133.	0
LAT = 30.68	TILT = 90:	953.	986.	864.	644.	474.	413.	426.	535.	741.	1070.	1067.	937.	0
	KT:	.44	.48	.50	.52	.52	.50	.47	.48	.49	.53	.49	.43	0
	AMB TEMP:	51.2	54.0	59.4	67.9	74.8	80.3	81.6	81.5	77.5	68.9	58.5	52.9	67.4
	HTG DEG DAYS:	451	337	221	40	0	0	0	0	0	39	211	385	1684
STA NO:13895	HORIZ INSOL:	752.	1013.	1341.	1729.	1897.	1972.	1841.	1746.	1468.	1262.	915.	719.	1388.
MONTGOMERY	TILT = LAT:	1046.	1276.	1492.	1693.	1683.	1674.	1598.	1639.	1556.	1564.	1280.	1047.	0
AL	TILT = LAT+15:	1092.	1287.	1431.	1527.	1447.	1407.	1360.	1450.	1459.	1563.	1336.	1108.	0
LAT = 32.30	TILT = 90:	887.	933.	855.	677.	503.	440.	459.	585.	784.	1086.	1072.	931.	0
	KT:	.42	.45	.49	.53	.53	.53	.51	.52	.50	.53	.49	.43	0
	AMB TEMP:	47.5	50.6	56.5	65.2	72.4	78.9	81.0	80.7	76.0	65.8	55.0	48.5	64.8
	HTG DEG DAYS:	556	419	299	76	8	0	0	0	0	93	306	512	2269
STA NO:13897	HORIZ INSOL:	580.	824.	1130.	1544.	1825.	1963.	1891.	1737.	1398.	1114.	711.	521.	1270.
NASHVILLE	TILT = LAT:	832.	1063.	1273.	1517.	1610.	1648.	1625.	1634.	1511.	1437.	1031.	774.	0
TN	TILT = LAT+15:	862.	1065.	1214.	1364.	1380.	1379.	1375.	1440.	1413.	1433.	1068.	809.	0
LAT = 36.12	TILT = 90:	716.	803.	776.	682.	556.	500.	523.	652.	826.	1048.	880.	689.	0
	KT:	.37	.40	.43	.48	.51	.52	.52	.52	.50	.51	.42	.36	0
	AMB TEMP:	38.3	41.0	48.7	60.1	68.5	76.6	79.6	78.5	72.0	60.9	48.4	40.4	59.4
	HTG DEG DAYS:	828	672	524	176	45	0	0	0	10	180	498	763	3696
STA NO:13923	HORIZ INSOL:	794.	1037.	1366.	1610.	1852.	2114.	2077.	1932.	1580.	1268.	919.	744.	1441.
SHERMAN	TILT = LAT:	1164.	1346.	1543.	1577.	1640.	1780.	1788.	1818.	1701.	1608.	1333.	1141.	0
TX	TILT = LAT+15:	1224.	1361.	1481.	1421.	1406.	1468.	1511.	1603.	1596.	1610.	1395.	1215.	0
LAT = 33.72	TILT = 90:	1020.	1010.	909.	664.	521.	471.	502.	656.	882.	1144.	1141.	1044.	0
	KT:	.46	.48	.50	.50	.51	.57	.57	.57	.55	.55	.51	.47	0
	AMB TEMP:	41.7	45.9	52.3	63.7	71.2	79.4	83.6	83.7	76.0	65.8	53.4	44.8	63.5
	HTG DEG DAYS:	722	535	411	114	13	0	0	0	0	90	353	626	2864
STA NO:13957	HORIZ INSOL:	762.	1038.	1342.	1613.	1886.	2065.	2014.	1877.	1554.	1303.	929.	731.	1426.
SHREVEPORT	TILT = LAT:	1069.	1318.	1496.	1575.	1673.	1748.	1741.	1764.	1657.	1628.	1308.	1073.	0
LA	TILT = LAT+15:	1118.	1331.	1434.	1422.	1438.	1464.	1475.	1559.	1555.	1631.	1367.	1137.	0
LAT = 32.47	TILT = 90:	911.	970.	860.	642.	505.	447.	475.	618.	836.	1138.	1100.	959.	0
	KT:	.43	.47	.49	.49	.52	.56	.55	.56	.53	.55	.50	.44	0
	AMB TEMP:	47.2	50.5	56.8	66.4	73.4	80.2	83.2	83.2	77.4	67.5	56.2	49.2	65.9
	HTG DEG DAYS:	552	416	291	65	5	0	0	0	0	70	278	490	2167
STA NO:13958	HORIZ INSOL:	865.	1125.	1429.	1605.	1834.	2072.	2106.	1931.	1606.	1333.	987.	825.	1476.
AUSTIN	TILT = LAT:	1184.	1397.	1576.	1563.	1634.	1765.	1826.	1814.	1696.	1620.	1339.	1175.	0
TX	TILT = LAT+15:	1243.	1414.	1515.	1413.	1410.	1482.	1547.	1605.	1594.	1621.	1399.	1250.	0
LAT = 30.30	TILT = 90:	997.	1003.	870.	603.	465.	412.	442.	586.	814.	1093.	1100.	1038.	0
	KT:	.46	.48	.50	.49	.51	.56	.58	.57	.54	.54	.50	.47	0
	AMB TEMP:	49.7	53.3	59.5	68.6	75.2	81.6	84.6	84.7	78.9	70.1	59.1	52.3	68.1
	HTG DEG DAYS:	483	344	223	44	0	0	0	0	0	39	205	399	1737
STA NO:13959	HORIZ INSOL:	833.	1096.	1428.	1612.	1774.	2112.	2130.	1958.	1601.	1301.	957.	803.	1467.
WACO	TILT = LAT:	1168.	1385.	1591.	1573.	1576.	1790.	1841.	1841.	1702.	1605.	1329.	1178.	0
TX	TILT = LAT+15:	1226.	1402.	1529.	1420.	1361.	1499.	1557.	1626.	1599.	1606.	1389.	1254.	0
LAT = 31.62	TILT = 90:	997.	1012.	901.	627.	478.	434.	466.	618.	842.	1105.	1109.	1057.	0
	KT:	.46	.48	.51	.49	.49	.57	.58	.58	.54	.54	.50	.47	0
	AMB TEMP:	47.0	50.9	57.2	67.3	74.5	81.9	85.6	85.7	78.9	69.1	57.5	49.8	67.1
	HTG DEG DAYS:	558	401	280	56	0	0	0	0	0	51	241	471	2058

STA NO:13960
DALLAS
TX
LAT = 32.85

HORIZ INSOL:	822.	1071.	1422.	1627.	1889.	2135.	2122.	1950.	1587.	1276.	936.	780.	1468.
TILT = LAT:	1187.	1376.	1601.	1591.	1674.	1801.	1829.	1834.	1699.	1598.	1334.	1180.	0
TILT = LAT+15:	1248.	1393.	1538.	1435.	1438.	1506.	1545.	1619.	1595.	1599.	1396.	1257.	0
TILT = 90:	1031.	1023.	929.	654.	512.	456.	489.	642.	864.	1121.	1131.	1073.	0
KT:	.47	.49	.52	.50	.52	.57	.58	.58	.54	.54	.51	.48	0
AMB TEMP:	45.4	49.4	55.8	66.4	73.8	81.6	85.7	85.8	78.2	68.0	55.9	48.2	66.2
HTG DEG DAYS:	608	437	314	71	0	0	0	0	0	55	284	521	2290

STA NO:13962
ABILENE
TX
LAT = 32.43

HORIZ INSOL:	924.	1183.	1576.	1843.	2037.	2209.	2139.	1956.	1598.	1315.	1008.	863.	1554.
TILT = LAT:	1358.	1537.	1790.	1810.	1804.	1862.	1845.	1840.	1707.	1645.	1445.	1323.	0
TILT = LAT+15:	1436.	1562.	1724.	1632.	1546.	1554.	1558.	1624.	1603.	1648.	1517.	1417.	0
TILT = 90:	1194.	1149.	1033.	717.	521.	451.	482.	635.	860.	1150.	1230.	1216.	0
KT:	.52	.53	.57	.56	.57	.59	.59	.58	.55	.56	.54	.52	0
AMB TEMP:	43.7	47.9	54.5	65.2	72.4	80.3	83.9	83.6	76.1	66.1	54.1	46.4	64.5
HTG DEG DAYS:	660	479	354	104	11	0	0	0	0	89	338	577	2610

STA NO:13963
LITTLE ROCK
AR
LAT = 34.73

HORIZ INSOL:	731.	1003.	1313.	1611.	1929.	2107.	2032.	1861.	1518.	1228.	847.	674.	1404.
TILT = LAT:	1081.	1317.	1490.	1581.	1704.	1769.	1747.	1751.	1639.	1576.	1238.	1039.	0
TILT = LAT+15:	1133.	1331.	1428.	1423.	1460.	1477.	1476.	1544.	1536.	1576.	1292.	1102.	0
TILT = 90:	948.	1000.	893.	683.	550.	488.	516.	660.	869.	1136.	1064.	950.	0
KT:	.44	.47	.49	.50	.54	.56	.56	.55	.53	.54	.48	.44	0
AMB TEMP:	39.5	42.9	50.3	61.7	69.8	78.1	81.4	80.6	73.3	62.4	50.3	41.6	61.0
HTG DEG DAYS:	791	619	470	139	21	0	0	0	5	143	441	725	3354

STA NO:13964
FORT SMITH
AR
LAT = 35.33

HORIZ INSOL:	744.	999.	1312.	1616.	1912.	2089.	2065.	1877.	1502.	1201.	851.	682.	1404.
TILT = LAT:	1125.	1326.	1498.	1588.	1688.	1752.	1772.	1768.	1626.	1550.	1267.	1077.	0
TILT = LAT+15:	1182.	1341.	1435.	1429.	1445.	1463.	1495.	1558.	1524.	1549.	1324.	1145.	0
TILT = 90:	999.	1017.	907.	696.	557.	498.	531.	675.	874.	1126.	1099.	996.	0
KT:	.46	.48	.49	.50	.53	.56	.57	.56	.53	.54	.49	.46	0
AMB TEMP:	39.0	43.3	50.3	62.2	70.1	78.0	82.2	81.4	74.0	63.2	50.4	41.5	61.3
HTG DEG DAYS:	806	608	471	132	17	0	0	0	0	135	438	729	3336

STA NO:13966
WICHITA FALLS
TX
LAT = 33.97

HORIZ INSOL:	862.	1123.	1472.	1763.	2017.	2221.	2166.	1969.	1602.	1291.	957.	799.	1520.
TILT = LAT:	1301.	1487.	1683.	1734.	1782.	1864.	1861.	1854.	1729.	1650.	1412.	1260.	0
TILT = LAT+15:	1375.	1510.	1618.	1562.	1525.	1553.	1569.	1634.	1623.	1653.	1482.	1348.	0
TILT = 90:	1159.	1132.	999.	723.	548.	480.	513.	671.	901.	1181.	1221.	1171.	0
KT:	.51	.52	.54	.54	.56	.60	.59	.59	.56	.56	.53	.51	0
AMB TEMP:	41.5	45.9	52.5	64.3	72.3	81.3	85.8	85.5	77.0	66.0	52.9	44.2	64.1
HTG DEG DAYS:	729	535	409	112	13	0	0	0	0	92	369	645	2904

STA NO:13967
OKLAHOMA CITY
OK
LAT = 35.40

HORIZ INSOL:	801.	1055.	1400.	1725.	1918.	2144.	2128.	1950.	1554.	1233.	901.	725.	1461.
TILT = LAT:	1239.	1419.	1614.	1702.	1693.	1795.	1824.	1839.	1690.	1601.	1363.	1171.	0
TILT = LAT+15:	1308.	1439.	1550.	1531.	1449.	1497.	1537.	1619.	1585.	1602.	1429.	1249.	0
TILT = 90:	1115.	1097.	982.	739.	560.	503.	538.	698.	909.	1167.	1194.	1094.	0
KT:	.49	.51	.53	.54	.53	.57	.58	.58	.55	.55	.52	.49	0
AMB TEMP:	36.8	41.3	48.2	60.4	68.3	76.8	81.5	81.1	73.0	62.4	49.2	40.0	59.9
HTG DEG DAYS:	874	664	532	180	36	0	0	0	12	148	474	775	3695

STA NO:13968
TULSA
OK
LAT = 36.20

HORIZ INSOL:	732.	979.	1306.	1603.	1822.	2021.	2030.	1865.	1473.	1164.	827.	659.	1373.
TILT = LAT:	1133.	1316.	1503.	1578.	1608.	1693.	1740.	1759.	1602.	1516.	1254.	1064.	0
TILT = LAT+15:	1191.	1331.	1440.	1419.	1377.	1415.	1468.	1548.	1500.	1515.	1310.	1132.	0
TILT = 90:	1017.	1020.	925.	707.	557.	507.	543.	691.	877.	1113.	1097.	992.	0
KT:	.46	.48	.50	.50	.51	.54	.56	.56	.52	.53	.49	.46	0
AMB TEMP:	36.6	41.2	48.3	60.6	68.8	77.3	82.1	81.4	73.3	62.9	49.4	39.8	60.2
HTG DEG DAYS:	880	666	528	176	28	0	0	0	10	143	468	781	3680

STA NO:13970
BATON ROUGE
LA
LAT = 30.53

HORIZ INSOL:	785.	1054.	1379.	1681.	1871.	1926.	1746.	1677.	1464.	1301.	920.	737.	1379.
TILT = LAT:	1057.	1300.	1519.	1640.	1666.	1647.	1526.	1574.	1537.	1580.	1237.	1027.	0
TILT = LAT+15:	1104.	1312.	1458.	1482.	1436.	1388.	1305.	1395.	1443.	1580.	1288.	1084.	0
TILT = 90:	878.	929.	841.	629.	472.	413.	427.	541.	746.	1068.	1008.	891.	0
KT:	.42	.46	.49	.51	.52	.52	.48	.49	.49	.53	.47	.42	0
AMB TEMP:	51.0	53.9	59.7	68.4	74.8	80.3	82.0	81.6	77.5	68.5	58.6	52.9	67.4
HTG DEG DAYS:	451	335	208	33	0	0	0	0	0	54	208	381	1670

STA NO:13985
DODGE CITY
KS
LAT = 37.77

HORIZ INSOL:	827.	1122.	1476.	1886.	2090.	2358.	2295.	2055.	1687.	1301.	894.	732.	1560.
TILT = LAT:	1401.	1614.	1763.	1885.	1837.	1953.	1953.	1948.	1889.	1788.	1454.	1299.	0
TILT = LAT+15:	1438.	1646.	1696.	1693.	1564.	1616.	1636.	1710.	1773.	1798.	1530.	1394.	0
TILT = 90:	1309.	1303.	1124.	859.	634.	566.	606.	783.	1068.	1363.	1315.	1257.	0
KT:	.55	.57	.58	.60	.58	.63	.63	.52	.61	.61	.56	.54	0
AMB TEMP:	30.8	35.2	41.2	54.0	64.0	73.7	79.2	78.1	68.9	57.9	42.8	33.4	54.9
HTG DEG DAYS:	1060	834	738	344	115	21	0	0	41	247	666	980	5046

STA NO:13994
ST. LOUIS
MO
LAT = 38.75

HORIZ INSOL:	627.	886.	1205.	1564.	1871.	2092.	2049.	1817.	1459.	1100.	718.	531.	1327.
TILT = LAT:	1013.	1232.	1410.	1550.	1645.	1739.	1747.	1717.	1619.	1488.	1134.	881.	0
TILT = LAT+15:	1062.	1244.	1348.	1390.	1404.	1447.	1469.	1509.	1514.	1487.	1182.	930.	0
TILT = 90:	924.	981.	904.	739.	610.	560.	593.	729.	933.	1131.	1010.	825.	0
KT:	.44	.46	.48	.50	.52	.56	.56	.55	.53	.53	.47	.41	0
AMB TEMP:	31.3	35.1	43.3	56.5	65.8	74.9	78.6	77.2	69.6	59.1	45.0	34.6	55.9
HTG DEG DAYS:	1045	837	682	272	103	10	0	0	35	224	600	942	4750

STA NO:13995
SPRINGFIELD
MO
LAT = 37.23

HORIZ INSOL:	684.	926.	1235.	1604.	1882.	2075.	2063.	1873.	1481.	1144.	775.	603.	1362.
TILT = LAT:	1074.	1258.	1426.	1585.	1657.	1732.	1763.	1769.	1625.	1514.	1191.	983.	0
TILT = LAT+15:	1128.	1270.	1364.	1424.	1417.	1444.	1485.	1555.	1521.	1513.	1243.	1042.	0
TILT = 90:	969.	983.	892.	728.	585.	530.	566.	715.	908.	1128.	1049.	918.	0
KT:	.45	.47	.48	.50	.52	.55	.56	.56	.53	.53	.48	.43	0
AMB TEMP:	32.9	37.0	44.0	56.5	65.1	73.6	77.8	77.1	69.3	59.0	45.5	36.0	56.1
HTG DEG DAYS:	995	784	660	275	94	10	0	6	35	227	585	899	4570

STA NO:13996
TOPEKA
KS
LAT = 39.07

HORIZ INSOL:	681.	941.	1257.	1642.	1915.	2126.	2128.	1910.	1516.	1147.	772.	584.	1385.
TILT = LAT:	1143.	1340.	1487.	1633.	1683.	1764.	1811.	1810.	1696.	1576.	1259.	1016.	0
TILT = LAT+15:	1205.	1357.	1424.	1465.	1435.	1466.	1520.	1588.	1587.	1578.	1318.	1081.	0
TILT = 90:	1060.	1080.	961.	781.	625.	570.	611.	767.	983.	1210.	1138.	971.	0
KT:	.48	.50	.50	.52	.54	.57	.58	.58	.56	.56	.51	.45	0
AMB TEMP:	28.0	33.4	41.2	54.5	64.5	73.5	78.2	77.2	68.2	57.6	42.9	31.8	54.3
HTG DEG DAYS:	1147	885	745	329	118	13	0	0	55	259	663	1029	5243

STA NO:14601
BANGOR
ME
LAT = 44.80

HORIZ INSOL:	455.	725.	1094.	1440.	1729.	1857.	1859.	1611.	1255.	839.	471.	379.	1143.
TILT = LAT:	856.	1131.	1370.	1452.	1511.	1526.	1571.	1533.	1449.	1225.	817.	749.	0
TILT = LAT+15:	897.	1142.	1308.	1296.	1283.	1267.	1316.	1340.	1349.	1218.	844.	792.	0
TILT = 90:	817.	962.	966.	794.	675.	623.	662.	768.	930.	991.	751.	735.	0
KT:	.41	.45	.48	.46	.49	.49	.51	.51	.50	.47	.39	.39	0
AMB TEMP:	0	0	0	0	0	0	0	0	0	0	0	0	0
HTG DEG DAYS:	0	0	0	0	0	0	0	0	0	0	0	0	0

STA NO:14607
CARIBOU
ME
LAT = 46.87

HORIZ INSOL:	419.	724.	1133.	1414.	1578.	1757.	1762.	1501.	1103.	688.	366.	311.	1063.
TILT = LAT:	861.	1214.	1485.	1439.	1374.	1433.	1485.	1428.	1275.	1004.	629.	642.	0
TILT = LAT+15:	905.	1231.	1421.	1282.	1160.	1193.	1242.	1246.	1183.	991.	644.	677.	0
TILT = 90:	839.	1064.	1086.	820.	657.	630.	669.	753.	843.	819.	575.	633.	0
KT:	.42	.49	.52	.48	.45	.47	.49	.48	.45	.41	.33	.36	0
AMB TEMP:	10.7	12.9	23.6	36.7	49.7	59.6	64.9	62.3	54.1	43.8	31.4	16.1	33.8

	HTG DEG DAYS:	1683	1459	1283	849	474	170	84	122	327	657	1008	1516	9632
STA NO:14732	HORIZ INSOL:	548.	795.	1118.	1457.	1690.	1802.	1784.	1583.	1280.	951.	593.	457.	1171.
NEW YORK CITY (LA GU	TILT = LAT:	914.	1130.	1324.	1445.	1483.	1498.	1520.	1493.	1421.	1299.	946.	784.	0
NY	TILT = LAT+15:	956.	1137.	1263.	1294.	1266.	1252.	1282.	1310.	1325.	1292.	980.	826.	0
LAT = 40.77	TILT = 90:	842.	914.	875.	726.	600.	550.	579.	682.	849.	1003.	844.	740.	0
	KT:	.41	.44	.46	.47	.47	.48	.49	.49	.48	.48	.41	.38	0
	AMB TEMP:	32.1	33.1	40.6	51.7	61.8	71.5	76.7	74.9	68.1	56.1	47.3	35.6	54.3
	HTG DEG DAYS:	1020	893	756	399	145	0	0	0	30	224	531	911	4909
STA NO:14733	HORIZ INSOL:	349.	546.	889.	1315.	1596.	1804.	1776.	1513.	1152.	784.	403.	283.	1034.
BUFFALO	TILT = LAT:	526.	729.	1034.	1303.	1397.	1491.	1507.	1428.	1285.	1071.	599.	429.	0
NY	TILT = LAT+15:	536.	720.	978.	1164.	1191.	1242.	1267.	1251.	1194.	1058.	608.	439.	0
LAT = 42.93	TILT = 90:	462.	576.	695.	689.	606.	583.	611.	691.	795.	836.	518.	382.	0
	KT:	.29	.32	.38	.43	.45	.48	.49	.47	.45	.42	.31	.26	0
	AMB TEMP:	23.7	24.4	32.1	44.9	55.1	65.7	70.1	68.4	61.6	51.5	39.8	27.9	47.1
	HTG DEG DAYS:	1280	1137	1020	603	321	58	12	33	138	419	756	1150	6927
STA NO:14734	HORIZ INSOL:	552.	793.	1109.	1449.	1687.	1795.	1760.	1565.	1273.	951.	596.	454.	1165.
NEWARK	TILT = LAT:	920.	1125.	1311.	1437.	1480.	1493.	1500.	1475.	1411.	1298.	950.	775.	0
NJ	TILT = LAT+15:	963.	1132.	1250.	1286.	1264.	1248.	1266.	1294.	1315.	1290.	984.	816.	0
LAT = 40.70	TILT = 90:	848.	909.	864.	721.	599.	548.	573.	674.	842.	1000.	846.	731.	0
	KT:	.42	.44	.45	.47	.47	.48	.48	.48	.48	.48	.42	.38	0
	AMB TEMP:	31.4	32.6	40.6	51.7	61.9	71.4	76.4	74.6	67.8	57.5	46.2	34.5	53.9
	HTG DEG DAYS:	1042	907	756	399	143	0	0	0	34	243	564	946	5034
STA NO:14735	HORIZ INSOL:	457.	688.	986.	1335.	1570.	1730.	1725.	1499.	1170.	817.	457.	356.	1066.
ALBANY	TILT = LAT:	775.	990.	1171.	1325.	1373.	1433.	1464.	1414.	1306.	1124.	713.	604.	0
NY	TILT = LAT+15:	806.	992.	1112.	1183.	1172.	1196.	1233.	1238.	1214.	1113.	730.	630.	0
LAT = 42.75	TILT = 90:	716.	810.	792.	697.	596.	565.	596.	682.	807.	879.	629.	564.	0
	KT:	.38	.40	.42	.44	.44	.46	.47	.46	.45	.43	.35	.33	0
	AMB TEMP:	21.5	23.5	33.4	46.9	57.7	67.5	72.0	69.6	61.9	51.4	39.6	25.9	47.6
	HTG DEG DAYS:	1349	1162	980	543	253	39	9	22	135	422	762	1212	6888
STA NO:14737	HORIZ INSOL:	528.	764.	1078.	1410.	1637.	1777.	1765.	1546.	1239.	926.	568.	430.	1139.
ALLENTOWN	TILT = LAT:	863.	1070.	1267.	1395.	1436.	1473.	1504.	1456.	1367.	1254.	888.	716.	0
PA	TILT = LAT+15:	901.	1075.	1207.	1249.	1226.	1237.	1269.	1278.	1273.	1246.	918.	751.	0
LAT = 40.65	TILT = 90:	789.	860.	833.	700.	585.	544.	573.	666.	814.	963.	786.	667.	0
	KT:	.40	.42	.44	.45	.46	.47	.48	.47	.46	.47	.40	.36	0
	AMB TEMP:	27.8	29.4	38.1	49.9	60.1	69.5	74.1	71.7	64.7	54.1	42.3	30.7	51.0
	HTG DEG DAYS:	1153	997	834	453	190	21	0	6	85	344	681	1063	5827
STA NO:14740	HORIZ INSOL:	478.	715.	979.	1315.	1568.	1686.	1649.	1422.	1154.	853.	497.	385.	1058.
HARTFORD	TILT = LAT:	795.	1016.	1147.	1299.	1374.	1400.	1403.	1335.	1276.	1164.	777.	650.	0
CT	TILT = LAT+15:	827.	1019.	1089.	1161.	1174.	1172.	1185.	1171.	1186.	1154.	798.	679.	0
LAT = 41.93	TILT = 90:	730.	826.	765.	673.	585.	545.	566.	637.	776.	903.	686.	607.	0
	KT:	.38	.41	.41	.43	.44	.45	.45	.44	.44	.44	.36	.34	0
	AMB TEMP:	24.8	26.8	35.6	47.7	58.3	67.8	72.7	70.4	62.8	52.6	41.3	28.2	49.1
	HTG DEG DAYS:	1246	1070	911	519	226	24	0	12	106	384	711	1141	6350
STA NO:14742	HORIZ INSOL:	385.	607.	940.	1296.	1574.	1729.	1721.	1475.	1122.	741.	375.	283.	1021.
BURLINGTON	TILT = LAT:	660.	880.	1133.	1291.	1374.	1425.	1456.	1394.	1266.	1033.	575.	470,	0
VT	TILT = LAT+15:	684.	879.	1075.	1151.	1170.	1187.	1223.	1219.	1175.	1020.	584.	485.	0
LAT = 44.47	TILT = 90:	611.	727.	785.	704.	621.	589.	620.	699.	804.	820.	504.	434.	0
	KT:	.34	.38	.41	.43	.45	.46	.47	.46	.44	.41	.30	.29	0

	AMB TEMP:	16.8	18.6	29.1	43.0	54.8	65.2	69.8	67.4	59.3	48.8	37.0	22.6	44.4
	HTG DEG DAYS:	1494	1299	1113	660	331	63	20	49	191	502	840	1314	7876
STA NO:14745	HORIZ INSOL:	460.	686.	974.	1317.	1582.	1705.	1675.	1455.	1140.	817.	463.	362.	1053.
CONCORD	TILT = LAT:	799.	999.	1160.	1307.	1384.	1410.	1421.	1371.	1273.	1136.	739.	636.	0
NH	TILT = LAT+15:	834.	1002.	1102.	1167.	1180.	1178.	1197.	1201.	1182.	1125.	759.	665.	0
LAT = 43.20	TILT = 90:	745.	823.	790.	695.	606.	566.	590.	670.	792.	895.	659.	601.	0
	KT:	.39	.41	.42	.43	.45	.45	.46	.45	.44	.44	.36	.34	0
	AMB TEMP:	20.6	22.6	32.3	44.2	55.1	64.7	69.7	67.2	59.5	49.3	38.0	24.8	45.6
	HTG DEG DAYS:	1376	1187	1014	624	315	58	16	45	182	487	810	1246	7360
STA NO:14751	HORIZ INSOL:	536.	771.	1083.	1411.	1652.	1805.	1764.	1551.	1267.	934.	579.	447.	1150.
HARRISBURG	TILT = LAT:	866.	1072.	1267.	1394.	1451.	1503.	1505.	1460.	1397.	1256.	896.	741.	0
PA	TILT = LAT+15:	903.	1076.	1207.	1248.	1240.	1257.	1270.	1282.	1302.	1247.	925.	778.	0
LAT = 40.22	TILT = 90:	788.	856.	827.	694.	583.	542.	566.	661.	828.	959.	790.	690.	0
	KT:	.40	.42	.44	.45	.46	.48	.48	.47	.47	.47	.40	.37	0
	AMB TEMP:	30.1	32.3	41.0	52.8	63.1	72.0	76.1	73.9	67.0	55.8	43.8	32.6	53.4
	HTG DEG DAYS:	1082	916	744	370	128	0	0	0	51	293	636	1004	5224
STA NO:14764	HORIZ INSOL:	450.	682.	970.	1304.	1567.	1712.	1659.	1461.	1158.	822.	459.	363.	1051.
PORTLAND	TILT = LAT:	794.	1004.	1162.	1295.	1370.	1414.	1407.	1378.	1302.	1159.	746.	655.	0
ME	TILT = LAT+15:	829.	1008.	1104.	1156.	1167.	1180.	1185.	1205.	1210.	1149.	767.	687.	0
LAT = 43.65	TILT = 90:	744.	833.	797.	695.	608.	573.	592.	680.	816.	919.	670.	625.	0
	KT:	.39	.41	.42	.43	.44	.46	.46	.46	.45	.45	.36	.35	0
	AMB TEMP:	21.5	22.9	31.8	42.7	52.7	62.2	68.0	66.4	58.7	49.1	38.6	25.7	45.0
	HTG DEG DAYS:	1349	1179	1029	669	381	106	27	55	200	493	792	1218	7498
STA NO:14765	HORIZ INSOL:	506.	739.	1032.	1374.	1655.	1775.	1695.	1499.	1209.	907.	538.	419.	1112.
PROVIDENCE	TILT = LAT:	855.	1055.	1220.	1361.	1450.	1473.	1443.	1411.	1343.	1252.	858.	726.	0
RI	TILT = LAT+15:	893.	1060.	1160.	1218.	1237.	1230.	1217.	1238.	1250.	1243.	886.	763.	0
LAT = 41.73	TILT = 90:	790.	858.	814.	701.	606.	559.	574.	665.	815.	975.	766.	686.	0
	KT:	.40	.42	.43	.45	.47	.47	.47	.46	.46	.47	.39	.37	0
	AMB TEMP:	28.4	29.4	36.9	47.3	56.9	66.4	72.1	70.4	63.4	53.7	43.3	31.5	50.0
	HTG DEG DAYS:	1135	997	871	531	259	36	0	10	93	350	651	1039	5972
STA NO:14768	HORIZ INSOL:	364.	560.	903.	1339.	1606.	1817.	1781.	1519.	1160.	782.	404.	281.	1043.
ROCHESTER	TILT = LAT:	566.	757.	1058.	1331.	1405.	1501.	1510.	1435.	1297.	1072.	605.	428.	0
NY	TILT = LAT+15:	580.	749.	1002.	1188.	1198.	1250.	1270.	1256.	1206.	1059.	615.	438.	0
LAT = 43.12	TILT = 90:	505.	603.	714.	706.	612.	583.	615.	695.	806.	838.	525.	382.	0
	KT:	.30	.33	.39	.44	.45	.48	.49	.47	.45	.42	.31	.26	0
	AMB TEMP:	24.0	24.8	33.0	46.1	56.5	66.9	71.2	69.3	62.3	52.3	40.5	28.3	47.9
	HTG DEG DAYS:	1271	1126	992	567	285	46	9	26	126	398	735	1138	6719
STA NO:14771	HORIZ INSOL:	385.	571.	890.	1324.	1578.	1778.	1758.	1504.	1165.	777.	399.	285.	1034.
SYRACUSE	TILT = LAT:	615.	779.	1039.	1314.	1380.	1469.	1491.	1419.	1305.	1064.	594.	439.	0
NY	TILT = LAT+15:	633.	772.	984.	1174.	1177.	1225.	1254.	1243.	1213.	1051.	602.	449.	0
LAT = 43.12	TILT = 90:	555.	623.	701.	697.	604.	580.	609.	690.	611.	831.	513.	393.	0
	KT:	.32	.34	.38	.44	.45	.47	.48	.47	.45	.42	.31	.27	0
	AMB TEMP:	23.6	24.6	33.2	46.5	56.8	66.9	71.5	69.7	62.8	52.5	41.0	28.1	48.1
	HTG DEG DAYS:	1283	1131	986	555	272	46	11	18	120	392	720	1144	6678
STA NO:14777	HORIZ INSOL:	455.	689.	991.	1339.	1591.	1760.	1746.	1513.	1199.	897.	490.	368.	1086.
WILKES-BARRE-SCRANTO	TILT = LAT:	723.	952.	1156.	1322.	1394.	1463.	1486.	1425.	1326.	1223.	742.	589.	0
PA	TILT = LAT+15:	748.	952.	1098.	1183.	1192.	1223.	1253.	1250.	1233.	1214.	760.	611.	0
LAT = 41.33	TILT = 90:	651.	762.	764.	676.	582.	551.	579.	665.	798.	946.	647.	538.	0

	KT:	.35	.39	.41	.43	.45	.47	.48	.47	.45	.46	.35	.32	0
	AMB TEMP:	26.0	27.3	36.0	48.5	58.9	67.9	72.2	70.0	62.9	52.6	40.8	29.1	49.4
	HTG DEG DAYS:	1209	1056	899	495	219	28	7	18	116	391	726	1113	6277
STA NO:14780	HORIZ INSOL:	560.	797.	1109.	1456.	1672.	1775.	1703.	1533.	1261.	956.	621.	475.	1160.
LAKEHURST	TILT = LAT:	913.	1113.	1300.	1441.	1468.	1480.	1455.	1442.	1388.	1287.	980.	799.	0
NJ	TILT = LAT+15:	954.	1119.	1239.	1291.	1235.	1239.	1230.	1267.	1293.	1279.	1016.	842.	0
LAT = 40.03	TILT = 90:	834.	891.	847.	712.	585.	535.	552.	651.	817.	983.	871.	750.	0
	KT:	.41	.43	.45	.47	.47	.47	.47	.47	.47	.47	.42	.39	0
	AMB TEMP:	0	0	0	0	0	0	0	0	0	0	0	0	0
	HTG DEG DAYS:	0	0	0	0	0	0	0	0	0	0	0	0	0
STA NO:14819	HORIZ INSOL:	507.	760.	1107.	1459.	1789.	2007.	1944.	1719.	1354.	969.	566.	402.	1215.
CHICAGO	TILT = LAT:	859.	1095.	1327.	1453.	1565.	1658.	1650.	1631.	1531.	1362.	923.	685.	0
IL	TILT = LAT+15:	897.	1102.	1266.	1300.	1335.	1377.	1385.	1429.	1428.	1357.	956.	718.	0
LAT = 41.78	TILT = 90:	794.	896.	891.	746.	643.	602.	629.	755.	938.	1070.	831.	644.	0
	KT:	.40	.43	.46	.48	.50	.53	.53	.53	.52	.50	.41	.35	0
	AMB TEMP:	24.3	27.4	36.8	49.9	60.0	70.5	74.7	73.7	65.9	55.4	40.4	28.5	50.6
	HTG DEG DAYS:	1262	1053	874	453	208	26	0	8	57	316	738	1132	6127
STA NO:14820	HORIZ INSOL:	388.	601.	922.	1350.	1661.	1843.	1828.	1583.	1240.	867.	466.	318.	1091.
CLEVELAND	TILT = LAT:	576.	797.	1061.	1334.	1474.	1529.	1554.	1494.	1378.	1174.	694.	474.	0
OH	TILT = LAT+15:	589.	790.	1005.	1193.	1258.	1275.	1309.	1310.	1283.	1164.	709.	486.	0
LAT = 41.40	TILT = 90:	502.	624.	698.	683.	608.	567.	598.	693.	832.	905.	600.	419.	0
	KT:	.30	.34	.38	.44	.47	.49	.50	.49	.47	.44	.33	.28	0
	AMB TEMP:	26.9	27.9	36.1	48.3	58.3	67.9	71.4	70.0	63.9	53.8	41.6	30.3	49.7
	HTG DEG DAYS:	1181	1039	896	501	244	40	9	17	95	354	702	1076	6154
STA NO:14821	HORIZ INSOL:	459.	677.	980.	1353.	1647.	1813.	1755.	1641.	1282.	945.	538.	387.	1123.
COLUMBUS	TILT = LAT:	693.	900.	1122.	1332.	1446.	1510.	1498.	1548.	1414.	1268.	805.	596.	0
OH	TILT = LAT+15:	714.	897.	1065.	1193.	1237.	1263.	1265.	1359.	1316.	1260.	826.	617.	0
LAT = 40.00	TILT = 90:	610.	703.	724.	662.	578.	541.	561.	691.	832.	967.	698.	536.	0
	KT:	.34	.37	.40	.43	.46	.48	.48	.50	.48	.47	.37	.31	0
	AMB TEMP:	28.4	30.3	39.2	51.2	61.1	70.4	73.6	71.9	65.2	54.2	41.7	30.7	51.5
	HTG DEG DAYS:	1135	972	800	418	176	13	0	8	76	342	699	1063	5702
STA NO:14822	HORIZ INSOL:	417.	680.	1000.	1399.	1716.	1866.	1835.	1576.	1253.	876.	478.	344.	1120.
DETROIT	TILT = LAT:	670.	966.	1186.	1392.	1503.	1543.	1558.	1489.	1409.	1218.	749.	562.	0
MI	TILT = LAT+15:	692.	967.	1127.	1244.	1280.	1285.	1309.	1305.	1313.	1209.	768.	583.	0
LAT = 42.42	TILT = 90:	605.	785.	798.	726.	633.	587.	616.	708.	867.	955.	662.	518.	0
	KT:	.34	.39	.42	.46	.48	.50	.50	.49	.48	.46	.36	.31	0
	AMB TEMP:	25.5	26.9	35.4	48.1	58.4	69.1	73.3	71.9	64.5	54.3	41.1	29.6	49.9
	HTG DEG DAYS:	1225	1067	918	507	238	26	0	11	80	342	717	1097	6228
STA NO:14826	HORIZ INSOL:	383.	636.	957.	1339.	1658.	1813.	1797.	1555.	1196.	829.	429.	309.	1075.
FLINT	TILT = LAT:	606.	897.	1132.	1330.	1451.	1498.	1524.	1470.	1342.	1150.	657.	492.	0
MI	TILT = LAT+15:	623.	896.	1075.	1188.	1236.	1248.	1281.	1287.	1248.	1140.	670.	508.	0
LAT = 42.97	TILT = 90:	544.	728.	767.	703.	625.	585.	616.	709.	832.	904.	575.	448.	0
	KT:	.32	.38	.41	.44	.47	.48	.49	.48	.46	.44	.33	.29	0
	AMB TEMP:	22.3	23.8	32.6	45.9	55.8	65.8	69.7	68.2	61.0	51.2	38.3	26.8	46.8
	HTG DEG DAYS:	1324	1154	1004	573	306	65	14	36	147	433	801	1184	7041
STA NO:14827	HORIZ INSOL:	455.	698.	982.	1361.	1672.	1842.	1787.	1594.	1274.	924.	516.	370.	1123.
FORT WAYNE	TILT = LAT:	713.	960.	1139.	1344.	1466.	1529.	1522.	1504.	1416.	1261.	789.	583.	0
IN	TILT = LAT+15:	737.	960.	1091.	1203.	1252.	1276.	1283.	1320.	1320.	1253.	810.	605.	0

LAT = 41.00	TILT = 90:	638.	766.	748.	682.	599.	560.	583.	691.	849.	973.	691.	530.	0
	KT:	.35	.39	.40	.44	.47	.49	.49	.49	.48	.47	.36	.31	0
	AMB TEMP:	25.3	27.6	36.5	49.3	59.6	69.5	73.0	71.3	64.5	53.6	40.2	28.6	49.9
	HTG DEG DAYS:	1231	1047	684	471	216	23	0	12	90	363	744	1128	6209
STA NO:14837	HORIZ INSOL:	515.	804.	1136.	1398.	1743.	1948.	1934.	1708.	1299.	911.	504.	389.	1191.
MADISON	TILT = LAT:	936.	1227.	1396.	1395.	1526.	1605.	1638.	1624.	1481.	1302.	832.	704.	0
WI	TILT = LAT+15:	983.	1241.	1334.	1247.	1298.	1333.	1374.	1421.	1380.	1297.	860.	741.	0
LAT = 43.13	TILT = 90:	887.	1031.	961.	738.	652.	615.	652.	778.	924.	1037.	753.	674.	0
	KT:	.43	.48	.49	.46	.49	.52	.53	.53	.50	.49	.39	.37	0
	AMB TEMP:	16.8	20.3	30.2	45.3	56.0	65.8	70.1	68.7	59.7	49.9	34.7	21.9	44.9
	HTG DEG DAYS:	1494	1252	1079	591	297	72	14	39	173	474	909	1336	7730
STA NO:14839	HORIZ INSOL:	479.	737.	1089.	1443.	1768.	1977.	1962.	1719.	1310.	908.	525.	378.	1191.
MILWAUKEE	TILT = LAT:	838.	1088.	1323.	1443.	1548.	1629.	1662.	1634.	1492.	1291.	873.	669.	0
WI	TILT = LAT+15:	876.	1096.	1262.	1283.	1317.	1352.	1393.	1430.	1391.	1285.	903.	702.	0
LAT = 42.95	TILT = 90:	784.	902.	905.	759.	657.	617.	655.	779.	928.	1025.	792.	635.	0
	KT:	.40	.43	.46	.47	.50	.53	.54	.53	.51	.48	.40	.35	0
	AMB TEMP:	19.4	22.5	31.4	44.7	54.2	64.5	69.9	69.2	61.1	51.0	36.5	24.2	45.7
	HTG DEG DAYS:	1414	1190	1042	609	348	90	15	36	140	440	855	1265	7444
STA NO:14847	HORIZ INSOL:	325.	603.	1029.	1383.	1688.	1811.	1835.	1523.	1049.	673.	332.	253.	1042.
SAULT STE. MARIE	TILT = LAT:	572.	935.	1308.	1401.	1473.	1483.	1547.	1449.	1196.	962.	528.	449.	0
MI	TILT = LAT+15:	591.	938.	1248.	1248.	1248.	1229.	1293.	1264.	1108.	949.	536.	466.	0
LAT = 46.47	TILT = 90:	533.	796.	944.	793.	689.	638.	684.	757.	783.	778.	470.	424.	0
	KT:	.32	.40	.47	.47	.48	.48	.51	.48	.43	.40	.30	.29	0
	AMB TEMP:	14.2	15.2	24.0	38.2	49.0	58.7	63.8	63.2	55.3	46.2	32.8	20.1	40.0
	HTG DEG DAYS:	1575	1394	1271	804	496	200	96	125	291	583	966	1392	9193
STA NO:14848	HORIZ INSOL:	416.	660.	993.	1387.	1723.	1922.	1852.	1666.	1291.	909.	497.	340.	1138.
SOUTH BEND	TILT = LAT:	645.	909.	1164.	1376.	1510.	1590.	1574.	1578.	1448.	1255.	769.	533.	0
IN	TILT = LAT+15:	663.	907.	1105.	1231.	1287.	1324.	1324.	1382.	1350.	1247.	790.	551.	0
LAT = 41.70	TILT = 90:	574.	727.	774.	707.	623.	586.	608.	732.	880.	978.	677.	483.	0
	KT:	.33	.37	.41	.45	.48	.51	.51	.51	.49	.47	.36	.30	0
	AMB TEMP:	24.0	26.3	35.3	48.1	58.4	68.6	72.3	71.0	63.8	53.4	39.6	28.2	49.1
	HTG DEG DAYS:	1271	1084	921	507	245	35	6	24	98	368	762	1141	6462
STA NO:14850	HORIZ INSOL:	311.	568.	1001.	1405.	1729.	1912.	1910.	1609.	1165.	754.	377.	257.	1083.
TRAVERSE CITY	TILT = LAT:	483.	810.	1228.	1413.	1511.	1571.	1614.	1531.	1327.	1065.	587.	409.	0
MI	TILT = LAT+15:	492.	806.	1169.	1261.	1283.	1302.	1351.	1338.	1233.	1053.	597.	419.	0
LAT = 44.73	TILT = 90:	429.	665.	860.	772.	675.	634.	674.	766.	849.	850.	518.	371.	0
	KT:	.28	.35	.44	.47	.49	.51	.53	.51	.46	.42	.31	.26	0
	AMB TEMP:	20.8	20.7	28.7	42.7	52.8	63.7	68.7	67.5	59.4	49.8	36.9	25.9	44.8
	HTG DEG DAYS:	1370	1240	1125	669	387	104	33	66	178	471	843	1212	7698
STA NO:14852	HORIZ INSOL:	385.	587.	890.	1278.	1586.	1759.	1734.	1506.	1194.	851.	457.	315.	1045.
YOUNGSTOWN	TILT = LAT:	566.	768.	1015.	1257.	1390.	1462.	1476.	1418.	1318.	1144.	670.	465.	0
OH	TILT = LAT+15:	577.	760.	960.	1124.	1189.	1222.	1246.	1244.	1226.	1133.	683.	476.	0
LAT = 41.27	TILT = 90:	491.	598.	664.	644.	580.	550.	576.	661.	793.	878.	576.	409.	0
	KT:	.30	.33	.37	.41	.45	.47	.48	.46	.45	.44	.33	.27	0
	AMB TEMP:	25.7	26.7	35.3	47.7	57.6	67.0	70.7	69.2	62.7	52.6	40.3	28.8	48.7
	HTG DEG DAYS:	1218	1072	921	519	258	42	9	22	118	384	741	1122	6426
STA NO:14858	HORIZ INSOL:	244.	484.	933.	1366.	1660.	1838.	1838.	1521.	1010.	671.	291.	192.	1004.
HOUGHTON	TILT = LAT:	379.	704.	1174.	1387.	1447.	1501.	1547.	1451.	1152.	978.	447.	297.	0

MI	TILT = LAT+15:	383.	698.	1116.	1234.	1226.	1243.	1293.	1265.	1066.	965.	450.	301.	0
LAT = 47.17	TILT = 90:	337.	587.	850.	796.	691.	656.	697.	769.	762.	799.	393.	266.	0
	KT:	.25	.33	.43	.47	.47	.49	.51	.49	.42	.40	.27	.23	0
	AMB TEMP:	0	0	0	0	0	0	0	0	0	0	0	0	0
	HTG DEG DAYS:	0	0	0	0	0	0	0	0	0	0	0	0	0
STA NO:14860	HORIZ INSOL:	346.	577.	920.	1359.	1646.	1847.	1833.	1455.	1201.	827.	416.	278.	1059.
ERIE	TILT = LAT:	499.	767.	1068.	[illegible]	1442.	1529.	1557.	1369.	1338.	1124.	606.	397.	0
PA	TILT = LAT+15:	506.	759.	1011.	1204.	1230.	1274.	1309.	1200.	1245.	1112.	614.	403.	0
LAT = 42.08	TILT = 90:	429.	603.	710.	699.	608.	578.	610.	653.	817.	871.	518.	344.	0
	KT:	.28	.33	.39	.44	.46	.49	.50	.45	.46	.43	.31	.25	0
	AMB TEMP:	25.1	25.2	32.9	44.8	54.6	64.6	68.7	67.5	61.4	51.6	40.1	29.1	47.1
	HTG DEG DAYS:	1237	1114	995	606	336	80	24	43	141	415	747	1113	6851
STA NO:14895	HORIZ INSOL:	428.	650.	964.	1357.	1668.	1839.	1787.	1596.	1272.	908.	505.	353.	1111.
AKRON-CANTON	TILT = LAT:	650.	872.	1113.	1340.	1463.	1528.	1522.	1505.	1413.	1230.	762.	543.	0
OH	TILT = LAT+15:	669.	868.	1056.	1199.	1249.	1275.	1283.	1321.	1317.	1221.	781.	560.	0
LAT = 40.92	TILT = 90:	574.	687.	729.	679.	597.	559.	582.	690.	846.	947.	663.	487.	0
	KT:	.33	.36	.40	.44	.47	.49	.49	.49	.48	.46	.35	.30	0
	AMB TEMP:	26.3	27.7	36.2	48.5	58.7	68.3	71.7	70.3	63.7	53.3	40.7	29.4	49.6
	HTG DEG DAYS:	1200	1044	893	495	231	33	9	16	101	369	729	1104	6224
STA NO:14898	HORIZ INSOL:	451.	725.	1104.	1439.	[illegible]	1908.	1889.	1622.	1218.	821.	465.	350.	1143.
GREEN BAY	TILT = LAT:	832.	1118.	1378.	1448.	1503.	1568.	1597.	1542.	1394.	1181.	790.	652.	0
WI	TILT = LAT+15:	871.	1128.	1317.	1293.	1276.	1301.	1337.	1349.	1297.	1172.	815.	685.	0
LAT = 44.48	TILT = 90:	790.	947.	968.	786.	667.	629.	664.	767.	889.	949.	721.	628.	0
	KT:	.40	.45	.48	.48	.49	.51	.52	.51	.48	.46	.38	.35	0
	AMB TEMP:	15.4	18.0	28.6	43.8	54.5	64.5	69.2	67.7	58.9	49.2	34.1	20.9	43.7
	HTG DEG DAYS:	1538	1316	1128	636	338	91	22	54	191	490	927	1367	8098
STA NO:14913	HORIZ INSOL:	389.	673.	1034.	1373.	1643.	1767.	1654.	1547.	1095.	725.	381.	292.	1064.
DULUTH	TILT = LAT:	769.	1099.	1326.	1392.	1432.	1446.	1562.	1475.	1264.	1074.	665.	581.	0
MN	TILT = LAT+15:	805.	1110.	1266.	1239.	1214.	1199.	1305.	1287.	1173.	1064.	683.	610.	0
LAT = 46.83	TILT = 90:	742.	955.	963.	793.	679.	632.	696.	776.	835.	881.	612.	567.	0
	KT:	.39	.45	.48	.47	.47	.47	.51	.49	.45	.43	.35	.34	0
	AMB TEMP:	8.5	12.1	23.5	38.6	49.4	59.0	65.6	64.1	54.4	45.3	28.4	14.4	38.6
	HTG DEG DAYS:	1751	1481	1287	792	464	194	67	104	318	611	1098	1569	9756
STA NO:14914	HORIZ INSOL:	415.	706.	1098.	1476.	1835.	1994.	2120.	1825.	1304.	874.	457.	337.	1203.
FARGO	TILT = LAT:	850.	1175.	1429.	1509.	1604.	1628.	1786.	1764.	1557.	1374.	873.	730.	0
ND	TILT = LAT+15:	893.	1189.	1367.	1345.	1356.	1343.	1486.	1538.	1451.	1373.	908.	774.	0
LAT = 46.90	TILT = 90:	828.	1027.	1043.	860.	748.	691.	774.	918.	1039.	1152.	826.	729.	0
	KT:	.42	.47	.51	.50	.52	.53	.59	.58	.54	.52	.42	.40	0
	AMB TEMP:	5.9	10.7	24.2	42.3	54.6	64.7	70.7	69.2	57.9	47.0	28.6	13.0	40.8
	HTG DEG DAYS:	1832	1520	1265	681	334	97	13	33	234	558	1092	1612	9271
STA NO:14918	HORIZ INSOL:	356.	663.	1046.	1444.	1716.	1853.	1921.	1618.	1121.	704.	346.	272.	1088.
INTERNATIONAL FALLS	TILT = LAT:	758.	1153.	1391.	1488.	1497.	1509.	1615.	1558.	1331.	1092.	636.	598.	0
MN	TILT = LAT+15:	795.	1168.	1330.	1324.	1265.	1246.	1345.	1357.	1236.	1083.	655.	631.	0
LAT = 48.57	TILT = 90:	743.	1025.	1039.	877.	735.	683.	748.	848.	907.	916.	594.	596.	0
	KT:	.40	.47	.50	.50	.49	.50	.53	.52	.47	.44	.34	.36	0
	AMB TEMP:	1.9	7.0	20.6	38.2	50.1	60.4	65.8	63.2	53.0	43.5	24.9	8.7	36.5
	HTG DEG DAYS:	1956	1624	1376	804	462	168	66	112	364	667	1203	1745	10547
STA NO:14920	HORIZ INSOL:	481.	765.	1101.	1426.	1713.	1905.	1900.	1666.	1242.	864.	494.	370.	1161.

LA CROSSE	TILT = LAT:	883.	1176.	1360.	1430.	1498.	1568.	1608.	1585.	1416.	1240.	836.	682.	0
WI	TILT = LAT+15:	926.	1188.	1298.	1277.	1273.	1302.	1348.	1386.	1318.	1233.	864.	717.	0
LAT = 43.87	TILT = 90:	838.	993.	945.	767.	655.	618.	657.	774.	894.	993.	762.	655.	0
	KT:	.42	.46	.48	.47	.48	.51	.52	.52	.49	.47	.39	.36	0
	AMB TEMP:	16.1	20.0	31.1	47.6	59.0	68.5	72.8	71.4	61.8	51.8	35.4	21.8	46.4
	HTG DEG DAYS:	1516	1260	1051	522	224	39	10	17	130	421	888	1339	7417
STA NO:14922	HORIZ INSOL:	464.	764.	1104.	1442.	1737.	1928.	1970.	1687.	1255.	860.	480.	353.	1170.
MINNEAPOLIS-ST. PAUL	TILT = LAT:	885.	1214.	1387.	1455	1518.	1582.	1664.	1610.	1450.	1267.	843.	679.	0
MN	TILT = LAT+15:	930.	1229.	1325.	1298.	1289.	1311.	1391.	1407.	1350.	1261.	873.	716.	0
LAT = 44.88	TILT = 90:	849.	1040.	980.	796.	680.	640.	693.	805.	933.	1030.	778.	660.	0
	KT:	.42	.48	.49	.48	.49	.51	.54	.53	.50	.48	.40	.37	0
	AMB TEMP:	12.2	16.5	28.3	45.1	57.1	66.9	71.9	70.2	60.0	50.0	32.4	18.6	44.1
	HTG DEG DAYS:	1637	1358	1138	597	271	65	11	21	173	472	978	1438	8159
STA NO:14923	HORIZ INSOL:	535.	812.	1119.	1459.	1754.	1969.	1939.	1715.	1357.	996.	595.	433.	1224.
MOLINE	TILT = LAT:	912.	1183.	1338.	1452.	1537.	1629.	1647.	1625.	1529.	1398.	974.	751.	0
IL	TILT = LAT+15:	955.	1194.	1276.	1294.	1310.	1355.	1383.	1424.	1427.	1394.	1012.	790.	0
LAT = 41.45	TILT = 90:	846.	971.	894.	740.	628.	590.	622.	746.	926.	1095.	880.	711.	0
	KT:	.42	.46	.46	.47	.49	.52	.53	.53	.51	.51	.43	.38	0
	AMB TEMP:	21.5	25.7	35.7	50.6	61.1	70.8	74.5	72.9	64.6	54.4	39.2	26.6	49.8
	HTG DEG DAYS:	1349	1100	908	436	184	20	0	11	79	344	774	1190	6395
STA NO:14925	HORIZ INSOL:	477.	753.	1082.	1410.	1695.	1902.	1909.	1662.	1250.	870.	494.	370.	1156.
ROCHESTER	TILT = LAT:	874.	1154.	1333.	1413.	1483.	1565.	1615.	1561.	1428.	1253.	838.	685.	0
MN	TILT = LAT+15:	916.	1165.	1271.	1261.	1251.	1300.	1353.	1383.	1329.	1246.	867.	721.	0
LAT = 43.92	TILT = 90:	829.	973.	926.	759.	651.	618.	660.	774.	902.	1005.	765.	660.	0
	KT:	.41	.46	.47	.47	.48	.51	.52	.52	.49	.48	.39	.36	0
	AMB TEMP:	12.9	16.9	27.8	44.5	56.2	66.0	70.1	68.6	59.3	49.6	32.6	18.9	43.6
	HTG DEG DAYS:	1615	1347	1153	615	292	78	21	35	185	485	972	1429	8227
STA NO:14931	HORIZ INSOL:	579.	859.	1165.	1538.	1876.	2121.	2085.	1828.	1416.	1061.	664.	481.	1306.
BURLINGTON	TILT = LAT:	987.	1248.	1391.	1533.	1645.	1753.	1770.	1735.	1596.	1488.	1100.	843.	0
IA	TILT = LAT+15:	1036.	1261.	1329.	1373.	1401.	1454.	1484.	1521.	1491.	1488.	1147.	891.	0
LAT = 40.78	TILT = 90:	918.	1021.	922.	767.	647.	602.	639.	776.	956.	1163.	999.	803.	0
	KT:	.44	.47	.48	.50	.53	.57	.57	.56	.53	.54	.46	.40	0
	AMB TEMP:	22.9	27.3	36.9	51.3	61.8	71.4	75.4	73.9	65.4	55.3	39.8	27.6	50.8
	HTG DEG DAYS:	1305	1056	871	416	172	16	0	8	70	320	756	1159	6149
STA NO:14933	HORIZ INSOL:	581.	861.	1181.	1557.	1868.	2125.	2097.	1828.	1434.	1068.	658.	487.	1312.
DES MOINES	TILT = LAT:	1024.	1278.	1428.	1558.	1637.	1753.	1778.	1738.	1630.	1528.	1121.	891.	0
IA	TILT = LAT+15:	1077.	1293.	1365.	1394.	1393.	1452.	1489.	1522.	1524.	1528.	1171.	945.	0
LAT = 41.53	TILT = 90:	963.	1057.	959.	791.	659.	617.	656.	793.	991.	1208.	1028.	861.	0
	KT:	.45	.49	.49	.51	.52	.57	.57	.56	.54	.55	.47	.42	0
	AMB TEMP:	19.4	24.2	33.9	49.5	60.9	70.5	75.1	73.3	64.3	54.3	37.8	25.0	49.0
	HTG DEG DAYS:	1414	1142	964	465	186	26	0	13	94	350	816	1240	6710
STA NO:14935	HORIZ INSOL:	661.	917.	1265.	1692.	1972.	2242.	2216.	1939.	1509.	1138.	739.	569.	1405.
GRAND ISLAND	TILT = LAT:	1189.	1363.	1537.	1701.	1729.	1848.	1878.	1847.	1719.	1630.	1276.	1075.	0
NE	TILT = LAT+15:	1258.	1383.	1473.	1524.	1470.	1528.	1570.	1618.	1608.	1635.	1339.	1148.	0
LAT = 40.97	TILT = 90:	1129.	1127.	1028.	848.	674.	623.	666.	822.	1035.	1287.	1178.	1052.	0
	KT:	.50	.51	.52	.55	.55	.60	.61	.59	.57	.58	.52	.48	0
	AMB TEMP:	22.3	27.7	35.5	49.9	60.7	70.7	76.3	75.0	64.4	53.7	38.2	27.0	50.1
	HTG DEG DAYS:	1324	1044	915	461	184	35	6	5	107	362	804	1178	6425

STA NO:14936	HORIZ INSOL:	488.	745.	1114.	1530.	1872.	2101.	2183.	1892.	1418.	988.	577.	405.	1276.
HURON	TILT = LAT:	926.	1155.	1391.	1549.	1638.	1722.	1843.	1817.	1663.	1493.	1063.	806.	0
SD	TILT = LAT+15:	973.	1166.	1328.	1383.	1388.	1423.	1536.	1588.	1553.	1494.	1110.	855.	0
LAT = 44.38	TILT = 90:	887.	979.	976.	837.	713.	668.	735.	887.	1065.	1222.	998.	793.	0
	KT:	.43	.46	.49	.51	.53	.56	.60	.59	.56	.55	.47	.41	0
	AMB TEMP:	12.5	17.9	29.0	45.8	57.0	67.1	73.7	72.1	60.7	49.6	32.4	19.2	44.8
	HTG DEG DAYS:	1627	1319	1116	576	273	72	9	13	169	482	978	1420	8054
STA NO:14940	HORIZ INSOL:	554.	836.	1168.	1519.	1835.	2114.	2084.	1833.	1405.	1010.	600.	443.	1288.
MASON CITY	TILT = LAT:	1035.	1291.	1444.	1527.	1660.	1737.	1764.	1750.	1622.	1485.	1059.	852.	0
IA	TILT = LAT+15:	1092.	1309.	1381.	1365.	1409.	1437.	1474.	1531.	1514.	1485.	1105.	904.	0
LAT = 43.15	TILT = 90:	990.	1091.	996.	804.	696.	647.	686.	831.	1015.	1196.	982.	832.	0
	KT:	.46	.50	.50	.50	.53	.56	.57	.57	.55	.54	.46	.42	0
	AMB TEMP:	14.2	18.5	29.0	45.7	57.4	67.2	71.3	69.9	60.2	50.5	33.6	20.1	44.8
	HTG DEG DAYS:	1575	1302	1116	579	265	64	13	31	165	457	942	1392	7901
STA NO:14943	HORIZ INSOL:	569.	842.	1170.	1578.	1901.	2124.	2122.	1845.	1421.	1038.	643.	469.	1310.
SIOUX CITY	TILT = LAT:	1035.	1273.	1431.	1587.	1666.	1748.	1797.	1759.	1629.	1507.	1125.	886.	0
IA	TILT = LAT+15:	1091.	1289.	1368.	1419.	1415.	1447.	1503.	1539.	1522.	1507.	1176.	941.	0
LAT = 42.40	TILT = 90:	983.	1064.	975.	821.	684.	634.	679.	819.	1006.	1204.	1041.	863.	0
	KT:	.46	.49	.49	.52	.54	.57	.58	.57	.55	.55	.48	.43	0
	AMB TEMP:	18.0	23.4	33.2	49.4	60.9	70.3	75.3	73.5	63.4	53.1	36.3	23.5	48.4
	HTG DEG DAYS:	1457	1165	986	474	189	33	0	10	113	378	864	1287	6953
STA NO:14944	HORIZ INSOL:	533.	802.	1152.	1543.	1894.	2100.	2150.	1845.	1410.	1005.	608.	441.	1290.
SIOUX FALLS	TILT = LAT:	1002.	1240.	1430.	1557.	1658.	1725.	1817.	1764.	1635.	1492.	1098.	867.	0
SD	TILT = LAT+15:	1056.	1255.	1367.	1391.	1407.	1426.	1517.	1542.	1527.	1493.	1147.	921.	0
LAT = 43.57	TILT = 90:	960.	1048.	992.	827.	704.	652.	709.	846.	1032.	1209.	1025.	852.	0
	KT:	.45	.48	.50	.51	.54	.56	.59	.57	.55	.55	.48	.43	0
	AMB TEMP:	14.2	19.4	30.0	46.1	57.7	67.6	73.3	71.8	60.9	50.2	33.1	20.0	45.4
	HTG DEG DAYS:	1575	1277	1085	567	259	65	10	18	165	465	957	1395	7838
STA NO:14991	HORIZ INSOL:	452.	746.	1090.	1426.	1681.	1872.	1886.	1621.	1196.	826.	451.	341.	1132.
EAU CLAIRE	TILT = LAT:	851.	1177.	1366.	1437.	1468.	1538.	1594.	1543.	1371.	1204.	768.	643.	0
WI	TILT = LAT+15:	892.	1190.	1304.	1282.	1247.	1276.	1334.	1348.	1275.	1196.	792.	675.	0
LAT = 44.87	TILT = 90:	813.	1006.	965.	787.	661.	627.	671.	774.	879.	973.	702.	621.	0
	KT:	.41	.47	.48	.48	.48	.50	.52	.51	.48	.46	.37	.35	0
	AMB TEMP:	11.7	15.4	27.3	44.5	56.2	66.1	70.5	68.4	58.7	48.7	32.0	18.0	43.1
	HTG DEG DAYS:	1652	1389	1169	615	293	65	14	37	202	505	990	1457	8388
STA NO:22010	HORIZ INSOL:	958.	1206.	1580.	1700.	1827.	2024.	2054.	1936.	1584.	1360.	1060.	903.	1516.
DEL RIO	TILT = LAT:	1314.	1494.	1747.	1655.	1632.	1732.	1789.	1819.	1663.	1636.	1429.	1281.	0
TX	TILT = LAT+15:	1386.	1516.	1684.	1498.	1409.	1457.	1518.	1611.	1563.	1638.	1498.	1367.	0
LAT = 29.37	TILT = 90:	1109.	1065.	947.	613.	450.	398.	424.	568.	781.	1088.	1170.	1132.	0
	KT:	.49	.51	.55	.51	.51	.55	.57	.57	.53	.54	.52	.50	0
	AMB TEMP:	50.8	55.7	62.6	72.0	78.2	84.3	86.7	86.1	80.2	71.2	59.6	52.3	70.0
	HTG DEG DAYS:	449	283	163	16	0	0	0	0	0	34	184	394	1523
STA NO:23023	HORIZ INSOL:	1081.	1383.	1839.	2192.	2430.	2562.	2389.	2210.	1844.	1522.	1176.	1000.	1802.
MIDLAND-ODESSA	TILT = LAT:	1627.	1833.	2118.	2165.	2142.	2142.	2052.	2081.	1989.	1935.	1724.	1565.	0
TX	TILT = LAT+15:	1735.	1876.	2050.	1951.	1822.	1770.	1723.	1833.	1872.	1950.	1824.	1689.	0
LAT = 31.93	TILT = 90:	1455.	1386.	1216.	809.	537.	431.	478.	672.	984.	1362.	1491.	1462.	0
	KT:	.60	.62	.66	.67	.68	.69	.66	.65	.63	.64	.62	.60	0
	AMB TEMP:	43.6	47.8	54.3	64.3	72.3	79.9	82.3	81.8	75.4	65.8	53.3	45.9	63.9
	HTG DEG DAYS:	663	482	349	98	0	0	0	0	0	81	356	592	2621

STA NO:23034
SAN ANGELO
TX
LAT = 31.37

HORIZ INSOL:	962.	1208.	1606.	1851.	2031.	2186.	2123.	1966.	1607.	1337.	1044.	895.	1568.
TILT = LAT:	1386.	1547.	1810.	1813.	1801.	1850.	1836.	1848.	1707.	1649.	1469.	1339.	0
TILT = LAT+15:	1467.	1572.	1744.	1637.	1545.	1547.	1553.	1632.	1604.	1652.	1542.	1434.	0
TILT = 90:	1207.	1140.	1023.	696.	501.	431.	461.	614.	839.	1134.	1237.	1219.	0
KT:	.52	.53	.57	.56	.56	.59	.58	.58	.54	.55	.54	.52	0
AMB TEMP:	46.4	50.4	57.1	67.2	74.5	81.6	84.7	84.5	76.8	67.2	55.5	48.3	66.2
HTG DEG DAYS:	577	413	287	74	0	0	0	0	0	73	298	518	2240

STA NO:23042
LUBBOCK
TX
LAT = 33.65

HORIZ INSOL:	1031.	1332.	1762.	2168.	2396.	2544.	2412.	2208.	1820.	1468.	1116.	935.	1766.
TILT = LAT:	1613.	1813.	2056.	2151.	2108.	2119.	2063.	2083.	1986.	1910.	1694.	1522.	0
TILT = LAT+15:	1721.	1855.	1988.	1937.	1792.	1750.	1729.	1833.	1868.	1924.	1792.	1641.	0
TILT = 90:	1468.	1403.	1221.	852.	578.	473.	518.	717.	1024.	1379.	1491.	1441.	0
KT:	.60	.62	.65	.67	.67	.68	.66	.66	.63	.64	.62	.59	0
AMB TEMP:	39.1	42.7	48.9	60.0	68.5	77.1	79.7	78.4	71.0	61.0	48.8	41.3	59.7
HTG DEG DAYS:	803	624	508	190	29	0	0	0	8	162	488	735	3545

STA NO:23043
ROSWELL
NM
LAT = 33.40

HORIZ INSOL:	1047.	1373.	1807.	2218.	2459.	2610.	2441.	2242.	1913.	1527.	1131.	952.	1810.
TILT = LAT:	1531.	1871.	2111.	2201.	2163.	2171.	2087.	2115.	2094.	1994.	1711.	1545.	0
TILT = LAT+15:	1741.	1917.	2042.	1982.	1837.	1789.	1749.	1860.	1972.	2012.	1810.	1667.	0
TILT = 90:	1483.	1448.	1249.	860.	575.	463.	512.	717.	1071.	1440.	1503.	1462.	0
KT:	.61	.63	.66	.68	.68	.70	.67	.66	.66	.66	.62	.60	0
AMB TEMP:	38.1	42.9	49.3	59.7	68.5	77.0	79.2	77.9	70.4	59.6	46.9	39.3	59.1
HTG DEG DAYS:	834	619	487	185	20	0	0	0	17	195	543	797	3697

STA NO:23044
EL PASO
TX
LAT = 31.80

HORIZ INSOL:	1125.	1480.	1909.	2364.	2601.	2682.	2450.	2284.	1987.	1639.	1244.	1031.	1900.
TILT = LAT:	1703.	1984.	2208.	2341.	2237.	2235.	2102.	2151.	2157.	2110.	1842.	1620.	0
TILT = LAT+15:	1820.	2037.	2140.	2109.	1938.	1639.	1763.	1894.	2034.	2134.	1955.	1751.	0
TILT = 90:	1529.	1510.	1265.	850.	532.	416.	474.	681.	1059.	1494.	1603.	1517.	0
KT:	.62	.66	.69	.72	.72	.72	.67	.67	.67	.68	.65	.61	0
AMB TEMP:	43.6	48.4	54.6	63.9	72.2	80.3	82.3	80.5	74.2	64.0	51.6	44.4	63.4
HTG DEG DAYS:	663	465	328	89	0	0	0	0	0	92	402	639	2678

STA NO:23047
AMARILLO
TX
LAT = 35.23

HORIZ INSOL:	960.	1244.	1631.	2019.	2212.	2393.	2280.	2103.	1760.	1404.	1033.	872.	1659.
TILT = LAT:	1552.	1726.	1916.	2007.	1947.	1993.	1950.	1986.	1938.	1864.	1611.	1474.	0
TILT = LAT+15:	1654.	1763.	1849.	1806.	1659.	1651.	1637.	1747.	1821.	1876.	1701.	1588.	0
TILT = 90:	1430.	1358.	1171.	845.	598.	511.	548.	733.	1037.	1375.	1434.	1412.	0
KT:	.59	.60	.61	.63	.62	.64	.62	.63	.62	.63	.60	.58	0
AMB TEMP:	36.0	39.7	45.6	56.5	65.6	74.6	78.7	77.6	69.8	59.5	46.3	38.5	57.4
HTG DEG DAYS:	899	708	601	275	61	10	0	0	20	206	561	822	4183

STA NO:23048
TUCUMCARI
NM
LAT = 35.18

HORIZ INSOL:	1009.	1297.	1712.	2098.	2314.	2484.	2349.	2164.	1829.	1443.	1073.	910.	1724.
TILT = LAT:	1650.	1815.	2026.	2090.	2035.	2064.	2006.	2045.	2022.	1926.	1687.	1555.	0
TILT = LAT+15:	1763.	1858.	1957.	1880.	1730.	1706.	1682.	1797.	1902.	1941.	1785.	1680.	0
TILT = 90:	1530.	1434.	1240.	873.	609.	511.	551.	747.	1080.	1424.	1509.	1499.	0
KT:	.62	.62	.64	.65	.64	.67	.64	.65	.64	.64	.62	.61	0
AMB TEMP:	37.0	41.1	46.7	56.9	65.6	75.1	78.4	76.7	69.6	58.7	46.2	38.6	57.6
HTG DEG DAYS:	868	669	567	260	57	7	0	0	20	217	564	818	4047

STA NO:23050
ALBUQUERQUE
NM
LAT = 35.05

HORIZ INSOL:	1017.	1342.	1768.	2228.	2538.	2679.	2489.	2290.	1972.	1547.	1134.	928.	1827.
TILT = LAT:	1659.	1887.	2098.	2227.	2227.	2214.	2120.	2167.	2198.	2091.	1804.	1587.	0
TILT = LAT+15:	1773.	1935.	2030.	2003.	1886.	1819.	1772.	1903.	2071.	2114.	1913.	1715.	0
TILT = 90:	1537.	1495.	1283.	914.	623.	502.	553.	772.	1168.	1556.	1623.	1530.	0
KT:	.62	.64	.66	.69	.71	.72	.68	.68	.69	.69	.65	.62	0
AMB TEMP:	35.2	40.0	45.8	55.8	65.3	74.6	78.7	76.6	70.1	58.2	44.5	36.2	56.8
HTG DEG DAYS:	924	700	595	282	58	0	0	0	7	218	615	893	4292

STA NO:23051	HORIZ INSOL:	962.	1241.	1652.	2040.	2222.	2418.	2284.	2097.	1802.	1434.	1028.	861.	1670.
CLAYTON	TILT = LAT:	1623.	1770.	1974.	2038.	1953.	2006.	1948.	1984.	2010.	1960.	1664.	1520.	0
NM	TILT = LAT+15:	1734.	1811.	1906.	1832.	1662.	1659.	1634.	1743.	1890.	1977.	1760.	1641.	0
LAT = 36.45	TILT = 90:	1521.	1418.	1236.	887.	627.	539.	575.	762.	1105.	1479.	1506.	1478.	0
	KT:	.62	.61	.63	.64	.62	.65	.62	.63	.64	.66	.62	.60	0
	AMB TEMP:	33.1	36.1	40.4	50.8	60.0	69.2	73.6	72.4	65.0	54.8	42.3	35.1	52.7
	HTG DEG DAYS:	989	809	763	431	172	38	0	5	73	324	681	927	5212
STA NO:23062	HORIZ INSOL:	840.	1127.	1530.	1879.	2135.	2351.	2273.	2044.	1727.	1301.	884.	732.	1568.
DENVER	TILT = LAT:	1550.	1709.	1890.	1894.	1873.	1938.	1929.	1946.	1980.	1871.	1539.	1420.	0
CO	TILT = LAT+15:	1655.	1748.	1821.	1698.	1591.	1601.	1613.	1705.	1859.	1884.	1625.	1531.	0
LAT = 39.75	TILT = 90:	1491.	1422.	1252.	908.	686.	610.	648.	830.	1167.	1471.	1429.	1411.	0
	KT:	.61	.61	.62	.60	.60	.63	.62	.62	.64	.64	.59	.59	0
	AMB TEMP:	29.9	32.8	37.0	47.5	57.0	66.0	73.0	71.6	62.8	52.0	39.4	32.6	50.1
	HTG DEG DAYS:	1088	902	868	525	253	80	0	0	120	408	768	1004	6016
STA NO:23063	HORIZ INSOL:	754.	1078.	1502.	1933.	2255.	2509.	2385.	2084.	1767.	1307.	869.	691.	1594.
EAGLE	TILT = LAT:	1338.	1612.	1845.	1951.	1979.	2063.	2021.	1985.	2030.	1878.	1500.	1308.	0
CO	TILT = LAT+15:	1421.	1644.	1777.	1750.	1677.	1697.	1686.	1739.	1907.	1892.	1582.	1407.	0
LAT = 39.65	TILT = 90:	1269.	1332.	1219.	931.	707.	621.	660.	841.	1195.	1476.	1387.	1290.	0
	KT:	.54	.58	.60	.62	.63	.67	.65	.63	.65	.64	.58	.55	0
	AMB TEMP:	18.0	23.3	31.1	41.9	51.3	58.9	65.9	63.7	55.6	44.8	30.9	20.3	42.2
	HTG DEG DAYS:	1457	1168	1051	693	425	190	43	79	285	626	1023	1386	8426
STA NO:23065	HORIZ INSOL:	789.	1056.	1424.	1829.	2062.	2357.	2319.	2046.	1643.	1268.	857.	695.	1529.
GOODLAND	TILT = LAT:	1405.	1558.	1726.	1836.	1810.	1945.	1968.	1945.	1862.	1796.	1457.	1301.	0
KS	TILT = LAT+15:	1494.	1588.	1660.	1647.	1539.	1607.	1645.	1705.	1746.	1806.	1535.	1398.	0
LAT = 39.37	TILT = 90:	1334.	1280.	1131.	875.	662.	602.	645.	821.	1088.	1400.	1340.	1278.	0
	KT:	.56	.56	.57	.58	.58	.63	.64	.62	.61	.62	.57	.55	0
	AMB TEMP:	27.6	31.5	36.3	48.7	58.9	69.1	75.8	74.1	64.3	52.8	38.5	30.1	50.6
	HTG DEG DAYS:	1159	938	890	489	216	55	0	0	108	387	795	1082	6119
STA NO:23066	HORIZ INSOL:	791.	1119.	1553.	1986.	2380.	2599.	2465.	2182.	1834.	1345.	918.	731.	1659.
GRAND JUNCTION	TILT = LAT:	1395.	1665.	1906.	2004.	2087.	2134.	2089.	2080.	2108.	1922.	1580.	1377.	0
CO	TILT = LAT+15:	1483.	1700.	1837.	1799.	1767.	1753.	1741.	1822.	1983.	1938.	1669.	1482.	0
LAT = 39.12	TILT = 90:	1321.	1372.	1250.	940.	716.	612.	656.	859.	1227.	1502.	1461.	1357.	0
	KT:	.56	.59	.62	.63	.67	.69	.67	.66	.67	.65	.60	.57	0
	AMB TEMP:	26.6	33.6	41.2	51.7	62.2	71.3	78.7	75.4	67.2	54.9	39.8	29.5	52.7
	HTG DEG DAYS:	1190	879	738	404	133	20	0	0	60	324	756	1101	5605
STA NO:23090	HORIZ INSOL:	945.	1281.	1693.	2133.	2452.	2666.	2478.	2252.	1934.	1479.	1047.	837.	1766.
FARMINGTON	TILT = LAT:	1603.	1855.	2040.	2140.	2151.	2196.	2106.	2137.	2186.	2050.	1722.	1484.	0
NM	TILT = LAT+15:	1712.	1901.	1971.	1924.	1822.	1803.	1758.	1875.	2058.	2072.	1824.	1601.	0
LAT = 36.75	TILT = 90:	1505.	1499.	1287.	932.	663.	549.	596.	812.	1209.	1560.	1569.	1444.	0
	KT:	.61	.64	.65	.67	.68	.71	.68	.68	.69	.68	.64	.59	0
	AMB TEMP:	28.6	35.0	40.6	49.7	59.5	67.9	75.0	72.6	64.6	52.9	39.2	30.1	51.3
	HTG DEG DAYS:	1128	840	756	465	184	36	0	6	67	375	774	1082	5713
STA NO:23129	HORIZ INSOL:	928.	1215.	1610.	1938.	2064.	2140.	2300.	2100.	1701.	1326.	1004.	847.	1598.
LONG BEACH	TILT = LAT:	1420.	1629.	1859.	1913.	1823.	1800.	1971.	1979.	1845.	1699.	1491.	1350.	0
CA	TILT = LAT+15:	1507.	1660.	1792.	1723.	1560.	1504.	1657.	1743.	1734.	1704.	1568.	1449.	0
LAT = 33.82	TILT = 90:	1276.	1250.	1104.	780.	551.	474.	518.	698.	957.	1216.	1295.	1263.	0
	KT:	.54	.56	.59	.60	.57	.57	.63	.62	.59	.58	.56	.54	0
	AMB TEMP:	54.2	55.5	57.2	60.6	64.1	67.3	72.2	73.3	71.8	66.9	60.6	55.5	63.3

	HTG DEG DAYS:	339	273	247	148	71	23	0	0	7	48	155	295	1606
STA NO:23153	HORIZ INSOL:	918.	1274.	1777.	2251.	2577.	2788.	2703.	2433.	2043.	1521.	1031.	827.	1846.
TONOPAH	TILT = LAT:	1625.	1907.	2202.	2283.	2259.	2283.	2284.	2330.	2364.	2191.	1769.	1545.	0
NV	TILT = LAT+15:	1737.	1957.	2132.	2051.	1907.	1866.	1894.	2039.	2230.	2221.	1876.	1671.	0
LAT = 38.07	TILT = 90:	1546.	1571.	1429.	1023.	713.	582.	641.	901.	1347.	1711.	1638.	1527.	0
	KT:	.62	.65	.70	.71	.72	.74	.74	.74	.74	.72	.65	.62	0
	AMB TEMP:	30.2	34.6	39.6	48.1	56.9	65.3	73.0	70.7	63.5	52.1	39.8	31.9	50.5
	HTG DEG DAYS:	1079	851	787	512	269	92	0	13	108	407	756	1026	5900
STA NO:23154	HORIZ INSOL:	820.	1141.	1606.	2009.	2311.	2513.	2447.	2230.	1935.	1408.	926.	723.	1672.
ELY	TILT = LAT:	1470.	1714.	1987.	2030.	2027.	2067.	2073.	2129.	2249.	2045.	1609.	1365.	0
NV	TILT = LAT+15:	1567.	1752.	1918.	1822.	1717.	1701.	1728.	1865.	2118.	2067.	1701.	1470.	0
LAT = 39.28	TILT = 90:	1402.	1419.	1310.	955.	709.	612.	659.	879.	1316.	1611.	1493.	1347.	0
	KT:	.58	.61	.64	.64	.65	.67	.67	.68	.71	.69	.61	.57	0
	AMB TEMP:	23.6	27.9	32.8	41.3	50.0	57.7	67.2	65.5	56.7	46.0	34.0	26.2	44.1
	HTG DEG DAYS:	1283	1039	998	711	470	241	23	62	265	589	930	1203	7814
STA NO:23155	HORIZ INSOL:	766.	1102.	1595.	2095.	2509.	2749.	2684.	2421.	1992.	1458.	942.	677.	1749.
BAKERSFIELD	TILT = LAT:	1172.	1497.	1872.	2088.	2202.	2266.	2276.	2296.	2231.	1959.	1443.	1072.	0
CA	TILT = LAT+15:	1234.	1521.	1805.	1878.	1865.	1857.	1892.	2014.	2102.	1976.	1516.	1139.	0
LAT = 35.42	TILT = 90:	1048.	1163.	1147.	878.	631.	507.	562.	811.	1196.	1456.	1272.	991.	0
	KT:	.47	.53	.60	.65	.70	.74	.73	.72	.70	.65	.55	.46	0
	AMB TEMP:	47.5	52.4	56.6	62.7	69.8	76.9	83.9	81.6	76.6	66.9	56.0	47.9	64.9
	HTG DEG DAYS:	543	353	266	140	22	0	0	0	0	55	276	530	2185
STA NO:23159	HORIZ INSOL:	914.	1236.	1685.	2133.	2454.	2655.	2424.	2157.	1920.	1465.	1015.	818.	1740.
BRYCE CANYON	TILT = LAT:	1592.	1817.	2055.	2150.	2152.	2184.	2059.	2048.	2189.	2070.	1711.	1499.	0
UT	TILT = LAT+15:	1700.	1861.	1986.	1931.	1821.	1792.	1719.	1796.	2061.	2093.	1812.	1619.	0
LAT = 37.70	TILT = 90:	1507.	1483.	1319.	962.	689.	575.	616.	812.	1236.	1599.	1573.	1472.	0
	KT:	.61	.63	.66	.67	.68	.71	.66	.65	.69	.69	.64	.60	0
	AMB TEMP:	19.8	23.2	28.7	37.7	46.2	46.2	61.6	59.9	52.9	42.8	30.7	22.4	40.0
	HTG DEG DAYS:	1401	1170	1125	819	563	330	128	176	363	688	1029	1321	9133
STA NO:23160	HORIZ INSOL:	1099.	1432.	1864.	2363.	2671.	2730.	2341.	2183.	1979.	1602.	1208.	996.	1872.
TUCSON	TILT = LAT:	1669.	1918.	2156.	2343.	2345.	2269.	2012.	2055.	2152.	2064.	1792.	1567.	0
AZ	TILT = LAT+15:	1782.	1967.	2088.	2111.	1983.	1864.	1692.	1811.	2029.	2085.	1900.	1691.	0
LAT = 32.12	TILT = 90:	1500.	1462.	1243.	860.	538.	418.	482.	672.	1066.	1465.	1560.	1466.	0
	KT:	.61	.64	.67	.72	.74	.73	.64	.64	.67	.67	.64	.60	0
	AMB TEMP:	50.9	53.5	57.6	65.5	73.6	82.1	86.3	83.8	80.1	70.1	58.5	52.0	67.8
	HTG DEG DAYS:	442	333	243	81	0	0	0	0	0	29	221	403	1752
STA NO:23161	HORIZ INSOL:	958.	1281.	1772.	2274.	2591.	2766.	2603.	2383.	2008.	1516.	1085.	876.	1843.
DAGGETT	TILT = LAT:	1529.	1775.	2100.	2274.	2273.	2281.	2213.	2256.	2240.	2032.	1695.	1463.	0
CA	TILT = LAT+15:	1628.	1815.	2031.	2046.	1922.	1869.	1845.	1980.	2111.	2052.	1794.	1576.	0
LAT = 34.87	TILT = 90:	1401.	1393.	1279.	924.	620.	490.	549.	786.	1185.	1503.	1512.	1396.	0
	KT:	.58	.61	.66	.70	.72	.74	.71	.71	.70	.67	.62	.58	0
	AMB TEMP:	47.3	52.0	56.7	64.3	72.3	80.1	87.3	85.5	79.2	68.1	55.5	48.0	66.4
	HTG DEG DAYS:	549	371	271	118	14	0	0	0	0	57	296	527	2203
STA NO:23169	HORIZ INSOL:	978.	1340.	1823.	2319.	2646.	2778.	2588.	2355.	2037.	1540.	1086.	881.	1864.
LAS VEGAS	TILT = LAT:	1637.	1930.	2206.	2335.	2319.	2284.	2197.	2235.	2305.	2126.	1764.	1544.	0
NV	TILT = LAT+15:	1749.	1981.	2137.	2099.	1957.	1870.	1830.	1960.	2174.	2151.	1870.	1667.	0
LAT = 36.08	TILT = 90:	1530.	1553.	1380.	981.	658.	524.	582.	817.	1255.	1608.	1601.	1498.	0
	KT:	.62	.65	.69	.72	.74	.74	.71	.71	.72	.70	.64	.61	0

	AMB TEMP:	44.2	49.1	54.8	63.8	73.3	82.3	89.6	87.4	80.1	67.1	53.3	45.2	65.8
	HTG DEG DAYS:	645	451	324	126	10	0	0	0	0	74	357	614	2601
STA NO:23174	HORIZ INSOL:	926.	1214.	1619.	1951.	2060.	2119.	2307.	2080.	1681.	1317.	1004.	849.	1594.
LOS ANGELES	TILT = LAT:	1422.	1631.	1873.	1928.	1819.	1783.	1977.	1960.	1824.	1688.	1496.	1359.	0
CA	TILT = LAT+15:	1509.	1663.	1806.	1736.	1556.	1490.	1661.	1726.	1713.	1692.	1574.	1459.	0
LAT = 33.93	TILT = 90:	1280.	1253.	1115.	788.	552.	475.	521.	696.	948.	1210.	1302.	1274.	0
	KT:	.54	.56	.60	.60	.57	.57	.63	.62	.58	.57	.56	.54	0
	AMB TEMP:	54.5	55.6	56.5	58.8	61.9	64.5	68.5	69.5	68.7	65.2	60.5	56.9	61.7
	HTG DEG DAYS:	331	270	267	195	114	71	19	15	23	77	158	279	1819
STA NO:23179	HORIZ INSOL:	985.	1353.	1825.	2317.	2652.	2791.	2541.	2278.	2015.	1537.	1124.	913.	1861.
NEEDLES	TILT = LAT:	1578.	1895.	2170.	2319.	2324.	2301.	2163.	2154.	2246.	2064.	1767.	1538.	0
CA	TILT = LAT+15:	1683.	1943.	2102.	2086.	1964.	1883.	1806.	1892.	2118.	2085.	1873.	1660.	0
LAT = 34.77	TILT = 90:	1450.	1495.	1321.	935.	618.	484.	547.	762.	1185.	1527.	1582.	1474.	0
	KT:	.60	.64	.68	.72	.74	.75	.70	.68	.70	.68	.64	.60	0
	AMB TEMP:	51.6	56.5	61.6	70.4	79.6	88.3	95.4	93.3	86.9	74.3	60.7	52.7	72.6
	HTG DEG DAYS:	421	261	150	42	0	0	0	0	0	10	163	381	1428
STA NO:23183	HORIZ INSOL:	1021.	1374.	1814.	2355.	2677.	2739.	2486.	2293.	2015.	1576.	1150.	932.	1869.
PHOENIX	TILT = LAT:	1584.	1875.	2121.	2346.	2347.	2269.	2124.	2164.	2220.	2073.	1749.	1505.	0
AZ	TILT = LAT+15:	1688.	1921.	2052.	2112.	1983.	1862.	1778.	1902.	2094.	2095.	1852.	1622.	0
LAT = 33.43	TILT = 90:	1435.	1451.	1256.	902.	577.	453.	513.	727.	1135.	1504.	1541.	1421.	0
	KT:	.59	.63	.67	.72	.74	.74	.68	.68	.70	.68	.63	.58	0
	AMB TEMP:	51.2	55.1	59.7	67.7	76.3	84.6	91.2	89.1	83.8	72.2	59.8	52.5	70.3
	HTG DEG DAYS:	428	292	185	60	0	0	0	0	0	17	182	388	1552
STA NO:23184	HORIZ INSOL:	1016.	1335.	1777.	2275.	2629.	2762.	2309.	2092.	1955.	1543.	1140.	927.	1813.
PRESCOTT	TILT = LAT:	1636.	1858.	2101.	2273.	2305.	2279.	1976.	1973.	2168.	2068.	1794.	1561.	0
AZ	TILT = LAT+15:	1747.	1904.	2032.	2045.	1949.	1863.	1659.	1736.	2042.	2090.	1902.	1686.	0
LAT = 34.65	TILT = 90:	1507.	1461.	1274.	917.	614.	484.	537.	716.	1142.	1528.	1606.	1497.	0
	KT:	.61	.63	.66	.70	.73	.74	.63	.62	.68	.68	.65	.61	0
	AMB TEMP:	37.1	40.5	44.3	52.0	60.4	69.1	75.5	73.0	68.1	57.2	45.8	38.6	55.1
	HTG DEG DAYS:	865	686	642	394	165	33	0	0	23	254	576	818	4456
STA NO:23185	HORIZ INSOL:	800.	1150.	1649.	2159.	2523.	2701.	2692.	2406.	1998.	1431.	912.	706.	1761.
RENO	TILT = LAT:	1438.	1741.	2058.	2199.	2212.	2212.	2272.	2308.	2343.	2100.	1590.	1336.	0
NV	TILT = LAT+15:	1531.	1781.	1989.	1974.	1867.	1811.	1883.	2019.	2209.	2125.	1680.	1437.	0
LAT = 39.50	TILT = 90:	1371.	1447.	1365.	1032.	748.	625.	684.	940.	1378.	1665.	1477.	1318.	0
	KT:	.57	.61	.66	.69	.71	.72	.74	.73	.74	.70	.61	.56	0
	AMB TEMP:	31.9	37.1	40.3	46.8	54.6	61.5	69.3	66.9	60.2	50.3	40.1	33.0	49.4
	HTG DEG DAYS:	1026	781	766	546	328	145	17	50	168	456	747	992	6022
STA NO:23188	HORIZ INSOL:	976.	1266.	1632.	1937.	2003.	2062.	2186.	2057.	1717.	1373.	1063.	904.	1598.
SAN DIEGO	TILT = LAT:	1465.	1676.	1866.	1907.	1773.	1744.	1882.	1936.	1851.	1738.	1554.	1415.	0
CA	TILT = LAT+15:	1556.	1709.	1800.	1719.	1520.	1461.	1588.	1707.	1739.	1744.	1637.	1520.	0
LAT = 32.73	TILT = 90:	1305.	1270.	1085.	753.	523.	451.	490.	663.	936.	1226.	1339.	1316.	0
	KT:	.55	.57	.59	.59	.56	.55	.60	.61	.59	.58	.57	.55	0
	AMB TEMP:	55.2	56.7	58.0	60.7	63.3	65.5	69.6	71.4	69.9	66.1	60.8	56.7	62.9
	HTG DEG DAYS:	314	237	219	144	79	52	6	0	16	43	140	257	1507
STA NO:23194	HORIZ INSOL:	985.	1327.	1780.	2283.	2595.	2712.	2347.	2141.	1928.	1513.	1119.	894.	1802.
WINSLOW	TILT = LAT:	1591.	1860.	2115.	2285.	2276.	2239.	2005.	2022.	2142.	2033.	1772.	1511.	0
AZ	TILT = LAT+15:	1697.	1906.	2046.	2056.	1924.	1838.	1681.	1778.	2017.	2053.	1879.	1630.	0
LAT = 35.02	TILT = 90:	1466.	1470.	1292.	932.	624.	499.	547.	737.	1138.	1507.	1591.	1449.	0

	KT:	.60	.63	.67	.71	.72	.73	.64	.64	.68	.67	.64	.59	0
	AMB TEMP:	32.6	39.1	44.8	53.7	62.7	71.8	78.3	76.1	69.5	57.3	43.2	33.8	55.3
	HTG DEG DAYS:	1004	725	626	348	124	14	0	0	19	252	654	967	4733
STA NO:23195	HORIZ INSOL:	1096.	1443.	1919.	2413.	2728.	2814.	2453.	2329.	2051.	1623.	1215.	1000.	1924.
YUMA	TILT = LAT:	1691.	1958.	2243.	2401.	2392.	2329.	2101.	2195.	2250.	2117.	1832.	1604.	0
AZ	TILT = LAT+15:	1807.	2010.	2175.	2162.	2019.	1906.	1761.	1931.	2123.	2142.	1944.	1734.	0
LAT = 32.67	TILT = 90:	1532.	1507.	1310.	891.	551.	420.	495.	712.	1127.	1521.	1609.	1514.	0
	KT:	.62	.65	.70	.74	.76	.76	.67	.69	.70	.69	.65	.61	0
	AMB TEMP:	55.4	59.4	63.9	71.2	78.7	85.8	93.7	92.8	87.1	75.9	63.5	56.3	73.7
	HTG DEG DAYS:	308	192	97	24	0	0	0	0	0	5	108	276	1010
STA NO:23230	HORIZ INSOL:	708.	1018.	1456.	1922.	2211.	2350.	2322.	2053.	1701.	1212.	822.	647.	1535.
OAKLAND	TILT = LAT:	1144.	1429.	1735.	1923.	1942.	1947.	1976.	1945.	1907.	1640.	1305.	1102.	0
CA	TILT = LAT+15:	1205.	1451.	1668.	1727.	1651.	1611.	1654.	1708.	1790.	1644.	1367.	1175.	0
LAT = 37.73	TILT = 90:	1046.	1140.	1105.	874.	655.	564.	608.	782.	1077.	1240.	1167.	1049.	0
	KT:	.47	.52	.57	.61	.62	.63	.64	.62	.62	.57	.52	.48	0
	AMB TEMP:	48.6	51.9	53.7	56.1	58.9	61.9	63.1	63.5	64.5	61.1	55.3	49.9	57.4
	HTG DEG DAYS:	508	367	350	270	193	114	80	74	59	135	291	468	2909
STA NO:23232	HORIZ INSOL:	597.	939.	1458.	2004.	2435.	2684.	2688.	2368.	1907.	1315.	782.	538.	1643.
SACRAMENTO	TILT = LAT:	939.	1319.	1755.	2017.	2135.	2202.	2271.	2263.	2191.	1843.	1256.	890.	0
CA	TILT = LAT+15:	982.	1335.	1688.	1811.	1807.	1805.	1884.	1981.	2062.	1855.	1315.	940.	0
LAT = 38.52	TILT = 90:	846.	1054.	1134.	931.	708.	597.	654.	897.	1260.	1423.	1128.	832.	0
	KT:	.41	.49	.58	.64	.68	.72	.74	.72	.70	.63	.50	.41	0
	AMB TEMP:	45.1	49.8	53.0	58.3	64.3	70.5	75.2	74.1	71.5	63.3	53.0	45.8	60.3
	HTG DEG DAYS:	617	426	372	227	120	20	0	0	5	101	360	595	2843
STA NO:23234	HORIZ INSOL:	708.	1009.	1455.	1920.	2226.	2377.	2392.	2116.	1742.	1226.	821.	642.	1553.
SAN FRANCISCO	TILT = LAT:	1139.	1411.	1730.	1920.	1955.	1969.	2033.	2008.	1956.	1659.	1299.	1087.	0
CA	TILT = LAT+15:	1199.	1432.	1664.	1724.	1661.	1628.	1699.	1762.	1837.	1663.	1360.	1158.	0
LAT = 37.62	TILT = 90:	1039.	1123.	1099.	870.	654.	564.	612.	798.	1102.	1253.	1159.	1031.	0
	KT:	.47	.51	.57	.61	.62	.63	.65	.64	.63	.58	.51	.47	0
	AMB TEMP:	48.3	51.2	53.0	55.3	58.3	61.6	62.5	63.0	64.1	61.0	55.3	49.7	56.9
	HTG DEG DAYS:	518	386	372	291	210	120	93	84	66	137	291	474	3042
STA NO:23244	HORIZ INSOL:	738.	1037.	1485.	1944.	2277.	2453.	2441.	2167.	1760.	1248.	843.	660.	1588.
SUNNYVALE	TILT = LAT:	1193.	1453.	1767.	1944.	1999.	2029.	2074.	2057.	1975.	1689.	1334.	1118.	0
CA	TILT = LAT+15:	1259.	1476.	1701.	1746.	1698.	1675.	1732.	1804.	1855.	1634.	1399.	1192.	0
LAT = 37.42	TILT = 90:	1093.	1156.	1120.	875.	658.	563.	611.	807.	1103.	1274.	1192.	1062.	0
	KT:	.49	.52	.58	.61	.64	.65	.67	.65	.63	.58	.52	.48	0
	AMB TEMP:	0	0	0	0	0	0	0	0	0	0	0	0	0
	HTG DEG DAYS:	0	0	0	0	0	0	0	0	0	0	0	0	0
STA NO:23273	HORIZ INSOL:	854.	1141.	1582.	1921.	2141.	2349.	2341.	2106.	1730.	1353.	974.	804.	1608.
SANTA MARIA	TILT = LAT:	1323.	1544.	1844.	1902.	1886.	1960.	2000.	1988.	1896.	1773.	1481.	1312.	0
CA	TILT = LAT+15:	1400.	1571.	1777.	1712.	1609.	1626.	1678.	1748.	1781.	1781.	1557.	1407.	0
LAT = 34.90	TILT = 90:	1193.	1195.	1118.	801.	582.	503.	544.	726.	1007.	1295.	1301.	1237.	0
	KT:	.52	.54	.59	.60	.60	.63	.64	.63	.61	.60	.56	.53	0
	AMB TEMP:	50.5	52.0	52.8	54.9	57.1	59.6	62.1	62.3	62.6	60.4	56.1	51.8	56.9
	HTG DEG DAYS:	450	364	378	303	245	167	112	102	94	159	270	409	3053
STA NO:24011	HORIZ INSOL:	467.	776.	1168.	1459.	1848.	2060.	2184.	1877.	1354.	908.	507.	373.	1248.
BISMARCK	TILT = LAT:	998.	1326.	1538.	1489.	1616.	1681.	1840.	1817.	1626.	1438.	1005.	839.	0
ND	TILT = LAT+15:	1055.	1348.	1474.	1327.	1366.	1386.	1530.	1584.	1518.	1439.	1050.	894.	0

LAT = 46.77	TILT = 90:	984.	1168.	1126.	847.	750.	705.	789.	940.	1085.	1207.	960.	646.	0
	KT:	.47	.52	.54	.50	.53	.55	.60	.60	.56	.54	.46	.43	0
	AMB TEMP:	8.2	13.5	25.1	43.0	54.4	63.8	70.8	69.2	57.5	46.8	28.9	15.6	41.4
	HTG DEG DAYS:	1761	1442	1237	660	339	122	18	35	252	564	1083	1531	9044
STA NO:24013	HORIZ INSOL:	384.	656.	1044.	1461.	1846.	1975.	2098.	1800.	1277.	850.	439.	310.	1178.
MINOT	TILT = LAT:	830.	1123.	1380.	1505.	1614.	1607.	1765.	1743.	1549.	1386.	891.	715.	0
ND	TILT = LAT+15:	874.	1137.	1319.	1340.	1362.	1325.	1467.	1522.	1444.	1386.	928.	760.	0
LAT = 48.27	TILT = 90:	818.	993.	1026.	881.	778.	711.	797.	939.	1058.	1181.	855.	722.	0
	KT:	.42	.46	.50	.51	.53	.53	.58	.58	.54	.53	.43	.40	0
	AMB TEMP:	7.9	12.8	23.6	41.1	52.8	62.0	68.8	67.2	56.2	46.1	27.9	14.7	40.1
	HTG DEG DAYS:	1770	1462	1283	717	384	150	27	70	286	586	1113	1559	9407
STA NO:24018	HORIZ INSOL:	766.	1068.	1433.	1771.	1995.	2258.	2230.	1966.	1667.	1242.	823.	671.	1491.
CHEYENNE	TILT = LAT:	1458.	1662.	1786.	1767.	1749.	1860.	1890.	1874.	1931.	1828.	1482.	1355.	0
WY	TILT = LAT+15:	1555.	1698.	1718.	1601.	1486.	1537.	1579.	1641.	1812.	1840.	1564.	1460.	0
LAT = 41.15	TILT = 90:	1412.	1402.	1208.	891.	653.	629.	672.	837.	1170.	1461.	1390.	1357.	0
	KT:	.59	.60	.59	.57	.56	.60	.61	.60	.63	.63	.58	.57	0
	AMB TEMP:	26.6	29.0	31.6	42.7	52.4	61.3	69.1	67.6	58.2	47.9	35.5	29.2	45.9
	HTG DEG DAYS:	1190	1008	1035	669	394	156	22	31	225	530	885	1110	7255
STA NO:24023	HORIZ INSOL:	692.	958.	1333.	1724.	1988.	2267.	2277.	1990.	1565.	1177.	759.	605.	1445.
NORTH PLATTE	TILT = LAT:	1274.	1448.	1639.	1736.	1743.	1866.	1929.	1898.	1795.	1708.	1333.	1178.	0
NE	TILT = LAT+15:	1351.	1472.	1573.	1555.	1481.	1542.	1611.	1662.	1681.	1715.	1400.	1262.	0
LAT = 41.13	TILT = 90:	1219.	1207.	1103.	868.	681.	630.	680.	845.	1085.	1357.	1237.	1165.	0
	KT:	.53	.54	.55	.56	.56	.60	.62	.61	.59	.60	.54	.52	0
	AMB TEMP:	23.4	28.1	34.3	47.8	58.3	68.0	74.3	73.0	62.3	51.0	36.2	26.8	48.6
	HTG DEG DAYS:	1290	1033	952	522	238	65	7	8	141	439	864	1184	6743
STA NO:24025	HORIZ INSOL:	530.	795.	1206.	1614.	1966.	2195.	2278.	1993.	1496.	1052.	623.	442.	1349.
PIERRE	TILT = LAT:	1039.	1258.	1532.	1643.	1722.	1797.	1923.	1920.	1770.	1616.	1179.	914.	0
SD	TILT = LAT+15:	1097.	1275.	1467.	1467.	1458.	1482.	1600.	1677.	1656.	1622.	1236.	974.	0
LAT = 44.38	TILT = 90:	1006.	1076.	1081.	886.	741.	687.	757.	932.	1137.	1331.	1117.	909.	0
	KT:	.47	.49	.53	.54	.56	.58	.63	.62	.59	.58	.51	.45	0
	AMB TEMP:	15.6	20.4	29.8	46.3	57.4	67.4	75.2	73.9	62.1	62.1	33.8	21.5	46.2
	HTG DEG DAYS:	1531	1249	1091	561	267	74	6	10	152	451	936	1349	7677
STA NO:24027	HORIZ INSOL:	735.	1089.	1530.	1944.	2344.	2574.	2547.	2240.	1833.	1306.	826.	651.	1635.
ROCK SPRINGS	TILT = LAT:	1409.	1729.	1945.	1983.	2056.	2106.	2151.	2155.	2170.	1973.	1519.	1330.	0
WY	TILT = LAT+15:	1501.	1769.	1876.	1777.	1737.	1727.	1785.	1884.	2041.	1993.	1604.	1433.	0
LAT = 41.60	TILT = 90:	1366.	1472.	1333.	992.	775.	677.	732.	952.	1331.	1597.	1433.	1335.	0
	KT:	.57	.62	.64	.63	.66	.69	.70	.69	.70	.67	.60	.57	0
	AMB TEMP:	19.2	23.4	28.9	40.1	50.4	58.9	68.2	66.1	56.4	44.7	30.7	22.6	42.5
	HTG DEG DAYS:	1420	1165	1119	747	453	198	18	49	269	629	1029	1314	8410
STA NO:24028	HORIZ INSOL:	676.	951.	1307.	1668.	1933.	2237.	2284.	2000.	1599.	1145.	723.	575.	1425.
SCOTTSBLUFF	TILT = LAT:	1275.	1464.	1619.	1681.	1694.	1840.	1932.	1912.	1855.	1681.	1288.	1141.	0
NE	TILT = LAT+15:	1353.	1489.	1554.	1505.	1440.	1520.	1612.	1673.	1738.	1687.	1352.	1222.	0
LAT = 41.87	TILT = 90:	1228.	1232.	1102.	856.	682.	641.	698.	868.	1138.	1346.	1200.	1132.	0
	KT:	.53	.54	.55	.54	.54	.60	.63	.62	.61	.59	.53	.51	0
	AMB TEMP:	24.9	29.5	34.3	46.2	56.5	65.9	73.7	71.6	61.2	50.2	36.2	27.6	48.2
	HTG DEG DAYS:	1243	994	952	564	280	91	0	8	160	459	864	1159	6774
STA NO:24029	HORIZ INSOL:	518.	788.	1205.	1537.	1883.	2156.	2329.	2006.	1502.	1005.	591.	441.	1330.
SHERIDAN	TILT = LAT:	1026.	1260.	1539.	1560.	1647.	1765.	1965.	1937.	1787.	1543.	1118.	934.	0

WY	TILT = LAT+15:	1083.	1277.	1475.	1393.	1396.	1456.	1633.	1691.	1672.	1546.	1171.	996.	0
LAT = 44.77	TILT = 90:	996.	1082.	1094.	850.	723.	687.	777.	948.	1156.	1272.	1058.	934.	0
	KT:	.47	.49	.53	.51	.53	.57	.64	.63	.60	.56	.49	.46	0
	AMB TEMP:	21.0	25.9	31.0	43.6	53.1	61.1	70.4	69.2	57.9	47.8	33.4	25.5	45.0
	HTG DEG DAYS:	1364	1095	1054	642	375	168	28	31	245	533	948	1225	7708
STA NO:24033	HORIZ INSOL:	486.	763.	1189.	1526.	1913.	2174.	2384.	2022.	1470.	987.	561.	421.	1325.
BILLINGS	TILT = LAT:	996.	1252.	1544.	1557.	1674.	1775.	2009.	1962.	1768.	1554.	1098.	935.	0
MT	TILT = LAT+15:	1052.	1269.	1480.	1389.	1416.	1462.	1667.	1712.	1653.	1558.	1150.	998.	0
LAT = 45.80	TILT = 90:	974.	1087.	1115.	867.	753.	712.	816.	985.	1164.	1298.	1048.	944.	0
	KT:	.46	.49	.54	.52	.54	.58	.66	.64	.59	.57	.49	.46	0
	AMB TEMP:	21.9	27.4	32.6	44.6	54.5	62.6	71.8	70.1	58.9	49.3	35.7	26.8	46.3
	HTG DEG DAYS:	1336	1053	1004	612	333	131	10	15	221	487	879	1184	7265
STA NO:24036	HORIZ INSOL:	420.	692.	1128.	1444.	1807.	2059.	2288.	1901.	1372.	905.	502.	363.	1240.
LEWISTOWN	TILT = LAT:	873.	1151.	1482.	1475.	1579.	1679.	1927.	1845.	1658.	1445.	1007.	823.	0
MT	TILT = LAT+15:	919.	1165.	1419.	1314.	1335.	1384.	1599.	1609.	1548.	1446.	1052.	877.	0
LAT = 47.05	TILT = 90:	854.	1006.	1087.	843.	741.	710.	824.	960.	1113.	1218.	965.	832.	0
	KT:	.43	.47	.52	.49	.52	.55	.63	.61	.57	.54	.46	.43	0
	AMB TEMP:	19.1	23.8	27.5	40.1	49.6	56.6	65.5	64.4	54.0	45.5	32.2	24.5	41.9
	HTG DEG DAYS:	1423	1154	1163	747	477	265	70	94	348	605	984	1256	8586
STA NO:24037	HORIZ INSOL:	457.	745.	1185.	1542.	1896.	2146.	2293.	1977.	1444.	961.	551.	399.	1300.
MILES CITY	TILT = LAT:	949.	1241.	1555.	1580.	1659.	1751.	1932.	1920.	1746.	1531.	1107.	905.	0
MT	TILT = LAT+15:	1000.	1259.	1491.	1410.	1402.	1442.	1604.	1674.	1632.	1535.	1160.	966.	0
LAT = 46.43	TILT = 90:	929.	1085.	1134.	891.	760.	719.	810.	981.	1162.	1287.	1063.	916.	0
	KT:	.45	.49	.54	.52	.54	.57	.63	.53	.59	.57	.49	.45	0
	AMB TEMP:	15.4	21.6	30.2	45.3	56.3	64.9	74.4	72.5	59.9	48.8	32.4	22.0	45.3
	HTG DEG DAYS:	1538	1215	1079	591	288	117	9	16	217	508	978	1333	7889
STA NO:24089	HORIZ INSOL:	683.	1014.	1441.	1847.	2204.	2501.	2535.	2225.	1750.	1219.	765.	594.	1565.
CASPER	TILT = LAT:	1362.	1642.	1850.	1889.	1937.	2044.	2138.	2150.	2089.	1672.	1453.	1264.	0
WY	TILT = LAT+15:	1450.	1678.	1782.	1691.	1634.	1678.	1774.	1879.	1963.	1888.	1533.	1360.	0
LAT = 42.92	TILT = 90:	1331.	1412.	1291.	979.	775.	703.	767.	983.	1313.	1533.	1382.	1276.	0
	KT:	.57	.60	.61	.61	.62	.67	.70	.59	.68	.65	.58	.55	0
	AMB TEMP:	23.2	26.8	31.0	42.7	52.7	61.9	71.0	69.6	58.7	47.7	33.9	26.2	45.4
	HTG DEG DAYS:	1296	1070	1054	669	368	147	13	17	229	536	933	1203	7555
STA NO:24090	HORIZ INSOL:	542.	827.	1229.	1589.	1887.	2131.	2223.	1963.	1518.	1064.	647.	476.	1341.
RAPID CITY	TILT = LAT:	1053.	1309.	1557.	1612.	1652.	1743.	1877.	1888.	1793.	1624.	1222.	995.	0
SD	TILT = LAT+15:	1112.	1328.	1492.	1442.	1401.	1444.	1564.	1643.	1678.	1630.	1282.	1063.	0
LAT = 44.05	TILT = 90:	1018.	1118.	1095.	864.	711.	668.	737.	909.	1145.	1333.	1157.	994.	0
	KT:	.47	.50	.53	.53	.53	.57	.61	.51	.60	.59	.52	.47	0
	AMB TEMP:	21.9	25.8	31.2	44.6	55.2	64.2	72.6	71.6	60.5	50.0	35.4	26.5	46.6
	HTG DEG DAYS:	1336	1098	1048	612	319	134	13	17	191	474	888	1194	7324
STA NO:24121	HORIZ INSOL:	689.	1034.	1463.	1900.	2303.	2534.	2623.	2316.	1893.	1322.	812.	617.	1626.
ELKO	TILT = LAT:	1249.	1581.	1821.	1926.	2020.	2077.	2214.	2227.	2233.	1964.	1438.	1190.	0
NV	TILT = LAT+15:	1323.	1613.	1753.	1726.	1709.	1706.	1836.	1947.	2102.	1983.	1515.	1276.	0
LAT = 40.83	TILT = 90:	1189.	1324.	1227.	948.	746.	653.	719.	955.	1349.	1574.	1341.	1175.	0
	KT:	.52	.57	.60	.61	.65	.68	.72	.71	.71	.67	.57	.52	0
	AMB TEMP:	23.2	29.2	35.0	43.5	51.9	59.6	69.5	67.0	57.6	46.9	34.8	25.9	45.4
	HTG DEG DAYS:	1296	1002	930	645	406	190	27	60	248	561	906	1212	7483
STA NO:24127	HORIZ INSOL:	639.	989.	1454.	1894.	2362.	2561.	2590.	2254.	1843.	1293.	788.	570.	1603.

SALT LAKE CITY	TILT = LAT:	1126.	1491.	1806.	1919.	2072.	2099.	2187.	2163.	2162.	1906.	1379.	1065.	0
UT	TILT = LAT+15:	1189.	1517.	1739.	1721.	1751.	1723.	1815.	1892.	2034.	1921.	1450.	1137.	0
LAT = 40.77	TILT = 90:	1061.	1240.	1215.	943.	756.	653.	713.	931.	1303.	1521.	1279.	1040.	0
	KT:	.48	.55	.60	.61	.66	.68	.71	.69	.69	.65	.55	.48	0
	AMB TEMP:	28.0	33.4	39.6	49.2	58.3	66.2	76.7	74.5	64.8	52.4	39.1	30.3	51.0
	HTG DEG DAYS:	1147	885	787	474	237	88	0	5	105	402	777	1076	5983
STA NO:24128	HORIZ INSOL:	691.	1028.	1472.	1967.	2362.	2569.	2678.	2343.	1907.	1322.	810.	618.	1648.
WINNEMUCCA	TILT = LAT:	1256.	1571.	1836.	2002.	2071.	2105.	2258.	2260.	2256.	1967.	1436.	1198.	0
NV	TILT = LAT+15:	1332.	1602.	1768.	1794.	1751.	1727.	1871.	1976.	2124.	1986.	1513.	1285.	0
LAT = 40.90	TILT = 90:	1198.	1315.	1239.	984.	760.	658.	725.	963.	1365.	1578.	1339.	1184.	0
	KT:	.52	.57	.61	.64	.66	.68	.73	.72	.72	.67	.57	.52	0
	AMB TEMP:	28.2	34.1	37.6	45.1	53.8	61.7	71.0	67.8	59.2	48.3	37.3	30.4	47.9
	HTG DEG DAYS:	1141	865	849	597	359	149	6	42	199	518	831	1073	6629
STA NO:24131	HORIZ INSOL:	485.	840.	1304.	1827.	2277.	2463.	2613.	2197.	1737.	1138.	628.	437.	1495.
BOISE	TILT = LAT:	880.	1315.	1659.	1874.	1997.	2012.	2202.	2126.	2091.	1746.	1149.	856.	0
ID	TILT = LAT+15:	922.	1334.	1593.	1676.	1686.	1652.	1823.	1857.	1964.	1756.	1203.	909.	0
LAT = 43.57	TILT = 90:	832.	1118.	1162.	986.	810.	714.	798.	995.	1330.	1433.	1078.	840.	0
	KT:	.41	.50	.56	.60	.64	.66	.72	.58	.68	.62	.49	.42	0
	AMB TEMP:	29.0	35.5	41.1	49.0	57.4	64.8	74.5	72.2	63.1	52.1	39.8	32.1	50.9
	HTG DEG DAYS:	1116	826	741	480	252	97	0	12	127	406	756	1020	5833
STA NO:24134	HORIZ INSOL:	490.	792.	1187.	1649.	2052.	2280.	2460.	2083.	1620.	1043.	594.	431.	1390.
BURNS	TILT = LAT:	893.	1220.	1482.	1674.	1798.	1867.	2076.	2003.	1924.	1563.	1064.	838.	0
OR	TILT = LAT+15:	936.	1234.	1419.	1497.	1522.	1539.	1723.	1754.	1804.	1566.	1111.	889.	0
LAT = 43.58	TILT = 90:	845.	1030.	1031.	886.	749.	685.	773.	949.	1221.	1271.	991.	821.	0
	KT:	.42	.48	.51	.55	.58	.61	.68	.65	.63	.57	.46	.42	0
	AMB TEMP:	25.2	31.0	36.1	44.2	52.2	59.0	68.4	66.1	58.2	47.3	35.8	27.9	46.0
	HTG DEG DAYS:	1234	952	896	624	402	205	30	68	226	549	876	1150	7212
STA NO:24137	HORIZ INSOL:	402.	688.	1128.	1485.	1883.	2045.	2287.	1897.	1352.	871.	480.	334.	1238.
CUT BANK	TILT = LAT:	911.	1215.	1530.	1536.	1647.	1663.	1925.	1853.	1667.	1449.	1036.	825.	0
MT	TILT = LAT+15:	962.	1233.	1466.	1368.	1389.	1368.	1595.	1614.	1556.	1452.	1085.	880.	0
LAT = 48.60	TILT = 90:	906.	1084.	1150.	905.	799.	736.	862.	1000.	1148.	1243.	1009.	844.	0
	KT:	.45	.49	.54	.52	.54	.55	.63	.61	.57	.55	.48	.44	0
	AMB TEMP:	16.2	22.4	26.8	39.5	49.6	56.5	64.4	62.6	53.2	44.1	29.7	21.4	40.5
	HTG DEG DAYS:	1513	1193	1184	765	477	267	82	125	368	648	1059	1352	9033
STA NO:24138	HORIZ INSOL:	527.	846.	1279.	1639.	1989.	2143.	2392.	2023.	1521.	1023.	602.	450.	1370.
DILLON	TILT = LAT:	1080.	1406.	1670.	1679.	1742.	1753.	2017.	1959.	1827.	1601.	1175.	990.	0
MT	TILT = LAT+15:	1142.	1431.	1604.	1499.	1473.	1446.	1674.	1710.	1710.	1607.	1233.	1059.	0
LAT = 45.25	TILT = 90:	1057.	1225.	1201.	922.	766.	694.	804.	970.	1193.	1332.	1122.	1000.	0
	KT:	.49	.54	.57	.55	.56	.57	.66	.64	.61	.58	.51	.48	0
	AMB TEMP:	20.2	25.5	29.6	41.1	50.4	57.5	66.4	64.6	54.7	45.0	31.8	23.9	42.6
	HTG DEG DAYS:	1389	1106	1097	717	453	238	54	85	325	620	996	1274	8354
STA NO:24143	HORIZ INSOL:	421.	720.	1170.	1489.	1848.	2101.	2329.	1933.	1379.	925.	498.	336.	1262.
GREAT FALLS	TILT = LAT:	900.	1235.	1564.	1529.	1615.	1712.	1962.	1882.	1678.	1507.	1018.	759.	0
MT	TILT = LAT+15:	948.	1253.	1500.	1363.	1365.	1409.	1626.	1640.	1567.	1511.	1065.	807.	0
LAT = 47.48	TILT = 90:	885.	1091.	1158.	882.	764.	729.	846.	988.	1135.	1281.	981.	765.	0
	KT:	.44	.50	.55	.51	.53	.56	.64	.62	.57	.56	.47	.41	0
	AMB TEMP:	20.5	26.6	30.5	43.4	53.3	60.8	69.3	67.4	57.3	48.3	34.6	26.5	44.9
	HTG DEG DAYS:	1380	1075	1070	648	367	162	18	42	260	524	912	1194	7652

STA NO:24144	HORIZ INSOL:	419.	709.	1146.	1487.	1860.	2040.	2334.	1933.	1412.	926.	521.	364.	1262.
HELENA	TILT = LAT:	847.	1168.	1497.	1519.	1627.	1665.	1967.	1872.	1705.	1468.	1034.	801.	0
MT	TILT = LAT+15:	890.	1182.	1433.	1354.	1376.	1374.	1631.	1632.	1593.	1470.	1081.	852.	0
LAT = 46.60	TILT = 90:	823.	1018.	1091.	860.	751.	697.	824.	962.	1137.	1232.	989.	804.	0
	KT:	.42	.47	.52	.51	.53	.54	.64	.61	.58	.55	.47	.42	0
	AMB TEMP:	18.1	25.4	30.6	42.7	52.2	59.2	67.9	66.2	55.5	45.3	31.7	23.3	43.2
	HTG DEG DAYS:	1454	1109	1066	669	401	194	33	57	304	611	999	1293	8190
STA NO:24149	HORIZ INSOL:	340.	609.	1020.	1435.	1842.	2015.	2336.	1931.	1435.	860.	413.	286.	1210.
LEWISTON	TILT = LAT:	610.	944.	1292.	1458.	1611.	1646.	1968.	1871.	1731.	1324.	731.	545.	0
ID	TILT = LAT+15:	632.	947.	1232.	1300.	1363.	1359.	1633.	1632.	1618.	1321.	754.	570.	0
LAT = 46.38	TILT = 90:	572.	804.	930.	823.	741.	687.	819.	957.	1151.	1100.	676.	525.	0
	KT:	.33	.40	.46	.49	.53	.54	.65	.61	.59	.51	.37	.32	0
	AMB TEMP:	31.2	38.1	42.9	50.3	58.1	65.0	73.4	71.5	63.3	51.8	40.5	34.8	51.7
	HTG DEG DAYS:	1048	753	685	441	232	84	0	17	124	409	735	936	5464
STA NO:24153	HORIZ INSOL:	312.	574.	982.	1382.	1783.	1933.	2327.	1881.	1358.	813.	410.	267.	1169.
MISSOULA	TILT = LAT:	553.	888.	1245.	1403.	1557.	1579.	1961.	1823.	1634.	1251.	746.	508.	0
MT	TILT = LAT+15:	570.	889.	1185.	1249.	1317.	1305.	1626.	1589.	1525.	1246.	771.	531.	0
LAT = 46.92	TILT = 90:	515.	756.	901.	801.	730.	676.	831.	946.	1093.	1041.	696.	490.	0
	KT:	.32	.39	.45	.47	.51	.52	.64	.60	.56	.49	.37	.31	0
	AMB TEMP:	20.8	27.2	33.3	43.9	52.2	58.9	66.6	65.0	55.3	44.1	32.3	24.7	43.7
	HTG DEG DAYS:	1370	1058	983	633	397	201	39	71	301	648	981	1249	7931
STA NO:24155	HORIZ INSOL:	348.	614.	1044.	1503.	1926.	2144.	2396.	1994.	1502.	908.	438.	293.	1259.
PENDLETON	TILT = LAT:	607.	931.	1313.	1529.	1685.	1752.	2019.	1932.	1811.	1391.	770.	538.	0
OR	TILT = LAT+15:	627.	933.	1253.	1364.	1425.	1444.	1675.	1686.	1694.	1390.	795.	561.	0
LAT = 45.68	TILT = 90:	564.	785.	937.	850.	755.	703.	816.	968.	1191.	1150.	710.	514.	0
	KT:	.33	.40	.47	.51	.55	.57	.66	.63	.61	.52	.38	.32	0
	AMB TEMP:	32.0	38.9	43.8	50.9	58.5	65.6	73.5	71.5	64.0	52.6	41.4	35.7	52.4
	HTG DEG DAYS:	1023	731	657	423	220	70	6	13	97	384	708	908	5240
STA NO:24156	HORIZ INSOL:	539.	882.	1371.	1820.	2280.	2480.	2600.	2239.	1769.	1203.	689.	477.	1529.
POCATELLO	TILT = LAT:	987.	1373.	1743.	1859.	2000.	2027.	2192.	2165.	2118.	1841.	1262.	933.	0
ID	TILT = LAT+15:	1038.	1395.	1676.	1664.	1689.	1665.	1816.	1892.	1990.	1855.	1325.	993.	0
LAT = 42.92	TILT = 90:	938.	1163.	1212.	964.	794.	700.	777.	993.	1332.	1506.	1186.	918.	0
	KT:	.45	.52	.58	.60	.64	.66	.71	.69	.68	.64	.52	.45	0
	AMB TEMP:	23.2	29.4	35.4	45.3	54.4	61.8	71.5	69.5	59.4	48.4	35.7	26.9	46.7
	HTG DEG DAYS:	1296	997	918	591	336	138	0	20	192	515	879	1181	7063
STA NO:24157	HORIZ INSOL:	315.	606.	1041.	1495.	1916.	2083.	2357.	1942.	1435.	841.	398.	255.	1224.
SPOKANE	TILT = LAT:	589.	983.	1357.	1538.	1679.	1696.	1986.	1893.	1765.	1339.	743.	500.	0
WA	TILT = LAT+15:	610.	989.	1296.	1370.	1417.	1396.	1645.	1649.	1650.	1337.	768.	522.	0
LAT = 47.63	TILT = 90:	558.	853.	998.	883.	792.	727.	857.	997.	1199.	1129.	698.	485.	0
	KT:	.33	.42	.49	.51	.55	.56	.65	.62	.60	.51	.38	.31	0
	AMB TEMP:	25.4	32.2	37.5	46.1	54.7	61.5	69.7	68.0	59.6	47.8	35.5	29.0	47.3
	HTG DEG DAYS:	1228	918	853	567	327	144	21	47	196	533	885	1116	6835
STA NO:24172	HORIZ INSOL:	804.	1165.	1657.	2165.	2555.	2750.	2784.	2484.	2027.	1451.	929.	714.	1791.
LOVELOCK	TILT = LAT:	1482.	1800.	2087.	2213.	2240.	2246.	2346.	2393.	2403.	2172.	1666.	1396.	0
NV	TILT = LAT+15:	1580.	1844.	2018.	1985.	1888.	1836.	1939.	2052.	2267.	2201.	1765.	1505.	0
LAT = 40.07	TILT = 90:	1424.	1511.	1399.	1054.	769.	643.	705.	983.	1431.	1740.	1563.	1389.	0
	KT:	.59	.63	.67	.70	.72	.73	.76	.76	.75	.72	.63	.58	0
	AMB TEMP:	28.9	35.2	40.1	48.5	57.5	65.6	14.3	71.3	62.7	51.2	38.4	30.8	50.4
	HTG DEG DAYS:	1119	834	772	495	255	86	0	17	126	428	798	1060	5990

STA NO:24215	HORIZ INSOL:	561.	857.	1250.	1756.	2186.	2436.	2577.	2213.	1735.	1155.	659.	505.	1491.
MOUNT SHASTA	TILT = LAT:	967.	1264.	1523.	1773.	1917.	1999.	2176.	2125.	2027.	1676.	1114.	927.	0
CA	TILT = LAT+15:	1015.	1278.	1459.	1588.	1624.	1645.	1806.	1859.	1904.	1682.	1162.	985.	0
LAT = 41.32	TILT = 90:	902.	1042.	1024.	888.	732.	655.	728.	933.	1234.	1332.	1018.	897.	0
	KT:	.43	.48	.52	.57	.61	.65	.71	.68	.66	.59	.47	.44	0
	AMB TEMP:	33.6	37.8	40.4	46.3	53.3	60.0	67.8	66.0	61.2	51.4	41.7	35.5	49.6
	HTG DEG DAYS:	973	762	763	561	371	178	37	34	145	422	699	915	5890
STA NO:24216	HORIZ INSOL:	570.	893.	1354.	1910.	2375.	2600.	2672.	2311.	1845.	1228.	706.	511.	1581.
RED BLUFF	TILT = LAT:	941.	1289.	1646.	1931.	2083.	2131.	2255.	2216.	2148.	1757.	1167.	890.	0
CA	TILT = LAT+15:	985.	1304.	1580.	1731.	1761.	1749.	1869.	1939.	2021.	1766.	1219.	942.	0
LAT = 40.15	TILT = 90:	864.	1049.	1090.	934.	742.	640.	702.	931.	1279.	1381.	1058.	847.	0
	KT:	.42	.48	.55	.61	.67	.69	.73	.71	.69	.61	.48	.42	0
	AMB TEMP:	45.2	50.0	53.2	59.5	67.4	75.5	82.3	79.9	75.3	65.0	53.7	46.4	62.8
	HTG DEG DAYS:	614	420	366	218	64	8	0	0	0	82	339	577	2688
STA NO:24225	HORIZ INSOL:	407.	737.	1133.	1639.	2034.	2278.	2475.	2121.	1589.	982.	504.	337.	1353.
MEDFORD	TILT = LAT:	644.	1072.	1375.	1653.	1782.	1870.	2090.	2038.	1852.	1404.	806.	544.	0
OR	TILT = LAT+15:	664.	1078.	1313.	1479.	1511.	1543.	1737.	1782.	1736.	1401.	830.	563.	0
LAT = 42.37	TILT = 90:	578.	880.	934.	852.	718.	658.	743.	923.	1148.	1115.	719.	498.	0
	KT:	.33	.43	.48	.54	.57	.61	.68	.66	.61	.52	.38	.31	0
	AMB TEMP:	36.6	41.3	44.8	50.2	57.3	64.3	71.7	70.4	64.4	53.4	43.5	37.7	53.0
	HTG DEG DAYS:	880	664	626	444	250	94	11	21	89	360	645	846	4930
STA NO:24227	HORIZ INSOL:	269.	503.	845.	1255.	1632.	1693.	1913.	1549.	1157.	636.	339.	222.	1001.
OLYMPIA	TILT = LAT:	438.	738.	1033.	1260.	1422.	1386.	1611.	1478.	1352.	906.	562.	375.	0
WA	TILT = LAT+15:	446.	733.	978.	1121.	1206.	1151.	1345.	1289.	1256.	891.	573.	385.	0
LAT = 46.97	TILT = 90:	396.	617.	738.	721.	678.	615.	716.	779.	898.	733.	508.	347.	0
	KT:	.27	.34	.39	.43	.47	.45	.53	.49	.48	.38	.31	.26	0
	AMB TEMP:	37.2	41.0	43.2	48.2	54.0	58.9	63.6	62.8	58.6	50.6	43.3	39.5	50.1
	HTG DEG DAYS:	862	672	676	504	341	197	89	103	198	446	651	791	5530
STA NO:24229	HORIZ INSOL:	310.	554.	895.	1308.	1663.	1772.	2037.	1674.	1217.	724.	388.	260.	1067.
PORTLAND	TILT = LAT:	505.	806.	1085.	1310.	1452.	1455.	1719.	1600.	1411.	1034.	639.	441.	0
OR	TILT = LAT+15:	517.	803.	1028.	1167.	1232.	1209.	1434.	1397.	1313.	1021.	654.	455.	0
LAT = 45.60	TILT = 90:	458.	669.	762.	730.	667.	615.	724.	813.	918.	832.	577.	410.	0
	KT:	.29	.36	.40	.44	.47	.47	.56	.53	.49	.42	.33	.28	0
	AMB TEMP:	38.1	42.8	45.7	50.6	56.7	62.0	67.1	66.6	62.2	53.8	45.3	40.7	52.6
	HTG DEG DAYS:	834	622	598	432	264	128	48	56	119	347	591	753	4792
STA NO:24230	HORIZ INSOL:	491.	775.	1190.	1683.	2080.	2287.	2446.	2069.	1584.	999.	572.	425.	1383.
REDMOND	TILT = LAT:	928.	1212.	1504.	1719.	1822.	1871.	2063.	1999.	1891.	1510.	1044.	856.	0
OR	TILT = LAT+15:	975.	1226.	1440.	1536.	1541.	1540.	1712.	1745.	1772.	1511.	1090.	909.	0
LAT = 44.27	TILT = 90:	888.	1031.	1059.	923.	772.	702.	789.	962.	1214.	1235.	978.	846.	0
	KT:	.43	.48	.52	.56	.59	.61	.67	.65	.63	.55	.46	.43	0
	AMB TEMP:	30.2	35.8	38.6	38.6	51.3	58.2	65.7	63.8	57.7	48.4	39.0	33.4	47.2
	HTG DEG DAYS:	1079	818	818	618	425	220	55	102	233	515	780	980	6643
STA NO:24232	HORIZ INSOL:	332.	588.	947.	1370.	1738.	1849.	2142.	1775.	1328.	769.	410.	277.	1127.
SALEM	TILT = LAT:	540.	856.	1151.	1375.	1519.	1519.	1808.	1700.	1551.	1099.	672.	467.	0
OR	TILT = LAT+15:	554.	854.	1093.	1227.	1289.	1261.	1507.	1485.	1446.	1088.	689.	483.	0
LAT = 44.92	TILT = 90:	490.	709.	804.	755.	680.	623.	737.	846.	1000.	882.	605.	434.	0
	KT:	.30	.37	.42	.46	.49	.49	.59	.56	.53	.43	.34	.29	0
	AMB TEMP:	38.8	42.9	45.2	49.8	55.7	61.2	66.6	66.1	61.9	53.2	45.2	40.9	52.3
	HTG DEG DAYS:	812	619	614	456	295	133	43	53	120	366	594	747	4852

STA NO:24233	HORIZ INSOL:	262.	495.	849.	1294.	1714.	1802.	2248.	1616.	1148.	656.	337.	211.	1053.
SEATTLE-TACOMA	TILT = LAT:	433.	735.	1048.	1306.	1495.	1472.	1893.	1550.	1348.	958.	573.	359.	0
WA	TILT = LAT+15:	441.	730.	993.	1162.	1265.	1218.	1572.	1351.	1252.	945.	585.	368.	0
LAT = 47.45	TILT = 90:	394.	618.	755.	754.	715.	651.	822.	823.	902.	784.	522.	333.	0
	KT:	.27	.34	.40	.44	.49	.48	.62	.52	.48	.40	.32	.26	0
	AMB TEMP:	38.2	42.3	44.1	48.7	54.9	59.8	64.5	63.8	59.6	52.2	44.6	40.5	51.1
	HTG DEG DAYS:	831	636	648	489	313	167	80	82	170	397	612	760	5185
STA NO:24243	HORIZ INSOL:	365.	666.	1122.	1598.	2009.	2169.	2358.	1975.	1483.	891.	444.	295.	1281.
YAKIMA	TILT = LAT:	689.	1074.	1459.	1645.	1759.	1769.	1987.	1919.	1806.	1394.	823.	579.	0
WA	TILT = LAT+15:	717.	1083.	1396.	1468.	1485.	1456.	1648.	1673.	1690.	1393.	853.	607.	0
LAT = 46.57	TILT = 90:	655.	928.	1061.	929.	801.	727.	830.	983.	1206.	1165.	771.	563.	0
	KT:	.36	.44	.51	.54	.57	.58	.65	.53	.61	.53	.40	.34	0
	AMB TEMP:	27.5	35.7	41.8	49.5	57.9	64.5	70.7	68.8	61.3	50.1	38.4	31.3	49.8
	HTG DEG DAYS:	1163	820	719	465	239	94	20	37	147	462	798	1045	6009
STA NO:24255	HORIZ INSOL:	283.	532.	918.	1345.	1760.	1820.	1981.	1593.	1173.	655.	357.	233.	1054.
WHIDBEY ISLAND	TILT = LAT:	523.	843.	1176.	1372.	1537.	1483.	1666.	1530.	1401.	984.	659.	457.	0
WA	TILT = LAT+15:	540.	844.	1118.	1220.	1298.	1226.	1387.	1333.	1303.	972.	679.	477.	0
LAT = 48.35	TILT = 90:	493.	727.	866.	806.	748.	670.	763.	830.	954.	816.	616.	443.	0
	KT:	.31	.38	.44	.47	.51	.49	.55	.51	.50	.41	.35	.30	0
	AMB TEMP:	0	0	0	0	0	0	0	0	0	0	0	0	0
	HTG DEG DAYS:	0	0	0	0	0	0	0	0	0	0	0	0	0
STA NO:24283	HORIZ INSOL:	529.	793.	1133.	1587.	1843.	1962.	1808.	1579.	1342.	936.	593.	470.	1214.
ARCATA	TILT = LAT:	879.	1133.	1349.	1586.	1616.	1625.	1539.	1489.	1503.	1280.	953.	824.	0
CA	TILT = LAT+15:	918.	1141.	1288.	1421.	1376.	1353.	1297.	1307.	1403.	1272.	988.	870.	0
LAT = 40.98	TILT = 90:	808.	920.	895.	795.	643.	581.	587.	665.	903.	989.	853.	784.	0
	KT:	.40	.44	.47	.51	.52	.52	.50	.48	.51	.47	.42	.40	0
	AMB TEMP:	0	0	0	0	0	0	0	0	0	0	0	0	0
	HTG DEG DAYS:	0	0	0	0	0	0	0	0	0	0	0	0	0
STA NO:24284	HORIZ INSOL:	439.	705.	1058.	1510.	1857.	1994.	2108.	1786.	1377.	893.	525.	381.	1219.
NORTH BEND	TILT = LAT:	755.	1041.	1287.	1519.	1626.	1641.	1782.	1704.	1589.	1279.	891.	694.	0
OR	TILT = LAT+15:	786.	1046.	1227.	1358.	1380.	1360.	1489.	1490.	1483.	1272.	923.	730.	0
LAT = 43.42	TILT = 90:	701.	864.	885.	805.	690.	629.	697.	817.	999.	1020.	814.	665.	0
	KT:	.37	.42	.45	.50	.52	.53	.58	.56	.54	.48	.41	.36	0
	AMB TEMP:	44.6	46.6	46.9	49.1	53.1	56.9	59.0	59.7	58.4	54.9	50.1	46.5	52.2
	HTG DEG DAYS:	632	515	561	477	369	243	188	168	201	313	447	574	4688
STA NO:25704	HORIZ INSOL:	231.	433.	716.	1033.	1180.	1182.	1120.	949.	759.	528.	308.	187.	719.
ADAK	TILT = LAT:	507.	734.	920.	1034.	1010.	955.	927.	870.	860.	830.	677.	454.	0
AK	TILT = LAT+15:	527.	734.	868.	914.	854.	797.	779.	754.	788.	818.	702.	477.	0
LAT = 51.88	TILT = 90:	496.	651.	699.	648.	557.	508.	504.	517.	603.	708.	657.	457.	0
	KT:	.32	.36	.37	.37	.34	.32	.31	.31	.34	.37	.37	.32	0
	AMB TEMP:	0	0	0	0	0	0	0	0	0	0	0	0	0
	HTG DEG DAYS:	0	0	0	0	0	0	0	0	0	0	0	0	0
STA NO:93037	HORIZ INSOL:	891.	1178.	1550.	1931.	2129.	2369.	2212.	2025.	1759.	1359.	944.	782.	1594.
COLORADO SPRINGS	TILT = LAT:	1610.	1762.	1892.	1942.	1869.	1957.	1881.	1923.	2002.	1933.	1621.	1484.	0
CO	TILT = LAT+15:	1721.	1803.	1824.	1743.	1589.	1617.	1576.	1687.	1881.	1949.	1714.	1603.	0
LAT = 38.82	TILT = 90:	1542.	1454.	1234.	907.	664.	590.	618.	800.	1157.	1505.	1498.	1470.	0
	KT:	.62	.62	.62	.61	.60	.63	.61	.61	.64	.65	.61	.60	0
	AMB TEMP:	28.6	31.3	35.3	46.2	55.5	64.6	70.7	69.1	60.9	50.5	37.5	31.0	48.4

	HTG DEG DAYS:	1128	944	921	564	301	103	9	13	155	456	825	1054	6473
STA NO:93044	HORIZ INSOL:	986.	1297.	1688.	2167.	2473.	2602.	2264.	2078.	1895.	1496.	1088.	893.	1744.
ZUNI	TILT = LAT:	1599.	1812.	1990.	2162.	2172.	2155.	1937.	1962.	2102.	2009.	1714.	1512.	0
NM	TILT = LAT+15:	1706.	1855.	1922.	1945.	1841.	1775.	1627.	1726.	1979.	2028.	1814.	1631.	0
LAT = 35.10	TILT = 90:	1476.	1430.	1215.	895.	620.	507.	544.	724.	1120.	1490.	1534.	1451.	0
	KT:	.60	.62	.63	.67	.69	.70	.62	.62	.67	.67	.63	.59	0
	AMB TEMP:	30.3	34.6	39.6	48.1	56.6	65.4	71.4	69.4	63.3	52.5	40.1	32.0	50.3
	HTG DEG DAYS:	1076	851	787	507	264	68	0	13	91	388	747	1023	5815
STA NO:93045	HORIZ INSOL:	1118.	1451.	1886.	2338.	2557.	2650.	2365.	2216.	1940.	1579.	1217.	1003.	1860.
TRUTH OR CONSEQUENCE	TILT = LAT:	1764.	1996.	2213.	2326.	2246.	2202.	2026.	2089.	2124.	2069.	1867.	1641.	0
NM	TILT = LAT+15:	1890.	2051.	2145.	2095.	1904.	1813.	1701.	1838.	2001.	2091.	1982.	1776.	0
LAT = 33.23	TILT = 90:	1616.	1552.	1307.	890.	573.	456.	507.	707.	1082.	1496.	1654.	1562.	0
	KT:	.64	.66	.69	.72	.71	.71	.65	.66	.67	.68	.66	.62	0
	AMB TEMP:	40.0	44.9	50.2	59.5	59.5	76.9	79.3	77.4	71.6	61.3	48.7	40.8	59.9
	HTG DEG DAYS:	775	563	459	188	19	0	0	0	5	144	489	750	3392
STA NO:93058	HORIZ INSOL:	894.	1172.	1564.	1956.	2162.	2434.	2312.	2102.	1779.	1361.	954.	782.	1623.
PUEBLO	TILT = LAT:	1584.	1725.	1898.	1964.	1899.	2010.	1965.	1996.	2018.	1913.	1610.	1449.	0
CO	TILT = LAT+15:	1691.	1764.	1830.	1763.	1614.	1659.	1644.	1751.	1896.	1928.	1702.	1563.	0
LAT = 38.28	TILT = 90:	1506.	1412.	1226.	903.	659.	583.	619.	811.	1153.	1478.	1480.	1426.	0
	KT:	.61	.60	.62	.62	.60	.65	.63	.64	.65	.65	.61	.59	0
	AMB TEMP:	30.1	34.7	40.0	51.7	61.1	70.7	76.4	74.5	66.2	54.5	40.8	33.0	52.8
	HTG DEG DAYS:	1082	848	775	405	148	28	0	0	55	335	726	992	5394
STA NO:93101	HORIZ INSOL:	947.	1236.	1610.	1928.	2070.	2194.	2363.	2155.	1737.	1357.	1026.	869.	1625.
EL TORO	TILT = LAT:	1451.	1658.	1856.	1903.	1829.	1844.	2024.	2032.	1887.	1740.	1527.	1389.	0
CA	TILT = LAT+15:	1541.	1691.	1790.	1714.	1564.	1538.	1698.	1788.	1773.	1747.	1607.	1492.	0
LAT = 33.67	TILT = 90:	1304.	1271.	1099.	773.	548.	473.	517.	706.	974.	1246.	1327.	1301.	0
	KT:	.55	.57	.59	.59	.58	.59	.65	.64	.60	.59	.57	.55	0
	AMB TEMP:	0	0	0	0	0	0	0	0	0	0	0	0	0
	HTG DEG DAYS:	0	0	0	0	0	0	0	0	0	0	0	0	0
STA NO:93104	HORIZ INSOL:	909.	1229.	1735.	2233.	2549.	2747.	2612.	2616.	1980.	1473.	1034.	841.	1830.
CHINA LAKE	TILT = LAT:	1470.	1720.	2070.	2238.	2235.	2263.	2218.	2491.	2221.	1994.	1635.	1431.	0
CA	TILT = LAT+15:	1564.	1757.	2001.	2013.	1891.	1854.	1847.	2180.	2093.	2012.	1728.	1541.	0
LAT = 35.68	TILT = 90:	1353.	1360.	1280.	936.	641.	515.	571.	855.	1198.	1490.	1465.	1373.	0
	KT:	.57	.60	.66	.70	.71	.73	.71	.78	.70	.66	.61	.57	0
	AMB TEMP:	0	0	0	0	0	0	0	0	0	0	0	0	0
	HTG DEG DAYS:	0	0	0	0	0	0	0	0	0	0	0	0	0
STA NO:93111	HORIZ INSOL:	927.	1220.	1636.	1951.	2018.	2055.	2118.	1935.	1608.	1296.	1006.	856.	1552.
POINT MUGU	TILT = LAT:	1432.	1647.	1899.	1929.	1782.	1731.	1820.	1822.	1738.	1661.	1509.	1384.	0
CA	TILT = LAT+15:	1521.	1679.	1832.	1737.	1525.	1448.	1536.	1606.	1631.	1665.	1588.	1486.	0
LAT = 34.12	TILT = 90:	1292.	1270.	1135.	792.	551.	474.	513.	666.	909.	1192.	1317.	1302.	0
	KT:	.55	.57	.61	.60	.56	.55	.58	.58	.56	.57	.56	.55	0
	AMB TEMP:	0	0	0	0	0	0	0	0	0	0	0	0	0
	HTG DEG DAYS:	0	0	0	0	0	0	0	0	0	0	0	0	0
STA NO:93129	HORIZ INSOL:	882.	1180.	1636.	2092.	2467.	2706.	2503.	2241.	1968.	1460.	992.	786.	1743.
CEDAR CITY	TILT = LAT:	1521.	1715.	1984.	2106.	2164.	2222.	2123.	2131.	2253.	2061.	1660.	1421.	0
UT	TILT = LAT+15:	1621.	1752.	1916.	1892.	1831.	1822.	1770.	1869.	2123.	2083.	1757.	1531.	0
LAT = 37.70	TILT = 90:	1433.	1392.	1272.	945.	690.	574.	622.	836.	1273.	1591.	1522.	1388.	0
	KT:	.59	.60	.64	.66	.69	.72	.68	.68	.71	.69	.62	.58	0

	AMB TEMP:	28.7	33.1	38.4	47.1	56.2	65.0	73.2	71.3	63.2	51.5	38.8	30.8	49.8
	HTG DEG DAYS:	1125	893	825	537	281	86	0	6	114	424	786	1060	6137
STA NO:93193	HORIZ INSOL:	657.	1012.	1566.	2093.	2484.	2733.	2685.	2423.	1985.	1429.	889.	574.	1711.
FRESNO	TILT = LAT:	1003.	1390.	1863.	2097.	2179.	2247.	2273.	2306.	2253.	1966.	1397.	905.	0
CA	TILT = LAT+15:	1049.	1408.	1796.	1885.	1844.	1841.	1888.	2021.	2123.	1983.	1467.	955.	0
LAT = 36.77	TILT = 90:	892.	1091.	1170.	917.	667.	547.	602.	854.	1246.	1490.	1245.	831.	0
	KT:	.42	.50	.60	.66	.69	.73	.73	.73	.71	.66	.54	.41	0
	AMB TEMP:	45.3	49.9	53.9	60.3	67.4	73.9	80.6	78.3	73.8	64.2	53.5	45.8	62.3
	HTG DEG DAYS:	611	423	344	182	51	9	0	0	0	90	345	595	2650
STA NO:93721	HORIZ INSOL:	587.	840.	1162.	1488.	1714.	1879.	1823.	1600.	1330.	998.	660.	499.	1215.
BALTIMORE	TILT = LAT:	942.	1165.	1359.	1471.	1507.	1567.	1558.	1505.	1464.	1333.	1031.	825.	0
MD	TILT = LAT+15:	985.	1173.	1298.	1319.	1289.	1310.	1315.	1324.	1367.	1326.	1071.	870.	0
LAT = 39.18	TILT = 90:	855.	927.	876.	712.	582.	538.	562.	661.	849.	1009.	913.	770.	0
	KT:	.42	.44	.46	.47	.48	.50	.50	.49	.49	.49	.44	.39	0
	AMB TEMP:	33.4	34.8	42.8	53.8	63.7	72.4	76.6	74.9	68.5	57.4	46.1	35.3	55.0
	HTG DEG DAYS:	980	846	688	340	110	0	0	0	27	250	567	921	4729
STA NO:93729	HORIZ INSOL:	686.	952.	1326.	1774.	1962.	2036.	1921.	1705.	1470.	1137.	873.	659.	1375.
CAPE HATTERAS	TILT = LAT:	1011.	1250.	1516.	1751.	1731.	1710.	1652.	1603.	1589.	1450.	1305.	1028.	0
NC	TILT = LAT+15:	1057.	1260.	1453.	1576.	1481.	1430.	1399.	1414.	1488.	1446.	1366.	1090.	0
LAT = 35.27	TILT = 90:	885.	951.	918.	755.	564.	492.	513.	627.	853.	1046.	1135.	944.	0
	KT:	.42	.46	.50	.55	.55	.55	.53	.51	.52	.51	.51	.44	0
	AMB TEMP:	45.3	45.8	50.6	58.9	67.0	74.3	78.0	77.5	73.7	65.2	56.0	47.7	61.7
	HTG DEG DAYS:	611	538	458	188	47	0	0	0	0	76	277	536	2731
STA NO:93734	HORIZ INSOL:	572.	815.	1125.	1459.	1718.	1901.	1818.	1617.	1340.	1004.	651.	481.	1208.
WASHINGTON-STERLING	TILT = LAT:	901.	1116.	1305.	1439.	1511.	1585.	1554.	1523.	1474.	1337.	1003.	776.	0
DC	TILT = LAT+15:	940.	1122.	1245.	1291.	1293.	1325.	1312.	1339.	1375.	1330.	1041.	814.	0
LAT = 38.95	TILT = 90:	812.	881.	836.	695.	579.	538.	557.	664.	851.	1009.	883.	716.	0
	KT:	.40	.43	.45	.46	.48	.51	.50	.49	.49	.49	.43	.37	0
	AMB TEMP:	32.1	33.8	41.8	53.1	62.6	71.1	75.3	73.6	66.9	55.9	44.7	34.0	53.7
	HTG DEG DAYS:	1020	874	719	357	131	5	0	0	43	291	609	961	5010
STA NO:93805	HORIZ INSOL:	877.	1138.	1479.	1823.	1936.	1883.	1748.	1675.	1493.	1318.	1009.	813.	1433.
TALLAHASSEE	TILT = LAT:	1207.	1418.	1639.	1782.	1723.	1612.	1529.	1573.	1568.	1600.	1377.	1156.	0
FL	TILT = LAT+15:	1269.	1436.	1576.	1610.	1482.	1361.	1307.	1395.	1472.	1601.	1441.	1229.	0
LAT = 30.38	TILT = 90:	1020.	1020.	906.	667.	475.	409.	425.	538.	758.	1080.	1136.	1020.	0
	KT:	.46	.49	.52	.55	.54	.51	.48	.49	.50	.54	.51	.46	0
	AMB TEMP:	52.6	54.8	60.3	67.9	74.8	80.0	81.1	81.1	78.1	69.3	58.9	53.2	67.7
	HTG DEG DAYS:	408	323	187	34	0	0	0	0	0	31	204	376	1563
STA NO:93814	HORIZ INSOL:	501.	738.	1027.	1399.	1672.	1837.	1771.	1634.	1312.	990.	589.	433.	1158.
CINCINNATI (COVINGTO	TILT = LAT:	754.	986.	1175.	1376.	1471.	1534.	1514.	1540.	1440.	1317.	881.	671.	0
OH	TILT = LAT+15:	779.	986.	1117.	1234.	1259.	1284.	1280.	1353.	1343.	1310.	908.	699.	0
LAT = 39.07	TILT = 90:	664.	769.	749.	669.	571.	530.	551.	672.	833.	994.	764.	607.	0
	KT:	.35	.39	.41	.45	.47	.49	.48	.50	.48	.48	.39	.34	0
	AMB TEMP:	31.1	33.3	41.7	53.9	63.2	72.1	75.6	74.4	67.8	56.8	43.8	33.7	54.0
	HTG DEG DAYS:	1051	888	722	341	138	9	0	0	44	271	636	970	5070
STA NO:93815	HORIZ INSOL:	489.	725.	1025.	1403.	1699.	1874.	1810.	1646.	1318.	969.	564.	408.	1161.
DAYTON	TILT = LAT:	754.	983.	1183.	1384.	1493.	1560.	1544.	1552.	1458.	1305.	856.	639.	0
OH	TILT = LAT+15:	780.	983.	1125.	1240.	1276.	1302.	1303.	1363.	1360.	1298.	882.	665.	0
LAT = 39.90	TILT = 90:	671.	775.	765.	685.	590.	549.	571.	691.	857.	996.	747.	580.	0

	KT:	.36	.39	.41	.45	.48	.50	.50	.50	.49	.48	.38	.33	0
	AMB TEMP:	28.1	30.4	39.0	51.4	61.6	71.3	74.6	73.0	66.3	55.5	41.8	30.9	52.0
	HTG DEG DAYS:	1144	969	806	413	166	13	0	7	63	307	696	1057	5641
STA NO:93817	HORIZ INSOL:	574.	823.	1151.	1501.	1783.	1983.	1920.	1735.	1403.	1087.	683.	499.	1262.
EVANSVILLE	TILT = LAT:	876.	1106.	1327.	1480.	1569.	1655.	1642.	1636.	1540.	1447.	1037.	785.	0
IN	TILT = LAT+15:	912.	1111.	1266.	1328.	1342.	1381.	1386.	1433.	1440.	1443.	1076.	824.	0
LAT = 38.05	TILT = 90:	778.	862.	838.	698.	579.	534.	560.	683.	875.	1086.	906.	718.	0
	KT:	.39	.42	.45	.47	.50	.53	.53	.52	.51	.52	.43	.37	0
	AMB TEMP:	32.6	35.9	44.3	56.7	65.7	74.7	77.8	76.2	69.1	58.2	44.9	35.3	56.0
	HTG DEG DAYS:	1004	815	653	263	95	5	0	0	34	236	603	921	4629
STA NO:93819	HORIZ INSOL:	496.	747.	1037.	1398.	1688.	1868.	1806.	1643.	1324.	977.	579.	417.	1165.
INDIANAPOLIS	TILT = LAT:	763.	1017.	1197.	1379.	1483.	1556.	1542.	1550.	1463.	1314.	882.	654.	0
IN	TILT = LAT+15:	790.	1018.	1139.	1235.	1268.	1300.	1301.	1361.	1365.	1307.	909.	682.	0
LAT = 39.73	TILT = 90:	678.	802.	773.	680.	584.	545.	567.	687.	858.	1001.	771.	595.	0
	KT:	.36	.40	.42	.45	.47	.50	.49	.50	.49	.48	.39	.33	0
	AMB TEMP:	27.9	30.7	39.7	52.3	62.2	71.7	75.0	73.2	66.3	55.7	41.7	30.9	52.3
	HTG DEG DAYS:	1150	960	784	387	159	11	0	5	63	302	699	1057	5577
STA NO:93820	HORIZ INSOL:	546.	780.	1100.	1479.	1747.	1897.	1850.	1685.	1362.	1044.	657.	486.	1219.
LEXINGTON	TILT = LAT:	818.	1032.	1258.	1457.	1538.	1586.	1584.	1587.	1490.	1377.	986.	756.	0
KY	TILT = LAT+15:	848.	1034.	1198.	1308.	1317.	1327.	1339.	1396.	1391.	1372.	1021.	792.	0
LAT = 38.03	TILT = 90:	719.	798.	792.	688.	572.	523.	548.	671.	845.	1029.	857.	688.	0
	KT:	.37	.40	.43	.47	.49	.51	.51	.51	.49	.49	.42	.36	0
	AMB TEMP:	32.9	35.3	43.6	55.3	64.7	73.0	76.2	75.0	68.6	57.8	44.6	35.5	55.2
	HTG DEG DAYS:	995	832	673	302	106	8	0	0	40	246	612	915	4729
STA NO:93821	HORIZ INSOL:	546.	789.	1102.	1467.	1720.	1904.	1838.	1680.	1361.	1042.	653.	488.	1216.
LOUISVILLE	TILT = LAT:	822.	1052.	1263.	1444.	1514.	1591.	1573.	1583.	1490.	1378.	982.	766.	0
KY	TILT = LAT+15:	852.	1055.	1204.	1296.	1296.	1331.	1329.	1392.	1392.	1373.	1017.	803.	0
LAT = 38.18	TILT = 90:	724.	817.	797.	685.	568.	526.	549.	672.	848.	1032.	854.	699.	0
	KT:	.37	.41	.43	.46	.48	.51	.50	.51	.49	.50	.42	.37	0
	AMB TEMP:	33.3	35.8	44.0	55.9	64.8	73.3	76.9	75.9	69.1	58.1	45.0	35.6	55.6
	HTG DEG DAYS:	983	818	661	286	105	5	0	0	35	241	600	911	4645
STA NO:93822	HORIZ INSOL:	585.	861.	1143.	1515.	1855.	2097.	2058.	1806.	1454.	1068.	677.	490.	1302.
SPRINGFIELD	TILT = LAT:	961.	1221.	1344.	1503.	1638.	1737.	1751.	1710.	1629.	1470.	1089.	828.	0
IL	TILT = LAT+15:	1007.	1233.	1283.	1347.	1396.	1443.	1470.	1500.	1523.	1468.	1135.	873.	0
LAT = 39.83	TILT = 90:	882.	985.	875.	737.	628.	581.	615.	748.	958.	1133.	978.	779.	0
	KT:	.42	.46	.46	.49	.52	.56	.56	.55	.54	.53	.46	.39	0
	AMB TEMP:	26.7	30.4	39.4	53.1	63.4	72.9	76.1	74.4	67.2	56.6	41.9	30.5	52.7
	HTG DEG DAYS:	1187	969	794	363	132	12	0	8	48	282	693	1070	5558
STA NO:93987	HORIZ INSOL:	794.	1069.	1376.	1624.	1867.	2055.	2006.	1864.	1531.	1349.	963.	768.	1439.
LUFKIN	TILT = LAT:	1090.	1337.	1523.	1584.	1660.	1747.	1740.	1751.	1618.	1663.	1328.	1102.	0
TX	TILT = LAT+15:	1140.	1350.	1462.	1431.	1429.	1466.	1476.	1549.	1519.	1666.	1388.	1169.	0
LAT = 31.23	TILT = 90:	917.	967.	855.	624.	483.	427.	454.	591.	795.	1142.	1103.	975.	0
	KT:	.43	.47	.49	.49	.52	.55	.55	.55	.52	.56	.50	.45	0
	AMB TEMP:	48.8	52.2	58.0	67.3	74.1	80.3	83.0	83.1	77.5	68.2	57.2	50.8	66.7
	HTG DEG DAYS:	509	371	256	56	0	0	0	0	0	52	256	440	1940
STA NO:94008	HORIZ INSOL:	388.	671.	1105.	1488.	1828.	2047.	2193.	1863.	1340.	877.	479.	334.	1218.
GLASGOW	TILT = LAT:	841.	1157.	1479.	1536.	1597.	1665.	1846.	1814.	1640.	1443.	1007.	799.	0
MT	TILT = LAT+15:	885.	1172.	1416.	1368.	1349.	1371.	1532.	1580.	1531.	1445.	1054.	851.	0

LAT = 48.22	TILT = 90:	829.	1024.	1103.	898.	770.	729.	825.	971.	1122.	1232.	975.	813.	0
	KT:	.42	.47	.52	.52	.52	.55	.61	.50	.56	.55	.47	.43	0
	AMB TEMP:	9.2	15.2	25.2	42.8	54.2	62.0	70.5	69.0	57.2	46.4	29.0	17.1	41.5
	HTG DEG DAYS:	1730	1394	1234	666	344	151	15	30	263	577	1080	1485	8969
STA NO:94224	HORIZ INSOL:	315.	545.	866.	1253.	1608.	1626.	1746.	1499.	1183.	713.	387.	261.	1000.
ASTORIA	TILT = LAT:	534.	804.	1050.	1253.	1402.	1335.	1473.	1424.	1374.	1030.	657.	461.	0
OR	TILT = LAT+15:	549.	801.	995.	1115.	1190.	1112.	1234.	1242.	1278.	1018.	674.	478.	0
LAT = 46.15	TILT = 90:	491.	671.	743.	706.	657.	586.	653.	740.	901.	835.	599.	434.	0
	KT:	.31	.36	.39	.42	.46	.43	.48	.47	.48	.42	.34	.29	0
	AMB TEMP:	40.6	43.6	44.4	47.8	52.3	56.5	60.0	60.3	58.4	52.8	46.5	42.8	50.5
	HTG DEG DAYS:	756	599	639	516	394	255	163	151	201	378	555	688	5295
STA NO:94701	HORIZ INSOL:	476.	710.	1016.	1326.	1620.	1817.	1749.	1486.	1260.	890.	503.	403.	1105.
BOSTON	TILT = LAT:	806.	1019.	1208.	1313.	1419.	1504.	1486.	1401.	1417.	1240.	803.	711.	0
MA	TILT = LAT+15:	840.	1023.	1149.	1173.	1210.	1254.	1251.	1227.	1320.	1232.	827.	747.	0
LAT = 42.37	TILT = 90:	745.	833.	814.	686.	605.	577.	596.	670.	871.	974.	716.	675.	0
	KT:	.38	.41	.43	.43	.46	.48	.48	.46	.48	.47	.37	.37	0
	AMB TEMP:	29.2	30.4	38.1	48.6	58.6	68.0	73.3	71.3	64.5	55.4	45.2	33.0	51.3
	HTG DEG DAYS:	1110	969	834	492	218	27	0	8	76	301	594	992	5621
STA NO:94725	HORIZ INSOL:	391.	620.	978.	1343.	1613.	1779.	1751.	1484.	1124.	736.	388.	294.	1042.
MASSENA	TILT = LAT:	693.	921.	1197.	1345.	1408.	1463.	1480.	1405.	1275.	1038.	619.	514.	0
NY	TILT = LAT+15:	720.	922.	1138.	1199.	1197.	1216.	1242.	1227.	1183.	1026.	632.	535.	0
LAT = 44.93	TILT = 90:	648.	759.	839.	739.	641.	607.	636.	711.	816.	829.	552.	484.	0
	KT:	.36	.39	.43	.45	.46	.47	.48	.47	.45	.42	.32	.31	0
	AMB TEMP:	14.5	16.7	27.6	42.2	54.1	64.3	64.3	66.7	59.2	48.5	35.9	20.1	43.2
	HTG DEG DAYS:	1566	1352	1159	684	350	78	22	57	192	512	873	1392	8237
STA NO:94728	HORIZ INSOL:	500.	721.	1037.	1364.	1636.	1710.	1688.	1483.	1214.	895.	533.	404.	1099.
NEW YORK CITY (CENTR	TILT = LAT:	807.	997.	1212.	1346.	1435.	1425.	1439.	1394.	1338.	1206.	818.	657.	0
NY	TILT = LAT+15:	839.	998.	1153.	1205.	1227.	1194.	1216.	1224.	1245.	1196.	841.	686.	0
LAT = 40.78	TILT = 90:	732.	796.	796.	680.	587.	534.	559.	644.	798.	924.	717.	607.	0
	KT:	.38	.40	.43	.44	.46	.46	.46	.45	.46	.45	.37	.34	0
	AMB TEMP:	32.2	33.4	41.1	52.1	62.3	71.6	76.6	74.9	68.4	58.7	47.4	35.5	54.5
	HTG DEG DAYS:	1017	885	741	387	137	0	0	0	29	209	528	915	4848
STA NO:94823	HORIZ INSOL:	424.	625.	943.	1317.	1602.	1762.	1689.	1510.	1209.	895.	505.	347.	1069.
PITTSBURGH	TILT = LAT:	631.	821.	1078.	1295.	1406.	1467.	1442.	1421.	1328.	1198.	750.	518.	0
PA	TILT = LAT+15:	648.	815.	1022.	1159.	1203.	1228.	1219.	1248.	1236.	1188.	768.	533.	0
LAT = 40.50	TILT = 90:	552.	638.	699.	652.	574.	539.	555.	650.	788.	914.	648.	458.	0
	KT:	.32	.34	.38	.42	.45	.47	.46	.46	.45	.45	.35	.29	0
	AMB TEMP:	28.1	29.3	38.1	50.2	59.8	68.6	71.9	70.2	63.8	53.2	41.3	30.5	50.4
	HTG DEG DAYS:	1144	1000	834	444	208	26	7	16	98	372	711	1070	5930
STA NO:94830	HORIZ INSOL:	435.	680.	997.	1384.	1717.	1878.	1849.	1616.	1276.	911.	498.	355.	1133.
TOLEDO	TILT = LAT:	685.	944.	1168.	1372.	1505.	1556.	1572.	1527.	1427.	1255.	767.	567.	0
OH	TILT = LAT+15:	707.	944.	1110.	1227.	1283.	1296.	1322.	1339.	1330.	1247.	788.	588.	0
LAT = 41.60	TILT = 90:	614.	757.	776.	704.	620.	576.	606.	710.	865.	977.	674.	517.	0
	KT:	.34	.39	.41	.45	.48	.50	.51	.50	.48	.47	.36	.31	0
	AMB TEMP:	24.8	27.1	35.8	48.4	58.8	68.9	72.3	70.8	63.8	53.0	39.6	28.0	49.3
	HTG DEG DAYS:	1246	1061	905	498	229	32	5	18	99	379	762	1147	6381
STA NO:94849	HORIZ INSOL:	362.	617.	1028.	1407.	1720.	1879.	1885.	1583.	1156.	743.	382.	270.	1086.
ALPENA	TILT = LAT:	622.	917.	1276.	1418.	1503.	1542.	1592.	1506.	1320.	1055.	609.	454.	0

MI	TILT = LAT+15:	643.	919.	1216.	1264.	1276.	1279.	1332.	1315.	1226.	1043.	621.	468.	0
LAT = 45.07	TILT = 90:	575.	767.	900.	780.	677.	632.	674.	760.	848.	845.	543.	420.	0
	KT:	.33	.39	.46	.47	.49	.50	.52	.50	.46	.42	.32	.28	0
	AMB TEMP:	17.8	18.3	26.2	40.1	50.5	60.9	65.5	64.2	56.3	47.3	34.9	23.4	42.1
	HTG DEG DAYS:	1463	1308	1203	747	455	150	75	110	265	549	903	1290	8518
STA NO:94860	HORIZ INSOL:	370.	648.	1014.	1412.	1755.	1956.	1914.	1673.	1262.	858.	446.	311.	1135.
GRAND RAPIDS	TILT = LAT:	572.	918.	1214.	1403.	1537.	1613.	1622.	1591.	1428.	1199.	691.	494.	0
MI	TILT = LAT+15:	586.	917.	1155.	1259.	1307.	1339.	1361.	1393.	1330.	1190.	707.	510.	0
LAT = 42.88	TILT = 90:	509.	746.	825.	741.	652.	612.	643.	759.	886.	945.	609.	450.	0
	KT:	.31	.38	.43	.46	.50	.52	.53	.52	.49	.46	.34	.29	0
	AMB TEMP:	23.2	24.5	33.1	46.5	57.1	67.4	71.5	70.0	62.4	52.0	38.7	27.4	47.8
	HTG DEG DAYS:	1296	1134	989	555	270	44	8	27	114	409	789	1166	6801
STA NO:94918	HORIZ INSOL:	634.	892.	1223.	1554.	1873.	2123.	2107.	1858.	1373.	1050.	644.	511.	1321.
NORTH OMAHA	TILT = LAT:	1144.	1331.	1484.	1559.	1642.	1751.	1787.	1763.	1549.	1490.	1082.	946.	0
NE	TILT = LAT+15:	1209.	1349.	1421.	1395.	1397.	1452.	1496.	1548.	1446.	1489.	1128.	1005.	0
LAT = 41.37	TILT = 90:	1086.	1104.	997.	783.	657.	613.	654.	801.	937.	1173.	986.	918.	0
	KT:	.49	.50	.51	.51	.53	.57	.58	.57	.52	.54	.46	.44	0
	AMB TEMP:	20.2	25.5	34.6	50.0	60.9	70.2	75.1	73.7	64.4	54.4	37.9	25.7	49.4
	HTG DEG DAYS:	1389	1106	942	456	166	33	7	10	99	342	813	1218	6601

Climatic data and horizontal solar flux are from V. Cinquemani et al., *Input Data for Solar Systems,* National Climatic Center, U.S. Department of Commerce, Asheville, N.C., 1978. Blank table entries indicate that data were unavailable from this source.

Solar flux on tilted surfaces (tilts equal to latitude, latitude $+15^\circ$, and vertical are used) was calculated by the method described in "Solar Radiation Calculations" in Chap. 2.

Tabulated solar flux values on tilted surfaces *do not include radiation reflected from the foreground.* To include this effect, add the term $\bar{H}_h \rho sm^2 (\beta/2)$, where $\bar{H}$ is the horizontal flux, ρ is the diffuse foreground reflectance, and β the surface tilt angle.

Units used are Btu/(day$\cdot$ft^2) for solar flux, °F for temperature, and °F-days for degree-days; $\bar{K}_T$ has no units. The degree-day basis is 65°F. For other units see Appendix 1.

Appendix 7 Physical Property Data for Mineral Wool, Industrial Felt, and Fiberglass Insulation

Insulation Data

Material[a]	Nominal density (lb/ft³)	Temperature limitation (°F)	Mean temperature thermal conductivity: °F	Mean temperature thermal conductivity: (Btu·in)/(h·ft²·°F)[b]	Federal specification compliance	Producer	Standard sizes (variable)	Cost[c] ($/bd ft)
					Mineral wool			
Insulation								
No. 10	10.0	1200	200	0.26	HH-I-558 B	Forty-Eight	2 ft × 4 ft (board)	0.125–0.14 (dist)[d]
			350	0.32	form A,	Insulations	THK: 1–3 in	Carload: 30,000 bd ft
			500	0.375	class 4		(½ in increments)	
LTR	8.0	1000	200	0.27	HH-I-558 B	Forty-Eight	2 ft × 4 ft (board)	0.105–0.115 (dist)
			350	0.32	form A,	Insulations	THK: 1–4 in	Carload: 30,000 bd ft
			500	0.385	class 4		(½ in increments)	
I-T	6.0	850	200	0.27	HH-I-558 B	Forty-Eight	2 ft × 4 ft (board)	0.095–0.10 (dist)
			350	0.34	form A,	Insulations	THK: 1–4 in	Carload: 30,000 bd ft
			500	0.45	class 3		(½ in increments)	
MT-board								
MT-10	10.0	1050	200	0.25	HH-I-558 B	Eagle-Picher	2 ft × 4 ft (board)	0.13–0.14 (dist)
			350	0.333	form A,		THK: 1–3 in	Carload: 36,000 bd ft
			500	0.445	class 1, 2, 3		(½ in increments)	
MT-8	8.0	1050	200	0.255	HH-I-558 B	Eagle-Picher	2 ft × 4 ft (board)	0.107–0.12 (dist)
			350	0.350	form A,		THK: 1–4 in	Carload: 36,000 bd ft
			500	0.470	class 1, 2, 3		(½ in increments)	
MT-6	6.0	1050	200	0.270	HH-I-558 B	Eagle-Picher	2 ft × 4 ft (board)	0.085–0.10 (dist)
			350	0.373	form A,		THK: 1–4 in	Carload: 36,000 bd ft
			500	0.495	class 1, 2, 3		(½ in increments)	

					Industrial felt[e]			
Thermafiber								
SF-234	8.0	1000	200 350 500	0.27 0.36 0.48	HH-I-558 B form A, class 1, 2, 3	United States Gypsum	THK: 1–2 in (N.J.) Length: 60 in THK: 1–2½ in (Ind.) Length: 48 in	0.131 (dist) 7,000–38,000 bd ft
SF-240	6.0	1000	200 350 500	0.27 0.37 0.50	HH-I-558 B form A, class 1, 2, 3	United States Gypsum	THK: 1–2½ in (Tex.) Length: 90 in THK: 1–3⅓ in (Ind.) Length: 48 in	0.095–0.113 (dist) 7,000–38,000 bd ft
SF-250	4.5	800	200 350 500	0.29 0.415 0.55	HH-I-558 B form A, class 1, 2	United States Gypsum	THK:1–4 in (Tex.) Length: 90 in THK: 1–5 in (Ind.) Length: 48 in	0.081–0.10 (dist) 7,000–38,000 bd ft
SF-252	4.0	800	200 350 500	0.30 0.435 0.59	HH-I-558 B form A, class 1, 2	United States Gypsum	THK: 1–4 in (Tex.) Length: 90 in THK: 1–5 in (Ind.) Length: 48 in	0.07–0.087 (dist) 7,000–38,000 bd ft
SF-256	3.5	600	200 350 500	0.33 0.47 0.62	HH-I-558 B form A, class 1, 2	United States Gypsum	THK: 1–4 in (Tex.) Length: 90 in THK: 1–6 in (Ind.) Length: 48 in	0.066–0.084 (dist) 7,000–38,000 bd ft
SF-260	3.0	500	200 350 500	0.35 0.50 0.65	HH-I-558 B form A, class 1, 2	United States Gypsum	THK: 1–4 in (Tex.) Length: 90 in THK: 1–6 in (Ind.) Length: 48 in	0.064–0.82 (dist) 7,000–38,000 bd ft
SF-270	2.5	400	200 350 500	0.39 0.56 –	No data provided	United States Gypsum	THK: 1–4 in (Tex.) Length: 90 in THK: 1–5 in (Ind.) Length: 48 in	0.06–0.078 (dist) 7,000–38,000 bd ft

(See footnotes on p. 419.)

Insulation Data (*continued*)

Material[a]	Nominal density (lb/ft³)	Temperature limitation (°F)	Mean temperature thermal conductivity °F	(Btu·in)/(h·ft²·°F)[b]	Federal specification compliance	Producer	Standard sizes (variable)	Cost[c] ($/bd ft)
Foamglas and Fiberglas insulation[f]								
Foamglas[g]	8.5	600	200 350 500	0.46[j] 0.58 0.74	HH-I-551 D (Fed) ASTMC 552-73	Pittsburgh Corning	1 ft × 1.5 ft (board) $1\frac{1}{2}$ ft × 2 ft (board) THX: $1\frac{1}{2}$–4 in ($\frac{1}{2}$ in inc)	0.22–0.24 (Corning)[h] Carload: 36,000 bd ft
Fiberglas[i]								
701	1.6	450	200 350	0.33 0.51	HH-I-558 B form A, class 1 HH-I-558 B, type 1 form B, class 7	Owens-Corning Fiberglas	2 ft × 4 ft (board) THK: $1\frac{1}{2}$–4 in ($\frac{1}{2}$ in inc)	0.07–0.08 (dist) Carload: 30,000–35,000 bd ft
703	3.0	450	200 350	0.30 0.41	HH-I-558 B form A, class 1, 2	Owens-Corning Fiberglas	2 ft × 4 ft (board) THK: 1–2 in ($\frac{1}{2}$ in inc)	0.14–0.15 (dist) Carload: 30,000–35,000 bd ft
705	6.0	450	200 350	0.27 0.38	HH-I-558 B form A, class 1, 2	Owens-Corning Fiberglas	2 ft × 4 ft (board) THK: 1–2 in ($\frac{1}{2}$ in inc)	0.25–0.27 (dist) Carload: 30,000–35,000 bd ft
Thermal insulating wool[j]								
Type I	1.25	1000	200 350 500	0.41 0.65 0.85	HH-I-558 B form B, type 1, class 6	Owens-Corning Fiberglas	Rolls Width: 2 or 3 ft THK: 2, 3, 4 in Length: 76, 52, 38 ft	0.04–0.06 (Corning) Half carload: 35,000 bd ft

Type II	2.4	1000	200	0.30[j]	HH-I-558 B	Owens-Corning	2 ft × 8 ft	0.08–0.09
			350	0.44	form B, type 1,	Fiberglas	2 ft × 4 ft (board)	(Corning)
			500	0.60	class 7, 8		THK: 1–3 in	0.14–0.15 (dist)
							($\frac{1}{2}$ in inc)	Carload: 35,000 bd ft
IS board[i]	4.0	800	200	0.30	HH-I-558 B	Owens-Corning	2 ft × 4 ft, 3 ft × 4 ft	0.10–0.13
			350	0.44	form A	Fiberglas	4 ft × 8 ft (board)	(Corning)
			500	0.61	class 3		THK: 1–6 in	0.18–0.20 (dist)
							($\frac{1}{2}$ in inc)	Carload: 35,000 bd ft

[a]All insulations listed will not cause or aggravate corrosion and will absorb less than 1% moisture. All insulations listed appear as semirigid board composed of silica base refractory fibers bonded with special binders for service in indicated temperature ranges.

[b]Units are consistently employed within the insulation industry. Conductivity measurements consider a test specimen 1 in thick and 1 ft^2 in normal area.

[c]Cost data current through October 30, 1975. Costs are based on carload purchases and include freight where necessary to move insulation to Houston.

[d]Cost from Houston distributors noted by (dist).

[e]Industrial felt is preformed mineral fiber felt, which will not cause or sustain corrosion. It absorbs less than 1% moisture by weight and is rated noncombustible. Insulation to be ordered in varying thicknesses and lengths; standard width of 24 in employed (see column 8).

[f]All codes in specification compliance column are federal specifications unless otherwise noted.

[g]Foamglas is an impermeable, incombustible, rigid insulation composed of completely sealed glass cells with no binder material. Its rigid form may allow for Foamglas being implemented as the collector box.

[h]Cost from Corning Houston warehouse noted by (Corning); cost from Houston distributor noted by (dist).

[i]Insulations are made of inorganic glass fibers preformed into semirigid to rigid rectangular boards (T1W I in blankets). Insulations will not accelerate or cause corrosion and will absorb less than 1% moisture (by volume).

[j]The units of the mean temperature thermal conductivity for Foamglas and Fiberglas insulation are (Btu·in)/(h·ft·°F).

Source: R. C. Ratzel and R. B. Bannerot, "Optimal Material Selection for Flat-Plate Solar Energy Collectors Utilizing Commercially Available Materials," presented at the ASME-AIChE National Heat Transfer Conference, 1976.

Appendix 8 Thermal Properties

Thermal Properties of Some Nonmetals

Material	Average temperature (°F)	k [Btu/(h·ft·°F)]	c [Btu/(lb$_m$·°F)]	ρ (lb$_m$/ft^3)	α (ft^2/h)
		Insulating materials			
Asbestos	32	0.087	0.25	36	~0.01
	392	0.12	–	36	~0.01
Cork	86	0.025	0.04	10	~0.006
Cotton fabric	200	0.046			
Diatomaceous earth, powdered	100	0.030	0.21	14	~0.01
	300	0.036	–		
	600	0.046	–		
Molded pipe covering	400	0.051	–	26	
	1600	0.088	–		
Glass wool					
Fine	20	0.022	–		
	100	0.031	–	1.5	
	200	0.043	–		
Packed	20	0.016	–		
	100	0.022	–	6.0	
	200	0.029	–		
Hair felt	100	0.027	–	8.2	
Kaolin insulating brick	932	0.15	–	27	
	2102	0.26	–		

Thermal Properties of Some Nonmetals (*continued*)

Material	Average temperature (°F)	k [Btu/(h·ft·°F)]	c [Btu/(lb$_m$·°F)]	ρ (lb$_m$/ft^3)	α (ft^2/h)
Kaolin insulating firebrick	392	0.05	–	19	
	1400	0.11	–		
Magnesia, 85%	32	0.032	–	17	
	200	0.037	–	17	
Rock wool	20	0.017	–	8	
	200	0.030	–		
Rubber	32	0.087	0.48	75	0.0024
		Building materials			
Brick					
Fire clay	392	0.58	0.20	144	0.02
	1832	0.95			
Masonry	70	0.38	0.20	106	0.018
Zirconia	392	0.84	–	304	
	1832	1.13	–		
Chrome brick	392	0.82	–	246	
	1832	0.96	–		
Concrete					
Stone	~70	0.54	0.20	144	0.019
10% moisture	~70	0.70	–	140	~0.025
Glass, window	~70	~0.45	0.2	170	0.013
Limestone, dry	70	0.40	0.22	105	0.017
Sand					
Dry	68	0.20	–	95	
10% water	68	0.60	–	100	
Soil					
Dry	70	~0.20	0.44	–	~0.01
Wet	70	~1.5	–	–	~0.03
Wood					
Oak ⊥ to grain*	70	0.12	0.57	51	0.0041
∥ to grain	70	0.20	0.57	51	0.0069
Pine ⊥ to grain	70	0.06	0.67	31	0.0029
∥ to grain	70	0.14	0.67	31	0.0067
Ice	32	1.28	0.46	57	0.048

*⊥, perpendicular; ∥, parallel.

Source: F. Kreith, *Principles of Heat Transfer*, Intext, New York, 1973.

Thermal Properties of Metals and Alloys

Material	k, [Btu/(h·ft·°F)] 32°F	212°F	572°F	932°F	c [Btu/(lb_m·°F)] 32°F	ρ (lb_m/ft^3) 32°F	α (ft^2/h) 32°F
				Metals			
Aluminum	117	119	133	155	0.208	169	3.33
Bismuth	4.9	3.9	–	–	0.029	612	0.28
Copper, pure	224	218	212	207	0.091	558	4.42
Gold	169	170	–	–	0.030	1203	4.68
Iron, pure	35.8	36.6	–	–	0.104	491	0.70
Lead	20.1	19	18	–	0.030	705	0.95
Magnesium	91	92	–	–	0.232	109	3.60
Mercury	4.8	–	–	–	0.033	849	0.17
Nickel	34.5	34	32	–	0.103	555	0.60
Silver	242	238	–	–	0.056	655	6.6
Tin	36	34	–	–	0.054	456	1.46
Zinc	65	64	59	–	0.091	446	1.60
				Alloys			
Admiralty metal	65	64					
Brass, 70% Cu, 30% Zn	56	60	66	–	0.092	532	1.14
Bronze, 75% Cu, 25% Sn	15	–	–	–	0.082	540	0.34
Cast iron							
Plain	33	31.8	27.7	24.8	0.11	474	0.63
Alloy	30	28.3	27	–	0.10	455	0.66
Constantan, 60% Cu, 40% Ni	12.4	12.8	–	–	0.10	557	0.22
18-8 Stainless steel,							
Type 304	8.0	9.4	10.9	12.4	0.11	488	0.15
Type 347	8.0	9.3	11.0	12.8	0.11	488	0.15
Steel, mild, 1% C	26.5	26	25	22	0.11	490	0.49

Source: F. Kreith, *Principles of Heat Transfer,* Intext, New York, 1973.

Appendix 9 P-Chart Coefficients

P-Chart Coefficients*

Station Name	State	With R-9 night insulation						Without night insulation						De-gree-days
		Direct gain		Masonry wall		Water wall		Direct gain		Masonry wall		Water wall		
		A	*B*	*A*	*B*	*A*	*B*	*A*	*B*	*A*	*B*	*A*	*B*	
Washington-Sterling	DC	0.459	0.242	0.534	0.172	0.725	0.120	0.261	0.204	0.370	0.125	0.414	0.106	5010
Birmingham	AL	0.549	0.180	0.584	0.141	0.813	0.097	0.455	0.142	0.534	0.091	0.586	0.085	2844
Prescott	AZ	0.597	0.269	0.611	0.224	0.977	0.127	0.580	0.175	0.645	0.119	0.900	0.084	4456
Sacramento	CA	0.465	0.299	0.488	0.232	0.601	0.184	0.453	0.206	0.493	0.136	0.577	0.121	2843
San Francisco	CA	0.543	0.343	0.528	0.299	0.705	0.213	1.316	0.094	0.717	0.127	0.933	0.103	3042
Denver	CO	0.555	0.305	0.596	0.239	0.924	0.141	0.481	0.197	0.538	0.142	0.643	0.118	6016
Hartford	CT	0.424	0.255	0.499	0.186	0.659	0.132	0.094	0.544	0.249	0.174	0.237	0.167	6350
Wilmington	DE	0.474	0.234	0.535	0.175	0.738	0.119	0.288	0.186	0.394	0.117	0.440	0.102	4940
Tallahassee	FL	0.618	0.147	0.626	0.120	0.959	0.073	0.468	0.153	0.622	0.078	0.688	0.076	1563
Atlanta	GA	0.537	0.190	0.567	0.151	0.833	0.095	0.420	0.155	0.511	0.096	0.567	0.088	3095
Boise	ID	0.392	0.439	0.439	0.322	0.509	0.278	0.279	0.336	0.340	0.221	0.376	0.199	5833
Chicago	IL	0.403	0.292	0.488	0.199	0.622	0.149	0.121	0.478	0.260	0.186	0.252	0.180	6127
Indianapolis	IN	0.412	0.258	0.500	0.175	0.647	0.129	0.124	0.424	0.262	0.168	0.253	0.165	5577
Des Moines	IA	0.421	0.306	0.500	0.216	0.664	0.152	0.130	0.515	0.260	0.214	0.255	0.207	6710
Wichita	KS	0.505	0.261	0.566	0.192	0.811	0.125	0.332	0.237	0.448	0.136	0.475	0.130	4687
Louisville	KY	0.454	0.230	0.525	0.164	0.720	0.112	0.227	0.240	0.347	0.130	0.354	0.126	4645
Shreveport	LA	0.562	0.170	0.590	0.133	0.884	0.083	0.483	0.140	0.584	0.082	0.599	0.086	2167
Caribou	ME	0.378	0.378	0.441	0.289	0.498	0.257	N-R	N-R	N-R	N-R	N-R	N-R	9632
Baltimore	MD	0.481	0.228	0.547	0.168	0.745	0.117	0.312	0.175	0.411	0.113	0.462	0.098	4729
Boston	MA	0.434	0.259	0.509	0.185	0.681	0.130	0.168	0.323	0.293	0.162	0.312	0.141	5621
Detroit	MI	0.401	0.265	0.469	0.196	0.590	0.148	0.075	0.798	0.223	0.198	0.195	0.214	6228
Minneapolis-St. Paul	MN	0.392	0.314	0.445	0.250	0.539	0.198	N-R	N-R	N-R	N-R	N-R	N-R	8159
Kansas City	MO	0.469	0.257	0.528	0.193	0.721	0.132	0.232	0.282	0.350	0.154	0.368	0.143	5357
Billings	MT	0.406	0.404	0.477	0.282	0.577	0.229	0.206	0.384	0.309	0.220	0.313	0.212	7265
Helena	MT	0.358	0.496	0.430	0.336	0.496	0.291	0.136	0.588	0.253	0.273	0.250	0.264	8190

North Omaha	NE	0.435	0.306	0.515	0.215	0.713	0.143	0.160	0.431	0.285	0.203	0.288	0.191	6601
Las Vegas	NV	0.583	0.239	0.562	0.212	0.792	0.141	0.552	0.184	0.640	0.111	0.717	0.105	2601
Concord	NH	0.419	0.274	0.482	0.210	0.612	0.158	N-R	N-R	0.213	0.224	0.185	0.234	7360
Newark	NJ	0.464	0.237	0.529	0.175	0.724	0.120	0.258	0.205	0.370	0.124	0.402	0.110	5034
Albuquerque	NM	0.580	0.257	0.621	0.198	0.968	0.117	0.615	0.147	0.644	0.107	0.793	0.088	4292
Farmington	NM	0.543	0.307	0.586	0.239	0.866	0.150	0.414	0.238	0.505	0.153	0.671	0.111	5713
Buffalo	NY	0.391	0.243	0.453	0.191	0.521	0.164	N-R	N-R	N-R	N-R	N-R	N-R	7285
New York City (La Guardia)	NY	0.477	0.227	0.539	0.170	0.727	0.119	0.273	0.196	0.380	0.121	0.414	0.108	4909
Rochester	NY	0.390	0.246	0.449	0.193	0.523	0.163	N-R	N-R	N-R	N-R	N-R	N-R	6719
Charlotte	NC	0.538	0.202	0.569	0.160	0.810	0.105	0.434	0.160	0.512	0.102	0.574	0.093	3218
Bismarck	ND	0.379	0.392	0.436	0.299	0.515	0.247	N-R	N-R	0.190	0.337	N-R	N-R	9044
Columbus	OH	0.411	0.247	0.487	0.177	0.625	0.130	0.095	0.551	0.243	0.172	0.222	0.178	5702
Burns	OR	0.402	0.436	0.472	0.302	0.559	0.254	0.226	0.385	0.319	0.230	0.353	0.199	7212
Salem	OR	0.385	0.373	0.437	0.267	0.494	0.241	0.224	0.381	0.308	0.213	0.347	0.185	4852
Harrisburg	PA	0.447	0.240	0.522	0.172	0.701	0.121	0.207	0.243	0.328	0.135	0.352	0.119	5224
Providence	RI	0.443	0.263	0.521	0.188	0.705	0.130	0.173	0.312	0.297	0.162	0.334	0.131	5972
Columbia	SC	0.558	0.186	0.600	0.141	0.855	0.093	0.503	0.138	0.584	0.087	0.611	0.088	2598
Pierre	SD	0.389	0.385	0.465	0.268	0.579	0.205	0.104	0.792	0.236	0.273	0.218	0.285	7677
Nashville	TN	0.483	0.197	0.543	0.145	0.749	0.099	0.279	0.199	0.393	0.111	0.403	0.110	3696
Lubbock	TX	0.574	0.243	0.622	0.183	0.996	0.104	0.594	0.150	0.640	0.103	0.804	0.084	3545
Salt Lake City	UT	0.444	0.381	0.500	0.277	0.623	0.217	0.437	0.228	0.389	0.193	0.443	0.167	5983
Richmond	VA	0.498	0.219	0.541	0.170	0.739	0.117	0.368	0.168	0.454	0.108	0.514	0.095	3939
Charleston	WV	0.439	0.227	0.511	0.162	0.674	0.116	0.203	0.251	0.320	0.134	0.330	0.126	4590
La Crosse	WI	0.402	0.294	0.461	0.227	0.579	0.170	N-R	N-R	0.198	0.256	N-R	N-R	7417
Cheyenne	WY	0.529	0.337	0.584	0.259	0.876	0.159	0.395	0.235	0.488	0.158	0.586	0.128	7255
Sheridan	WY	0.412	0.405	0.484	0.285	0.610	0.217	0.194	0.406	0.306	0.224	0.314	0.210	7708

*A and B are used in Chapter 7 to predict passive solar heating performance. Table courtesy of Solar Energy Design Corp. of America, P.O. Box 67, Ft. Collins, Colorado. For other sites contact this company.

Appendix 10 Factors for Simplified Active System Sizing Method

Collector factors and location factors are numbers derived from computer-simulated performance of approximately 30,000 solar heating systems. The results of these computer simulations indicate that performance depends mainly on (1) the amount of solar energy available together with local climatic conditions (location factor), and (2) the collector's effectiveness for capturing that energy (collector factor). By combining these factors, you can estimate how much heat a given collector will provide in a given location as described in Chap. 8.

(*a*) Collector Factors

		Collector type					
		Air			Liquid		
Collector orientation	Collector* tilt angle	Selective surface and single glazing	Single† glazing	Double† glazing	Selective surface and single glazing	Single† glazing	Double† glazing
South	Roof	7.4	6.6	6.8	8.5	7.7	7.8
	Optimal	8.2	7.3	7.5	9.3	8.6	8.6
	Wall	6.8	6.0	6.1	8.1	7.3	7.2
	Wall with reflector	8.0	7.2	7.3	9.3	8.5	8.4
Southeast or south-west	Roof	7.2	6.3	6.5	8.2	7.0	7.5
	Optimal	7.7	6.8	7.0	8.8	8.0	8.0
	Wall	6.3	5.5	5.6	7.5	6.6	6.6
	Wall with reflector	7.5	6.7	6.8	8.7	7.9	7.9
East or west	Roof	6.5	5.7	5.9	7.5	6.7	6.8
	Optimal	6.4	5.6	5.8	7.4	6.6	6.7
	Wall	5.0	4.3	4.5	5.9	5.2	5.3
	Wall with reflector	6.4	5.6	5.8	7.5	6.6	6.7

*Roof, standard 4/12 pitch roof; optimal, latitude of location where collector array is to be installed plus 12°; wall, vertical wall (90°); wall with reflector, vertical wall with attached horizontal polished reflector 90° to wall.
†No selective surface.

(*b*) Location Factors

State	City	Small	Medium	Large
Alabama	Birmington	5.7	7.2	6.7
	Mobile	7.8	9.7	11.1
	Montgomery	6.7	8.4	9.9
Alaska	Annette	2.3	3.5	4.7
	Homer	0.8	1.3	1.8
	Juneau	0.7	1.2	1.6
	King Salmon	0.7	1.1	1.5
	Kodiak	0.9	1.4	2.0
	Matanuska	1.9	2.9	3.9
	Yakutat	0.6	1.0	1.4
Arizona	Page	5.5	7.5	9.2
	Phoenix	9.6	11.7	12.7
	Prescott	6.2	8.3	10.1
	Tucson	9.6	11.6	12.7
	Winslow	5.7	7.7	9.5
	Yuma	11.5	12.9	13.4

(*b*) Location Factors (*continued*)

State	City	Small	Medium	Large
Arkansas	Fort Smith	5.2	6.7	8.2
	Little Rock	5.2	6.7	8.1
California	Arcata	4.6	6.6	8.3
	Bakersfield	7.0	8.6	9.9
	China Lake	4.6	5.5	6.3
	Daggett	7.8	9.7	11.1
	Davis	6.1	7.7	9.1
	El Toro	9.2	11.3	12.5
	Fresno	6.6	8.3	9.6
	Inyokern	8.3	10.3	11.7
	Long Beach	9.6	11.7	12.7
	Los Angeles	9.5	11.7	12.7
	Mt. Shasta	4.2	5.7	7.2
	Needles	9.5	11.2	12.4
	Oakland	7.2	9.3	10.8
	Pasadena	9.6	11.7	12.7
	Point Muga	6.8	9.6	11.3
	Red Bluff	6.2	7.9	9.2
	Riverside	9.5	11.6	12.7
	Sacramento	6.3	7.9	9.3
	San Diego	10.1	12.1	13.0
	San Francisco	7.1	9.2	10.7
	San Jose	7.2	9.1	10.6
	Santa Maria	7.6	9.9	11.5
	Sunnyvale	7.5	9.5	11.0
Colorado	Boulder	4.4	6.2	7.9
	Colorado Springs	4.5	6.3	8.0
	Denver	4.6	6.4	8.1
	Eagle	3.6	5.2	6.6
	Grand Junction	4.6	6.2	7.7
	Grand Lake	3.1	4.7	6.3
	Pueblo	4.9	6.7	8.3
Connecticut	Hartford	2.7	3.7	4.7
Delaware	Wilmington	3.7	5.0	6.3
District of Columbia	Washington	3.6	4.8	6.0
Florida	Apalachicola	8.9	10.7	12.0
	Daytona Beach	10.7	12.3	13.1
	Gainesville	10.4	12.1	13.0
	Jacksonville	9.1	11.0	12.2
	Orlando	11.5	12.9	13.4
	Pensacola	8.5	10.5	11.8
	Tallahassee	8.4	10.3	11.6
	Tampa	11.6	12.9	13.4
Georgia	Atlanta	5.5	7.0	8.5
	Augusta	6.2	7.9	9.4
	Griffin	6.4	8.1	9.7
	Macon	6.7	8.4	10.0
	Savannah	7.3	9.1	10.6

(*b*) Location Factors (*continued*)

State	City	Small	Medium	Large
Idaho	Boise	3.9	5.3	6.5
	Lewiston	2.1	2.7	3.1
	Pocatello	3.5	4.9	6.1
	Twin Falls	3.5	4.9	6.2
Illinois	Chicago	3.0	4.1	5.1
	Lemont	3.3	4.5	5.6
	Moline	3.0	4.1	5.2
	Peoria	3.4	4.6	5.7
	Springfield	3.5	4.6	5.8
Indiana	Evansville	3.8	5.0	6.2
	Fort Wayne	2.7	3.6	4.6
	Indianapolis	3.1	4.1	5.1
	South Bend	2.6	3.6	4.4
Iowa	Ames	3.2	4.4	5.7
	Burlington	2.0	2.5	3.0
	Des Moines	3.2	4.3	5.4
	Mason City	2.8	3.9	5.0
	Sioux City	3.1	4.2	5.4
Kansas	Dodge City	4.8	6.5	8.0
	Goodland	2.5	3.2	3.8
	Manhattan	4.2	5.6	7.0
	Topeka	4.1	5.5	6.8
	Wichita	4.7	6.3	7.8
Kentucky	Covington	3.0	3.9	4.7
	Lexington	3.6	4.8	5.9
	Louisville	3.7	4.8	6.0
Louisiana	Baton Rouge	7.8	9.6	11.0
	Lake Charles	7.9	9.6	11.0
	New Orleans	8.4	10.2	11.6
	Shreveport	6.9	8.6	10.1
Maine	Bangor	1.6	2.0	2.5
	Caribou	2.0	2.8	3.7
	Portland	2.5	3.5	4.5
Maryland	Annapolis	4.2	5.6	7.0
	Baltimore	3.6	4.8	5.9
	Patuxent River	4.2	5.7	7.1
	Silver Hill	4.5	6.0	7.4
Massachusetts	Amherst	2.8	3.9	5.0
	Blue Hill	3.1	4.3	5.5
	Boston	3.1	4.2	5.3
	Lynn	3.1	4.3	5.4
	Natick	3.3	4.5	5.7
Michigan	Alpena	2.1	3.0	3.8
	Detroit	2.6	3.5	4.4
	East Lansing	2.7	3.6	4.6
	Flint	2.4	3.2	4.1

(*b*) Location Factors (*continued*)

State	City	Small	Medium	Large
	Grand Rapids	2.4	3.3	4.1
	Houghton	1.3	1.8	2.2
	Lansing	2.7	3.7	4.7
	Sault Ste. Marie	1.9	2.8	3.6
	Traverse City	2.2	3.0	3.8
Minnesota	Duluth	2.0	2.9	3.8
	International Falls	1.9	2.8	3.6
	Minneapolis	2.4	3.4	4.3
	Rochester	2.4	3.3	4.2
	St. Cloud	2.7	3.8	4.9
Mississippi	Jackson	6.5	8.2	9.7
	Meridian	6.4	8.0	9.5
Missouri	Columbia	3.7	5.0	6.2
	Kansas City	3.8	5.1	6.4
	St. Louis	4.0	5.3	6.5
	Springfield	4.2	5.6	7.0
Montana	Billings	3.3	4.7	6.0
	Cut Bank	1.6	2.1	2.6
	Dillon	2.9	4.2	5.4
	Glasgow	2.5	3.6	4.6
	Great Falls	2.9	4.2	5.3
	Helena	2.8	4.0	5.1
	Lewiston	2.7	3.9	5.1
	Miles City	2.9	4.0	5.1
	Missoula	2.4	3.3	4.2
	Summit	2.2	3.4	4.5
Nebraska	Grand Island	3.6	5.0	6.3
	Lincoln	3.9	5.3	6.7
	North Omaha	3.3	4.5	5.8
	North Platte	3.7	5.2	6.6
	Scotts Bluff	3.7	5.2	6.7
Nevada	Elko	3.8	5.3	6.7
	Ely	4.1	5.8	7.5
	Las Vegas	7.9	10.0	11.4
	Love Lock	4.9	6.7	8.3
	Reno	5.0	6.9	8.6
	Tonopah	5.1	7.0	8.8
	Winnemucca	4.3	5.9	7.5
	Yucca Flats	3.6	4.4	5.1
New Hampshire	Concord	2.4	3.4	4.3
New Jersey	Atlantic City	4.4	5.9	7.4
	Lakehurst	3.5	4.7	5.9
	Newark	3.5	4.8	6.0
	Trenton	4.0	5.5	6.8
New Mexico	Albuquerque	6.0	8.0	9.7
	Clayton	3.1	4.0	4.9

(*b*) Location Factors (*continued*)

State	City	Small	Medium	Large
	Farmington	5.0	6.9	8.6
	Roswell	6.4	8.4	10.2
	Truth or Consequences	4.3	5.3	6.3
	Tucumcari	3.6	4.5	5.4
	Zuni	5.0	7.0	8.8
New York	Albany	2.5	3.4	4.3
	Binghampton	2.0	2.8	3.6
	Buffalo	2.2	3.0	3.8
	Ithaca	2.6	3.6	4.6
	Massena	2.2	3.1	4.0
	New York	3.4	4.5	5.6
	New York (La Guardia)	2.3	2.8	3.3
	Rochester	2.4	3.3	4.2
	Schenectady	2.5	3.5	4.5
	Syracuse	2.3	3.0	3.8
North Carolina	Asheville	4.6	6.2	7.7
	Cape Hatteras	5.6	7.1	8.5
	Charlotte	5.4	7.0	8.5
	Cherry Point	6.3	8.0	9.6
	Greensboro	4.9	6.5	7.9
	Raleigh	5.0	6.6	8.1
North Dakota	Bismarck	2.5	3.6	4.6
	Fargo	2.3	3.2	4.1
	Minot	2.4	3.3	4.3
Ohio	Akron	2.7	3.6	4.5
	Cincinnati	3.2	4.3	5.4
	Cleveland	2.5	3.4	4.2
	Columbus	2.8	3.8	4.7
	Dayton	3.0	4.0	4.9
	Put-In-Bay	2.9	3.9	4.8
	Toledo	2.7	3.6	4.5
	Youngstown	2.4	3.3	4.1
Oklahoma	Oklahoma City	5.3	6.9	8.4
	Stillwater	5.4	7.1	8.6
	Tulsa	5.0	6.5	8.0
Oregon	Astoria	3.4	4.9	6.2
	Burns	3.1	4.3	5.4
	Corvallis	4.2	5.8	7.3
	Medford	3.9	5.3	6.5
	North Bend	4.6	6.5	8.1
	Pendleton	3.8	5.0	6.2
	Portland	3.2	4.3	5.3
	Redmond	3.4	4.7	5.9
	Salem	3.3	4.5	5.5
Pennsylvania	Allentown	2.9	4.0	5.0
	Avoca	2.6	3.5	4.4
	Erie	2.3	3.2	4.0

(*b*) Location Factors (*continued*)

State	City	Small	Medium	Large
	Harrisburg	3.3	4.4	5.5
	Philadelphia	3.6	4.8	6.0
	Pittsburgh	2.6	3.5	4.4
	Scranton	1.8	2.3	2.7
	State College	2.9	3.9	4.9
Rhode Island	Newport	3.6	4.9	6.3
	Providence	2.9	4.0	5.1
South Carolina	Charleston	6.9	8.7	10.2
	Columbia	6.3	8.0	9.6
	Greenville	5.6	7.2	8.7
South Dakota	Huron	2.7	3.8	4.8
	Pierre	3.1	4.2	5.4
	Rapid City	3.3	4.6	5.9
	Sioux Falls	2.9	4.0	5.1
Tennessee	Chattanooga	4.7	6.1	7.4
	Knoxville	4.7	6.1	7.4
	Memphis	5.2	6.6	8.0
	Nashville	4.4	5.6	6.8
	Oak Ridge	4.5	5.8	7.1
Texas	Abilene	7.0	8.9	10.5
	Amarillo	5.8	7.7	9.5
	Austin	7.8	9.6	11.1
	Big Springs	7.3	9.3	10.9
	Brownsville	11.1	12.5	13.2
	Corpus Christi	10.1	11.7	12.7
	Dallas	6.6	8.3	9.8
	Del Rio	6.3	7.4	8.3
	El Paso	7.8	10.0	11.5
	Forth Worth	6.7	8.4	10.0
	Houston	8.1	9.8	11.1
	Kingsville	9.8	11.4	12.5
	Laredo	8.2	9.4	10.4
	Lubbock	6.3	8.3	10.0
	Lufkin	7.3	9.2	10.7
	Midland	7.3	9.3	10.9
	Port Arthur	8.1	9.9	11.3
	San Angelo	7.5	9.4	11.0
	San Antonio	8.3	10.1	11.5
	Sherman	4.0	4.7	5.5
	Waco	7.4	9.2	10.7
	Wichita Falls	6.2	8.0	9.6
Utah	Bryce Canyon	2.7	3.7	4.6
	Cedar City	4.7	6.5	8.2
	Salt Lake City	4.2	5.7	7.1
Vermont	Burlington	2.0	2.8	3.5
Virginia	Mt. Weather	3.7	5.1	6.5
	Norfolk	5.1	6.7	8.1

(*b*) Location Factors (*continued*)

State	City	Small	Medium	Large
	Richmond	4.5	5.9	7.3
	Roanoke	4.3	5.8	7.2
Washington	Olympia	2.8	4.0	4.9
	Prosser	4.0	5.3	6.4
	Pullman	3.5	4.8	5.9
	Richland	3.8	4.9	6.0
	Seattle	3.1	4.2	5.1
	Spokane	2.8	3.8	4.7
	Tacoma	3.2	4.3	5.3
	Whidbey Island	3.1	4.4	5.5
	Yakina	3.4	4.5	5.5
West Virginia	Charleston	3.3	4.4	5.4
	Huntington	2.5	3.0	3.5
	Parkersburg	3.4	4.5	5.5
Wisconsin	Eau Claire	2.3	3.2	4.1
	Green Bay	2.4	3.4	4.3
	La Crosse	2.6	3.5	4.4
	Madison	2.6	3.6	4.6
	Milwaukee	2.6	3.7	4.7
Wyoming	Casper	3.8	5.4	7.0
	Cheyenne	3.9	5.6	7.2
	Lander	4.1	5.9	7.6
	Laramie	3.5	5.2	6.8
	Rock Springs	3.7	5.3	6.9
	Sheridan	3.1	4.4	5.7
Canada				
Alberta	Calgary	1.4	2.0	2.5
	Edmonton	2.7	3.9	5.1
	Lethbridge	2.8	4.0	5.0
British Columbia	Vancouver	2.7	3.7	4.6
	Victoria	2.3	3.0	3.6
Manitoba	Churchill	1.7	2.7	3.8
	Winnipeg	2.1	3.1	4.0
New Brunswick	Moncton	2.2	3.2	4.2
	St. John	1.6	2.2	2.7
Newfoundland	St. John	1.9	2.8	3.8
Nova Scotia	Halifax	1.6	2.3	2.8
Ontario	Hamilton	1.8	2.3	2.8
	Kapuskasing	1.7	2.4	3.2
	London	1.7	2.2	2.6
	Ottawa	2.4	3.4	4.3
	Toronto	2.5	3.5	4.4
	Windsor	1.8	2.3	2.8
Quebec	Montreal	2.1	3.0	3.8
	Quebec City	1.4	1.9	2.3
Saskatchewan	Saskatoon	1.2	1.6	2.0

(*c*) Compound Interest

Interest rate (%)	Years 2	3	4	5	7	10	15	20	25	30
6	23	33	43	52	69	91	119	140	156	168
6.5	23	33	42	51	68	89	115	135	149	159
7	22	33	42	51	67	87	112	130	142	151
7.5	22	33	42	50	66	85	109	125	136	142
8	22	32	41	50	65	83	105	120	130	137
8.5	22	32	41	49	64	81	102	116	125	131
9	22	32	40	49	63	80	99	112	120	125
9.5	22	31	40	48	62	79	97	108	115	120
10	22	31	40	47	61	76	94	104	111	115
10.5	22	31	39	47	60	75	91	101	107	110
11	22	31	39	46	59	73	89	98	103	106
11.5	22	31	39	46	58	72	86	95	99	102
12	21	30	38	45	57	70	84	92	96	98
12.5	21	30	38	45	56	69	82	89	93	95
13	21	30	38	44	56	68	80	86	90	91
13.5	21	30	37	44	55	66	78	84	87	88
14	21	30	37	43	54	65	76	81	84	85
14.5	21	29	37	43	53	64	74	79	81	83
15	21	29	36	43	52	63	72	77	79	80
16	21	29	36	42	51	60	69	73	75	75
17	21	28	35	41	50	58	66	69	71	71
18	20	28	35	40	48	56	63	66	67	67
19	20	28	34	39	47	54	60	63	64	64
20	20	27	33	38	46	53	58	60	61	61

Appendix 11 Solar Information Sources

Solar Energy Research Institute (SERI)
1617 Cole Boulevard
Golden, Colorado 80401
(303) 231-1415
(800) 525-5555 (alcohol fuel inquiries from the U.S. outside Colorado)
(800) 332-8339 (alcohol fuel inquiries from Colorado outside Denver)
(303) 231-7303 (alcohol fuel inquiries from Denver only)

Areas of Interest: Manages and coordinates research and development and works with all levels of government to promote application of solar energy technologies.

Resources: Over 75,000 solar and solar-related books, journals, directories, proceedings, technical reports, and other documents; access to SEIDB data bases and over 140 public and private computerized data bases.

Information Services: (free) Answers inquiries and makes referrals, provides on-site use of library, provides computerized information searching, conducts workshops and seminars. Operates the National Alcohol Fuels Information Center.

Note: For information on solar heating and cooling, contact the National Solar Heating and Cooling Information Center (see listing below).

Publications: (free) *Pocket Solar Energy Information Data Bank, Solar*

Abstracted from the *1980 Solar Energy Information Locator*, Solar Energy Research Institute, Golden, Colo., 1980.

Energy Information Locator, Facts About Gasohol, State Solar Educational Programs and Courses, Analysis Methods for Photovoltaic Applications and many other brochures, reading lists, reference guides, directories; microfiche copies of SERI technical reports and papers.

For a complete publications listing, contact SERI for free copies of the *SEIDB Publications Bulletins: Solar Awareness* (introductory information) or *Technical Information.*

Operated for the U.S. Department of Energy.

Mid-America Solar Energy Complex (MASEC)

8140 26th Avenue South
Bloomington, Minnesota 55420
(612) 853-0400

Serves: Illinois, Indiana, Iowa, Missouri, Nebraska, North Dakota, Kansas, Michigan, Minnesota, Ohio, South Dakota, and Wisconsin.

Areas of Interest: Regional commercialization of solar technologies; conservation integral to solar applications.

Resources: Books, journals, reports, proceedings, directories, resource data, legislation, periodicals, newsletters, bibliographies, and solar equipment brochure file.

Information Services: (free) Answers inquiries, conducts seminars and workshops, provides technical assistance, provides on-site use of library, conducts computerized information searching at MASEC offices. Information services also available through state solar offices within the MASEC service region (see state sources listed below).

Publications: *MASEC Newsletter* (monthly), *MASEC Yellow Pages*, and various technical reports.

Operated for the U.S. Department of Energy.

Northeast Solar Energy Center (NESEC)

470 Atlantic Avenue
Boston, Massachusetts 02110
(617) 292-9250

Serves: Connecticut, Maine, Massachusetts, New Hampshire, Vermont, New Jersey, New York, Rhode Island, and Pennsylvania.

Areas of Interest: Accelerating commercialization of solar technologies; current emphasis on domestic solar hot water, space heating, passive architecture, wood energy, small wind, and industrial process heat.

Resources: Over 5000 books, journals, directories, reports, and educational materials on all aspects of solar and energy conservation; industry activity and state resource files.

Information Services: (free) Answers inquiries and provides on-site use of library; conducts conferences and seminars for consumers, manufacturers, designers, installers, bankers, architects, legislators, and others.

Publications: (free) *NESEC Update* (monthly newsletter). *Barriers and Incentives to Solar Energy Development*, brochures, audiovisuals, and other materials.
Operated for the U.S. Department of Energy.

Southern Solar Energy Center (SSEC)
61 Perimeter Park
Atlanta, Georgia 30341
(404) 458-8765

Serves: Alabama, Arkansas, Delaware, District of Columbia, Florida, Georgia, Kentucky, Louisiana, Maryland, MIssissippi, North Carolina, Oklahoma, Puerto Rico, South Carolina, Tennessee, Texas, Virginia, Virgin Islands, and West Virginia.
Areas of Interest: Solar energy commercialization.
Resources: Technical staff support collection (emphasis on reports and journals; limited collection of general interest items).
Information Services: (free) Answers general inquiries.
Publications: *SSEC News* and selected Department of Energy publications.
Operated for the U.S. Department of Energy.

Western Solar Utilization Network (Western SUN)
Pioneer Park Building
715 SW Morrison
Portland, Oregon 97204
(503) 241-1222

Serves: Alaska, Arizona, California, Colorado, Hawaii, Idaho, Montana, Nevada, New Mexico, Oregon, Utah, U.S. Pacific Territories, Washington, and Wyoming.
Areas of Interest: Commercialization of solar technologies to meet the needs of the Western region.
Information Services: (free) Answers inquiries; conducts seminars and workshops, provides technical assistance to citizens, businesses, industry, and government; other activities. Information services also available through state solar offices within the Western SUN service region (see state sources listed below).
Publications: Under development—to include regional studies of energy demand and market analysis for passive solar and industrial process heat, homeowners solar decision-making guides for several typical Western climate regions, teachers' aids, consumer guide to anemometers, and other publications.
Operated for the U.S. Department of Energy.

National Solar Heating and Cooling Information Center (NSHCIC)
P. O. Box 1607
Rockville, Maryland 20850
(800) 523-2929 (Continental U.S. outside Pennsylvania)
(800) 462-4983 (Pennsylvania only)
(800) 523-4700 (Alaska and Hawaii only)

Areas of Interest: Active, passive, and hybrid solar heating and cooling of residential and commercial buildings; solar hot-water systems.

Resources: Articles, books, journals, reports, proceedings. Access to energy-related computerized files, including the SEIDB data bases.

Information Services: (free) Answers inquiries, makes referrals to other information sources, and provides information on federal solar heating and cooling demonstration programs for residential and commercial buildings.

Publications: (free) Data compilations, bibliographies, fact sheets, source sheets, and pamphlets.

Operated for the U.S. Department of Housing and Urban Development and the U.S. Department of Energy.

STATE SOURCES

Where can you find out what your state is doing to conserve energy and promote the use of solar resources? In most states, these programs are handled by several agencies. On the following pages we list state offices and contacts responsible for solar energy and energy conservation programs.

State energy offices, in most states, are responsible for carrying out state energy policies and coordinating programs.

Energy extension services (EES) are state-level offices, funded and coordinated by the U.S. Department of Energy, designed to actively promote energy conservation and solar energy at the local level. Families, small businesses, community groups, and local governments are now receiving information and technical assistance through energy extension service programs. (Because the EES program was recently broadened to include all states and U.S. territories, activities in some states may still be under development.)

Solar offices, which coordinate solar energy programs, have been created in most states with funding from the regional solar energy centers. On the following pages we list state solar offices that can provide information to the public.

Some states have created solar policy advisory offices and energy research and development centers. These organizations are referenced below, but are described separately because of their diverse functions.

State Sources

State or territory	State energy offices	Energy extension services (EES)	State solar offices and representatives
Alabama	Alabama Energy Management Board 3734 Atlanta Highway Montgomery, Alabama 36130 (205)832-5010 (800) 572-7226 (State Solar Hotline)	Alabama Energy Extension Service 313 Ross Hall Auburn University Auburn, Alabama 36830 (205) 826-4718	See: Alabama Energy Management Board
Alaska	Alaska State Energy Office Division of Power and Energy Mackay Building, 7th Floor 338 Denali Street Anchorage, Alaska 99501 (907) 272-0527	Alaska Energy Extension Service Division of Power and Energy MacKay Building, 7th Floor 338 Denali Street Anchorage, Alaska 99501 (907) 274-8655	Alaska State Solar Representative Division of Power and Energy MacKay Building, 7th Floor 338 Denali Street Anchorage, Alaska 99501 (907) 276-0508
Arizona	Arizona Office of Economic Planning and Development Energy Conservation Office 1700 West Washington, Room 504 Phoenix, Arizona 85007 (602) 255-3303 (800) 352-5499 (State Energy Hotline)	Energy Extension Service Arizona Solar Energy Commission 1700 West Washington, Room 502 Phoenix, Arizona 85007 (602) 255-3682	Arizona State Solar Representative Arizona Solar Energy Commission 1700 West Washington, Suite 502 Phoenix, Arizona 85007 (602) 255-3682
Arkansas	Arkansas State Energy Office Commerce Department 3000 Kavanaugh Little Rock, Arkansas 72205 (501) 371-1379 (800) 482-1122 (State Energy Hotline)	Energy Extension Service Arkansas State Energy Office 3000 Kavanaugh Little Rock, Arkansas 72205 (501) 371-1370 Contact: Paul F. Levy, Director	See: Arkansas State Energy Office

State Sources (*continued*)

State or territory	State energy offices	Energy extension services (EES)	State solar offices and representatives
California Also see: California Office of Appropriate Technology (page 21): Solar CAL Office (page 26)	California Energy Commission 1111 Howe Avenue Sacramento, California 95825 (916) 920-6430 (800) 952-5670 (State Solar Hotline)	California Energy Extension Service Office of Appropriate Technology 1530 Tenth Street Sacramento, California 95814 (916) 445-1803 (916) 322-8901	California State Solar Representative 1111 Howe Avenue Sacramento, California 95825 (916) 920-7623 Contact: Larry Levin or Landon Williams California State Solar Representative 6022 West Pico Boulevard Los Angeles, California 90035 (213) 852-5135
Colorado Also see: Colorado Energy Research Institute (page 22)	Colorado Office of Energy Conservation 1600 Downing Street, 2nd Floor Denver, Colorado 80218 (303) 839-2507 (800) 234-2105 (State Energy Hotline)	Energy Extension Service Colorado Office of Energy Conservation 1600 Downing Street, 2nd Floor Denver, Colorado 80218 (303) 839-2507	Colorado State Solar Representative Colorado Office of Energy Conservation 1600 Downing Street, 2nd Floor Denver, Colorado 80218 (303) 839-2507 (303) 839-2186
Connecticut	Connecticut Office of Policy and Management Energy Division 80 Washington Street Hartford, Connecticut 06115 (203) 566-3394 (800) 842-1648 (State Energy Hotline)	Connecticut Energy Extension Service Office of Policy and Management 80 Washington Street Hartford, Connecticut 06115 (203) 566-5803	Energy Division Connecticut Office of Policy and Management 80 Washington Street Hartford, Connecticut 06115 (203) 566-3394

Delaware	Delaware Energy Office 114 West Water Street P.O. Box 1401 Dover, Delaware 19901 (302) 678-5644 (800) 282-8616 (State Energy Hotline)	Energy Extension Service Delaware Energy Office 114 West Water Street P.O. Box 1401 Dover, Delaware 19901 (302) 678-5644	See: Delaware Energy Office
District of Columbia	District of Columbia Office of Planning and Development 409 District Building 1350 E Street N.W. Washington, D.C. 20004 (202) 727-6365	Energy Extension Service 1420 New York Avenue N.W., 2d Floor Washington, D.C. 20005 (202) 727-1804	See: District of Columbia Office of Planning and Development
Florida Also see: Florida Solar Energy Center (page 22)	Florida Governor's Energy Office 301 Bryant Building Tallahassee, Florida 32301 (904) 488-6764 (800) 432-0575 (State Solar Hotline)	Energy Extension Service Florida Governor's Energy Office 301 Bryant Building Tallahassee, Florida 32301 (904) 488-6143	See: Florida Governor's Energy Office
Georgia	Georgia Office of Energy Resources 270 Washington Street S.W., Suite 615 Atlanta, Georgia 30334 (404) 656-5176	Energy Extension Service Georgia Office of Energy Resources 270 Washington Street S.W., Room 615 Atlanta, Georgia 30334 (404) 656-5176	See: Georgia Office of Energy Resources
Hawaii	Hawaii State Energy Office Department of Planning and Economic Development 1164 Bishop Street, Suite 1515 Honolulu, Hawaii 96813 (808) 548-4150	Energy Extension Service Hawaii State Energy Office Department of Planning and Economic Development 1164 Bishop Street, Suite 1515 Honolulu, Hawaii 96813 (808) 548-4080	Hawaii State Solar Representative Center for Science Policy and Technology Assessment P.O. Box 2359 Honolulu, Hawaii 96804 (808) 548-4195

State Sources (*continued*)

State or territory	State energy offices	Energy extension services (EES)	State solar offices and representatives
Idaho	Idaho Office of Energy State Capitol Building Boise, Idaho 83720 (208) 334-3800 (800) 632-5954 (State Energy Hotline)	Energy Extension Service Idaho Office of Energy State Capitol Building Boise, Idaho 83720 (208) 334-3800	Idaho State Solar Representative State Capitol Building Boise, Idaho 83720 (208) 334-3800 (208) 342-5435
Illinois	Illinois Institute of Natural Resources Divisions of Solar Energy and Conservation 325 West Adams Street, Room 300 Springfield, Illinois 62706 (217) 785-2431 (800) 424-9122 (State Energy Hotline)	Energy Extension Service Illinois Institute of Natural Resources 325 West Adams Street, Room 300 Springfield, Illinois 62706 (217) 785-2800	Illinois Institute of Natural Resources Alternative Energy Division 325 West Adams Street, Room 300 Springfield, Illinois 62706 (217) 785-2431
Indiana	Indiana Energy Office Indiana Department of Commerce 440 North Meridian Street Indianapolis, Indiana 46204 (317) 232-8940	Energy Extension Service Energy Group Indiana Department of Commerce 440 North Meridian Street Indianapolis, Indiana 46204 (317) 232-8940	Energy Group Indiana Department of Commerce 440 North Meridian Street Indianapolis, Indiana 46204 (317) 232-8940
Iowa	Iowa Energy Policy Council Lucas Building, 6th Floor Capitol Complex Des Moines, Iowa 50319 (515) 281-4420 (800) 523-1114 (State Energy Hotline)	Iowa Energy Extension Service 110 Marston Hall Iowa State University Ames, Iowa 50011 (515) 294-4266	Iowa Energy Policy Council Capitol Complex Des Moines, Iowa 50319 (515) 281-8071
Kansas	Kansas Energy Office 214 West 6th Street Topeka, Kansas 66603 (913) 296-2496 (800) 432-3537 (State Energy Hotline)	Kansas Energy Extension Service 214 West 6th Street Topeka, Kansas 66603 (913) 296-2496	Kansas Energy Office 214 West 6th Street Topeka, Kansas 66603 (913) 296-2496

Kentucky	Kentucky Department of Energy Capitol Plaza Tower Frankfort, Kentucky 40601 (502) 564-7416 (800) 372-2978 (State Energy Hotline)	Kentucky Energy Extension Service P.O. Box. 11888 Ironworks Pike Lexington, Kentucky 40578	See: Kentucky Department of Energy
Louisiana	Louisiana Division of Natural Resources Department of Research and Development P.O. Box 44156 Baton Rouge, Louisiana 70804 (504) 342-4592	Energy Extension Service Louisiana State University Cooperative Extension Service Knapp Hall Baton Rouge, Louisiana 70803 (504) 342-4500	See: Louisiana Division of Natural Resources, Department of Research and Development
Maine	Maine Office of Energy Resources 55 Capitol Street Augusta, Maine 04330 (207) 289-3811	Energy Extension Service Maine Office of Energy Resources 55 Capitol Street Augusta, Maine 04330 (207) 289-3811	Maine Office of Energy Resources 55 Capitol Street Augusta, Maine 04330 (207) 289-3811
Maryland	Maryland Energy Policy Office 301 West Preston Street, Suite 1302 Baltimore, Maryland 21201 (301) 383-6810 (800) 492-5903 (State Energy Hotline)	Energy Extension Service Maryland Energy Policy Office 301 West Preston Street, Suite 1302 Baltimore, Maryland 21201 (301) 383-6810	See: Maryland Energy Policy Office
Massachusetts Also see: Massachusetts Solar Action Office (page 23)	Massachusetts Office of Energy Resources 73 Tremont Street, Room 700 Boston, Masschusetts 02108 (617) 727-4732 (800) 922-8265 ("Energyphone")	Energy Extension Service Massachusetts Office of Energy Resources 73 Tremont Street Boston, Massachusetts 02108 (617) 727-5064	Massachusetts Office of Energy Resources 73 Tremont Street, Room 840 Boston, Massachusetts 02108 (617) 727-7297
Michigan	Michigan Energy Administration P.O. Box 30228 Lansing, Michigan 48909 (517) 373-6430 (800) 292-4704 (State Energy Hotline)	Michigan Energy Extension Service 6520 Mercantile Way, Suite 1 Lansing, Michigan 48910 (517) 373-0480	Michigan Energy Administration P.O. Box 30228 Lansing, Michigan 48909 (517) 373-6430

State or territory	State energy offices	Energy extension services (EES)	State solar offices and representatives
Minnesota	Minnesota Energy Agency 980 American Center Building 150 East Kellogg Boulevard St. Paul, Minnesota 55101 (612) 296-5120 (800) 652-9747 (State Energy Hotline)	Energy Extension Service Minnesota Energy Agency 980 American Center Building 150 East Kellogg Boulevard St. Paul, Minnesota 55101 (612) 296-8898	Minnesota Energy Agency 980 American Center Building 150 East Kellogg Boulevard St. Paul, Minnesota 55101 (612) 296-4737
Mississippi	Mississippi Office of Energy 455 North Lamar, Suite 228 Jackson, Mississippi 39201 (601) 981-5099	Mississippi Energy Extension Center Mississippi State University P.O. Box 5406 Mississippi State, Mississippi 39762 (601) 325-3136	See: Mississippi Office of Energy
Missouri	Missouri Division of Energy P.O. Box 176 Jefferson City, Missouri 65102 (314) 751-4000 (800) 392-0717 (State Energy Hotline)	Energy Extension Service Missouri Divison of Energy P.O. Box 176 Jefferson City, Missouri 65102 (314) 751-4000	Missouri Division of Energy P.O. Box 176 Jefferson City, Missouri 65102 (314) 751-4000
Montana	Montana Department of Natural Resources and Conservation Energy Division 32 South Ewing Street Helena, Montana 59601 (406) 449-3940	Energy Extension Service Montana Department of Natural Resources and Conservation Energy Division 32 South Ewing Street Helena, Montana 59601 (406) 449-3780	Montana State Solar Representative Montana Department of Natural Resources and Conservation Energy Division 32 South Ewing Street Helena, Montana 59601 (406) 449-3940
Nebraska	Nebraska Energy Office 301 Centennial Mall Lincoln, Nebraska 68509 (402) 471-2867	Energy Extension Service W-181 Nebraska Hall University of Nebraska Lincoln, Nebraska 68588 (402) 472-3181	University of Nebraska W-191 Nebraska Hall Lincoln, Nebraska 68588 (402) 472-3414

Nevada	Nevada Department of Energy 1050 East William, Suite 405 Carson City, Nevada 89701 (702) 885-5157	Energy Extension Service Nevada Department of Energy 400 West King Carson City, Nevada 89710 (702) 885-5157	Nevada State Solar Representative Energy Conservation and Planning Department of Energy 1050 East William, Suite 405 Carson City, Nevada 89701 (702) 885-5157 Nevada State Solar Representative Desert Research Institute 1500 Buchanan Boulevard Boulder City, Nevada 89005 (702) 293-4217
New Hampshire	New Hampshire Governor's Council on Energy 2½ Beacon Street Concord, New Hampshire 03301 (603) 271-2711 (800) 852-3466 (State Energy Hotline)	Energy Extension Service New Hampshire Governor's Council on Energy 2½ Beacon Street Concord, New Hampshire 03301 (603) 271-2711	New Hampshire Governor's Council on Energy 2½ Beacon Street Concord, New Hampshire 03301 (603) 271-2711
New Jersey	New Jersey Department of Energy Office of Alternate Technology 101 Commerce Street Newark, New Jersey 07102 (201) 648-6293 (800) 492-4242 (State Energy Hotline)	Energy Extension Service New Jersey Department of Energy 101 Commerce Street Newark, New Jersey 07102 (201) 648-3900	Office of Alternate Technology New Jersey Department of Energy 101 Commerce Street Newark, New Jersey 07102 (201) 648-6293
New Mexico Also see: New Mexico Energy Institute (page 24); New Mexico Solar Energy Institute (page 25)	New Mexico Energy and Minerals Department P.O. Box 2770 Santa Fe, New Mexico 87503 (505) 827-2472 (800) 432-6782 (State Energy Hotline)	New Mexico Energy Extension Service P.O. Box 00 Santa Fe, New Mexico 87501 (505) 827-2386	New Mexico State Solar Representative Energy Conservation and Management Division P.O. Box 2770 Santa Fe, New Mexico 87501 (505) 827-5621

State Sources (*continued*)

State or territory	State energy offices	Energy extension services (EES)	State solar offices and representatives
New York	New York State Energy Office Agency Building 2, 8th Floor Empire State Plaza Albany, New York 12223 (518) 474-7016 (800) 342-3722 (State Energy Hotline)	Energy Extension Service New York State Energy Office 2 Rockefeller Plaza Albany, New York 12223 (518) 474-4083	New York State Energy Office Agency Building 2, 8th Floor Albany, New York 12223 (518) 474-7016 (518) 473-8667
North Carolina	North Carolina Department of Commerce Energy Division Dobbs Building Raleigh, North Carolina 27611 (919) 733-2230 (800) 662-7131 (State Energy Hotline)	Energy Extension Service North Carolina Department of Commerce Energy Division Dobbs Building Raleigh, North Carolina 27611 (919) 733-2230	See: North Carolina Department of Commerce, Energy Division
North Dakota	North Dakota Energy Office 1533 North 12th Street Bismarck, North Dakota 58501 (701) 224-2250	Energy Extension Service North Dakota Office of Energy Management and Coordination 1533 North 12th Street Bismarck, North Dakota 58501 (701) 224-2250	North Dakota Energy Office 1533 North 12th Street Bismarck, North Dakota 58501 (701) 224-2250
Ohio	Ohio Department of Energy 30 East Broad Street, 34th Floor Columbus, Ohio 43215 (614) 466-6797 (800) 282-9234 (State Energy Hotline)	Energy Extension Service Ohio Department of Energy 30 East Broad Street, 34th Floor Columbus, Ohio 43215 (614) 466-6747	Ohio Department of Energy 30 East Broad Street, 34th Floor Columbus, Ohio 43215 (614) 466-8277

Oklahoma	Oklahoma Department of Energy 4400 North Lincoln Boulevard, Suite 251 Oklahoma City, Oklahoma 73105 (405) 521-3441	Energy Extension Service Oklahoma Department of Energy 4400 North Lincoln Boulevard, Suite 251 Oklahoma City, Oklahoma 73105 (405) 521-2995 (405) 521-3941	See: Oklahoma Department of Energy
Oregon	Oregon Department of Energy Labor and Industries Building Salem, Oregon 97310 (503) 378-4128 (800) 452-7813 ("Access 800" State Govt. Hotline)	Energy Extension Service 114 Dearborn Hall Oregon State University Corvallis, Oregon 97331 (503) 754-3004	Oregon State Solar Representative Oregon Department of Energy Labor and Industries Building Salem, Oregon 97310
Pennsylvania	Pennsylvania Governor's Energy Council 1625 North Front Street Harrisburg, Pennsylvania 17102 (717) 783-8610 (800) 882-8400 (State Energy Conservation Hotline)	Energy Extension Service Pennsylvania Governor's Energy Council 1625 North Front Street Harrisburg, Pennsylvania 17102 (717) 783-8610	Pennsylvania Governor's Energy Council 1625 North Front Street Harrisburg, Pennsylvania 17102 (717) 783-8610
Puerto Rico	Puerto Rico Office of Energy Minillas Governmental Center North Building Office, Postal Stop 22 P.O. Box 41089, Minillas Station Santurce, Puerto Rico 00940 (809) 726-3636	Energy Extension Service Puerto Rico Office of Energy Minillas Governmental Center North Building Office, Postal Stop 22 P.O. Box 41089, Minillas Station Santurce, Puerto Rico 00940 (809) 726-0196 (809) 726-5055	See: Puerto Rico Office of Energy
Rhode Island	Rhode Island Governor's Energy Office 80 Dean Street Providence, Rhode Island 02903 (401) 277-3773 (will accept collect calls from state residents)	Energy Extension Service Rhode Island Governor's Energy Office 80 Dean Street Providence, Rhode Island 02903 (401) 277-3370	Energy Capability and Management Rhode Island Governor's Energy Office 80 Dean Street Providence, Rhode Island 02903 (401) 277-3374

State Sources (*continued*)

State or territory	State energy offices	Energy extension services (EES)	State solar offices and representatives
South Carolina	South Carolina Department of Energy Resources 1122 Lady Street, Suite 1120 Columbia, South Carolina 29201	Energy Extension Service Project South Carolina State Board for Technical and Comprehensive Education 1429 Senate Street Columbia, South Carolina 29201 (803) 758-5794	See: South Carolina Department of Energy Resources
South Dakota	South Dakota Office of Energy Policy Capital Lake Plaza Pierre, South Dakota 57501 (605) 773-3604 (800) 592-1865 ("Tie Line" State Govt. Hotline)	Energy Extension Service South Dakota Office of Energy Policy Capital Lake Plaza Pierre, South Dakota 57501 (605) 773-3603	South Dakota State Energy Office Capitol Lake Plaza Pierre, South Dakota 57501 (605) 773-3603
Tennessee	Tennessee Energy Office 226 Capital Boulevard, Suite 707 Nashville, Tennessee 37219 (615) 741-2994 (800) 342-1340 (State Energy Hotline)	Tennessee Energy Extension Service 226 Capitol Boulevard, Suite 615 Nashville, Tennessee 37219 (615) 741-6677	See: Tennessee Energy Office
Texas	Texas Energy and Natural Resources Advisory Council 411 West 13th Street Austin, Texas 78701 (512) 475-5588	Energy Extension Service Texas Energy and Natural Resources Advisory Council Executive Building 411 West 13th Street, Suite 804 Austin, Texas 78701 (512) 475-5407	See: Texas Energy and Natural Resources Advisory Council

Utah	Utah Energy Office Empire Building, Suite 101 231 East 400 South Salt Lake City, Utah 84111 (801) 533-5424 (800) 662-3633 (State Energy Hotline)	Energy Extension Service Utah Energy Office Empire Building, Suite 101 231 East 400 South Salt Lake City, Utah 84111 (801) 533-5424	Utah State Solar Representative Utah Energy Office Empire Building, Suite 101 231 East 400 South Salt Lake City, Utah 84111 (801) 533-5424
Vermont	Vermont Energy Office State Office Building Montpelier, Vermont 05602 (802) 828-2393 (800) 642-3281 (State Energy Hotline)	Energy Extension Service State Office Building Montpelier, Vermont 05602 (802) 828-2393	Vermont State Energy Office State Office Building Montpelier, Vermont 05602 (802) 828-2393
Virginia	Virginia State Office of Emergency and Energy Services Energy Division 310 Turner Road Richmond, Virginia 23225 (804) 745-3245 (800) 552-3831 (State Energy Hotline)	Energy Extension Service Virginia State Office of Emergency and Energy Services Energy Division 310 Turner Road Richmond, Virginia 23225 (804) 745-3245	See: Virginia State Office of Emergency and Energy Services, Energy Division
Virgin Islands	Virgin Islands Energy Office P.O. Box 90 St. Thomas, Virgin Islands 00801 (809) 774-0750	Energy Study College of the Virgin Islands St. Thomas, U.S. Virgin Islands 00801 (809) 774-1251 ext. 249	See: Virgin Islands Energy Office
Washington	Washington State Energy Office 400 East Union, 1st Floor Olympia, Washington 98504 (206) 753-2417	Washington Energy Extension Service Cooperative Extension Service AG-Phase II Washington State University Pullman, Washington 99164 (509) 335-2511	Washington State Solar Representative SMT-F515 University of Washington Seattle, Washington 98122 (206) 543-1249 Washington State Solar Representative P.O. Box 295 Winthrop, Washington 98862 (509) 996-2451

State Sources (*continued*)

State or territory	State energy offices	Energy extension services (EES)	State solar offices and representatives
West Virginia	West Virginia Fuel and Energy Office 1262½ Greenbrier Street Charleston, West Virginia 25311 (304) 348-8860 (800) 642-9012 (State Energy Hotline)	Energy Extension Service West Virginia Fuel and Energy Office 1262½ Greenbrier Street Charleston, West Virginia 25311 (304) 348-8860	See: West Virginia Fuel and Energy Office
Wisconsin	Wisconsin Office of State Planning and Energy One West Wilson Street, Room 201 Madison, Wisconsin 53702 (608) 266-8234	Wisconsin Energy Extension Service University of Wisconsin Extension 432 North Lake Street, Room 435 Madison, Wisconsin 53706 (608) 263-1662 (608) 263-7950	Wisconsin Solar Energy Office One West Wilson Street, Room 201 Madison, Wisconsin 53702 (608) 266-9861
Wyoming	Wyoming Energy Conservation Office 320 West 25th Street Cheyenne, Wyoming 82002 (307) 777-7131 (800) 442-8334 (State Solar Referral Hotline) (800) 442-2744 (State Govt. Hotline)	Wyoming Energy Extension Service University Station Box 3295 Laramie, Wyoming 82071 (307) 766-3362 (800) 442-6783 (State Energy Extension Service Hotline)	Wyoming State Solar Representative Rocky Mountain Institute of Energy and Environment P.O. Box 3965 Laramie, Wyoming 82071 (307) 766-6760

REGIONAL DEPOSITORIES OF U.S. DOCUMENTS

Do you need to find a report that was produced for the federal government? Many public and university libraries in major cities and educational centers maintain extensive collections of government documents available for your use. There are two principal depository systems that include solar energy documents.

Federal (GPO) depository library system. Created by Congress in 1895, this system is intended to offer free public access to U.S. documents. Each depository library has one printed copy of every unclassified federal publication, filed under the GPO classification system.

U.S. Department of Energy report collections. Many libraries purchase microfiche collections of all U.S. Department of Energy reports. In addition to microfiche reading equipment, most of these libraries have reader-printers or photocopy centers that reproduce enlarged printed copies on demand (charges for these services vary). Reports can be found by using the *Energy Research Abstracts Index*. The reference librarian can help you locate specific documents if you are unfamiliar with the report cataloguing system.

Regional Depositories

State	Library	GPO depository	Department of energy report collection
Alabama	Auburn University Library Auburn, Alabama		•
	Johnson Environment and Energy Center		•
	University of Alabama Huntsville, Alabama		
	Auburn University at Montgomery Library Montgomery, Alabama	•	
	University of Alabama Library Tuscaloosa, Alabama	•	
	Tuskegee Institute Library Tuskegee, Alabama		•
Arizona	University of Arizona Library Tucson, Arizona	•	•
California	University of California Library Davis, California		•
	University of California at Los Angeles Library Los Angeles, California		•

Regional Depositories (*continued*)

State	Library	GPO depository	Department of energy report collection
California	California State Library Sacramento, California	•	
	University of California at Santa Barbara Library Santa Barbara, California		•
Colorado	Colorado State Library University of Colorado Boulder, Colorado	•	
Connecticut	Connecticut State Library Hartford, Connecticut	•	
District of Columbia	Library of Congress Washington, D.C.		•
Florida	University of Florida Library Gainesville, Florida	•	•
Georgia	University of Georgia Library Athens, Georgia	•	
	Georgia Institute of Technology Library Atlanta, Georgia		•
Hawaii	University of Hawaii Library Honolulu, Hawaii		•
	University of Hawaii Library Laie, Hawaii	•	
Idaho	University of Idaho Library Moscow, Idaho	•	
	Idaho State University Library Pocatello, Idaho		•
Illinois	Illinois State Library Springfield, Illinois	•	
	University of Illinois Library Urbana, Illinois		•
Indiana	Indiana State Library Indianapolis, Indiana	•	
	Purdue University Library Lafayette, Indiana		•

Regional Depositories (*continued*)

State	Library	GPO depository	Department of energy report collection
Iowa	Iowa State University Library Ames, Iowa		•
	University of Iowa Library Iowa City, Iowa	•	
Kansas	Watson Library University of Kansas Lawrence, Kansas	•	
	Kansas State University Library Manhattan, Kansas		•
Kentucky	Margaret I. King Library University of Kentucky Lexington, Kentucky	•	•
Louisiana	Louisiana State University Library Baton Rouge, Louisiana	•	
	Louisiana Technical University Library Ruston, Louisiana	•	
Maine	Raymond H. Fogler Library University of Maine Orono, Maine	•	
Maryland	Johns Hopkins University Library Baltimore, Maryland		•
	McKeldin Library University of Maryland College Park, Maryland	•	•
Massachusetts	Boston Public Library Boston, Massachusetts	•	
	Massachusetts Institute of Technology Library Cambridge, Massachusetts		•
	Worcester Polytechnic Institute Library Worcester, Massachusetts		•
Michigan	University of Michigan Library Ann Arbor, Michigan		•
	Detroit Public Library Detroit, Michigan	•	

Regional Depositories (*continued*)

State	Library	GPO depository	Department of energy report collection
Michigan	Michigan State Library Lansing, Michigan	•	
Minnesota	Wilson Library University of Minnesota Moorhead, Minnesota	•	
Mississippi	University of Mississippi Library Oxford, Mississippi	•	
	Mississippi State University Library State College, Mississippi		•
Missouri	University of Missouri Columbia, Missouri		•
	Linda Hall Library Kansas City, Missouri		•
Montana	University of Montana Library Missoula, Montana	•	
Nebraska	Nebraska Publications Clearinghouse Nebraska Library Commission Lincoln, Nebraska	•	
Nevada	University of Nevada Library Reno, Nevada	•	
New Jersey	Newark Public Library Newark, New Jersey	•	
	Princeton University Library Princeton, New Jersey		•
New Mexico	University of New Mexico Library Albuquerque, New Mexico		•
	Zimmerman Library University of New Mexico Hobbs, New Mexico	•	
New York	New York State Library Albany, New York	•	•
	State University of New York Library Albany, New York		•
	Cornell University Library Ithaca, New York		•

Regional Depositories (*continued*)

State	Library	GPO depository	Department of energy report collection
	Columbia University Library New York, New York		•
	University of Rochester Library Rochester, New York		•
	Syracuse University Library Syracuse, New York		•
North Carolina	University of North Carolina Library Chapel Hill, North Carolina	•	
	North Carolina State University Library Raleigh, North Carolina		•
North Dakota	North Dakota State University Library Fargo, North Dakota	•	
Ohio	University of Cincinnati Library Cincinnati, Ohio		•
	Cleveland Public Library Cleveland, Ohio		•
	Ohio State Library Columbus, Ohio	•	
	University of Toledo Library Toledo, Ohio		•
Oklahoma	University of Oklahoma Library Norman, Oklahoma		•
	Oklahoma Department of Libraries Oklahoma City, Oklahoma	•	
Oregon	Portland State University Library Portland, Oregon	•	
Pennsylvania	State Library of Pennsylvania Harrisburg, Pennsylvania	•	
	Carnegie Library Pittsburgh, Pennsylvania		•
	Pennsylvania State University Library University Park, Pennsylvania		•
Puerto Rico	University of Puerto Rico Library San Juan, Puerto Rico		•

Regional Depositories (*continued*)

State	Library	GPO depository	Department of energy report collection
South Carolina	University of South Carolina Library Columbia, South Carolina		•
Texas	Texas State Library Austin, Texas	•	
	University of Texas Library Austin, Texas		•
	Texas A & M University Library College Station, Texas		•
	Rice University Library Houston, Texas		•
	Texas Tech University Library Lubbock, Texas	•	
Utah	Merrill Library and Learning Resources Center Utah State University Logan, Utah	•	
	University of Utah Library Salt Lake City, Utah		•
Virginia	Alderman Library University of Virginia Charlottesville, Virginia	•	•
Washington	Washington State Library Olympia, Washington	•	
	Washington State University Library Pullman, Washington		•
	University of Washington Library Seattle, Washington		•
West Virginia	West Virginia University Library Morgantown, West Virginia	•	•
Wisconsin	University of Wisconsin Library Madison, Wisconsin		•
	Milwaukee Public Library Milwaukee, Wisconsin	•	

Appendix 12 Energy Conservation Checklist

A CHECKLIST OF ENERGY CONSERVATION OPPORTUNITIES, RANKED IN PRIORITY ACCORDING TO CLIMATIC CONDITIONS

This checklist of energy-saving opportunities includes some items that subsume others. Some seem to border on the obvious, yet many contemporary buildings are testimony to the need for even seemingly obvious measures.

The items are ranked in priority and coded to the following climatic features: For winter, A indicates a heating season of 6000 degree-days or more, B a heating season of 4000 to 6000 degree-days, and C a heating season of 4000 degree-days or less. The numeral 1 following these letters indicates sun 60 percent of daylight time or more and wind 9 miles per hour or more, 2 indicates the sun condition but not the wind condition, 3 indicates the wind condition without the sun condition, and 4 the absence of either condition.

For summer, the letter D indicates a cooling season or more than 1500 hours at 80°F, E 600 to 1500 hours at the same temperature, and F less than 600 hours. The numeral 1 indicates a dry climate of 60 percent relative humidity or less and 2 indicates 60 percent or more humidity.

Guidelines that are independent of climate are not rated in priority columns and are marked with asterisks.

Reproduced from *AIA Journal*, courtesy of Dubin-Bloome Associates.

Site

	Priority 1	Priority 2	Priority 3	N/A
1. Use deciduous trees for summer sun shading and wind break for buildings up to three stories.	A1	A2	A4	C4
	D1	D2	E1	F
2. Use conifer trees for summer and winter sun shading and wind breaks	C4	C1	C2	A2
	D1	D2	E1	F
3. Cover exterior walls and/or roof with earth and planting to reduce heat transmission and solar gain	A1	A2	A4	C4
	D1	D2	E1	F
4. Shade walls and paved areas adjacent to building to reduce indoor-outdoor temperature differential.	C2	C1	C3	A2
	D1	D2	E1	F
5. Reduce paved areas and use grass or other vegetation to reduce outdoor temperature buildup.	C2	C1	C3	A2
	D1	D2	E1	F
6. Use ponds, water fountains, to reduce ambient outdoor air temperature around building.	C2	C1	C3	A4
	D1	E1	D2	F
7. Collect rainwater for use in building.	*			
8. Locate building on site to induce airflow effects for natural ventilation and cooling.	C2	C1	C3	A4
	F	E1	E2	D2
9. Locate buildings to minimize wind effects on exterior surfaces.	A4	A1	B4	C2
	F	E2	E1	D1
10. Select site with high air quality (least contaminated) to enhance natural ventilation.	C2	C1	C3	A4
	F	E1	E2	D2
11. Select a site with year-round ambient wet- and dry-bulb temperatures close to and somewhat lower than those desired within the occupied spaces.	*			
12. Select a site that has topographic features and adjacent structures that provide breaks.	A4	A1	B4	C3
	F	E2	E1	D2
13. Select a site that has topographic features and adjacent structures that provide desirable shading.	C2	C1	B2	A1
	D2	D1	E2	F
14. Select a site that allows optimum orientation and configuration to minimize yearly energy consumption.	*			
15. Select a site to reduce specular heat reflection from water.	C2	C1	B2	A4
	D2	D1	E2	F
16. Use sloping site to bury building partially or use earth berms to reduce heat transmission and solar radiation.	A4	A1	A3	C2
	D1	D2	E1	F
17. Select a site that allows occupants to use public transportation systems.	*			

Building

	Priority 1	Priority 2	Priority 3	N/A
1. Construct building with minimum exposed surface area to minimize heat transmission for a given enclosed volume.	A4	A1	A3	C2
	D1	D2	E1	F
2. Select building configuration to give minimum north wall to reduce heat losses.	A4	A1	A3	C2
3. Select building configuration to give minimum south wall to reduce cooling load.	D1	D2	E1	F
4. Use building configuration and wall arrangement (horizontal and vertical sloping walls) to provide self-shading and windbreaks.	A4	A1	B4	C3
	D1	D2	E1	
5. Locate insulation for walls, roofs, and floors over garages at the exterior surface.	A4	A3	A1	C2
	D1	D2	E1	F
6. Construct exterior walls, roof, and floors with high thermal mass with a goal of 100 pounds per cubic foot.	A4	A1	A3	C3
	D1	D2	E1	F

	Priority			
	1	2	3	N/A
7. Select insulation to give a composite U factor from 0.06 when outdoor winter design temperatures are less than 10°F to 0.15 when outdoor design conditions are above 40°F.				
8. Select U factors from 0.06 where sol-air temperatures are above 144°F up to a U value of 0.3 with sol-air temperatures below 85°F.				
9. Provide vapor barrier on the interior surface of exterior walls and roof of sufficient impermeability to provide condensation.	*			
10. Use concrete slab on grade for ground floors.	A4	A1	A3	C2
	D1	D2	E1	F
11. Avoid cracks and joints in building construction to reduce infiltration.	A4	A1	A3	
	D2	E2	D1	
12. Avoid thermal bridges through exterior surfaces.	A4	A1	A3	C3
	D2	D1	E2	F
13. Provide textured finish to external surfaces to increase film coefficient.	A4	A1	B4	C2
14. Provide solar control for walls and roof in the areas where similar solar control is desirable for glazing.	D2	D1	E2	A
15. Consider length and width aspects for rectangular buildings as well as other geometric forms in relation to building height and interior and exterior floor areas to optimize energy conservation.	A4	A1	A3	C2
	D1	D2	E1	F
16. To minimize heat gain in summer due to solar radiation, finish walls and roofs with a light-colored surface having a high emissivity.	D1	D2	E1	
17. To increase heat gain due to solar radiation on walls and roofs, use a dark-colored finish having a high absorptivity.	A1	A2	A4	C2
18. Reduce heat transmission through roof by one or more of the following:				
a. Insulation.	A4	A1	A3	C3
	D1	D2	E1	F
b. Reflective surfaces.	C2	C1	C3	A4
	D1	D2	E1	
c. Roof spray.	D1	E1	F	
d. Roof pond.	D1	E1	F	
e. Sod and planting.	A4	A1	A3	C2
	D1	D2	E1	F
f. Equipment and equipment rooms located on roof.	A4	A1	A3	
	D1	D2	E1	
g. Double roof with ventilated space between.	D1	D2	E1	F
19. Increase roof heat gain when reduction of heat loss in winter exceeds heat gain increase in summer:				
a. Use dark-colored surfaces.	A2	A1	B2	B1
b. Avoid shadows.	A2	A1	B2	B1
20. Insulate slab on grade with both vertical and horizontal perimeter insulation under slab.	A	B	C	
21. Reduce infiltration quantities by one or more of the following:				
a. Reduce building height.				
b. Use impermeable exterior surface materials.	A4	A1	A3	C4
c. Reduce crackage area around doors, windows, etc.	D2	E2	D1	F

	Priority			
	1	2	3	N/A
d. Provide all external doors with weather stripping.				
e. Where operable windows are used, provide them with sealing gaskets and cam latches.				
f. Locate building entrances on downwind side and provide wind break.				
g. Provide all entrances with vestibules; where vestibules are not used, provide revolving doors.				
h. Provide vestibules with self-closing weather-stripped doors to isolate them from the stairwells and elevator shafts.	A4 D2	A1 E2	A3 D1	C4 F
i. Seal all vertical shafts.				
j. Locate ventilation louvers on downwind side of building and provide wind breaks.				
k. Provide break at intermediate points of elevator shafts and stairwells for tall buildings.				
22. Provide wind protection by using fins, recesses, etc., for any exposed surface having a U value greater than 0.5.	A4	A1	B4	C2
23. Do not heat parking garages.	*			
24. Consider the amount of energy required for the protection of materials and their transport on a life-cycle energy usage basis.	*			
25. Consider use of the insulation type that can be most efficiently applied to optimize the thermal resistance of the wall or roof; e.g., some types of insulation are difficult to install without voids or shrinkage.	*			
26. Protect insulation from moisture originating outdoors, since volume decreases when wet. Use insulation with low water absorption and one that dries out quickly and regains its original thermal performance after being wet.	*			
27. Where sloping roofs are used, face them south for greatest heat gain benefit in the winter.	A1	A2	B1	C4
28. To reduce heat loss from windows, consider one or more of the following:				
a. Use minimum ratio of window area to wall area.				
b. Use double glazing.				
c. Use triple glazing.				
d. Use double reflective glazing.	A4	B4	C4	
e. Use minimum percentage of double glazing on north wall.	A1 A2	B1 B2	C1 C2	
f. Manipulate east and west walls so that windows face south.	A3	B3	C3	
g. Allow direct sun on windows November through March.				
h. Avoid window frames that form a thermal bridge.				
i. Use operable thermal shutters that decrease the composite U value to 0.1.				
29. To reduce heat gains through windows, consider the following:				
a. Use minimum ratio of window area to wall area.				
b. Use double glazing.				
c. Use triple glazing.	D1	E1	F	
d. Use double reflective glazing.	D2	E2	F	
e. Use minimum percentage of double glazing on the south wall.				

	Priority 1	2	3	N/A
f. Shade windows from direct sun April through October.				
30. To take advantage of natural daylight within the building and reduce electrical energy consumption, consider the following:				
a. Increase window size but do not exceed the point where yearly energy consumption, due to heat gains and losses, exceeds the saving from use of natural light.				
b. Locate windows high in wall to increase reflection from ceiling, but reduce glare effect on occupants.	C2 C1	B2 B1	A2 A2	
c. Control glare with translucent draperies operated by photocells.	C3 C4	B3 B4	A3 A4	
d. Provide exterior shades that eliminate direct sunlight but reflect light into occupied spaces.	F	E	D	
e. Slope vertical wall surfaces so that windows are self-shading and walls below act as light reflectors.				
f. Use clear glazing. Reflective or heat-absorbing films reduce the quantity of natural light transmitted through the window.				
31. To allow use of natural light in cold zones where heat losses are high energy users, consider operable thermal barriers.	A4	A1	B4	C3
32. Use permanently sealed windows to reduce infiltration in climatic zones where this is a large energy user.	A1 D1	A4 D2	B1 E1	C3 F
33. Where codes of regulations require operable windows and infiltration is undesirable, use windows that close against a sealing gasket.	A1 D1	A4 D2	B1 E1	C3 F
34. In climatic zones where outdoor air conditions are suitable for natural ventilation for a major part of the year, provide operable windows.	C2 F	C3 E1	C1 E2	A4 D2
35. In climate zones where outdoor air conditions are close to desired indoor conditions for a major portion of the year, consider the following:				
a. Adjust building orientation and configuration to take advantage of prevailing winds.				
b. Use operable windows to control ingress and egress of air through the building.				
c. Adjust the configuration of the building to allow natural cross-ventilation through occupied spaces.	F	E1	E2	D2
d. Use stack effect in vertical shafts, stairwells, etc., to promote natural airflow through the building.				

Planning

	1	2	3	N/A
1. Group services rooms as a buffer and locate at north wall to reduce heat loss or at south wall to reduce heat gain, whichever is the greater yearly energy user.	A4 D1	A1 D2	A3 E1	C2 F
2. Use corridors as heat transfer buffers and locate against external walls.	A4 D1	A1 D2	A3 E1	C2 F
3. Locate rooms with high process-heat gain (computer rooms) against outside surfaces that have the highest exposure loss.	A4	A1	A3	C2
4. Landscaped open planning allows excess heat from interior spaces to transfer to perimeter spaces that have heat loss.	*			

	Priority 1	2	3	N/A
5. Group rooms such that the same ventilating air can be used more than once, by operating in cascade through spaces in decreasing order of priority, i.e., office-corridor-toilet.	*			
6. Reduced ceiling heights reduce the exposed surface area and the enclosed volume. They also increase illumination effectiveness.	*			
7. Increased density of occupants (less gross floor area per person) reduces the overall size of the building and yearly energy consumption per capita.	*			
8. Spaces of similar function located adjacent to each other on the same floor reduce elevator use.	*			
9. Offices frequented by the general public located on the ground floor reduce elevator use.	*			
10. Equipment rooms located on the roof reduce unwanted heat gain and heat loss through the surface. They can also allow more direct duct and pipe runs, reducing power requirements.	A4 D1	A3 D2	B4 E1	C2
11. Windows planned to make beneficial use of winter sunshine should be positioned to allow occupants to move out of direct solar radiation.	*			
12. Deep ceiling voids allow use of larger duct sizes with low pressure drop and reduce HVAC requirements.	*			
13. Processes that have temperature and humidity requirements different from normal physiological needs should be grouped together and served by one common system.	*			
14. Open planning allows more effective use of lighting fixtures. The reduced area of partitioned walls decreases light absorption.	*			
15. Judicious use of reflective surfaces such as sloping white ceilings can enhance the effect of natural lighting and increase the energy saved yearly.	*			

Ventilation and Infiltration

	1	2	3	N/A
1. To minimize infiltration, balance mechanical ventilation so that supply air quantity equals or exceeds exhaust air quantity.	A D	B E	C F	
2. Take credit for infiltration as part of the outdoor air requirements for the building occupants and reduce mechanical ventilation accordingly.	A D	B E	C F	
3. Reduce outdoor air requirements [CFM (cubic feet per minute) per occupant] to the minimum, considering the task they are performing, room volume, and periods of occupancy.	A D	B E	C F	
4. If odor removal requires more than 2000 ft^3/min exhaust and a corresponding introduction of outdoor air, consider recirculating through activated carbon filter.	C D	B E	C F	
5. Where outdoor conditions are close to but less than indoor conditions for major periods of the year and the air is clean and free from offensive odors, consider use of natural ventilation when yearly energy trade-offs with other systems are favorable.	C2 F	C1 E1	C3 E2	A2 D2
6. Exchange heat between outdoor air, intake, and exhaust air by using heat pipes, thermal wheels, runaround systems, etc.	A D	B E	C F	

	Priority			
	1	2	3	N/A
7. In areas subjected to high humidities, consider latent heat exchange in addition to sensible heat exchange.	D2	E2	F	
8. Provide selective ventilation as needed; i.e., 5 ft^3/min per occupant for general areas and increased volumes for areas of heavy smoking or odor control.	*			
9. Transfer air from "clean" areas to more contaminated areas (toilet rooms, heavy smoking areas) rather than supplying fresh air to all areas regardless of function.	*			
10. Provide controls to shut down all air systems at night and weekends except when used for economizer cycle cooling.	A	B	C	
11. Reduce the energy required to heat or cool ventilation air from outdoor conditions to interior design conditions by considering the following:				
a. Reduce indoor air temperature setting in winter and increase in summer.	A	B	C	
	D	E	F	
b. Provide outdoor air direct to perimeter of exhaust hoods in kitchens, laboratories, etc. Do not cool this air in summer or heat over 50°F in winter.	A	B	C	
	D	E	F	

Heating, Ventilation, and Air Conditioning

	1	2	3	N/A
1. Use outdoor air for sensible cooling whenever conditions permit and when recaptured heat cannot be stored.	*			
2. Use adiabatic saturation to reduce temperature of hot, dry air to extend the period of time when "free cooling" can be used.	D1	E1	F	
3. In the summer when the outdoor air temperature at night is lower than the indoor temperature, use full outdoor air ventilation to remove excess heat and precool structure.	D1	E1	F	
4. In principle, select the air-handling system that operates at the lowest possible air velocity and static pressure. Consider high-pressure systems only when other trade-offs such as reduced building size are an overriding factor.	*			
5. To enhance the possibility of using waste heat from other systems, design air-handling systems to circulate sufficient air to enable cooling loads to be met by a 60°F air supply temperature and heating loads to be met by a 90°F air temperature.	*			
6. Design HVAC systems so that the maximum proportion of heat gain to a space can be treated as an equipment load, not as a room load.	*			
7. Schedule air delivery so that exhaust from primary spaces (offices) can be used to heat or cool secondary spaces (corridors).	*			
8. Exhaust air from center zone through lighting fixtures and use this warmed exhaust air to heat perimeter zones.	*			
9. Design HVAC systems so that they do not heat and cool air simultaneously.	*			
10. To reduce fan horsepower, consider the following:				
a. Design duct systems for low pressure loss.	*			
b. Use high-efficiency fans.	*			

	Priority			
	1	2	3	N/A
c. Use low-pressure-loss filters concomitant with contaminant removal.	*			
d. Use one common air coil for heating and cooling.	*			
11. Reduce or eliminate air leakage from duct work.	*			
12. Limit use of reheat to a maximum of 10 percent of gross floor area and then consider its use only for areas that have atypical fluctuating internal loads, such as conference rooms.	*			
13. Design chilled water systems to operate with as high a supply temperature as possible—suggested goal: 50°F. (This allows higher suction temperatures at the chiller with increased operating efficiency.)	*			
14. Use modular pumps to give varying flows that can match varying loads.	*			
15. Select high-efficiency pumps that match load. Do not oversize.	*			
16. Design piping systems for low pressure loss and select routes and locate equipment to give shortest pipe runs.	*			
17. Adopt as large a temperature differential as possible for chilled water systems and water-heating systems.	*			
18. Consider operating chillers in series to increase efficiency.	*			
19. Select chillers that can operate over a wide range of condensing temperatures and then consider the following:				
a. Use double-bundle condensers to capture waste heat at high condensing temperatures and use directly for heating or store for later use.	*			
b. When waste heat cannot be used directly or stored, operate chiller at lowest condensing temperature compatible with ambient outdoor conditions.	*			
20. Consider chilled water storage systems to allow chillers to operate at night when condensing temperatures are lowest.	*			
21. Consider use of double-bundle evaporators so that chillers can be used as heat pumps to upgrade stored heat for use in unoccupied periods.	*			
22. Consider use of gas or diesel engine drive for chillers and large items of ancillary equipment and collect and use waste heat for absorption cooling, heating, and/or domestic hot water.	*			
23. Locate cooling towers or evaporative coolers so that induced air movement can be used to provide or supplement garage exhaust ventilation.	*			
24. Use modular boilers for heating and select units so that each module operates at optimum efficiency.	*			
25. Extract waste heat from boiler flue gas by extending surface coils or heat pipes.	*			
26. Select boilers that operate at the lowest practicable supply temperature while avoiding condensation within the furnaces.	*			
27. Use unitary water/air heat pumps that transport heat energy from zone to zone via a common hydronic loop.	*			
28. Consider use of thermal storage in combination with unit heat pumps and a hydronic loop so that excess heat during the day can be captured and stored for use at night.	*			

	Priority			
	1	2	3	N/A
29. Consider use of both water/air and air/air heat pumps if a continuing source of low-grade heat, such as a lake or river is near the building.	*			
30. Consider direct use of solar energy via a system of collectors for heating in winter and absorption cooling in summer.	*			
31. Minimize requirements for snow melting to those that are absolutely necessary and, where possible, use waste heat for this service.	*			
32. Provide all outside air dampers with accurate position indicators and ensure that dampers are airtight when closed.	*			
33. If electric heating is contemplated, consider use of heat pumps in place of direct resistance heating; by comparison, they consume one-third of the energy per unit output.	*			
34. Consider use of spot heating and/or cooling in spaces having large volume and low occupancy.	*			
35. Use electric ignition in place of gas pilots for gas burners.	*			
36. Consider use of a total energy system if the life-cycle costs are favorable.	*			
Lighting and Power				
1. a. Use natural illumination in areas where effective when a net energy conservation gain is possible vis-à-vis heating and cooling loads.	C F	B E	A D	
b. Provide exterior reflectors at windows for more effective internal illumination.	*			
2. Consider a selective lighting system in regard to the following:				
a. Reduce the wattage required for each specific task by review of user needs and method of providing illumination.				
b. Consider only the amount of illumination required for the specific task, considering the duration and character and user performance required as per design criteria.	*			
c. Group similar tasks together for optimum conservation of energy per floor.	*			
d. Design switch circuits to permit turning off unused and unnecessary light.	*			
e. Illuminate tasks with fixtures built into furniture and maintain low-intensity lighting elsewhere.	*			
f. Consider use of polarized lenses to improve quality of lighting at tasks.	*			
g. Provide timers to automatically turn off lights in remote or little-used areas.	*			
h. Use multilevel ballasts to permit varying the lumen output for fixtures by adding or removing lamps when tasks are changed in location or requirements.	*			
i. Arrange electrical systems to accommodate relocatable luminaires that can be removed to suit changing furniture layouts.	*			
j. Consider use of ballasts that can accommodate sodium	*			

	Priority 1	2	3	N/A
metal-halide bulbs interchangeably with other lamps.				
3. Consider use of high-frequency lighting to reduce wattage per lumen output. Additional benefits are reduced ballast heat loss into the room and longer lamp life.	*			
4. Consider use of landscape office planning to improve lighting efficiency. Approximately 25 percent less wattage per foot-candle required for open planning versus partitions.	*			
5. Consider use of light colors for walls, floors, and ceilings to increase reflectance, but avoid specular reflection.	*			
6. Lower ceilings or mounting height of luminaires to increase level of illumination with less wattage.	*			
7. Consider dry heat-of-light systems to improve lamp performance and reduce heat gain to space.	*			
8. Consider wet heat-of-light system to improve lamp performance and reduce heat gain to space and refrigeration load.	*			
9. Use fixtures that give high-contrast rendition factor at task.	*			
10. Provide suggestions to GSA for analysis of tasks to increase use of high-contrast material that requires less illumination.	*			
11. Select furniture and interior appointments that do not have glossy surfaces or give specular reflections.	*			
12. Use light spills from characteristic areas to illuminate noncharacteristic areas.	*			
13. Consider use of greater contrast between tasks and background lighting, such as 8 to 1 and 10 to 1.	*			
14. Consider washers and special illumination for features such as plants, murals, etc., in place of overhead space lighting to maintain proper contrast ratios.	*			
15. For horizontal tasks or duties, consider fixtures whose main light component is oblique and then locate for maximum equivalent sphere illumination footcandles on task.	*			
16. Consider using 250-W mercury vapor lamps and metal-halide lamps in place of 500-W incandescent lamps for special applications.	*			
17. Use lamps with higher lumens per watt input, such as:				
a. One 8-ft fluorescent lamp versus two 4-ft lamps.	*			
b. One 4-ft fluorescent lamp versus two 2-ft lamps.	*			
c. U-tube lamps versus two individual lamps.	*			
d. Fluorescent lamps in place of all incandescent lamps except for very close task lighting, such as at a typewriter paper holder.	*			
18. Use high utilization and maintenance factors in design calculations and instruct users to keep fixtures clean and change lamps earlier.	*			
19. Avoid decorative flood-lighting and display lighting.	*			
20. Direct exterior security lighting at entrances and avoid illuminating large areas adjacent to building.	*			
21. Consider switches activated by intruder devices rather than permanently lit security lighting.	*			
22. If already available, use street lighting for security purposes.				

	Priority			
	1	2	3	N/A
23. Reduce lighting requirements for hazards by:				
a. Using light fixtures close to and focused on hazard.	*			
b. Increasing contrast of hazard; i.e., paint stair treads and risers white with black nosing.	*			
24. Consider the following methods of coping with code requirements:				
a. Obtaining variance from existing codes.	*			
b. Changing codes to just fulfill health and safety functions of lighting by varying the qualitative and quantitative requirements to specific application.	*			
25. Consider use of a total energy system integrated with all other systems.	*			
26. Where steam is available, use turbine drive for large items of equipment.	*			
27. Use heat pumps in place of electric resistance heating and take advantage of the favorable coefficient of performance.	*			
28. Match motor sizes to equipment shaft power requirements and select to operate at the most efficient point.	*			
29. Maintain power factor as close to unity as possible.	*			
30. Minimize power losses in distribution system by:				
a. Reducing length of cable runs.	*			
b. Increasing conductor size within limits indicated by life-cycle costing.	*			
c. Use high-voltage distribution within the building.	*			
31. Match characteristics of electric motors to characteristics of machine driven.	*			
32. Design and select machinery to start in an unloaded condition to reduce starting torque requirements (e.g., pumps against closed valves).	*			
33. Use direct drive whenever possible to eliminate drive train losses.	*			
34. Use high-efficiency transformers (these are good candidates for life-cycle costing).	*			
35. Use liquid-cooled transformers and captive waste heat for beneficial use in other systems.	*			
36. In canteen kitchens, use gas for cooking rather than electricity.	*			
37. Use micro-wave ovens.	*			

Index

Page numbers in italic indicate figures, tables, or charts.